Flame Retardants for Textile Materials

Textile Institute Professional Publications

Series Editor: The Textile Institute

Care and Maintenance of Textile Products Including Apparel and Protective Clothing ·
Rajkishore Nayak, Saminathan Ratnapandian

Radio Frequency Identification (RFID) Technology and Application in Fashion and Textile Supply Chain
Rajkishore Nayak

The Grammar of Pattern
Michael Hann

Standard Methods for Thermal Comfort Assessment of Clothing
Ivana Špelić, Alka Mihelić Bogdanić, Anica Hursa Sajatovic

Fibres to Smart Textiles: Advances in Manufacturing, Technologies, and Applications
Asis Patnaik and Sweta Patnaik

Flame Retardants for Textile Materials
Asim Kumar Roy Choudhury

Textile Design: Products and Processes
Michael Hann

For more information about this series, please visit: www.crcpress.com/Textile-Institute-Professional-Publications/book-series/TIPP

Flame Retardants for Textile Materials

By

Asim Kumar Roy Choudhury

CRC Press
Taylor & Francis Group
Boca Raton London New York

CRC Press is an imprint of the
Taylor & Francis Group, an **informa** business

First edition published 2020
by CRC Press
6000 Broken Sound Parkway NW, Suite 300, Boca Raton, FL 33487-2742

and by CRC Press
2 Park Square, Milton Park, Abingdon, Oxon, OX14 4RN

© 2021 Taylor & Francis Group, LLC

CRC Press is an imprint of Taylor & Francis Group, LLC

International Standard Book Number-13: 978-0-367-14556-9 (Hardback)
International Standard Book Number-13: 978-0-367-53352-6 (Paperback)
International Standard Book Number-13: 978-0-429-03231-8 (eBook)

Visit the [companion website/eResources]: [insert comp website/eResources URL]

Contents

Preface vii
Author Biography xi

Chapter 1   Fire Hazards and Associated Terminology 1

Chapter 2   Flammability 39

Chapter 3   Inherent FR Fibers 129

Chapter 4   Flame Retardants 165

Chapter 5   Halogen-Based FRs 199

Chapter 6   Phosphorous-Based FRs 223

Chapter 7   Intumescent FRs (IFRs) 291

Chapter 8   Nanomaterial-based FRs 315

Chapter 9   Flame Retardancy of Synthetic Fibers 353

Chapter10   Flame Retardants and the Environment 383

Index 407

Preface

The words *fire* and *flame* have beencurses for millions of people throughout the ages all across the Globe. Those who have lost their homes, belongings, and relatives by fire cannot forget the these events for the rest of their lives. In many cases, fires are silent killers, killing people during sleep. With tremendous efforts by firemen and firefighters, fires may extinguish in time, only leading to realization that many unfortunate individuals have lost their lives, or have been injured, by fire. The most common natural textile materials (namely, cotton, flax, and jute), wood, and many household materials are cellulosic in nature. All of them burn quickly, spread rapidly, and release toxic gases. People have realized this since ancient times, and the flame retardancy concept has been applied using borax and other flame retardant (FR) materials. With the advent of synthetic fibers and polymers, this problem was intensified due to their poor absorbency, caused by hydrophobicity. They also melt easily, and the dripping of melt drops results in severe injury to the burn victim. The period from 1960 to 1980 saw the development of many of well-established flame retardant materials. During the last few decades, the knowledge about the toxicity and environmental impact of chemicals has rapidly grown, and people have become more aware of potential dangers associated with FRs. In February 2003, the Restriction of Hazardous Substances Directive (RoHS) was adopted by the European Union. This was followed by banning many FRs, mainly halogen- and halogenated-phosphosphorous FRs by various countries. The researchers put their best efforts to find eco-friendly substitutes and a large number of research works came into light. A good number of books on flame retardancy have been published in the last two decades, but most of them are devoted to specific or limited fields of flame retardancy. I came across a very large number of research publications on various chemicals and substrates. Hence, I decided to write a book covering broader topics. In this book, flame and fire retardancy of textiles and various nontextile materials (e.g., plastics, resins) are discussed which may help researchers to find newer FRs for the textile materials or vice-versa.

This book consists of 10 chapters. Chapter 1 discusses the hazards caused by fire from a historical perspective. From ancient times until the present, many cities in all parts of the world are ruined by fire; fire hazards are very common in cities and thousands of people are burned and die every year. Extinguishing fires is the job of the fireman or firefighter. Most textile materials are flammable and continue burning, even if they are taken away from fire or flame. Moreover, people who are rescued from fire die because of severe burns from burned garments; inhaling toxic gases released by burning; melting and dripping of polymers; and suffocation due to oxygen shortage. Various fire-related aspects, such as combustion, ignition, charring, and flammability are discussed in this chapter.

Chapter 2 discusses thermal and flammability properties and their variations among various natural and manmade textiles. Flammability test methods measure how easily materials ignite, how quickly they burn, and how they react when burned. A large number of flammability tests are in use, and may be classified into five groups: ignition tests (positioning test samples in vertical, horizontal, and inclined position); reaction to fire tests (how easily fire grows and spreads); application-based tests, i.e. performance of firefighters' clothing; radiant energy tests, i.e., testing on manikins in flash fire scenarios; and scientific assessment of thermal and flammability parameters such as limiting oxygen index (LOI), which measures minimum % oxygen in air required to initiate combustion, and is the most simple, effective and popular measure to express flammability. Various standard methods of flammability tests are discussed very elaborately.

Chapter 3 describes fibers and polymers that are self- or inherent flame retardants (IFR). They do not need any further treatment to protect from fire. Their fire retardancy property is durable and can prevent hazards that are caused during finishing. Wool is naturally flame retardant, while man-made fibers (including synthetics) can be made FR by adding FR chemicals during fiber spinning or by copolymerization. Most important IFR fibers are aramid fibers and polyvinyl chloride polymers.

In Chapter 4, a variety of flame retardants are described. Flame retardant finishes are chemicals which are added to combustible materials to render them resistant to ignition. Various FRs are classified according to characteristics such as chemical structure, and durability. Mineral-, halogen-, phosphorous-, nitrogen-, and silicon-based char-forming intumescent, reactive, and hybrid organic-inorganic FRs are described and their operating principles are explained.

Chapter 5 is devoted to the most economic, most popular, and at the same time most controversial halogen-based FRs. They are widely used in consumer products because of their low impact on other material properties, and the low loading levels required to meet the required flame retardancy. However, halogen-based FRs have raised concerns due to their persistency, their bioaccumulation on living organisms, and their potential toxic effects on human health. As a result, most of them are banned or awaiting substitution by more eco-friendly FRs.

Eco-friendly and versatile, phosphorus-based FRs are described in Chapter 6. Inorganic phosphorous derivatives, mostly nondurable or semidurable, entails primarily phosphoric acid and its ammonium salts. Organophosphorous FRs include aliphatic and aromatic phosphines, phosphine oxides, phosphites, phosphates, phosphinites, phosphinates, phosphonate esters, and phosphonium salts; they promote char formation and act in condensed mode. Nitrogen acts as a synergist in some cases, and some P-N-Si compounds are popular as FRs. These compounds are successfully used, both as additives and as reactive flame retardants for a wide variety of polymer-based systems, namely cotton, rayon, wool, polyester, polyamide, polyacrylic, epoxy resin, polyurethane, and polystyrene. They have also wide applications in nontextile sectors, such as resin, and plastics.

Chapter 7 is devoted to intumescent FRs (IFRs). Researchers showed that the sustainable materials obtained from natural resources can char on burning and form protective layer(s) to make a barrier between substrate and flame/

burning gases. These intumescent FRs (IFRs) are economic, efficient, and easily applicable on various substrates such as textile fibers, resins, and foam. The intumescent behavior resulting from a combination of charring and foaming of the surface of the burning polymers is being widely developed for fire retardance because it is characterized by a low environmental impact. Research work in intumescence is very active. New commercial molecules, as well as new concepts, have appeared.

Chapter 8 examines nanocomposites, The composites are made from two or more constituent materials;at least one of the phases shows dimensions in the nanometer range. These are high-performance materials that exhibit unusual property combinations and unique design possibilities, and are thought of as the materials of the 21st century. Fire retardant, carbon-based nanomaterials are made from graphene, carbon nanotubes (CNTs), and carbon black (CB). Layered aluminosilicates, also popularly described as clays, are one such type of filler; they are responsible for a revolutionary change in polymer composite synthesis, as well as for transforming polymer composites into fire retardant polymer nanocomposites.

Chapter 9 discusses flame retardancy of synthetic fibers. In the absence of functional groups, synthetic fibers are less prone to charring. Furthermore, hydrophobicity and melting are the two disadvantages of making synthetic fibers flame-resistant. To address these problems, back-coating and intumescent FRs are alternative ways to make FR synthetic fibers. The thermally-stable FRs can be added in melt or solutions of polymer before spinning, or may be applied as back-coating. Various FRs suitable for synthetic fibers and their methods of application are discussed in this chapter.

Finally, Chapter 10 explores environmental aspects of FRs. While FRs could ensure the production of fire safety products, many of them are not safe to human beings. There are more than 175 different types of FRs in the market, which contain bromine, chlorine, phosphorus, nitrogen, boron, and antimony compounds or their combinations of inorganic and organic origins. Flame retardant products do not easily obtain eco-labels. The introduction of novel, sustainable, natural-based, intumescent FR systems represents a major scientific and technological challenge. This is expected to make a breakthrough in the production of flame-retarded polymer materials that would follow the principles of *eco-designing*.

Who will read this book? Students who read portions of this book will gain a basic understanding of principles and issues related to fire retardancy, the knowledge on how FR materials and associated application methods changed with time and how their performances can be tested in different flaming environments. Researchers in one application field may find how a FR product used in other fields can be developed for their own applications. Developers, including quality assurance professionals, will find a variety of techniques which can fulfill the FR requirements of their products per specific national requirements that are dictated by prevailing national laws. Technical managers will find a coherent approach to prevent loss from burning and improve FR quality of their products. Therefore, a diverse reading audience should benefit from the contents of this book.

I would like to thank all those who helped to make this book possible. My special thanks to Professor Richard Murray, Chairman, Textile Institute Publication committee who inspired me to write a complete textbook on flame retardant applications in textile materials. I also wish to thank the reviewers who gratefully gave so much of their time to review each chapter, as well as those at CRC Press to make this project successful.

Asim Kr. Roy Choudhury

Author Biography

Dr. Asim Kumar Roy Choudhury is presently working as principal, KPS Institute of Polytechnic, Hooghly (W.B.) India. He retired from the post of Professor and HOD (Textile), Government College of Engineering and Textile Technology, Serampore (W.B.), India. He has over 35 years of experience in textile coloration and finishing in academia and in industry, he has written numerous research papers and acted as reviewer for several reputable international journals and presented papers in several international conferences. He is a fellow and silver medalist of the Society of Dyers and Colourists, Bradford, UK. He is also a member of the editorial board of the Textile Institute Book Series.

Dr. Choudhury's first book, *Modern Concept of Color and Appearance*, was published in the year 2000 and the second book, *Textile Preparation and Dyeing*, was published jointly in the year 2006 by Science Publishers, New Hampshire, USA and Oxford & IBH Publishing Co. Pvt. Ltd. New Delhi. Two books–*Principles of Colour and Appearance Measurement, Volume1: Object Appearance, Colour Perception and Instrumental Measurement* and *Volume 2: Visual Measurement of Colour, Colour Comparison and Management*–were published by Woodhead (UK) in the 2014. Dr. Choudhury also authored *Principles of Textile Finishing*, published by Elsevier in the year 2017. In addition, he has contributed chapters in 27 edited books, published by reputed publishers, such as CRC Press, Woodhead, Elsevier, Springer, Pan Stanford, Apple, and Academic Press. Future project: Principles of Textile Printing (CRC Press).

1 Fire Hazards and Associated Terminology

1.1 INTRODUCTION

About 100,000 people around the world die in fires each year. Fires account for ~ 1% of the GDP within EU. Human beings have been afraid of fire since they started making buildings from wood rather than from stone. In fact, firing infernos have been so common throughout history that nearly every major city in the world has been largely burnt to the ground at some point. Some of these cities have been burned on multiple occasions. Constantinople burned many times between 406 and 1204 only to be, like a damaged ants' or termites' nest, rebuilt each time, thereby setting the stage for the next great inferno. During historical wars, many of these fires were man-made, but most of them were due to natural calamity combined with poor construction methods, excessive use of flammable building materials, and/or the human incapability to fight really large blazes. Some of these fire incidences will be remembered forever because of their size and their dominant roles in shaping historical events. Which fires are these? The list of the top ten most destructive, most famous, or most historically significant non-war-related infernos in history are as follows (Danelek, 2011):

1. Rome (64 AD),
2. London (1212),
3. London (1666),
4. Chicago (1871),
5. Boston (1872),
6. Peshtigo, Wisconsin (1871),
7. San Francisco (1906),
8. Halifax, Nova Scotia (1917),
9. Tokyo, Japan (1923), and
10. Texas City, Texas (1947).

Among common household materials, wood is very fire-prone; it is made of cellulose, the same component of cotton and many other textile materials. These textile materials are, therefore, also fire-prone. Recently (2012–2013), about 800 people were injured in fires in garment and textile factories in Bangladesh, most of which were unreported, according to the data compiled by international labor campaigners. The high number of casualties raises concerns about the slow pace of change in this politically unstable southern Asian state, where more than 1,130 people died in a garment factory building.

1

The building caught fire on November 24, 2012, and collapsed in April 2013. The tragedy was the worst industrial accident anywhere in the world for a generation (Burke, 2013).

The material mainly responsible for the development of the fire in 25% of all dwelling fires, and the item first ignited in 26% of all dwelling fires in 2017–18 was "Textiles, upholstery and furnishings" (The Home Office, 2018). The former caused 46% of all fire-related fatalities in dwellings. This proportion is normally higher (64% in 2016–17).

The particular hazard posed by burning textiles, especially those based on natural cellulosic fibers such as cotton, jute and flax (linen), was recognised during early civilisations and such salts as alum had been used because they reduce their ignitability, and thereby confer flame retardancy. A major problem arises because most of the polymers on which textile materials are based are organic and thus flammable. In the United Kingdom alone, some 800–900 deaths and roughly 15,000 injuries result from fire each year. Most of the deaths are caused by inhalation of smoke and toxic combustion gases, with carbon monoxide being the most common cause, whilethe injuries result from exposure to the heat that evolves from fires.

The Home Office (UK) has responsibility for fire services in England. The vast majority of statistics produced by the Home Office are for England, but some clearly marked, tables are for the United Kingdom and are separated by national data. In the past, the Department for Communities and Local Government (which previously had responsibility for fire services in England) produced releases and tables for Great Britain and at times the UK.

The annual UK Fire Statistics 2018 (The Home Office, 2018) contains some of the most comprehensive documents available and provides information that is perhaps representative of a European country with a population of about 55 million. For every million people in England, there were 6.0 fire-related fatalities in 2017–18.

Smokers' materials (such as lighters, cigarettes, cigars, or pipe tobacco) were the source of ignition in 7% of accidental dwelling fires and in 9% of accidental dwelling fire nonfatal casualties in 2018–19. In contrast, smokers' materials were the source of ignition in 20% of fire-related fatalities in accidental dwelling fires in 2017–18.

The most common cause of death for fire-related fatalities in 2017–18 (where the cause of death was known) was "overcome by gas or smoke", given as 30% (99 fire-related fatalities) of fire-related fatalities. This was lower as a proportion compared with 2016–17 (38%). This was followed by "burns alone" (24%; 80 fire-related fatalities) and the combination of "burns and overcome by gas and smoke" (15%; 50 fire-related fatalities) in 2017–18.

Flame retardancy is an important characteristic of textile materials that protects consumers from unsafe apparel. Firefighters and emergency personnel require protection from flames. Floor coverings, upholstery, and draperies also need protection from fire, especially when used in public buildings. The military and airlines industries have multiple needs with respect to fire-retardancy (Schindler and Hauser, 2004).

Horrocks (2011) reviewed the state of the art for the different commercially available flame retardants (FRs) for textile materials, during the following periods.

1.1.1 1950–1980

The fire safety issue and federal regulations in European countries and in the United States dictated the development of new, effective flame retardant chemicals to reduce the fire hazards and meet the product flammability standards. The first patent on organo-phosphorus FRs for cellulosic textiles (i.e., cotton) was accepted during this period. Inherently FR synthetic fibers bearing aromatic structures were also developed during this "golden period" of flame retardant research.

1.1.2 1980–Late 1990s

This was a lean period in research on FRs.

1.1.3 2000–Onward

Phosphorus-based char-former flame retardant additives were developed during this period. The efforts were made to find the possibility of replacing bromine derivatives with other less toxic and efficient products. The outstanding potential of nanotechnology for conferring flame retardant features to fibers and fabrics was discovered. The preformed nanoparticle suspensions, single nanoparticles or nanoparticle assemblies, and hybrid organic–inorganic structures were proved prospective roles in FRs (Yu et al., 2013).

In 2013, the world consumption of flame retardants was more than 2 million tonnes. The construction sector is the most commercially important application area, requiring for instance, flame retardants for pipes and cables made of plastics. In 2008, the United States, Europe and Asia consumed 1.8 million tonnes, worth US\$4.20–4.25 billion. According to Ceresana, the market for flame retardants is increasing due to rising safety standards world-wide and the increased use of flame retardants. It is expected that the global flame retardant market will generate US\$5.8 billion. In 2010, the Asia–Pacific region was the largest market for flame retardants, accounting for approximately 41% of global demand, followed by North America and Western Europe (Ceresana, 2019).

1.2 FIRE HAZARDS IN TEXTILE INDUSTRY

The textile industry produces materials made of various natural and artificial fibers. It is one of the oldest and most important branches of industry. However, textile materials themselves are flammable. In addition, the textile industry deals with numerous flammable materials and chemicals. The chances of catching fire are very high along the entire textile production chain and easily cause fires and dust explosions.

A fire may break out at any place including houses, residential and commercial buildings, restaurants, cinemas, sports stadiums, jungles, industries, and mills (particularly those like textile mills and industries where raw material or finished products are combustible). In the case of a fire, there may be several reasons related to its initiation and propagation.

Statistics have clearly indicated that textile products have been significantly involved in fire hazards involving human lives. Such products include clothing (including oversuits, undergarments, work wear, and suiting), bedsheets, floor coverings, upholstered textiles in seats, bedding, and home furnishings. Human beings are always in close contact with these textile materials.

Since textiles made from natural fibers are flammable or combustible, they can provide a means of initiating fire. Actual fire cases had shown that textile and clothing were the main items causing injury and death to human lives.

Public concern over the fire-retarding textiles for the protection of human lives and property appeared in the form of legislation. An early example is seen in the United States. The Flammable Fabric Act, 1953, was the first major piece of such legislation.

In the 1970s and 1980s, there were significant discussions on textile flammability legislations in the United States and Western Europe. These were directed atcontrol of fire hazards to lives and other valuables.

In 1988, there were 15,080 textile-implicated fires in the United Kingdom, and textile-related fires caused 4,000 casualties and 495 deaths. The United Kingdom upholstery furniture safety regulations were also introduced in 1988 to exercise a control for reducing textile flammability hazards. The legislations employed in the United Kingdom covers nightwear, upholstered furniture, and toys, and specify how products are to be tested for safety purposes and for evaluating performance, labeling for safe use, and the nature of materials used. An equally important subject, along with the enforcements of legislation is the standard testing procedure designed for evaluating the degree or level of flame retardancy offered by the product for general and specific purposes.

A standard testing procedure usually incorporates the overall conditions that would be experienced by the textile item in an actual fire environment. Presently, several textile flammability testing procedures are available that can be used for a variety of textile products under specified application conditions. These products are apparel, upholstery, building materials, plastic toys, folding portable cots and, car racing suits.

British standard flammability tests are available for products such as curtain, carpets, and bedding, but statutory legislation does not exist for the use of these products in the domestic market.

It is a conventional practice for a retailer to specify a particular test to fulfill the requirements of safety and protection. The Consumer Protection Act 1975, the United Kingdom requires that all products sold in the United Kingdom must be fit for their purpose. In the United States, the Consumer Product Safety Act, 1972 is designed to protect the public from hazardous products.

The Consumer Safety Commission (CPSC), in conjunction with industry, has power to produce standards for product testing to protect masses from hazards.

This commission has recently amended the standard for the flammability of clothing textiles as originally issued in 1953. This is a voluntary standard for assessing the flammability risk of clothing textiles in terms of the ease of ignition and the speed of flame spread. This standard is aimed to reduce danger of injury and loss of life by introducing on-national-basis methods of testing and rating the flammability of textile products used for clothing (Uddin, 2019).

In the textile industry, nearly all materials being used are flammable to some degree. Some fire-prone substances are listed next (HSE, 2019):

- Loose materials, e.g., fabric offcuts or open layers of wadding–low density fibers burn very easily.
- Deposits of fluff and dust (fly)–dust on light fittings is a particular risk. Cotton fly is very hazardous when it is on fire.
- Oily fibers, such as contaminated wool or cotton; oil results from the spinning process.
- Rough, raw edges on rolls or bales–bales tend to burn on the surface and smoulder underneath–deep-seated smouldering in bales is almost impossible to put out from the outside.
- High piles of stock, especially if close together, can increase the speed at which a fire spreads.
- Traditional textile mills, constructed using a high amount of wood and with the presence of fly, means that a fire can spread rapidly.
- Flammable liquids that ignite easily or oxidizing agents that may make an existing fire more intense by fueling it with oxygen.

Some suggested precautionary steps to be followed to avoid fire hazards are listed next (HSE, 2019):

- Good housekeeping–cleaning up fluff and dust regularly, especially high ledges;
- Keeping offcuts in bins, preferably metal;
- Minimum storage in workrooms;
- Indirect heating in workrooms;
- Restricting smoking areas;
- Controlling heated work areas; and
- Storing raw materials and finished goods systematically with proper spacing–not randomly on the floor.

Some fire-prone areas in textile production units are discussed next (HSE, 2019).

1.2.1 CARPET MAKING

This involves the manufacture and storage of latex foam and rubber underlay and foam carpets. These can burn to produce enough smoke to classify the material as a highly flammable solid.

- A high level of sprinkler protection is needed where foam-backed carpets are stored.
- Traditional wool or nylon hessian-backed carpet is not particularly flammable.

1.2.2 SPINNING

During opening and carding, foreign bodies in fibers and baled raw material can come into contact with rotating metal parts of machinery and produce sparks or frictional heat. Natural fibers are more likely than synthetic fibers to contain foreign bodies.

Opening rags is a vigorous process and it is highly likely that they will contain foreign bodies, such as coins and metal buttons, that may cause a spark.

The spread of fire from opening machinery through ducting can be high; the spread of fire through the fiber delivery and trash recovery systems is also a fairly high risk. Automatic fire detection in a ducted system is essential.

Traditional spinning causes deposits of fly, and, if contaminated by oil, can be particularly flammable

1.2.3 WEAVING

The main hazard is ignition of fly by electrical faults, usually insulation failures caused by mechanical vibration. Modern looms are less susceptible to vibration.

Effective controls include good housekeeping and good maintenance of electrical systems and machinery.

1.2.4 FINISHING PROCESSES

These are processes that alter the physical characteristics of the cloth, either by

- Physical means, e.g., raising or milling, or
- Chemical means, e.g., crease resistance.

Processes involving a naked flame, e.g., flame bonding can cause smoldering.

The stenters used for thermal bonding are a common source of fires–smoldering in the finished reel of material can develop into a fire later. Also, if the material stops in the stenter, it is important for the heat supply to cut off automatically. Thermostats can also fail causing overheating.

Gas singeing, i.e., burning of projected fibers from the fabric surface by open flame may cause fire hazard.

Some causes for explosion are:

- Wool spinning: wool dust can cause explosions. Good housekeeping is essential and the dust in the carding machines should be controlled.
- Flocking: Ground flock (rather than precision cut) from mainly cotton, acrylic and nylon fibers, gives a higher risk of explosion. If

dispersed into the atmosphere, e.g., when cleaning down, it can cause an explosion and/or fire.

Burn injuries continue to be one of the leading causes of unintentional death and injury in the United States. Between 2011 and 2015, approximately 486,000 fire or burn injuries were seen at Emergency Departments. In 2016 alone, there were 3,390 civilian deaths from fires, which includes 2,800 deaths from residential structure fires, 150 deaths from non-residential structure fires 355 from vehicle fires, and 85 from outside and unclassified fires other than structure or vehicle fires. One civilian fire death occurs every 2 hours 35 minutes. The lifetime odds of a U.S. resident dying from exposure to fire, flames or smoke are 1 in 1,498.

The primary causes of burn injury include fire-flame, scalds, contact with hot object, electrical and chemicals.

(ABA, 2002)

1.2.5 BURN INJURY

A burn is an injury to the skin or other organic tissue primarily caused by heat radiation, radioactivity, electricity, friction, or contact with chemicals. Thermal (heat) burns occur when some or all of the cells in the skin or other tissues are destroyed by:

- Hot liquids (scalds),
- Hot solids (contact burns), or
- Flames (flame burns).

Burns are a global public health problem, accounting for an estimated 180,000 deaths annually. The majority of these occur in low- and middle-income countries and almost two thirds occur in Africa and Southeast Asia.

In many high-income countries, burn death rates have been decreasing, and the rate of child deaths from burns is currently over seventimes higher in low- and middle-income countries than in high-income countries.

Nonfatal burns are a leading cause of morbidity, including prolonged hospital-ization, disfigurement, and disability, often with resulting stigma and rejection.

- Burns are among the leading causes of disability-adjusted life-years (DALYs) lost in low- and middle-income countries.
- In 2004, nearly 11 million people worldwide were burned severely enough to require medical attention.

The burn statistics of some countries are as follows (WHO, 2018):

- In India, over 1,000,000 people are moderately or severely burned every year.
- Nearly 173,000 Bangladeshi children are moderately or severely burned every year.

- In Bangladesh, Colombia, Egypt, and Pakistan, 17% of children with burns have a temporary disability and 18% have a permanent disability.
- Burns are the second most common injury in rural Nepal, accounting for 5% of disabilities.
- In 2008, over 410,000 burn injuries occurred in the United States, with approximately 40,000 requiring hospitalization.

1.3   FIRE

A fire is defined as any combustion that is not under control. The development of fire can be subdivided into four different phases: ignition, propagation, development, and decline. Since the flames are still contained during the first two phases and the ambient temperatures are changing, the risk of damage can be relatively limited. The limit of this risk is tied to the duration of these two phases, which is determined by the geometry and ventilation of the area, and the amount of contact between the combustible source, the oxygen in the air, and the ignition. Up to the point when the flashover is reached, the mix of inflammable gases propagates the flames very quickly. The average temperature rises (over 1,200°C) and all combustible material burns and the fire increases. The decline or extinguishing phase begins after the maximum temperature is reached. The fire is considered extinguished when the ambient temperature drops around 300°C.

Fire is the rapid oxidation of a material in the exothermic chemical process of combustion, releasing heat, light, and various reaction products (NWCG, 2009). Slower oxidative processes, such as rusting or digestion, are not included by this definition.

Fire generates heat because the conversion of the weak double bond in molecular oxygen, O_2 to the stronger bonds in the combustion products carbon dioxide and water releases energy (418 kJ per 32 g of O_2); the bond energies of the fuel play only a minor role here (Schmidt-Rohr, 2015). At a certain point in the combustion reaction, called the ignition point, flames are produced. The flame is the visible portion of the fire. Flames consist primarily of carbon dioxide, water vapor, oxygen and nitrogen. If a large quantity of heat is generated, the gases may become ionized to produce plasma (Helmenstine, 2009). The color of the flame and the fire's intensity depend on the substances alight, and any impurities outside.

Fire is an important process that affects ecological systems around the globe. The positive effects of fire include stimulating growth and maintaining various ecological systems.

Fire has several negative effects, such as hazard to life and property, atmospheric pollution, and water contamination. If fire removes protective vegetation, heavy rainfall may lead to an increase in soil erosion by water (Morris and Moses, 1987). Also, when vegetation is burned, its nitrogen is released into the atmosphere, unlike elements such as potassium and phosphorus, which remain in the ash and are quickly recycled into the soil. This loss of nitrogen caused by a fire produces a long-term reduction in the fertility of the soil, which only slowly recovers as nitrogen is fixed from the atmosphere by lightning and by leguminous plants such as clover.

Fire has been used by humans in rituals; in agriculture, for clearing land; for cooking, generating heat and light, signaling, propulsion purposes, smelting, forging, incineration of waste, and cremation; and as a weapon or mode of destruction.

1.4 BURNING PROCESS

Fire initiation and propagation in textiles are mainly due to the formation of various gases and liquids during burning. The flame and heat resistance of textiles is concerned with the flammability of such materials, i.e., whether flammable or nonflammable, and the ability of these materials to reduce the transfer of heat from a high-temperature source, either by direct contact (conduction/convection) or via radiation. The required flame and heat resistance of a textile product dependon its end-uses in particular applications within a given textiles sector.

The burning process of textiles involves the release of heat, decomposition of the material, combustion, and propagation of the flame. The decomposition of the material is explained as the breakdown of the hydrogen bonds that make up the composition of the fabric. The fabric is broken down into gaseous liquid and solid composites, which further fuel the combustion process.

The burning of material is a complex phenomenon. It involves processes such as heat transferand thermal decomposition. For synthetic fibers, the thermoplastic behavior adds to the effect. While burning of the cellulosic textile materials, the combustible vapor is generated, and char is formed.

Carbon and oxygen react to form carbon monoxide (CO); it is an exothermic reaction and energy liberated is 26.4 kcal. The char becomes ash in an afterglow process by conversion of CO to CO_2 in the presence of excess oxygen. This reaction is also an exothermic reaction and the energy involved is 94.3 kcal, almost four times than that involved in CO formation. Because ignition, shrinkage, melting, dripping, and afterglow are involved, high energy and large amounts of heat evolve during burning.

When the textile material is heated, chemical and physical changes occur, depending on the temperature and chemical composition of the material. Thermoplastic fibers soften at glass transition temperature (T_g) and subsequently melt at melting temperature (T_m). At some higher temperature called pyrolysis temperature (T_p), both thermoplastic and nonthermoplastic materials chemically decompose or pyrolyze into lower molecular weight fragments and continue through the combustion temperature (T_c). For thermoplastic fibers, T_g and/or T_m are lower than T_p and/or T_c, while for non-thermoplastic fibers, T_g and/or T_m are higher than T_p and/or T_c.

In case of nonthermoplastic natural fibers, pyrolysis and combustion start before softening and melting. Thermoplastic synthetic fibers melt and drip away from the flame before pyrolysis and combustion temperatures are reached. However, if the melt does not shrink away from the flame front, pyrolysis and combustion temperatures are eventually reached and ignition occurs. FR cotton and inherent FR synthetic fibers (e.g., Nomex, Kevlar, PBI) can offer protection to the wearer because they do not shrink away from the flame. Thermoplastic fibers pass the ignition test by shrinking away from the flame. In reality, however, the wearer is exposed to direct heat and thereby suffers burning by contact of the body with the molten mass (Tomasino, 1992).

The thermal processes and combustion products of organic products occur in a progressive and definable cycle as shown in Figure 1.1. When heat is applied, the temperature of the fiber increases until the pyrolysis temperature (T_p) is reached. At this temperature the fiber undergoes irreversible chemical changes, producing nonflammable gases (carbon dioxide, water vapor, and higher oxides of nitrogen and sulfur), flammable gases (carbon monoxide, hydrogen and many oxidizable organic molecules), tars (liquid condensates) and carbonaceous char. As the temperature continues to rise, the tars also pyrolyze, producing more flammable and nonflammable gases and char. When combustion temperature (T_C) is reached, the flammable gases combine with oxygen in the process called combustion, which is a series of gas-phase free radical reactions. These highly exothermic reactions generate large amounts of heat and light. The generated heat provides additional thermal heat for the pyrolysis process to continue. More and more flammable gases and consequently, higher and higher amounts of heat are generated causing devastating effects. In the case of burning of textiles, the speed or rate of heat release is more important than the amount of generated heat (Schindler and Hauser, 2004). An important factor in combustion is the Limiting Oxygen Index

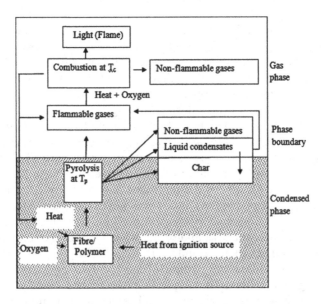

FIGURE 1.1 Combustion cycle for fibres and polymers.

(LOI), which is the percentage of oxygen in the fuel mix needed to support combustion. The higher this number is, the more difficult the combustion is.

Thermal decomposition precedes combustion and ignition of the material. Combustion is an exothermic process that requires three components, namely heat, oxygen, and fuel. When left unchecked, combustion becomes self-catalyzing and will continue until the oxygen, the fuel supply, or excess heat is depleted.

After combustion, the polymer may degrade without flame, burst into flame, or change physically by melting, shrinking, or charring. The combustion products may further be decomposed and ignited. In other words, on heating, a polymer may be liquefied with or without decomposition, be converted into carbon-residues (chars), or release combustible or noncombustible gases. The thermal decomposition products determine the flammability of polymers. The role of flame retardants (FRs) is to inhibit the formation of combustible products and/or to alter the normal distribution of decomposition products originating from the original material.

1.5 PYROLYSIS

The decomposition of materials due to fire is called pyrolysis or thermolysis. All textile fibers in their natural form are inherently fire retardants. Almost all known fibers have a high flash point or melting point. However, when the surrounding temperature reaches the flash point temperature of fibers, they catch fire. Cellulose such as cotton is solid and has an appreciably low vapor pressure. They do not burn but decompose into flammable fragments, which generate heat. This heat further decomposes the cellulose to carry on the decomposition process. Thermal decomposition of cellulose leads to the formation of products such as liquids, tar, and solid materials. Bond rupture, bond reformation, volatilization, and many exothermic reactions occur simultaneously.

Pyrolysis is the thermal decomposition of materials at elevated temperatures in an inert atmosphere (IUPAC, 2009). It involves the change of chemical composition and is irreversible. The word is coined from the Greek-derived elements *pyro* (fire) and *lysis* (separating).

Pyrolysis is most commonly used to the treatment of organic materials. It is one of the processes involved in charring wood (InnoFireWood's website, 2019). In general, pyrolysis of organic substances produces volatile products and leaves a solid residue enriched in carbon char. Extreme pyrolysis, which leaves mostly carbon as the residue, is called carbonization.

The aforementioned process is used heavily in the chemical industry, for example, to produce ethylene, many forms of carbon, and other chemicals from petroleum, coal, and even wood, to produce coke from coal. Inspirational applications of pyrolysis would convert biomass into syngas and biochar, waste plastics back into usable oil, or waste into safely disposable substances.

Pyrolysis differs from other processes such as combustion and hydrolysis in that it usually does not involve the addition of other reagents such as oxygen (O_2, in combustion) or water (in hydrolysis) (Cory et al., 2009). In practice, it is often not practical to achieve completely oxygen- or water-free conditions, especially as pyrolysis is often conducted on complex mixtures. This term has also been applied to the decomposition of organic material in the presence of superheated water or steam (hydrous pyrolysis), for example, in the steam cracking of oil. Pyrolysis has been assumed to take place during catagenesis, the conversion of buried organic matter to fossil fuels.

The possibility of extinguishing a polymer flame depends on the mechanism of thermal decomposition of the polymer. Whereas ignition of a polymer correlates primarily with the initial temperature of decomposition, steady combustion is related to the tendency of the polymer to yield a char, which is produced at the expense of combustible volatile fragments. Therefore, the dependence of steady combustion on the amount of char seems to be simple, and in an early study, it was established that the oxygen index shows a very good correlation with the char yield (Van Klevelen, 1975). In reality, char also serves as a physical barrier for heat flux from the flame to the polymer surface, as well as a diffusion barrier for gas transport to the flame (Levchik and Wilkie, 2000). Therefore, the contribution of the char can be more significant than is expected from a simple reduction in combustible gases.

Four general mechanisms are important for thermal decomposition of polymers (Hirschler, 2000):

1. Random chain scission, in which the polymer backbone is randomly split into smaller fragments;
2. Chain-end scission, in which the polymer depolymerizes from the chain ends;
3. Elimination of pendant groups without breaking of the backbone; and
4. Cross-linking.

Only a few polymers decompose predominantly through one mechanism; in many cases, a combination of two or more mechanisms is in effect. For example, polyethylene and polypropylene tend primarily to decompose via random chain scission, which in the case of polyethylene, is also accompanied by some cross-linking. Poly(methyl methacrylate) and polystyrene tend to depolymerize, poly(vinyl chloride) primarily undergoes elimination of pendant groups (dehydrochlorination), and polyacrylonitrile crosslinks.

In terms of flammability, random scission and depolymerization polymers are usually more flammable than polymers that cross-link or remove pendant groups. Cross-linking (Wilkie et al., 2001) leads to precursors of char and as a result, to lower flammability. Elimination of pendant groups results in double bonds, which can also give cross-links or lead to aromatization.

In general, polymers with aromatic or heterocyclic groups in the main chain are less combustible than polymers with an aliphatic backbone (Aseeva and Zaikov, 1986). Polymers with short flexible linkages between aromatic rings tend to cross-link and char. These polymers are thermally stable and show relatively good flame retardancy. For example, bisphenol A–based polycarbonate, phenol formaldehyde resins, and polyimides are self-extinguishing and show either a V-2 or V-1 rating in the UL-94 test. On the other hand, polymers with relatively long flexible (aliphatic) linkages are still relatively combustible despite aromatics in the backbone. Examples of these polymers are poly(ethylene terephthalate), poly(butylene terephthalate), polyurethanes, and bisphenol A–based epoxy resin.

1.6 COMBUSTION

Combustion, or burning, is a high-temperature exothermic redox chemical reaction between a fuel (the reductant) and an oxidant, usually atmospheric oxygen that produces oxidized, often gaseous products, in a mixture termed smoke. In thermodynamics, the term *exothermic process* (exo-: "outside") describes a process or reaction that releases energy from the system to its surroundings, usually in the form of heat, but also in a form of light (e.g., a spark, flame, or flash), electricity (e.g., a battery), or sound (e.g., explosion heard when burning hydrogen).

In complete combustion, the reactant burns in oxygen, and produces a limited number of products. When a hydrocarbon burns in oxygen, the reaction primarily yields carbon dioxide and water. When elements are burned, the products are primarily the most common oxides. Carbon yields carbon dioxide, sulfur yields sulfur dioxide, and iron yields iron (III) oxide. The combustion of methane, a hydrocarbon, is as in Equation 1.1.

$$CH_4 + 2O_2 \rightarrow CO_2 + 2H_2 \qquad (1.1)$$

Combustion in a fire produces a flame, and the heat produced can make combustion self-sustaining. Combustion is often a complicated sequence of elementary radical reactions. Solid fuels, such as wood and coal, first undergo endothermic pyrolysis to produce gaseous fuels; their combustion then supplies the heat required to produce more of them. Combustion is often hot enough that either incandescent light glows or a flame is produced.

Combustion is a type of chemical process in which a substance reacts rapidly with oxygen releasing heat. The original substance is called the fuel, and the source of oxygen is called the oxidizer. The fuel can be a solid, aliquid, or a gas, e.g., for aeroplane propulsion the fuel is usually a liquid.

Burning or combustion is a chemical process that occurs when oxygen combines/reacts with a substance producing sufficient heat and light (exothermic reaction) to cause ignition. The chemical process is called oxidation. The materials are oxidized continuously until they are exposed to an oxidizing agent (e.g., air) or directly to oxygen. At normal temperatures, the rate at which oxidation occurs is slow, and the heat generated is negligibly small, and is naturally conducted away from the material by the immediate environment. The oxidation rate increases with the increase of temperature, more and more heat releases, and pyrolysis takes place at a temperature specific to the material, i.e., the material decomposes by the action of heat.

Combustion means burning. It is an exothermic process that requires three components to start a chemical chain reaction, namely:

1. heat,
2. oxygen, and
3. a suitable fuel.

The combustion is self-catalyzing and unless controlled, combustion continues as long as the oxygen, the fuel or the excess heat remain. For the combustion process to take place, fuel, oxygen, and an ignition heat source are required. For example, in acampfire, wood is the fuel, the surrounding air provides the oxygen, and a match or lighter is the ignition heat source. Increasing any of these elements increases the fire's intensity, while eliminating any one of them causes the process to stop. If the campfire is smothered with water or dirt, for example, the oxygen can no longer get to the heat and fuel, and it goes out.

1.6.1   FUEL

Fuel is the substance that burns during the combustion process. All chemical fuels contain potential energy; this is the amount of energy that is released during a chemical reaction. The quantity of energy released by a substance during burning is known as the heat of combustion. Each fuel has a specific energy density, or megajoules (MJs) of energy produced per kilogram (kg) of the substance; methane, for example, has an energy density of 55.5 MJ/kg, meaning that it can supply more energy than sulfur, having an energy density of 9.16 MJ/kg.

A wide variety of substances may be used as fuels, but hydrocarbons are among the most common. Some examples of fuels are methane, propane, gasoline, and jet fuel. All fossil fuels, including coal and natural gas, are hydrocarbons. Other substances that are commonly used as fuels include hydrogen, alcohol, and biofuels such as wood.

Combustion, or burning, is a redox chemical reaction that releases heat. A fuel (the reductant) reacts with an oxidant (e.g., atmospheric oxygen), thereby producing mixed oxidized gaseous products called smoke. Combustion in a fire produces a flame, and the heat produced can make combustion self-sustaining. Combustion is often a complicated sequence of elementary radicals. Wood and some other solid fuels first undergo endothermic pyrolysis producing gaseous fuels, which on combustion, supply the heat required to produce more fuels. Combustion is often hot enough that either light glows or a flame is produced. A simple example can be seen in the combustion of hydrogen and oxygen into water vapor. The reaction is commonly used for fuelling rocket engines. This reaction releases 242 kJ/mol of heat and reduces the enthalpy accordingly at constant temperature and pressure (Equation 1.2).

$$2H_2(g) + O_2(g) \rightarrow 2H_2O(g) \qquad (1.2)$$

Combustion of an organic fuel in the air always releases heat because the double bond in O_2 is much weaker than other double bonds or pairs of single bonds, and therefore, the formation of the stronger bonds in the combustion products of CO_2 and H_2O results in the release of energy. The bond energies in the fuel are not significant, as they are similar to those in the combustion products; for example, the sum of the bond energies of CH_4 is nearly the same as that of CO_2. The heat of combustion is about -418 kJ per mole of O_2 consumed in a combustion reaction and can be estimated from the elemental composition of the fuel (Schmidt-Rohr, 2015).

In order for a fire to start or be sustained, a fuel, an oxidizer, and an ignition source must be present. If one of the three components is eliminated, then there cannot be a fire (or explosion).

Fuel, a flammable or combustible material, in combination with a sufficient quantity of an oxidizer such as oxygen gas or another oxygen-rich compound (though nonoxygen oxidizers exist), is exposed to a source of heat or ambient temperature above the flash point for the fuel–oxidizer mix. The fire tetrahedron or fire pyramid (Figure 1.2) adds a fourth component-chemical chain reaction–as a necessity in the prevention and control of fires. The free radicals formed during combustion are important intermediates in the initiation and propagation of the combustion reaction. Fire suppression materials that scavenge these free radicals are able to sustain a rate of rapid oxidation that produces a chain reaction. Fire cannot exist without all of these elements in place and in the right proportions. For example, a flammable liquid would start burning only if the fuel and oxygen are in the right proportions. Some fuel-

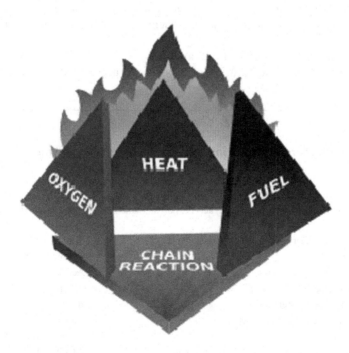

FIGURE 1.2 Fire Tetrahedron (https://en.wikipedia.org/wiki/File:Fire_tetrahedron.svg, (free to use).

oxygen mixes may require a catalyst, a substance that is not consumed when added, in any chemical reaction during combustion, but which enables the reactants to combust more readily.

Once ignited, a chain reaction must take place whereby fires can sustain their own heat by the further release of heat energy in the process of combustion and may propagate, provided there is a continuous supply of an oxidizer and fuel.

1.7   COMBUSTION PRODUCTS

The gaseous products released during combustion of polymers and fibers are shown in Table 1.1 (Lewin and Sello, 1975).

1.8   IGNITION

During burning, when the flashpoint is reached, runaway exothermic reactions are triggered. This is accompanied by the appearance of a flame or glowing zone. This phenomenon is known as ignition. The time interval between the onset of heating and ignition is called ignition time.

Ignition of fabrics (such as apparel, upholstery, and bedding materials) subjected to open flames is a topic of much relevance in understanding and controlling the initiation of unwanted fires. The ignitability of materials is of basic importance when fire initiation and developments are analyzed. For example, in order to predict the burning behavior of fabrics, it is crucial to understand the role played by various physical and chemical properties in determining:

- Whether ignition would occur, and
- If it does occur, the duration of exposure to accomplish it.

TABLE 1.1

Gaseous products released during combustion of organic polymers

Polymer/Fibers	Gases
All organic polymers	CO, CO_2
Nitrogen-containing polymers (wool, silk, acrylic, polyurethanes, amino resins etc.)	NO, NO_2, NH_3, HCN
Wool, vulcanised rubber, sulphur containing polymers	SO_2, H_2S, COS, CS_2
Cellulosic fibers	Formic and acetic acid
Wood, cotton and paper	Acrolein
Polyolefin and others	Alkanes, alkenes
Polystyrene, PVC, polyesters	Benzene
Wood, cotton, paper, phenolic resins	Aldehydes
Phenolic resins	Phenol, formaldehyde
PVC, PTEE, and other halogenated flame retardants	HCl, HF, HBr

Fabrics that are considered inherently noncombustible are made from what are termed high-performance fibers (HPFs), either inorganic fibers such as glass or ceramic fibers, or fibers spun from thermally resistant synthetic polymers, typically fibers such as Kevlar and Nomex, that are made from aramid polymers. HPFs have very strong bonds and require high-heat energy to break them.

One must understand the mechanisms and conditions that lead to a sustained appearance of a flame in the gas phase when a combustible solid is heated by an external source. Two types of ignition are possible under these conditions: spontaneous (auto) and piloted (forced). This depends on whether the ignition occurs with or without the aid of an external pilot such as a spark or a flame. From the fire research perspective, piloted ignition is more important because:

- It occurs at a lower threshold;
- It is the mechanism responsible for fire growth; and
- In practice, it is usually impossible to exclude all possible external pilot sources.

1.8.1 AUTOIGNITION

As bonds break, the bond-fragments can form combustible substances liberated as a gas, depending on the initial chemical composition of the base material. The amount of gas liberated increases with temperature, and when its ignition temperature is reached (forced ignition or auto-ignition) burning occurs. However, there must be sufficient oxygen present to combine with the gas molecules to generate the amount of heat that will raise the temperature to the point of ignition.

The lowest temperature at which a substance spontaneously (in the absence of external ignition sources such as flame or spark) ignites in a normal atmosphere is known as the autoignition temperature or kindling point of the particular substance. This temperature is required to supply the activation energy needed for combustion. The temperature at which a chemical ignites decreases as the pressure increases or oxygen concentration increases. This is applicable to a combustible fuel mixture. If the heat released is sufficient to sustain or increase the oxidation rate, then burning continues until the material is consumed. While burning continuously, more and more heat releases; the temperature may reach a level causing neighboring flammable materials to ignite; and flashover may occur.

1.8.3 PILOTED IGNITION

The earliest known scientific investigation into piloted ignition of wood was carried out by Bamford et al. (1946). They measured the time in which flaming would persist upon removal of the pilot heat source called the *ignition threshold*. They postulated a critical lower limit pyrolysate mass flow rate criterion for sustained ignition. The sustained ignition is possible if the pyrolysate mass flux at the fuel surface is less than 2.5×10^{-4} g/cm^2/s.

A comprehensive experimental and theoretical works of spontaneous and piloted ignition of cellulosic solids was reported by Akita (1959). He presented evidence that the ignition occurs due to some thermal phenomenon directly pertaining to the exposed surface itself. Formation of a combustible gas mixture in the proximity of the exposed surface in itself is a necessary and sufficient condition for piloted ignition (since an external heat source already exists). Such a condition is necessary but not sufficient to ensure spontaneous ignition. This sufficiency is fulfilled by a thermal condition at the attainment of a temperature above 500°C by the exposed surface.

Martin (1964) published his work on ignition based on the following:

- While the internal temperature profile is considerably influenced by pyrolysis, a critical exposed surface temperature criterion describes the onset of ignition.
- The persistence of ignition depends not on any unique composition of such a pyrolysis product mixture, but upon the continued outflow of flammable pyrolysis products.
- The exposed surface is completely pyrolyzed long before ignition.

Thomas and Dry Dale (1987) defined critical ignition temperature as the surface temperature of material at which ignition occurs. Furthermore, the piloted ignition temperature can be defined as the lowest temperature at which the ignition of the decomposition products gives rise to sustained burning at the surface. It is similar to the fire point of a combustible liquid, but differs in what is referred to as a surface temperature, rather than as a bulk temperature.

1.8.2 PROCESS OF IGNITION

The ability to control fire dramatically changed the habits of early humans. Making fire to generate heat and light made it possible for people to cook food, simultaneously increasing the variety and availability of nutrients, and reducing disease by killing organisms in the food. The heat produced also helped people stay warm in cold weather, enabling them to live in cooler climates. In addition, fire kept nocturnal predators at bay. Evidence of cooked food is found from 1.9 million years ago (Bowman et al., 2009), although there is a theory that fire could have been used in a controlled fashion about 1 million years ago. Early humans harnessed fire as early as 1 million years ago, much earlier than previously thought, which suggests evidence unearthed in a cave in South Africa (Krajick, 2011). Evidence became widespread around 50,000 to 100,000 years ago, suggesting regular use from this time; resistance to air pollution started to evolve in human populations at a similar point in time. The use of fire became progressively more sophisticated, for example, to create charcoal and to control wildlife from tens of thousands of years ago.

Fire has also been used for centuries as a method of torture and execution, as evidenced by deaths by burning, as well as by torture devices such as the iron boot, which could be filled with water, oil, or even lead, and then be heated over an open fire to the agony of the wearer.

Setting fuel a flame releases usable energy. Wood was a prehistoric fuel, and is still viable today. In power plants, the use of fossil fuels, such as petroleum, natural gas, and coal, supplies the vast majority of the world's electricity today.

1.9 LIMITING OXYGEN INDEX (LOI)

Minimum percent of oxygen in the environment that sustains burning under specified test conditions. In other terms, it is the content of oxygen in an oxygen-nitrogen mixture that keeps the sample at the limit of burning (Equation 1.3).

$$LOI = \frac{O_2}{H_2 + O_2} \times 100 \qquad (1.3)$$

Tesoro (1978) defined several terms relating to flammability, a few are discussed next.

1.10 CHARRING

Charring is a chemical process of incomplete combustion of certain solids when subjected to high heat. The resulting residue matter is called char. By the action of heat, charring removes hydrogen and oxygen from the solid, so that the remaining substance is composed primarily of carbon. Polymers such as thermoset, as well as most solid organic compounds such as wood or biological tissue, exhibit charring behavior (Chylek et al., 2015).

Charring means partially burning to blacken the surface. Charring can result from naturally occurring processes such as fire; it is also a deliberate and controlled reaction used in the manufacturing of certain products. The mechanism of charring is a part of the normal burning of certain solid fuels such as wood. During normal combustion, the volatile compounds created by charring are consumed at the flames within the fire or released to the atmosphere, while combustion of char can be seen as glowing red coals or embers that burn without the presence of flames.

Coke and charcoal are both produced by charring, whether on an industrial scale or through normal combustion of coal or wood. Normal combustion consumes the char, as well as the gases produced in its creation, while industrial processes seek to recover the purified char with minimal loss to combustion. This is accomplished by either burning the parent fuel (wood or coal) in a low-oxygen environment or by heating it to a high temperature without allowing combustion to occur. In industrial production of coke and charcoal, the volatile compounds that are driven off during charring are often captured for use in other chemical processes.

Charring is an important process in the combustion ignition of solid fuels and in smouldering. In construction of heavytimbered wood buildings, the predictable formation of char is used to determine the fire rating of supporting timbers and is an important consideration in fire protection engineering.

Charring of organic materials starts at temperatures considerably lower than that of soot formation. Burning of food during cooking (e.g., the production of nicely black toast) is an example of low-temperature charring. At temperatures above about 300°C, most of the organic materials undergo a slight thermal decomposition; hydrogen and other noncarbon elements are stripped from carbon chains and rings and the carbon condenses into a graphite like structure. The density of black porous residuum depends on the mass ratio of carbon to other elements in the original material.

Charring of polymers proceeds through various stages (Levchik and Wilkie, 2000):

1. Cross-linking,
2. Aromatization,
3. Fusion of aromatics, and
4. Graphitization.

The ability of a polymer to perform in one or several of these stages leading to char formation depends primarily on the polymer structure. However, this performance can be improved significantly by the use of flame retardants. Although many polymers tend to cross-link at early stages of thermal decomposition, this does not necessarily result in char formation. Char is formed only if the cross-linked polymer contains aromatic fragments and/or conjugated double bonds and is prone to aromatization during thermal decomposition (Wilkie et al., 2001). Fused aromatic rings in the char tend to assemble into small stacks, which are precursors of graphite. These pregraphitic domains are embedded in the amorphous char. This type of char, called turbostratic char, is usually formed at 600°C to 900°C, temperatures typically found on the surface of burning polymers. Char that contains more pregraphitic domains is more stable to thermal oxidation and therefore less likely to burn away and expose the polymer surface to the heat of the flame. On the other hand, highly graphitized chars are rigid and may have cracks, which do not retard diffusion of combustible materials to the flame. The best-performing char would be amorphous uncracked char with requisite pregraphitic domain content (Levchik, 2007).

1.11 SMOLDERING

Combustion occurs without flame and without prior flaming combustion, but usually with incandescence and smoke. Smoldering is the slow, low-temperature, flameless form of combustion, sustained by the heat that evolves when oxygen directly attacks the surface of a condensed-phase fuel. Many solid materials can sustain a smoldering reaction, including coal, cellulose, wood, tobacco, synthetic foams, charring polymers (including polyurethane foams), and some types of dust. Common examples of smoldering phenomena are the initiation of residential fires on upholstered furniture by weak heat sources and the persistent combustion of biomass behind the flaming front of wildfires.

The fundamental difference between smoldering and flaming combustion is that smoldering occurs on the surface of the solid, rather than in the gas phase. Smoldering is a surface phenomenon but can propagate to the interior of a porous fuel if it is permeable to flow. The characteristic temperature and heat released during smoldering are low compared to those in the flaming combustion. Smoldering propagates in a creeping fashion, around 0.1 mm/s (0.0039 in/s), which is about ten times slower than flames spread over a solid. In spite of its weak combustion characteristics, smoldering is a significant fire hazard. Smoldering emits toxic gases (e.g., carbon monoxide) at a higher yield than flaming fires and leaves behind a significant amount of solid residue. The emitted gases are flammable and could later be ignited in the gas phase, triggering the transition to flaming combustion.

1.12   AFTERGLOW

Glowing combustion in a material after cessation (natural or induced) of flame. The behavior of thermoplastic material towards the flame/heating has a different story. Fabrics made of synthetic materials that exhibit melting and surface involvement in afterglow is different from those exhibiting the same in flaming. Prevention method of afterglow is also different. Afterglow is mainly due to the burning of remnant char, which forms as a result of lack of oxygen in surrounding atmosphere. Carbon and oxygen react to form carbon monoxide (CO); this is an exothermic reaction, and energy liberated is 26.4 kcal. The char becomes ash in afterglow process by conversion of CO to CO_2 in the presence of excess oxygen.

1.13   SELF-EXTINGUISHING MATERIALS

Self-extinguishing materials are incapable of sustained combustion in air under the specified test conditions after removal of the external source of heat.

Hensel (2011) said flame retardant is good, but self-extinguishing is better for electrical installations. Only *self-extinguishing* material provides added safety with respect to fire protection. All Hensel products made from thermoplastic are flame retardant and self-extinguishing and fulfill the glow wire tests at least at 750°C or even at 960°C. The proof of self-extinguishing characteristics should be carried out using a glow wire test at a temperature of 750°C. The glow wire test at 750°C should prove that a flame would extinguish itself within 30 s of the energy source being switched off, i.e., it does not continue burning or catch fire.

In the Russian aviation industry, in flame resistance assessment, the determining indices (according to the industry standard OST 1090094-79) are the duration of combustion and the length of the burned part of the polymer specimen (measuring 50 x 290 mm) during heating by a gas burner with a flame temperature of 840°C. According to the given indices, materials are classified into low-burning, self-extinguishing, slow-burning, and burning (combustion time >15 s, length of burnt part of specimen >170 mm) (Petrova et al., 2014).

The airplane trims should satisfy the following requirements: as a minimum they should be self-extinguishing (after exposure to a gas burner flame for 12 s (60 s), the duration of residual burning should not exceed 15 s, and the length of burn-through should not exceed 152 mm); they should also have limited smoke formation (the specific optical density of smoke within 4 min should be no more than 200 units) (Barbot'ko, 2010; Mikhailin, 2011).

1.14   SMOKE

Smoke is a fine dispersion of carbon, other solids, and liquids resulting from incomplete combustion; particles are not individually visible, but cause opacity due to scattering and/or absorption of visible light.

Smoke is a collection of airborne solids, liquid particulates, and gases that are emitted when a material undergoes combustion or pyrolysis, together with the quantity of air that is entrained or otherwise mixed into the mass. It is commonly an unwanted by-product of fires (including stoves, candles, oil lamps, and fireplaces), but may also be used for pest control (fumigation), communication (smoke signals), defensive and offensive capabilities in the military (smoke screen), cooking, and smoking (tobacco, marijuana (drug). It is used in rituals in which incense, sage, or resin is burned to produce a smell for spiritual purposes. Smoke is sometimes used as a flavoring agent, and as a preservative for various foodstuffs. Smoke is also a component of internal combustion engine exhaust gases, particularly diesel exhaust.

Smoke inhalation is the primary cause of death in victims of indoor fires. The smoke kills by a combination of thermal damage, poisoning and lung irritation caused by carbon monoxide, hydrogen cyanide, and other combustion products.

Smoke is an aerosol (or mist) of solid particles and liquid droplets that are close to the ideal range of sizes for Mie scattering of visible light. A smoke cloud does not obstruct an image, but thoroughly scrambles it.

Polymers are a significant source of smoke. Aromatic side groups, e.g., in polystyrene, enhance generation of smoke. Aromatic groups integrated in the polymer backbone produce less smoke, likely due to significant charring. Aliphatic polymers tend to generate the least smoke, and are non-self-extinguishing. However presence of additives can significantly increase smoke formation. Phosphorus-based and halogen-based flame retardants decrease production of smoke. Higher degree of cross-linking between the polymer chains has such an effect as well (Van Krevelen and Nijenhuis, 2009).

1.15   FLAME

A flame (from Latin flamma) is the visible, gaseous part of a fire. It is caused by a highly exothermic reaction taking place in a thin zone. Very hot flames are hot enough to have ionized gaseous components of sufficient density to be considered plasma.

There are different methods of distributing the required components of combustion to a flame. In a diffusion flame, oxygen and fuel diffuse into each other; the flame occurs where they meet. As a result, the flame speed is limited by the rate of diffusion. In a diffusion flame, combustion takes place at the flame surface only, where the fuel meets oxygen in the right concentration; the interior of the flame contains unburnt fuel. This is opposite to combustion in a premixed flame.

In a premixed flame, the oxygen and fuel are premixed beforehand, which results in a different type of flame. Candle flames (diffusion flames) operate through evaporation of the fuel, which rises in a laminar flow of hot gas that then mixes with surrounding oxygen and combusts.

A flame is a mixture of reacting gases and solids emitting visible, infrared, and sometimes ultraviolet light, the frequency spectrum of which depends on the chemical composition of the burning material and intermediate reaction products. In many cases, such as the burning of organic matter, for example wood, or the incomplete combustion of gas, incandescent solid particles called soot produce the familiar red-orange glow of *fire*. This light has a continuous spectrum. Complete combustion of gas has a dim blue color due to the emission of single-wavelength radiation from various electron transitions in the excited molecules formed in the flame. Usually oxygen is involved, but hydrogen burning in chlorine also produces a flame, producing hydrogen chloride (HCl). Among many other possible combinations producing flames are fluorine and hydrogen, and hydrazine and nitrogen tetroxide. Hydrogen and hydrazine/unsymmetrical dimethylhydrazine (UDMH) flames are similarly pale blue, while burning boron and its compounds, evaluated in the mid-20th century as a high-energy fuel for jet and rocket engines, emits intense green flame, leading to its informal nickname of "Green Dragon".

The chemical kinetics occurring in the flame are very complex and typically involve a large number of chemical reactions and intermediate species, most of them radicals. For instance, a well-known chemical kinetics scheme, GRI-Mech (2007) uses 53 species and 325 elementary reactions to describe combustion of biogases.

The glow of a flame is complex. Black-body radiation is emitted from soot, gas, and fuel particles, though the soot particles are too small to behave like perfect blackbodies. There is also photon emission by de-excited atoms and molecules in the gases. Much of the radiation is emitted in the visible and infrared bands. Flame color depends on several factors, the most important typically being black-body radiation and spectral band emission, with both spectral line emission and spectral line absorption playing smaller roles. In hydrocarbon flames, the most common type of flame, the most important factor determining color is oxygen supply and the extent of fuel-oxygen premixing, which determines the rate of combustion, and thus the temperature and reaction paths, thereby producing different color hues.

The dominant color in a flame changes with temperature. The photo of the forest fire in Canada is an excellent example of this variation. Near the ground, where most burning occurs, the fire is either white, the hottest color possible for organic material in general, or yellow. Above the yellow region, the color changes to orange, which is cooler, then to red, which is cooler still. Above the red region, combustion no longer occurs, and the uncombusted carbon particles are visible as black smoke.

The common distribution of a flame under normal gravity conditions depends on convection, as soot tends to rise to the top of a general flame, as in a candle in normal gravity conditions, making it yellow. In microgravity or zero gravity (NASA, 2010), such as an environment in outer space, convection no longer occurs, and the flame becomes spherical, with a tendency to become more blue and more efficient (although it may go out if not moved steadily, because the CO_2 from combustion does not disperse as readily in microgravity, and tends to smother the flame). There are several possible explanations for this difference, of which the most likely is that the temperature is sufficiently evenly distributed so that soot is not formed and complete combustion occurs (NASA, 2007). Experiments by NASA reveal that diffusion flames in microgravity allow more soot to be completely oxidized after they are produced than diffusion flames on Earth, because of a series of mechanisms that behave differently in microgravity when compared to normal gravity conditions (www. derose.com, 2019). These discoveries have potential applications in applied science and industry, especially those concerning fuel efficiency.

In combustion engines, various steps are taken to eliminate a flame. The method depends mainly on whether the fuel is oil, wood, or a high-energy fuel such as jet fuel.

1.15.1   FLAME TEMPERATURES

It is true that objects at specific temperatures do radiate visible light. Objects whose surface is at a temperature above approximately 400°C (752°F) glow, emitting light at a color that indicates the temperature of that surface. It is a misconception that one can judge the temperature of a fire by the color of its flames or the sparks in the flames. For many reasons, chemically and optically, these colors may not match the red/orange/yellow/white heat temperatures on the chart.

1.15.2   ADIABATIC FLAME TEMPERATURE

Adiabatic means no loss of heat to the atmosphere. An adiabatic process occurs without transfer of heat or mass of substances between a thermodynamic system and its surroundings. In the study of combustion, there are two types of adiabatic flame temperature, depending on how the process is completed. These are constant volume and constant pressure, describing the temperature that the combustion products theoretically reach if no energy is lost to the outside environment. The constant volume adiabatic flame temperature is the temperature that results from a complete combustion process that occurs without any work, heat transfer or changes in kinetic or potential energy. Its temperature is higher than the constant pressure process because none of the energy is utilized to change the volume of the system (i.e., generate work).

TABLE 1.2

Adiabatic flame temperature of some fuel and oxidizer pairs

Fuel and oxidizer pairs	Adiabatic flame temperature
Oxy-dicyanoacetylene	4,990°C (9,000°F)
Oxy-acetylene	3,480°C (6,300 °F)
Oxyhydrogen	2,800°C (5,100 °F)
Air-acetylene	2,534°C (4,600 °F)
Bunsen burner (air-natural gas)	1,300 to 1,600°C (2,400 to 2,900°F) (Begon et al., 1996)
Candle (air-paraffin)	1,000°C (1,800°F)

The *adiabatic flame temperature* of a given fuel and oxidizer pair (Table 1.2) (Fire Wikipedia, accessed 9.2.2019) indicates the temperature at which the gases achieve stable combustion. This is the maximum temperature that can be achieved for given reactants. Heat transfer, incomplete combustion, and dissociation all result in lower temperature. The maximum adiabatic flame temperature for a given fuel and oxidizer combination occurs with a stoichiometric mixture (correct proportions such that all fuel and all oxidizer are consumed).

A fire-resistance rating typically means the duration for which a passive fire protection system can withstand a standard fire resistance test. This can be quantified simply as a measure of time, or it may entail a host of other criteria, involving other evidence of functionality or fitness for purpose.

1.16 FLAMMABILITY

Flammability is the tendency of a material to burn with a flame. Flammable materials are those that ignite more easily than other materials, whereas those that are harder to ignite or burn less vigorously are combustible.

Lewin (1985) explains flammability as the tendency of a material to burn with a flame. Indeed the flammability of textiles is a measurement of the ease with which fabric is able to be ignited and how effectively it burns. Kasem and Rouette (1972) stated that the ignitability of the fabric as well as the combustibility are the indicators of fabric flammability characteristics. The combustibility of the fabric is stated as the rate at which the flame (or the afterglow) is able to propagate. Ignition of the fabric is described as a more complex phenomenon by Backer et al. (1976). Ignition involves the transfer of heat with thermal decomposition governed by fluid mechanics and chemical kinetics. Exothermic reactions are triggered as the ignition temperature of the fabric is reached. The reactions accompanied by a flame or glowing of any sort is termed ignition.

The degree of flammability or combustibility in air depends largely upon the chemical composition of the subject material, as well as the ratio of mass to surface area. For example, finely divided wood dust can undergo explosive combustion and produce a blast wave. A piece of paper (made from wood) catches on fire quite easily. A heavy oak desk is much harder to ignite, even though the wood fiber is the same in all three materials.

Common sense (and indeed scientific consensus until the mid-1700s) would seem to suggest that material *disappears* when burned, as only the ash is left. In fact, there is an increase in weight because the combustible material reacts (or combines) chemically with oxygen, which also has mass. The original mass of combustible material and the mass of the oxygen required for combustion equals the mass of the combustion products (ash, water, carbon dioxide, and other gases). Antoine Lavoisier, "father of modern chemistry" (August 26, 1743–May 8, 1794), a French nobleman and one of the pioneer chemists in these early insights, stated that nothing is lost, nothing is created, everything is transformed, which would later be known as the law of conservation of mass. Lavoisier used the experimental fact that some metals gained mass when they burned to support his ideas.

Historically, flammable, inflammable and combustible meant capable of burning. The word "inflammable" came through French from the Latin inflammāre = "to set fire to", where the Latin preposition "in-" means "in" as in "indoctrinate", rather than "not" as in "invisible" and "ineligible".

The word "inflammable" may be erroneously thought to mean "nonflammable". The erroneous usage of the word "inflammable" is a significant safety hazard. Therefore, since the 1950s, efforts to put forward the use of "flammable" in place of "inflammable" have been accepted by linguists, and it is now the accepted standard in American English and British English. Antonyms of "flammable/inflammable" include: nonflammable, noninflammable, incombustible, noncombustible, not flammable, and fireproof.

Flammable applies to materials that ignite more easily than other materials, and thus, are more dangerous and more highly regulated. Less easily ignited less-vigorously burning materials are combustible. For example, in the United States, flammable liquids, by definition, have a flash point below 100°F (38°C),whereas combustible liquids have a flash point above 100°F (38°C). Flammable solids are solids that are readily combustible, or may cause or contribute to fire through friction. Readily combustible solids are powdered, granular, or pasty substances that easily ignite by brief contact with an ignition source, such as a burning match, and spread flame rapidly. The technical definitions vary between countries so the United Nations created the Globally Harmonized System of Classification and Labeling of Chemicals, which defines the flash point temperature of flammable liquids as between 0° and 140°F (−17.8° and 60°C) and combustible liquids between 140°F (60°C) and 200 F (93°C) (United Nations, 2011).

Flammability is the ability of a substance to burn or ignite, causing fire or combustion. The degree of difficulty required to cause the combustion of a substance is quantified through fire testing. Internationally, a variety of test protocols exist to quantify flammability. The ratings achieved are used in building codes, insurance requirements, fire codes and other regulations governing the use of building materials as well as the storage and handling of highly flammable substances inside and outside of structures and in surface and air transportation. For instance, changing occupancy by altering the flammability of the contents requires the owner of a building to apply for a building permit to make sure that the overall fire protection design basis of the facility can take the change into account.

The factors affecting the flammability of apparel are:

- Fiber content,
- Fiber construction,
- Fabric weight,
- Fabric construction,
- Fabric surface texture,
- Moisture content,
- Presence of additives or contaminants in the fabric,
- Garment design,
- Effect of components used in apparel, and
- Laundering of the apparel after use.

1.17 FLAME-RESISTANT VS. FLAME-RETARDANT

Various terms are used to express the way a fabric reacts when in contact with a flame. If negligibly affected it is said to be flame-proof or fire-proof; if it ignites but self-extinguishes on removal from the flame, it is called fire-resistant–difficult to burn; if the material does not burn but can melt and/or decompose at high temperatures, it is referred to as flame retardant, noncombustible or incombustible – not capable of igniting and burning. A very important aspect of flameproof/fireproof textiles is that they are thermally stable; they do not readily burn or shrink when exposed to a flame or intense heat (heatproof/heat-resistant).

Flame-resistant fabrics are made from materials that are inherently nonflammable–the materials have flame resistance built into their chemical structures. Fabrics made with these types of materials are designed to prevent the spread of fire and do not melt or drip when in close proximity to a flame. Because flame-resistant fabrics are not usually made from 100% flame-resistant materials, they burn, but do so very, very slowly, and are often self-extinguishing. The most important function of these materials and fabrics is to prevent the further spread of fire.

Fire-resistant is a synonym for flame-resistant. It means exactly the same thing, and it is correct to use them interchangeably.

With fire-resistant clothing, the promise is not that the garments will never catch fire. They are designed to resist igniting, and generally fulfill this purpose in all but the most extreme situations. The great strength of flame-resistant garments, however, is that they prevent fires from spreading. Even if the garments do catch fire, they almost always extinguish themselves quickly.

Fire-retardant or flame retardant fabrics are those that have undergone chemical treatment to acquire some of the same properties that flame-resistant fabrics inherently have. As a result of these chemical procedures, flame retardant fabrics become self-extinguishing and slow-burning. Any type of fabric may be used, but must undergo this treatment before it can be considered flame retardant.

If an employee works in environments where heat, fire, or electrical injuries are a real possibility, the odds are good that this employee should be wearing flame-resistant clothing.

Not all fire-resistant clothing is made from the same fabrics. There are multiple different choices available, and no choice is perfect. Each comes with different benefits and hazards. Each organization is best served by choosing the fabric that will be most suited to its needs and working environment. There are three broad categories of workers who should wear flame-resistant clothing for protection, based on the type of hazard to which the workers would be exposed while completing their work. The three primary hazards are (Ashley, 2017):

- Electric arc: People who are exposed to this hazard include electricians, as well as certain utility workers and others.
- Flash fire: This category includes pharmaceutical and chemical workers, as well as those who work in refineries and other industries.
- Combustible dust: This category includes workers in food processing plants, and in the paper and pulp industry, among others.

Here are a few of the common fibers with inherent flame-resistant qualities commonly used to create FR clothing (Ashley, 2017).

- Modacrylic: These are the most popular and common option available today. These fibers are often used as part of a blend to create several different flame-resistant fabrics. These various combinations of fibers work together to create fabrics that can easily stand up to several types of standards and regulations.
- Nomex: This is another type of fiber that has inherent flame-resistant qualities. Unlike modacrylic fibers, Nomex can create FR garments on its own. It doesn't have to be a standalone, however. It can also be combined with other materials such as Kevlar.
- Kevlar: These fibers are certainly flame-resistant, but have many other additional properties such as high strength. Kevlar can create flame-resistant clothing, as well as many other different items. When used to make FR clothing, Kevlar is often combined with Nomex.

Each type of flame-resistant fabric will come with its own pros and cons. Kevlar, for instance, is extremely heavy-duty, but consequently comes with a high price tag. There are no specific flame-resistant clothing dangers, however, and all are designed to protect the wearer from hazardous heat-based conditions.

Fire-resistant clothing is also referred to as FRC by industry people. The garments underneath the flame-resistant clothing have a significant impact on the safety and the effectiveness of the FRC clothing. Whenever someone wear these garments, he or she should always take care to wear only nonmelting garments underneath them.

There are two primary reasons for this caution. The first is that by doing this, essentially a second layer of FR protection is added. Even if the first layer of outerwear gets damaged or burned, a second layer would protect the wearer. The layer of air insulation between the two layers also helps keep the wearer safe.

Flame retardant fabrics are chemically treated to be slow-burning or self-extinguishing when exposed to an open flame. These fabrics can be made from any material, but they must be treated with special chemicals to qualify as flame retardant.

The term flame retardants (FRs) subsume a diverse group of chemicals which are added to manufacture materials, such as plastics and textiles, and surface finishes and coatings. Flame retardants are activated by the presence of an ignition source and are intended to prevent or slow the further development of ignition by a variety of different physical and chemical methods. They may be added as a copolymer during the polymerization of a polymer, mixed with polymers at an molding or extrusion process or, particularly for textiles, applied as a topical finish (EPA, 2005). Mineral flame retardants are typically additive while organohalogen and organophosphorus compounds can be either reactive or additive.

The most important difference between flame-resistant and flame retardant fabrics lies in how each is made. Without a special chemical application, a fabric does not qualify as flame retardant. Similarly, without being made of certain nonflammable fibers, a fabric will not qualify as fire resistant.

The special protective wear is called fire-resistant clothing/fire-resistant garments/fire-resistant apparel, or even personal protective equipment which also includes further accessories such as gloves, helmets, or boots. *FR* refers to the flame-resistant or fire-resistant and heat-resistant properties of the clothing, by virtue of which a fabric is able to resist burning or melting and even self-extinguish once the source of ignition or fire is removed from it.

Fire-resistant clothing is basically made from two kinds of fabric that are differentiated as *treated* and *inherent* FR fabrics. The former type quiet simply includes the natural or synthetic fabrics that are later treated with a combination of flame retardant chemicals to give them the flame-resistant properties. However, in the case of inherent fabrics, the fabric itself is made from fibers that have flame-resistant qualities and this resistance is ingrained into the molecular structure of the fiber by engineering retardant compounds into a permanent chemical change inside the hollow core of a fiber, creating an all-new fabric with FR characteristics (Islam, 2018).

Inherently flame retardant (FR) fibers may be of three types, namely:

1. Inherent thermally stable chemical structure (e.g., the polyaramids or other aromatic structures),
2. Flame-retardant additives incorporated during the production of manmade fibers (e.g., FR viscose), or
3. Produced by the synthesis of conventional fiber forming polymers with flame retardant comonomers (e.g., FR polyester).

The word *inherent* was not originally a textile or FR term. Its definition varies slightly from source to source, but the common thrust is "by its very nature, built-in, implicit," while "treated" is usually defined as chemical engineering to impart properties not previously present. Nature provides very few FR fibers; the most well-known of these is asbestos, which is obviously not in common use in protective apparel.

Conversely, all flame-resistant fibers in common use today for industrial protective apparel are engineered by humans, using chemistry, to be flame-resistant. What is important is not how the engineering was accomplished; what matters is that the engineering was accomplished, correctly and consistently, so that a garment is flame-resistant weeks later, months later, and years later, regardless of how many times it is laundered (Margolin, 2012).

1.18 FLAME RETARDANTS (FRS)

Normal textile materials are finished with flame retardant agents in the form of surface treatments, coatings, or functional finishes which become an integral part of the fiber structure.

Flame retardant materials were first produced around 400 BC, but the need for them was not realized much until the 17th century. In 1638, the idea of reducing the risk of fire in theaters originated in Paris with fireproofing of plaster and clay, thus beginning the process of creating flame-resistant materials. If humans intervene with chemistry to treat naturally flammable fibers, they could prevent potential harm. Therefore, the process of making things flame-resistant became a priority from this point on.

In 1735, Jonathan Wyld of England patented a flame-retarding mixture of alum, ferrous sulfate, and borax. The first systematic attempt to make textiles flame-resistant was made in 1821 by the eminent chemist Gay Lussac, who developed a flame-retarding finish for hemp and linen fabrics that contained various ammonium salts, with or without borax. The salts first broke down into a nonflammable vapor when they were heated up, while borax was low-melting and formed a glassy layer on fabrics. This was yet another step toward making today's textiles flame-resistant. By the 20th century, other scientists perfected the same method by incorporating stannic oxide into fabrics to make them flame retardant. Stannic oxide, also known as tin oxide, is an off-white, powdery product that is produced thermally from high-grade tin metal.

These techniques were used to make natural fibers fire-resistant. Once synthetic materials started dominating the market, cotton producers needed to come up with a better way to promote their products, or they would not have been able to survive. The Army Quartermaster Corps' were in demand for flame-resistant clothing and the research regarding fire-resistant fabrics increased to a large extent. Advanced technologies of the 20th century allowed scientists to start the process of chemically modifying the cellulose molecules on both the surfaces and within cotton fibers. To keep this special process commercially viable, the scientists needed to work hard to find a chemical combination that still kept the cotton's strength and durability while keeping the cost competitive (Caitlin, 2015).

TABLE 1.3
Early historical developments of flame retardants (WHO, 1997)

Alum used to reduce the flammability of wood by the Egyptians.	About 450 BC
The Romans used a mixture of alum and vinegar on wood.	About 200 BC
Mixture of clay and gypsum used to reduce flammability of theater curtains	1638
Mixture of alum, ferrous sulfate, and borax used on wood and textiles by Jonathan Wyld in Britain	1735
Alum used to reduce flammability of balloons	1783
GayLussac reported a mixture of $(NH_4)_3PO_4$, NH_4Cl and borax to be effective on linen and hemp	1821
Perkin described a flame retardant treatment for cotton using a mixture of sodium stannate and ammonium sulfate	1912

The chemical nature of textile substrates is highly diversified. Hence, the field of flame retardancy is multidisciplinary and complex as well. The flame-retarding chemicals and formulations are also numerous and include halogen, phosphorous, nitrogen, antimony, sulphur, boron, and other elements in many forms and combinations. Flame-retarding treatments require the application of a relatively large quantity of chemicals; e.g., 10%–30% of the weight of the material. Hence, the aesthetic properties (softness, stiffness, luster, handle, drape etc.), physical properties (washability, soil repellence and soil release, static charge accumulation), tensile properties, creasing, and pilling properties of textile materials may change.

A brief historical development of FRs through the ages is listed in Table 1.3.

1.19   CLASSIFICATION OF FLAME RETARDANTS

Flame retardants (FRs) are mainly of two types:

1. Additive flame retardants, and
2. Reactive flame retardants.

They can be further separated into several different classes (Van der Veen et al., 2012):

a. Minerals such as:

- Aluminum hydroxide (ATH),
- Magnesium hydroxide (MDH),
- Huntite and hydromagnesite,
- Various hydrates,
- Red phosphorus, and
- Boron compounds, mostly borates.

b. Organohalogen compounds. This class includes organochlorines such as chlorendic acid derivatives chlorinated paraffins and organobromines such as:

- Decabromodiphenyl ether (decabde),
- Decabromodiphenyl ethane (a replacement for decabde),
- Polymeric brominated compounds such as brominated polystyrenes, brominated carbonate oligomers (BCOs), brominated epoxy oligomers (BEOs),
- Tetrabromophthalic anyhydride,
- Tetrabromobisphenol A (TBBPA), and
- Hexabromocyclododecane (HBCD).

c. Organophosphorus compounds. This class includes organophosphates such as:

- Triphenyl phosphate (TPP),
- Resorcinol bis(diphenylphosphate) (RDP),
- Bisphenol A diphenyl phosphate (BADP), and
- Tricresyl phosphate (TCP),
- Phosphonates such as dimethyl methylphosphonate (DMMP), and
- Phosphinates such as aluminium diethyl phosphinate.

d. Halogenated phosphorus compounds:

- Brominated: tris(2,3-dibromopropyl) phosphate (tris), and
- Chlorinated: tris(1,3-dichloro-2-propyl) phosphate (TDCPP) and tetrakis(2-Chlorethyl)dichloroisopentyl diphosphate (V6) (Weil and Levchik, 2015).

Most but not all halogenated flame retardants are used in conjunction with a synergist to enhance their efficiency. Antimony trioxide is widely used, but other forms of antimony, such as the pentoxide and sodium antimonate are also used.

The mineral flame retardants mainly act as additive flame retardants and do not become chemically attached to the surrounding system. Most of the organohalogen and organophosphate compounds also do not react permanently to attach themselves into their surroundings, but further work is now underway to graft further chemical groups onto these materials to enable them to become integrated without losing their retardant efficiency. This will also make these materials nonemissive into the environment. Certain new nonhalogenated products with these reactive and nonemissive characteristics have been coming onto the market since 2010, because of the public debate about flame retardant emissions. Some of these new reactive materials have even received US-EPA approval for their low environmental impacts.

1.20 SYNERGISM AND ANTAGONISTIC

Synergism and antagonism are the interactions of two or more substances or agents to produce a combined effect greater or smaller than the sum of their separate respective effects.

Synergistic effects are observed in the vapor phase, as well as in condensed-phase, active FR systems. Their modes of action in the various systems are different and involve a wide variety of interactions. They involve chemical interactions between the FR additive and the synergist, between the synergist and the polymer or between all three: polymer, FR additive and synergist, between two synergists and the FR additive, and even between two polymers in a polymer blend. In some cases, the synergist itself is not a flame-retarding agent and becomes active only in the presence of an FR additive. This is the case with halogen-based additives and antimony trioxide. In other cases such as the bromine-phosphorus synergism, both additives are active flame retardants. In certain cases the interaction between the ingredients in a formulation brings about a decrease in the flame retardancy parameters and is thus antagonistic. This is encountered in the case of the application of phosphorus derivatives together with nitriles (Khanna and Pearce, 1978).

Lewin (1999) stated that the term synergism, in the FR terminology is a poorly defined term. Strictly speaking, it refers to the combined effect of two or more additives, which is greater than that predicted on the basis of the additivity of the effect of the components. The term *synergistic effectivity* (SE) (Lewin and Sello, 1975) is meant to serve as a general tool for characterizing and comparing synergistic systems. It is defined as the ratio of the FR effectivity (EFF) of the flame retardant additive plus the synergist to the EFF of the additive without synergist. EFF is defined as the increment in oxygen index (OI) for 1% of the flame retardant element. The differentiation between the synergistic systems and catalytic phenomena observed in FR technology is not straightforward. Some synergistic systems could easily be classified as catalytic and vice versa. The catalyst is highly effective at a low concentration in the formulation. The number of FR catalytic systems reported in the literature is, until now relatively small, but the interest in catalytic approaches appears to be growing.

1.21 FUTURE TRENDS

The Center of Fire Statistics (CFS) of International Association of Fire and Rescue Services (CTIF) presents its latest report №23 (CTIF, 2018), containing fire statistics of 27–57 countries during the years 1993–2016 representing 0.9–3.8 billion inhabitants of the Earth, depending on the year of reporting. In these countries 2.5–4.5 million fires and 21–62 thousand fire deaths were reported to fire services annually, depending on the year. The statistics clearly shows that fire hazards are not being abated in spite of all human efforts. Fire not only cause loss of human lives and our valuable wealth, but also causes tremendous air pollution. The biggest health threat from smoke is from fine particles. These microscopic particles can get into the eyes and respiratory system, where they can cause health problems such as burning eyes, runny nose, and illnesses such as bronchitis. Fine particles also can aggravate chronic heart and lung diseases. In addition, burning of various materials may generate toxic gases. To make textile and other substances less prone to burning, various flame retardant substances are used. In recent years, however, it has been observed that many of them, especially halogens, are very efficient, but not environment friendly.

It is a true challenge to find suitable and efficient substitutes at cheaper prices. The methods of application of flame retardants has also been revolutionized in recent years.

REFERENCES

Akita K. (1959). Studies on the mechanism of ignition of wood, Report by Fire Research Institute, Japan, 9(1–2).

American Burn Association (ABA). (2002). Burn injury fact sheet, https://ameriburn.org/wp-content/uploads/2017/12/nbaw-factsheet_121417-1.pdf, accessed on 29.4.2020.

Aseeva R.M. and Zaikov G.E. (1986). *Combustion of Polymer Materials.* Carl Hanser Verlag, Munich, Germany, p. 149.

Ashley. (2017). What is FR clothing? Your guide to flame-resistant clothing, www.degemmill.com/flame-resistant-clothing/, October 19, 2017, accessed on 7.2.2019.

Backer S., Tesoro G.C., Toong T.Y., and Moussa N.A. (1976). *Textile Fabric Flammability.* The Massachusetts Institute of Technology, Cambridge, MA.

Bamford C.H., Crank J., and Malon D.H. (1946). The combustion of wood. Part I, *Proceedings of the Cambridge Philosophical Society*, 42, 166.

Barbot'ko S.L. (2010). Ways of ensuring the fire safety of aviation materials, *Ross. Khim. Zh. (Zh. Ross. Khim. Obshch. Im. D.I. Mendeleeva)*, 54(1), 121–126.

Begon M., Harper J.L., and Townsend C.R. (1996). *Ecology: Individuals, Populations and Communities*, 3rd edition. Wiley-Blackwell, USA, ISBN-13: 978-0632038015.

Bowman D.M.J.S., Balch J.K., Artaxo P., Bond W.J., Carlson J.M., Cochrane M.A., D'Antonio C.M., DeFries R.S., Doyle J.C., Harrison S.P., Johnston F.H., Keeley J. E., Krawchuk M.A., Kull C.A., Marston J.B., Moritz M.A., Prentice I.C., Roos C. I., Scott A.C., Swetnam T.W., van der Werf G.R., and Pyne S.J. (2009). Fire in the Earth system, *Science*, 324(5926), 481–484. Bibcode:2009Sci..324.481B. DOI: 10.1126/science.1163886. PMID 19390038.

Burke J. (2013). Bangladesh factory fires: Fashion industry's latest crisis, https://www.theguardian.com/world/2013/dec/08/bangladesh-factory-fires-fashion-latest-crisis, December 8, 2013 13.58 GMT.

Caitlin A. (2015). Brief history of FR, http://workingperson.me/, accessed on 24.2.2015.

Ceresana. (2019). Flame retardants market report, www.ceresana.com/, accessed on 7.2.2019.

Chylek P., Jennings S.G., and Pinnick R. (2015). *AEROSOLS | Soot, Encyclopedia of Atmospheric Sciences*, 2nd edition, Academic Press, USA, pp. 86–91. DOI: 10.1016/B978-0-12-382225-3.00375-3.

Cory A.K., Loloee R., Wichman I.S., and Ghosh R.N. (2009). Time resolved measurements of pyrolysis products from thermoplastic Poly-Methyl-Methacrylate (PMMA), *ASME 2009 International Mechanical Engineering Congress and Exposition*, FL, November 13–19, 2009.

CTIF. (2018). Report 23, world fire statistics 2018, www.ctif.org/world-fire-statistics, accessed on 14.2.2019.

Danelek J. (2011). Top 10 most famous fires in history, March 18, 2011, https://www.top tenz.net/top-10-most-famous-fires-in-history.php

EPA (U.S. Environmental Protection Agency). (2005). Environmental profiles of chemical flame-retardant alternatives for low-density polyurethane foam (report), EPA 742-R-05-002A. Retrieved April 4, 2013.

GRI-Mech. (2007). Gregory P. Smith; David M. Golden; Michael Frenklach; Nigel W. Moriarty; Boris Eiteneer; Mikhail Goldenberg; C. Thomas Bowman; Ronald K. Hanson; Soonho Song; William C. Gardiner, Jr.; Vitali V. Lissianski; Zhiwei Qin. GRI-Mech 3.0. Archived from the original on 29 October 2007, http://combus tion.berkeley.edu/gri-mech/

Helmenstine A.M. (2009). What is the state of matter of fire or flame? Is it a liquid, solid, or gas? https://www.thoughtco.com/what-state-of-matter-is-fire-604300, accessed on 29.4.2020.

Hensel F.G. (2011). The elektro tipp, www.hensel-electric.de/, accessed on 8.2.2019.

Hindersinn R.R. (1990). Historical aspects of polymer fire retardance, *ACS Symposium Series*, 425, 87–96. DOI: 10.1021/bk-1990-0425.ch007.

Hirschler M.M. (2000). Chemical aspects of thermal decomposition, in: A.F. Grand and C.A. Wilkie, Eds., *Fire Retardancy of Polymeric Materials*. Marcel Dekker, New York, pp. 27–79.

Horrocks A.R. (2011). Flame retardant challenges for textiles and fibers: New chemistry versusinnovatory solutions, *Polymer Degradation and Stability*, 96, 377–392.

HSE. (2019). Fire and explosion, Health and Safety Executive, www.hse.gov.uk/textiles/fire-explosion.htm, accessed on 13.2.2019.

InnoFireWood website. (2019). Burning of wood at the Wayback Machine, http://virtual.vtt.fi/accessed on 8.2.2019.

Islam M. (2018). Fire-proof clothing versus fire-resistant clothing, https://garmentsmerch andising.com/fire-resistant-vs-fire-proof-clothing/, accesed on 29.4.2020.

IUPAC. (2009). *Compendium of Chemical Terminology*. International Union of Pure and Applied Chemistry, p. 1824, https://goldbook.iupac.org, accessed on 8.2.2019.

Kasem M.A. and Rouette H.K. (1972). *Flammability and Flame Retardancy of Fabrics: A Review Swiss Federal Institute of Technology*. Technomic Publishing Co. Ltd., Westport, CT.

Khanna Y.P. and Pearce E.M. (1978). *Flame Retardant Polymeric Materials*. M. Lewin, S.M. Atlas, and E.M. Pearce, Eds., Vol. 2, Plenum, New York, p. 43.

Krajick K. (2011). Farmers, flames and climate: Are we entering an age of 'mega-fires'?– State of the planet, https://blogs.ei.columbia.edu/2011/11/16/farmers-flames-and-cli mate-are-we-entering-an-age-of-mega-fires/accessed on 5. 2.2019.

Levchik S. and Wilkie C.A. (2000). Char formation, in: A.F. Grand and C.A. Wilkie, Eds., *Fire Retardancy of Polymeric Materials*. Marcel Dekker, New York, pp. 171–215.

Levchik S.V. (2007). Introduction to flame retardancy and polymer flammability, in: A. B. Morgan and C.A. Wilkie, Eds., *Flame Retardant Polymer Nanocomposites*. John Wiley & Sons, Inc, New York, pp. 1–29.

Lewin M. (1985). *Flame Retardance of Fabrics, Handbook of Fiber Science Technology*. Marcel Decker, New York.

Lewin M. (1999). Synergistic and catalytic effects in flame retardancy of polymeric materials–An overview, *Journal of Fire Sciences*, 17, 3–19.

Lewin M. and Sello S.B. (1975). *Flame Retardant Polymeric Materials*. M. Lewin, S. M. Atlas, and E.M. Pearce, Eds., Vol. 1, Plenum, New York, p. 19.

Margolin S.M. (2012). Flame resistant clothing: Get 'the facts,' not 'the story', https://ohsonline.com/, Dec 11, 2012, accessed on 7.2.2019.

Martin S.B. (1964). Ignition of organic materials by radiation, in Fire Research Abstracts and Reviews–6.

Mikhailin Y.A. (2011). Flame resistance indices of polymer materials and methods of determining them, *Polymer Materials*, 8, 32–34.

Morris S.E. and Moses T.A. (1987). Forest fire and the natural soil erosion regime in the Colorado front range, *Annals of the Association of American Geographers*, 77(2), 245–254. DOI: 10.1111/j.1467-8306.1987.tb00156.x.

NASA. (2007). Experiment results archived 2007-03-12 at the Wayback Machine, National Aeronautics and Space Administration, April 2005.

NASA. (2010). Spiral flames in microgravity archived 2010-03-19 at the Wayback Machine, CFM-1 experiment results archived 2007-09-12 at the Wayback Machine, LSP-1, National Aeronautics and Space Administration, April, 2015.

NWCG (National Wildfire Coordinating Group). (2009). Glossary of Wildland Fire Terminology. November. Retrieved 2008-12-18.

Petrova G.N., Perfilova D.N., Rumyantseva T.V., and Beider E.Y. (2014). Self-extinguishing thermoplastic elastomers, *International Polymer Science and Technology*, 41(5), 98–116.

Schindler W.D. and Hauser P.J. (2004). *Chemical Finishing of Textiles*. Woodhead, Cambridge, England.

Schmidt-Rohr K. (2015). Why combustions are always exothermic, yielding about 418 kJ per mole of O_2, *Journal of Chemical Education*, 92(12), 2094–2099. DOI: 10.1021/acs.jchemed.5b00333.

Tesoro G.C. (1978). Chemical modification of polymers with flame-retardant compounds, *Journal of Polymer Science: Macromolecular Reviews*, 13(1), 283–353.

The Home Office (UK). (2018). Detailed analysis of fires attended by fire and rescue services, England, April 2017 to March 2018, The Government Statistical Office, UK, ISSN 0143 6384, 1998, 6th September, ISBN: 978-1-78655-701-8, www.gov.uk/government/collections/fire-statistics#contents, accessed on 11.2.2019.

Thomas H.E. and Dry Dale D.D. (1987). Flammability of plastics. I. Ignition temperatures, *Fire and Materials*, 11, 163–172.

Tomasino C. (1992). *Chemistry & Technology of Fabric Preparation & Finishing*. College of Textiles, NCSU, USA.

Uddin F. (2019). Flammability hazards in textiles, www.pakissan.com/english/issues/flammability.hazards.in.textiles.shtml, accessed on 13.2.2019.

United Nations. (2011). *Globally Harmonized System of Classification and Labeling of Chemicals*, 4th revised edition. New York and Geneva, www.unece.org/fileadmin/DAM/trans/danger/publi/ghs/ghs_rev04/English/ST-SG-AC10-30-Rev4e.pdf, accessed on 29.4.2020.

Van der Veen I. and de Boer J. (2012). Phosphorus flame retardants: Properties, production, environmental occurrence, toxicity and analysis, *Chemosphere*, 88(10), 1119–1153. Bibcode:2012Chmsp.88.1119V. DOI: 10.1016/j.chemosphere.2012.03.067. PMID 22537891.

Van Klevelen D.W. (1975). Some basic aspects of flame resistance of polymeric materials, *Polymer*, 16, 615–620.

Van Krevelen D.W. and Nijenhuis K.T. (2009). Properties of polymers: Their correlation with chemical structure; their numerical estimation and prediction from additive group contributions. Elsevier, p. 864.

Weil E.D. and Levchik S.V. (2015). *Flame Retardants for Plastics and Textiles: Practical Applications*. Carl Hanser Verlag, Munich, p. 97.

WHO. (1997). *Flame Retardants: A General Introduction*, A report on Environmental Health Criteria 192, www.inchem.org/documents/ehc/ehc/ehc192.htm.

WHO. (2018). Burn 6 March 2018, www.who.int/news-room/fact-sheets/detail/burns, accessed on 21.2.2019.

Wilkie C.A., Levchik S.V., and Levchik G.F. (2001). Is there a correlation between cross-linking and thermal stability?, in: S. Al-Malaika, A. Golovoy, and C.A. Wilkie, Eds., *Specialty Polymer Additives: Principles and Application.* Blackwell Science, Oxford, England, pp. 359–374.

www.derose.com (2019). Flame temperatures, www.derose.net/steve/resources/engtables/flametemp.html, accessed on 5.2.2019.

Yu H.D., Regulacio M.D., Ye E., and Han M.Y. (2013). Chemical routes to top-down nanofabrication, *Chemical Society Reviews*, 2, 6006–6018. DOI: 10.1039/c3cs60113g.

2 Flammability

2.1 INTRODUCTION

Flammability is the ability of a substance to burn or ignite, causing fire or combustion. The degree of difficulty required to cause the combustion of a substance is quantified through fire testing. Internationally, a variety of test protocols exist to quantify flammability. The ratings achieved are used in building codes, insurance requirements, fire codes, and other regulations governing the use of building materials, as well as in the storage and handling of highly flammable substances inside and outside of structures and in surface and air transportation. For instance, changing occupancy by altering the flammability of the contents requires the owner of a building to apply for a building permit to make sure that the overall fire protection design basis of the facility can take such a change into account.

In the United States, the fire code (also fire prevention code or fire safety code) is a model code adopted by the state or local jurisdiction, and enforced by fire prevention officers within municipal fire departments. It is a set of rules prescribing minimum requirements to prevent fire and explosion hazards arising from storage, handling, or use of dangerous materials, or from other specific hazardous conditions. The fire code complements the building code; it is aimed primarily at preventing fires, ensuring that necessary training and equipment are available, and that the original design basis of the building, including the basic plan set out by the architect, is not compromised. The fire code also addresses inspection and maintenance requirements of fire protection equipment in order to maintain optimal active fire protection and passive fire protection measures.

Flammable materials ignite more easily than other materials, are thus more hazardous, and must be handled with care. Combustible materials are less easily ignited and burn less vigorously. In the United States, flammable liquids, by definition, have a flash point below 100°F (38°C), whereas combustible liquids have a flash point above 100°F (38°C). Flammable solids are solids that are readily combustible, or may cause or contribute to fire through friction. Readily combustible solids are powdered, granular, or pasty substances that easily ignite by brief contact with an ignition source, such as a burning match, and spread flame rapidly. Technical definitions vary between countries, so the United Nations created the Globally Harmonized System of Classification and Labeling of Chemicals (OHSA, 2006), which defines the flash point temperature of flammable liquids as between 0°F (−17.8°C) and 140°F (60°C) and combustible liquids between 140°F (60°C) and 200°F (93°C).

All common apparel textiles ignite in air if exposed to a flame or other heat source of adequate intensity for a sufficient period. Once ignited, large differences in their burning behavior become apparent, and hence, the intensity of injury caused to the human skin varies with different textile materials. However, it is important to understand the flammability behavior of individual fibers in order to predict the flammability behavior of combinations of fibers in fabric form and its effect on burn severity. The action of heat on a textile material gives rise to a complex and interrelated series of physical and chemical effects. Clearly, the burning process of even simple fabrics is not that simple and several reviews exist in which fabric flammability is related to fabric properties such as fiber type, fiber blend ratio, and mass per unit area and fabric construction.

2.1.1 FACTORS FOR FLAMMABILITY

The following three principal groups of fabric properties that dictate their flammability characteristics:

- physical properties
- chemical properties
- thermal properties

The physical properties of a fabric include its mass, construction, and configuration. The ease of ignition and linear burning rate are directly proportional to the mass per unit area. The denser and heavier the fabric is, the longer it takes to ignite and the longer it takes to burn. The fabric construction refers to the surface smoothness. A tightly configured fabric reduces the level of oxygen available to support combustion. A knitted textile is more porous; hence, it ignites and burns faster than a woven fabric.

The major variables that influence the burning of fabrics are (CPSC, 1975; Belshaw and Jerram, 1986):

- fiber type and content,
- fabric mass,
- fabric structure, and
- geometry of garment and configuration.

The weave pattern, yarn twist, and possibly, the linear density of yarns, also alter the burning rates of cellulosic and thermoplastic materials to some extent. The dynamic process of fabric burning is also influenced by the surrounding conditions such as temperature, relative humidity, oxygen content, rate of airflow, and orientation of fabric; for example, ignition at the base of a vertically oriented fabric provides the highest burning rate for a given sample. The effect of garment fit and garment configuration complicates burning behavior.

Bhatnagar (1974) points out that burning is a chemical reaction; therefore, the chemical properties of any fabric are able to influence its flammability characteristics. Chemical properties of a fabric are determined by the fibers used in fabrication.

Professor Pailthorpe (2000) noted that textiles with higher concentrations of carbon are less reactive to fire, compared to those with lower concentrations.

The thermal properties of fabric suggested by Reeves and Drake (1971) are the textile's ability to absorb heat. The ability of a textile to absorb heat can command its ignition temperature and burning rate.

The flammability characteristics of the fabric are affected by the conditions under which it is tested (Reeves and Drake, 1971). The moisture level of the atmosphere, and hence, the ability of the fabric to absorb the moisture, determine its burning ability.

Textile materials have very high fiber surface to mass ratios and hence tend to ignite easily and burn faster than other materials. Different fabrics exhibit different rate of ignition and ease of ignition, depending on fabric mass. The heavier the fabric, the longer it takes to ignite, compared toa light sheer fabric made of the same material. Surface characteristics, of course, have a bearing on this factor. A loose pile fabric usually ignites more easily than compact smooth surface fabric (Bennett, 1973). The raised fibers have a much larger exposed surface and can ignite easily with a rapid flash of fire across the fabric surface. In some cases, these surface flashes may cause the entire fabric to burn, but in others, the surface flash may not produce enough heat to ignite the base fabric.

Ignition sources in actual fires include malfunction of electrical heaters and hot plates, open flames, fuel-fired objects (stoves and fireplaces), and smoking materials (cigarettes, matches, lighters). Ignition can be caused by three usual methods of heat transfer, namely:

1. conduction: solid-phase conduction, if the fabric is in actual contact with the hot source.
2. convection: fire spread by convection from flaming sources
3. radiations: predominates for electrical heaters and hot plates

Self-extinguishing is flame spread-resistance. In general, the product or material is self-extinguishing if the flame is extinguished within a short time after the flame source is removed from the product or material. *Self-extinguishing* is a property of an object, meaning that the object can catch fire, but the flame would be extinguished without any influence on the object.

The usual criteria for self-extinguishment are char length, after-flame and afterglow time, and the presence of melt drip are:

- Char length is the length of the fabric area that is destroyed by the flame, measured from the point of flame application. The destruction can be due to charring or melting.
- After-flame time is the time during which the flame on the fabric continues to burn after the ignition flame is removed and is determined visually.
- Similarly, afterglow time is the time until the glow persists on the specimen after the disappearance of flame.
- Residual flame time is the time during which burning fragments falling from the fabric burn on the bottom of the cabinet. Such fragments, often called melt drip, have limiting injury potential, but can ignite other garments.

2.2   FLAMMABILITY OF TEXTILE FIBERS

Flammability depends on the composition of the yarn or fabric, i.e., which fibers it is made of. Flammability properties of different fibers are listed in Table 2.1.

It has been established that the size and depth of burn injuries are related to heat transfer, rather than the rate of flame-spread. Heavier fabrics produce more heat per unit area, although the flame spreads at a slower rate through them. The flammability characteristics of commercially available fibers vary widely.

Ignition of fabrics by flame is much more rapid at a cut edge than at the surface. This appears to be due to the higher temperature at the edges caused by the limited heat transfer by conduction away from the heat source. It has been suggested that surface ignition is a more practical test of fire resistance for a moderate risk environment than is *edge ignition* because garments with cut edges are not common. A significantly smaller amount of FR is needed to prevent surface ignition of PET-cotton blends, compared with edge ignition.

Both edge and surface ignition of cotton and blended polyester–cotton fabrics increases as the mass of the fabric increases. Surface ignition time per unit mass of fabric exceeds edge ignition time. Surface ignition time for cotton, as well for polyester–cotton fabric is twice that of edge ignition, and the difference is much greater for heavier fabrics.

TABLE 2.1

Flame characteristics of different textile fibers

Textile Fibers	Flame characteristics
Natural fibres	
Cotton	Burns readily with ash char formation and afterglow
Wool	Burns readily and form hair burning smell
Manmade (regenerated)	
Rayon	Burns readily with ash char formation and afterglow
Manmade (synthetic)	
Acetate	Burns and melts ahead of the flame
Triacetate	Burns readily and melts
Polyamide 6	Melts and forms a hard bead. Supports combustion with difficulty
Polyamide 66	Melts and forms a hard bead. Does not readily supports combustion
Polyester	Burns readily with soot
Acrylic	Burning readily with melting and sputtering
Polyethylene	Burns slowly
Polypropylene	Burns slowly
Nomex	Combustion with difficulty
PTFE	Does not easily burn

2.2.1 Propagation of Flame

In the case of cotton and cotton–PET blends, the fabric weight plays a dominant role in flame propagation. Other important parameters are the spacing between fabric and skin stimulant, the orientation of the sample and the direction of burning. For horizontal burning, a linear relationship exists between the flame propagation rate and the fabric weight, whereas a parabolic dependence was found for the vertical up-burning. In case of cotton–PET blends, the flame propagation is faster than for cotton, but the shape of the curves is similar. The flame velocities are higher by an order of magnitude for the vertical burning, compared with horizontal orientation.

The flame propagation rate of synthetic fabrics tends to be slow because of melting and shrinking away ahead of the flame; it depends on fiber composition, fabric structure, and the proximity of a nonthermoplastic fabric. Lightweight thermoplastic fabrics do not sustain combustion. Pure nylon or polyester heavyweight fabrics burn, giving out considerable black smoke. A small percentage of FRs can prevent the dripping of molten materials.

Besides fabric structural effects, Miller and Goswami (1971) have analyzed the effect of various yarn parameters on the burning behavior of fabrics. They found that no-twist yarns show higher burning rate values. The yarns with twist from 0 to 1.37 turns per cm decrease the mass burning rate by up to 40%. Thus, the low flammability of the twisted yarns can be attributed to the reduction of voids in the yarn. This has two effects; the first is the reduced access of fiber surfaces to air, and hence, reduced combustion rates; the second is the greater thermal conductivity across fibers in the yarn cross section, which could be expected to increase burning rates. The first effect appears to be the predominant one.

The burning of polyester/cotton blended textiles is far more intense and hazardous than expected from the average flammability of the individual blend component. Cotton begins to decompose thermally at about 350°C, whereas polyester decomposes at between 420°C–447°C. Flammable pyrolysis vapors are more readily formed from the cotton component, which ignites in the presence of sufficient oxygen. However, due to the thermoplastic behavior of polyester it shrinks and then melts (above 260°C) before the pyrolysis of cotton occurs. This melting of polyester envelops the surface of the polyester and cotton fibers developing carbonaceous char, which prevents any shrinkage of the blended fabric, keeping it away from an approaching flame or igniting source (Kolhatkar, 2006).

Moreover, the cellulosic carbonaceous char not only supports the molten polyester, but also acts as a wick into the flame source, thus enhancing the fuel supply within the flame zone (Horrocks, 1989). This so called scaffolding effect causes extremely intense burning of polyester–cotton blended fabrics which combine the high flame temperature of cotton and the ability to adhere to the victim due to the presence of molten polymer. Several workers have suggested that such burning blend combinations are even synergistic in their behavior, and that the rate of heat release from polyester–cotton blends is

proportional to the cotton content, but the rate of burning is controlled by
polyester content (Miller et al., 1976; Horrocks, 1989). In the case of regenerated fibers such as acetate and triacetate, woven fabrics often ignite and burn
vigorously, while knitted fabrics tend to shrink from the flames and do not
ignite and burn as readily.

Protein fibers such as silk and wool are more difficult to ignite. It takes
twice as much heat to ignite wool as cotton and wool extinguishes more
quickly in moving air (Mehta and Martin, 1975). Because of their nitrogen
contents,(and in the case of wool, sulfur content), both silk and wool fibers
have low intrinsic flammability with Limiting Oxygen Index (LOI) values of
23% and 25%, respectively. Wool is in the class of fibers that char and burn,
but do so to a lesser degree and at a slower rate than the other fibers in this
class, namely, acrylics and cellulosics.

2.3 THERMAL PROPERTIES OF TEXTILES

Various thermal parameters of textile fibers such as glass transition temperature
(T_g), melting point (T_m), ignition/pyrolysis temperature (T_p), combustion
temperature (T_c), maximum flame temperature, and LOI are shown in
Table 2.2 (Reeves et al., 1974). The average heat of combustion of textile
fibers determined by Bomb Calorimeter is also shown (Babrauskas, 1981).

All fabrics burn, but some burn more readily than others. Untreated,
natural fibers such as cotton, linen, and silk burn more readily than wool.
Wool is more difficult to ignite and it burns with a low flame velocity. In
spite of natural inherent flame retardancy, woolen fabrics burn easily due
to open fabric structure and the presence of dyes and finishes. Cellulosic
fibers, namely, cotton and rayon, burn readily with afterglow and with the
formation of char. Synthetic fabrics such as polyamide, acrylic, and polyester resist ignition. However, once ignited, these fibers melt. The hot, sticky
melts are of very high temperature and can cause deep and severe skin
burns. One major concern regarding such synthetics as polyester is that
they have a tendency to shrink at relatively low temperatures. In clothing,
this may be extremely dangerous, e.g., synthetic fiber underwear can shrink
onto the body when the outerwear burns, thereby removing the previously
present insulating layer of air. Polyvinyl chloride and polyvinylidene chloride fibers do not support combustion, polyamide 6, polyamide 6,6 and
polypropylene melt slowly without burning, while polyester and acrylic
melt and burn readily.

Blended fabrics may be more hazardous because, in most cases, they combine
a high rate of burning, with the problem of melting resulting in more severe skin
injuries. Loosely woven fabrics burn more easily than tightly woven fabrics. The
surface texture with loose or fluffy piles ignites more readily than fabric with
hard, tight surfaces. Garment designs also affect flammability – clothing that fits
closely to the body is less likely to stray into a flame source and if ignited, tends
to self-extinguish. Floating extra fabrics add to the hazards.

TABLE 2.2

Various thermal parameters of textile fibers

Fiber	T_g	T_m	T_p	T_c (°C)	MFT*	AHC*	LOI*
Natural fibers							
Cotton			350	400	860	4,330	18.4
Rayon			350	420	850	3,446	18.7
Wool			245	600	941	4,920	25.2
Synthetic fibers							
Acetate				475	960		
Triacetate				540	885		
Polyamide 6	50	215	431	450	875		20–21.5
Polyamide 6,6	50	265	403	530	–	6,926	20–20.1
Polyester	85	255	420–427	450–480	697	6,170	20–21
Acrylic	100	220	290	560	855	7,020	18.2
Modacrylic	80	240	273	690			29–30
Poly- propylene	-20	165	469	550–570	839	11,600	18.6
PTFE	126	327	400	560			95
Nomex	275	375	410	500			28.5–30
Kevlar	340		590	550			29
PBI	400		500	500			40–42

Temperature (T) in °C, *MFT – Maximum Flame Temperature (°C), *AHC-
Average Heat of Combustion (cal/g), *LOI – Limiting Oxygen Index
Ref.: Lewin and Sello, 1984, *Cullis and Herschler, 1981.

2.3.1 FLAME TEMPERATURE

Flame temperature indicates sensitively the interactions in the materials during burning. Miller and Martin (1975) extensively reviewed the fabric flame temperatures of blended fabrics. They observed that the flame temperature of some multicomponent fibrous systems is the same as that of the hotter burning components. For example, 100% cotton fabric burns at 1000° C and 100% polyester burns at 850°C, but the polyester–cotton 50–50 blend fabric burns at about 950°C, indicating that no additional chemical reactions are occurring. There are some blends, which burn at an average flame temperature of the two components, suggesting chemical interaction between the components.

Table 2.3 shows flame temperature of various double fabric layers (Miller and Martin, 1975). The table shows that the cotton-blended fabrics burn with much higher temperatures than those of the thermoplastic components of the blend. Carter et al. (1973) further studied the effect of blend ratio on flame temperatures of polyester–cotton fabrics and found that fabrics with higher polyester content burn with lower flame temperature.

TABLE 2.3

Flame Temperatures (°C) of Double Layer Fabrics (Horizontal Burning) at 21% O_2

Serial no.	Types of Blend	Flame temperature
1	Cotton–acrylic	976
2	Cotton–cotton	974
3	Cotton–polyester	950
4	Polyester–polyester	649
5	Acrylic–acrylic	910
6	Nylon–nylon	860
7	Acrylic–nylon	942
8	Cotton–nylon	902

In Table 2.3, the blends of serial numbers 1–6 are noninteracting and therefore burn with higher flame temperatures, while the last two blends are interacting and hence, the flame temperature are lower (average of two components).

In general, the more carbon and hydrogen are present in the chemical structure, the more heat emission will occur when material burns. Thus, many synthetics have a potential to give off more heat when they burn than do an equivalent amount of a cellulosic material; of course, the rate of heat release will vary. Moreover, the melting residue of polyester and nylon fibers holds heat and cools very slowly to form a beadlike plastic residue. These melting residues are at very high temperatures and can cause severe skin injury because they shrink and tend to stick to the skin. Increasing oxygen supply usually promotes burning, but does not necessarily increase the heat emission. However, the flame temperature does increase with increase in oxygen content (Kolhatkar, 2006).

2.3.2 FLAME SPREAD

Flame spread is influenced by various clothing structural factors such as the presence of belts, ties, cuffs and collars and tight fitting areas, since they act as fire stops. Loose fitting or flowing garments exhibit the so-called chimney effect, which makes the flame spread more rapidly up vertical to fabrics. Moreover, full-length nightdresses burn more vigorously than knee length garments of the same fabric. The flame spread is particularly rapid in the lightweight fabrics selected for nightwear if the more flammable fibers are present.

Moussa et al. (1973) studied flame-spreading mechanisms for different textile materials. His findings are as follows:

- For cellulosic fabrics, the most important physical process is the transfer of heat from the flame to the virgin material and the creation and subsequent diffusion of the combustible vapors to the fuel flame.

- The flame-spreading mechanisms are not affected by the fine structure of the cellulose.
- The materials of different porosities and of different interlacing geometries are found to have the same flame-spreading speed.
- For thermoplastic fabrics, shrinking, melting, and dripping play important roles in the flame-spreading mechanism. High shrinking levels may even extinguish the flame. Dripping can cause a reduction or increase in flame size, a fire jump, and/or a change in the flow field around the fabric.
- For blended fabrics, both the process characteristics of cellulose and thermoplastics are observed.
- For fuzzy cellulosic materials, a flashing flame consumes the outer layer of fuzz whereas the interior layer may smolder instead of burning.

2.4 LOI AND FLAMMABILITY

As discussed earlier the limiting oxygen index is the minimum concentration (%) of oxygen in the atmosphere surrounding a polymeric object that will support combustion of the polymer. The relation between LOI and flammability of different textile fibers is as follows:

- LOI ≤ 19: Easy ignition, rapid burning (cotton, acrylic, viscose, polypropylene)
- 19 ≤ LOI ≤ 22: Normal ignition and burning behavior (polyester, polyamide 6, polyamide 6,6)
- LOI around 25: Almost ignition resistant (wool)
- LOI > 26: Flame retardant (modacrylic, meta-aramid, para-aramid)
- LOI > 30: Flame resistant under severe conditions e.g., heavy air ventilation (melamine/phenol formaldehyde, polybenzimidazole or PBI, PTEE etc.)
- LOI around 100: Not burning even in pure oxygen only melting (glass and ceramic fibers)

The LOI values of some common polymers are given in Table 2.4

2.5 FLAMMABILITY TESTS

Nearly every country has its own set of textile fire testing standard methods in order to relate to the special local social and technical factors prevailing in that country. In addition, the test methods are defined by a number of national and international bodies such as air, land, and sea transport authorities, insurance organizations, and governmental departments relating to industry, defense, and health in particular. A brief overview of the various and many test methods that were available up to 1989 is given by Horrocks et al. (1989), and a more recent list of tests specifically relating to interior textiles has been published by Trevira GmbH in 1997 (Trevera, 1997).

TABLE 2.4

The LOI values of some common polymers

Polymer	limiting oxygen index (LOI)
Standard Polyurethane foam	16.5
PMMA (Perspex)	17.3
Poly(propylene)	17.4
Poly(styrene)	17.8
Plywood	23.0
Nylon 6.6	24–29
Polycarbonate	25–44
Nomex	28.5
Phenolic	26–64
PVC (unplasticized)	45–49
(PTFE)	95

The burning processes of materials such as textiles are complex and difficult to quantify in terms of their ignition and post-ignition behavior, because they are *thermally thin*, with a high specific volume and oxygen accessibility and hence difficult to rank, relative to other polymeric materials. Most common textile flammability tests are currently based on easiness of ignition and/or burning rate behavior, which can be easily quantified for fabrics and composites in varying geometries. Few, however, yield quantitative and fire science-related data, unlike the often maligned oxygen index methods. LOI, while it proves to be a very effective indicator of ease of ignition, has not achieved the status of an official test within the textile arena. For instance, it is well known that in order to achieve a degree of fabric flame retardancy sufficient to pass a typical vertical strip test, an LOI value of at least 26%–27% is required, which must be measurable in a reproducible fashion. However, because the sample ignition occurs at the top to give a vertically downward burning geometry, this is considered to be unrepresentative of the ignition geometry in the real world. Furthermore, the exact LOI value is influenced by fabric structural variables for the same fiber type and is not single-valued for a given fiber type or blend. However, it finds significant use in developing new flame retardants and optimizing levels of application to fibers and textiles (Horrocks and Price, 2001).

Ideally, all practical tests should be based on quite straightforward principles which transform into a practically simple and convenient-to-use test method. Observed parameters, such as time-to-ignition, post-ignition – after-flame times, burning rates, and nature of damage and debris produced should be measured with an acceptable and defined degree of accuracy. The measurement should be reproducible.

Furthermore, for similarly flame-retarded fabrics, the length of the damaged or char length can show semiquantitative relationships with the level of flame retardancy as determined by methods such as LOI.

As textile materials are used in more complex and demanding environments, the associated test procedures become more complex. This is especially the case for protective clothing when the garment and its components have to function not only as a typical textile material, and also be resistant to a number of agencies including heat and flame.

Flammability test methods can be divided into two general subdivisions, namely:

- Small-scale testing (or bench scale testing)
- Large-scale testing

Small-scale tests use a small section of the garment while large-scale tests use the entire garment.

2.5.1 SMALL-SCALE

Small-scale tests are performed on a bench top and are usually not very complicated. Fabric properties are the general concern in these tests. A small section of fabric is cut away from a large roll of fabric or a garment. These are also known as flaming exposure test methods; typically a flame is impinged on the fabric and subsequent flame-spread time is measured, along with the total distance of charring. These methods are generally inexpensive and easy to run, leading to a high level of repeatability. System behavior is considered to be an indication of the performance of the entire system. Small-scale tests are generally preferred since they are simple and inexpensive to run, compared to their full-scale counterparts.

The small-scale test methods for fabric flammability may be divided into two general categories based on the test setup. In the first category of tests, the sample of clothing is exposed to an ignition or pilot flame. The post-pilot time during which the material burns and the char lengths are the usual results sought in these tests. The second category of tests involves subjecting the material to a known heat flux or temperature and measuring the heat flux or temperature on the opposite side of the material. Usually the heat flux or temperature measured is correlated to burn data to determine the extent of burns that human skin might experience under the test conditions.

2.5.2 LARGE-SCALE

Large-scale testing is more complicated than the bench top tests and involve the use of the entire fabric system, or the whole garment. This type of testing generally provides a more detailed analysis of a fabric's performance, at a price greater than small-scale tests.

This type of testing is advantageous because the entire system is tested, instead of merely a sample. The behavior of the garment depends upon its configuration, as well as the materials in a fire scenario.

The critical test parameters for determining clothing exposure can be divided into three groups, namely (Fay, 2002):

* duration of exposure,
* intensity and,
* type of heat transfer involved.

The duration of exposure is a difficult variable to analyze because the behavior of people is difficult to predict. The time after which a fabric no longer needs to provide protection from a fire, is known as time to failure. Using this parameter instead of exposure time for a fabric is useful as the time to failure is a constant value for a given scenario, while the exposure time can vary. Knowing the time to failure of a fabric in various scenarios allows for a good comparison of how different fabrics perform.

In different fire scenarios, the time to failure can vary significantly. For example, a fabric may have a higher resistance to flame impingement than to radiant heat transfer, and as, such the time to failure would be greater in the test by flame impingement.

It is important to define the range of different fire scenarios to which a fabric or garment might be exposed. This range of fires can then be evaluated to determine a reasonable fire configuration in which the garment's performance can be evaluated. Exposing a garment to a fire scenario that is beyond the range of plausible fire environments does not seem an appropriate use of resources, because the garment is being tested beyond its intended use.

Convective and radiative heat transfer also plays a significant role in evaluating a fabric's performance as these two methods of heat transfer are the most likely exposure to be faced from a fire. It is also plausible that a fire that cannot be seen could be significant enough to cause thermal exposure to an individual nearby or in an adjacent corridor. However, accurately predicting the modes of heat transfer from a fire to a target is difficult to accomplish because small changes in the orientation of clothing, type of fuel burning, or the ventilation in a room could have significant impact on the heat transfer (Leblanc, 1998).

Determining the flammability potential of a fabric requires an understanding of a number of tests because flammability tests and specifications vary with the end use of the textile article. Some tests have been developed as research tools to quantify the retardancy value of finishes and fibers; LOI is one that is used often. Other tests have been developed to assess the flammability hazard of fabrics. These emulate actual in-service conditions that the textile is liable to encounter. Some of the variables are:

1. The way in which the heat source is presented as it is being ignited, i.e., vertical, horizontal, or 45-degree angle. Upward fire spread is far faster than downward and horizontal flame spread and, hence, upward fire spread is adopted as a better means of assessing the fire hazard of a fabric.
2. The temperature of the heat source.

3. Char length, after-flame and afterglow, and melt drip are some of the specifications of the specific test.

Horrocks (2011) presents a thorough review of the flammability tests. The test methods for evaluation of flammability and flame resistance issued by governments, institutes, societies, and administrative agencies are large. Many of the test methods are similar and differ only in minor details. They can be differentiated into two basic groups:

1. Comparison of the flammability of clothing textiles
2. Measurement of the degree of fire resistance of flame retardant materials

Recently, tests for the heat released during burning and the estimation of the smoke hazard and of toxicity of the products of combustion came into prominence. The source of ignition (burners and gas), number, size, and moisture content of the sample, as well as dimensions of the combustion chamber are meticulously described.

The whole area of flammability testing is very large and very much diversified. Hence, it is convenient to classify those into a few groups. One possible way of classification may be as follows:

1. Ignition tests (e.g., vertical, horizontal and inclined strip burning tests and UL94)
2. Reaction-to-fire tests (e.g., cone calorimetry, single burning item test)
3. Application-based (often composite) tests (e.g., EN 469 Firefighter's' clothing)
4. Radiant energy tests (e.g., ISO 506 manikin test)
5. Scientific methods for mechanistic elucidation (e.g., thermal analysis, LOI, microcombustion calorimetry).

Some of the commonly used flammability tests are listed in Table 2.5 (Schindler and Hauser, 2004).

2.5.3   IGNITION TESTS

Most of the tests address ignition, afterglow, char area or length upon burning, burning time of a particular length of fabric, mass loss, rate of surface spread of flame. The most important aspects are:

- Ignition
- Spread of flame
- Burn injury

Ignition takes place when atmospheric oxygen interacts with volatile combustible products ejected during decomposition of heated polymers. For thermoplastic fabrics, additionally, fabric shrinks and melts away from the flame and drips.

TABLE 2.5

Brief Specifications of a few commonly known flammability tests

Standard test methods	Recommended for	Short description
• ASTM D1230–17 • 16 CFR 1610	General apparel	Fabric ($2''$x $6''$) at 45 degrees (inclined), 16-mm butane burner, flame contact (FC) 1 s. The time required for the flame to propagate 5 in (127 mm) is recorded. Normal – burning time 3.5 s or more, rapid and intense burning – burning time less than 3.5 s.
ASTM D6413/15 D6413M – 15	Vertical flame resistance of textiles in general	Fabric ($12''$x $3''$), Vertical, FC 12 s. Flame height: 1.5 in. The bottom edge is ¾ in. above the burner's top. Measure after-flame and afterglow
ASTM D6545–18 16 CFR 1615/1616	Children's sleepwear (more stringent)	Fabric ($10''$x $3.5''$), Vertical, FC 3 s. burner with at least 97% pure methane. Subsequent flaming period (S) and char length (l) are measured. Fails if S = 10 s, l = 7 in (17.8 cm)
ASTM D2859-16 (16 CFR 1630 AND 1631)	Textile floor covering material	Fabric ($9''$x $9''$), horizontal, A 1588 methenamine tablet no. placed in the centre of the specimen, ignited, and the shortest distance between the charred area of the floor covering and the inside edge of the specified frame is measured, the charred area should not exceed one inch of the edge of the hole.
NFPA 701-77/NFPA 1971	Tent, upholstery, tarpaulin/protective clothing	Test 1. Specimen (150 mm x 400 mm), vertical, 100 mm flame, contact time (CT): 45s. After burn time not exceeding 2 s, %weight loss are measured. Test 2. Specimen (denser than 700 g/m^2), 1.2 m long, vertical flame of 280 mm±12 mm, CT: 2 minutes. Flaming combustion, after-burn, and afterglow time are measured.
ASTM F 1358–16 FTMS 5903.1	Flame impingement on materials used in protective clothing Not primarily designated for flame resistance.	Suspended over a gas flame, FC: 3 s. If no ignition, further expose to gas flame for 12 s. In FTMS 5903.1 – vertical position After-burn, and afterglow time, char length are measured.
BS 5852, part 1 & 2 EN 1021, EN 597	Ignitability of Upholstery against smoldering cigarettes and matches	Must be soaking-resistant at 40°C, then horizontally and vertically fixed on a minichair (supported foamed PU), tested by 7 ignition methods. FC = 40–70 s. If no flaming or progressive smoldering is observed on both cover and interior material, the test is recorded as no ignition and the material passes the test.

(Continued)

TABLE 2.5 (Cont.)

Standard test methods	Recommended for	Short description
ISO 6940-2004/6941	It determines the ease of ignition, as defined in ISO 4880; and measures flame spread time (see ISO 6941) of vertically oriented specimens.	A defined flame from a specified burner is applied for 10 s to the surface or the bottom edge of vertically-oriented textile specimens. The flame spread times in seconds for the flame front to travel between marker threads positioned adjacent to the surface of the test specimen and located at three distances from the igniting flame, are recorded.
DIN 54,333, part 1, 1981	Burning behavior of all types of fabrics, including coated, bonded, laminated, flocked	less severe test Size: 340 mm x 70 mm specimen in a horizontal position 40 mm flame height Perpendicular burner 15 s edge ignition Flame application 20 mm above the lower edge Burning time between marked places or until the flame extinguishes
UL94 of Underwriters Laboratories Inc. (UL)	Flammability of Plastic Materials	UL 94 contains 94HB, 94V, 94VTM, 94-5V, 94HBF, 94HF and radiant panel. The 94HB test is the horizontal burn method, while Methods 94V and 94VTM are used for more stringent vertical burning. The 94-5V test is for products that are not movable or are fixed to a conduit system. The 94HBF and HF tests are applicable for nonstructural foam, i.e., acoustical foam materials. Radiant panel test (ASTM, E162) determines the flame spread of a material exposed to fire.

Sponsoring organizations:
CPSC – Consumer Product Safety Commission, United States
NFPA – National Firefighters Protection Association, United States
ASTM – American Society for Testing Materials, United States
CFR – United States Government, Office of the Federal Registrar, 1995
BS – British Standards Institution
ISO – International Standards Organisation
DIN – Deutsches Institut für Nordmung

100% polyamide and polyester fabrics do not ignite when in contact with electric burners, irrespective of temperature, whereas cotton fabrics ignite at 600°C under similar conditions.

Different performance requirements and government regulations have led to the development of numerous test methods for evaluating the flame retardancy of textiles. Different test methods are developed with the vertical, horizontal, and inclined arrangement of the samples, with and without air ventilation.

Flammability test methods measure how easily materials ignite, how quickly they burn and how they react when burned. The materials are placed over a Bunsen burner, either vertically or horizontally, depending on the specification. During a vertical flammability test, a material is observed for the length of time it burns after the igniting flame is removed, how much of the specimen burns and whether it drips flaming particles. In contrast, horizontal flammability tests observe if the material continues to burn after the test flame is removed, and then calculate the rate at which the specimen burns.

The orientation of the specimen has a significant effect on its propensity for flame spread. The following configurations are ordered according to their likelihood of supporting flame propagation over the surface:

- Vertical specimen ignited at the bottom edge;
- Horizontal specimen facing down ignited at the underside;
- Horizontal specimen facing up ignited on the top side;
- Lateral specimen ignited at one end; and
- Vertical specimen ignited at the top edge.

This means, for example, that it is much easier to sustain upward flame spread over the surface of a vertical specimen ignited at the bottom edge than it is to support flame spread over the surface of the same specimen exposed to the same heat source from the top down.

Flammability testing of mattresses and furniture is an essential part of ensuring fire safety for consumers. These flammability tests occur in a burn room that contributes to the measurement of heat release, smoke release and opacity, combustive gas release and total mass loss. This type of flammability test ignites mattresses or furniture with a lit cigarette or a large open-flame to determine how a product will react in cases of accidental and intentional fires.

A few important flammability testing methods are discussed next.

2.5.4 INCLINED PLANE FLAMMABILITY TEST

ASTM D1230-17 (ASTM, 2019, Vol. 07.01) Standard Test Method for Flammability of Apparel Textiles, and 16 CFR 1610, covers textile fabrics intended to be used as apparel (other than children sleepwear or protective clothing).

Most textile materials can be evaluated using this test with the following exceptions: children's sleepwear, protective clothing, hats/gloves, footwear, and interlining fabric. Two factors are measured, namely:

1. Ease of ignition (how fast the sample catches fire).
2. Flame spread time (The time required for a flame to spread a certain distance).

All fabrics made of natural and regenerated cellulose, as well as many others made from natural or man-made fibers, are combustible. The potential danger to the wearer depends on such factors as ease of ignition, flame spread time and amount of heat released, and design of the garment. The test measures two such factors, namely, ease of ignition and flame spread time.

2.5.5 WORKING PRINCIPLE

The test should be conducted under standard atmosphere, maintaining standard temperature and moisture. Samples are mounted in a frame and held in a special apparatus at an angle of 45 degrees, (shown schematically in Figure 2.1 and photograph in Figure 2.2). A standardized flame is applied to the surface near the lower end for specified amount of time (e.g., 1 s). The flame travels up the length of the fabric (usually 5 in.) to a trigger string, which drops a weight to stop the timer when burned through. The time required for the flame travel the length of the fabric and break the trigger string is recorded, as well as the fabric's physical reaction(s) at the ignition point.

Equipment and tools:

- 45° flammability tester
- Precut fabric sample(s)
- Sample frame with four clips
- Brushing device (for piled or napped fabrics only)
- 50/3 mercerized sewing thread
- Double-sided tape (for slippery fabrics only)
- Sparker

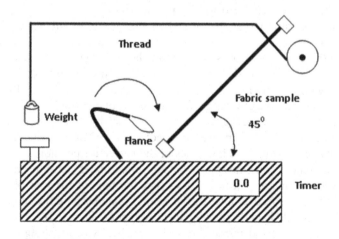

FIGURE 2.1 Schematic diagram of a inclined flammability tester.

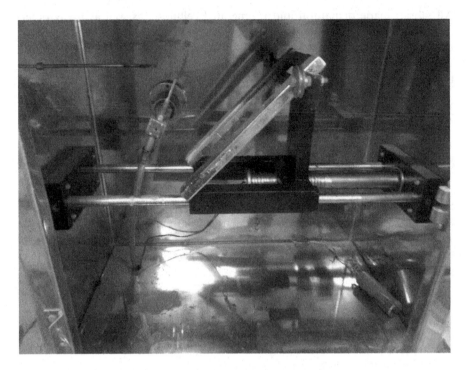

FIGURE 2.2   Inclined flammability tester (photograph).

Sample Preparation

1. The samples (both warp and weft directions) of 6.5"x 2"are cut.
2. All fabrics are oven dried for 30 min. at 105°C
3. All fabrics should then be placed in desiccator for at least 15 min. before testing
4. The individual specimen is clamped in the specimen holder, which consists of two frames – the lower one is small and the top one is longer. The specimen is inserted in the frame so that the bottom edge of the specimen coincides exactly with the lower edge of the longest (top) frame. The two halves of the frame are held together by bulldog clip. In some instruments, U Type 2 plates are used; one of them has five projected pins, while the other has five holes in the respective positions. The specimen is placed on the plate having pins under tension and then the second plate is placed on it so that the pins of the first plate pass through the respective holes of the second plate keeping the sandwiched specimen firmly held disallowing its movement (Figure 2.3)
5. The piled or napped fabric should be brushed with brushing device to raise the surface fibers

FIGURE 2.3 Specimen holders for inclined flammability test.

Preparing the Flammability Tester

- Main power switch is OFF.
- Timer is set at zero.
- The rack is moved to right, using lever arm located in front panel. The sample holder is placed in an instrument sample rack, so that the longest frame is on top.
- The rack is moved to the left until the sample comes in contact with the L–shaped locating arm. The burner tip now remains 5/16 in. away from the face of the specimen.
- The U–shaped glass manometer is filled with water to a convenient level
- Stop cord of 50/3 mercerized cotton sewing thread is threaded through instrument.

Performing the Test

1. The main power switch is turned ON.
2. The suitable auto impingement time (1, 5, 10 s, or manual) is selected. The control valve of the fuel supply is opened for 5 min. for the air to

be driven out. The gas is ignited and the flame height is adjusted to 16 mm (5/8 in.) from tip to the opening in the gas nozzle.

3. On pressing start button, the impingement is automatic and the flame is applied for the selected period.
4. The time required for the flame to proceed up to the fabric at a distance of 127 mm (5 in.) is recorded. It is also noted whether the base of a raised-surface fabric ignites, chars or melts.
5. The timer starts automatically – starting upon application of the flame and ending when the weight is released by the burning of the stop cord.
6. The results are recorded.
7. When testing is done, the power switch is turned OFF. The gas supply is also turned off.

Results

1. The arithmetic mean flame-spread time of all the specimens, usually 6 or 12 in number, is calculated. Faster the burning (less burning time), more chance for labeling as flammable and thereby subject to rejection.
2. If the mean time is less than 3.5 seconds, or any of the specimens do not burn, an additional set of specimens is tested.

The following three classes are used by the Consumer Product Safety Commission (CPSC), USA to interpret results for a similar test:

Class I

• For the textiles with and without raised fiber surface having an average time of flame spread in the test of ≥ 7 s and ≥ 3.5 s, respectively are considered by the trade to be generally acceptable for apparel.
• The fabrics those do not burn or burn with a surface flash (in less than 7 seconds) in which the base fabric is not affected by the flame are also accepted.

Class II
These textiles are considered by the trade to have flammability characteristics for apparel intermediate between Class I and Class III fabrics are limited to the following:

• The textiles having a raised fiber surface that have an average time of flame spread in the test of 4 to 7 seconds and in which the base fabric is ignited, charred or melted.

Class III
These textiles are considered by the trade to be unsuitable for apparel and are limited to the following:

- The textiles that do not have a raised fiber surface that have an average time of flame spread in the test of less than 3.5 seconds.
- The textiles having a raised fiber surface that have an average time of flame spread in the test of less than 4 seconds in which the base is ignited, charred, or melted.

2.5.6   VERTICAL FLAMMABILITY TEST OF TEXTILES OTHER THAN CHILDREN SLEEPWEAR

The concept of flammability assessment are opposite in inclined and vertical tests. In inclined test, the time required to burn a certain length of specimen (5 in.) is measured, while in vertical test, the after-flame time is measured. In the first case, shorter time means faster burning i.e. more flammable specimen, while in second case, shorter time means the flame extinguishes quickly i.e. less flammable specimen.

ASTM D6413/15 and D6413M/15 methods (ASTM, 2019, section 07.02) method determines the response of textiles to a standard ignition source, deriving measurement values for after-flame time, afterglow time, and char length. The vertical flame resistance, as determined by this test method, only relates to a specified flame exposure and application time. This test method maintains the specimen in a static, draft-free, vertical position and does not involve movement except that resulting from the exposure.

A specimen is positioned vertically above a controlled flame and exposed for a specified period of time. The test cabinet is made of a galvanized sheet or suitable metal with inner side painted black to facilitate viewing of specimen and flame. The test cabinet shall be set up in a laboratory hood so that combustion gases can be removed (Figure 2.4).

- Five lengthwise and five widthwise test specimens measuring $3''$(76 mm) by $12''$(300 mm) are cut and consider the long direction as the direction of the test.
- For patterned fabric, the specimens should be a representative sample of the pattern.
- The test specimen is suspended from the clip so that it hangs vertically while its lower end is held by the horizontal clamp plates. The plates shall be so fixed that not more than 9 mm of the specimen on either side (width wise) is covered by them.
- In most cases, the specimens are held by U-shaped frames (Figure 2.5). Some fabrics, e.g. woven aramid, thermoplastic fabrics often show longer char length as compared to that when the fabrics are suspended freely. Others show reverse trend.
- The specimen is held with holder using four clamps two on each side and near the top and bottom of the holder.
- The specimen holder containing specimen is inserted into the test cabinet.

FIGURE 2.4   Vertical Flammability Tester.

- The hood ventilation is turned off.
- The burner is moved so that it is under the flame height indicating rod. The nut near the bottom of the burner is rotated to shut off the air supply to the gas burner.
- The needle valve is adjusted to give a luminous flame of 38 mm (1.5″) high measured by the flame height indicating rod.
- The gas burner assembly is moved towards the hanging specimen. The burner position should be such that the burner mouth is 19 mm (0.75″) below the middle of the lower edge of the specimen.
- The flame is applied for 12 ±0.2 s and then the burner is slide back away from specimen.
- Any evidence of melting or dripping is noted.
- The flame then is removed, and the time till the specimen continues to flame after 12 s exposure (after-flame time) is noted.

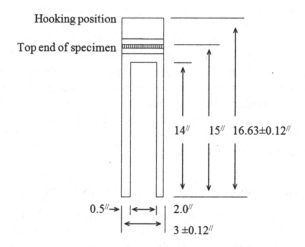

FIGURE 2.5 Specimen Sample Holder for ASTM Vertical Test.

- If the specimen continued to glow after (or without) the after-flame ceases, afterglow time is measured.
- The sample holder is removed from the cabinet. The hood ventilation is turned on to clear the test cabinet of fumes and smoke. The specimen is allowed to cool.

The char length is measured under a specified force as follows:

- The specimen is folded along a line through the peak of the highest charred area and parallel to the sides of the specimen.
- The specimen is punctured with the hook approx. 0.25″ from the bottom edge and the side edge of the specimen.
- A suitable weight (100, 200, 300 and 475 g for untreated fabrics weighing 68 to 203, 203 to 508, 508 to 780, over 780 g/m² respectively) is attached to the hook that will result in the appropriate tearing force.
- The corner of the specimen on the opposite bottom fabric edge, from where the hook and weight are attached, is grasped.
- The specimen is raised upward in a smooth continuous motion until the total tearing force is supported by the specimen. The fabric tear in the charred area of the specimen is noted.
- The end of tear is marked with a line across the width of the specimen and perpendicular to the fold line.
- The char length is measured along the undamaged area of the specimen to the nearest 3 mm.

Most of the tests (e.g., ASTM D6413-11) recommended for self-extinguishing fabrics specify vertical specimen ignited at the bottom edge. The methods work well for determining the efficiency of FR treatments on cotton and wool as they burn upwards. Thermoplastic fabrics, however, tend to shrink and melt ahead of flame. They may burn when ignited on the surface. In most cases, the specimens are held by U-shaped frames. Some fabrics, e.g., woven aramid, thermoplastic fabrics often show longer char length as compared to that when the fabrics are suspended freely. Others show a reverse trend (Lewin et al., 1982).

Most tests are conducted in a cabinet, with vent openings at the top and bottom. In most cases, the samples are preconditioned under standard textile conditions, 21°C and 65% relative humidity or oven dried at 105°C followed by cooling in desiccators for various period of time.

2.5.7 Vertical Flammability Test for Children's Sleepwear

The children's sleepwear standard became effective in 1972 and includes any garment (sizes 0-6X) worn primarily for sleeping or activities related to sleeping, such as nightgowns, pajamas, and robes, but excludes diapers and underwear. Fabrics intended or promoted for children's sleepwear must also meet this standard.

The vertical test confirming US federal regulations 16 CFR 1615 and 1616 and ASTM D 6545–18 (ASTM, 2019, section 07.02), states that a textile used in children's sleepwear must be tested to assess the flame resistance in its original state and after 50 laundering at 60± 3°C and drying circles.

The test requires five specimens (Size: 254 x 89 mm or 10″x 3.5″), which are to be conditioned, individually hung, vertically in a cabinet and exposed to a gas flame (38-mm or 1.5″) methane flame along the bottom edge for 3.0 seconds. The specimens cannot have an average char length greater than seven inches; no single specimen can have a char length of ten inches.

The extent of flame spread is determined by measuring the distance from the lower edge of the specimen to the point at which the specimen ceases to tear when subjected to a specified tearing load. The garment fails the test if the flaming period lasts more than 10 seconds and/or the char length is 7 in. (17.8 cm) or greater. The samples must be preconditioned at 20± 2°C for a minimum of 4 h.

2.5.8 Horizontal Flammability Test for Floor Covering Materials

The test measures the response or flame resistance to burning when a methenamine tablet (or pill) is ignited on a carpet or rug floor covering test sample under controlled conditions. The size of the burn hole or distance the carpet burns beyond the ignition point is measured. Small carpets and rugs must be tested by, but are not required to pass, the pill test. If they do not pass, they must be labeled as flammable.

The ASTM D 2859–16 (ASTM, 2019, Section 04.07) Standard Test Method for Ignition Characteristics of Finished Textile Floor Covering Materials and Standard for the Surface Flammability of Carpets and Rugs, CFR Part 1630, FF 1–70 (CPSC, 2019) became effective in 1971 and includes carpets or rugs which have one dimension greater than 6 feet and a surface area greater than 24 square feet, but excludes linoleum, vinyl tile, and asphalt tile.

Most horizontal methods (ASTM D 2859–16 and 16 CFR 1630 and 1631) for self-extinguishment are redesigned for mattresses and carpets and sometimes for open fabrics like netting. In US Standard test (FF-1-70) (Figure 2.6), Carpet specimen of 229 x 229 mm (9 in. square) is placed in an enclosure having a circular opening (9 in. dia.) at the top. A methenamine pill is placed in the center of the specimen and ignited. The pill burns up to 120 sec. A steel flattening plate with a 203-mm (8 in. dia.) cut-out is placed on the specimen. The carpet passes if, in case of seven out of eight specimens, the char does not extend to within 25 mm of the flattening plate. If a flame retardant has been used, testing only takes place after washing as specified.

In British standard test, a hexagonal nut is heated to 900°C and placed on the specimen for 30 sec. the radius of the charred area is measured on other sides of the specimen. The radius of the chars 35, 40–70, and 70 mm are considered low, medium, and high, respectively.

The standard for the Flammability of Mattresses and Mattress Pads (FF 4–72, Amended) CFR Part 1632 became effective in 1973 and includes ticking filled with a resilient material intended or promoted for sleeping upon, including mattress pads. Pillows, boxsprings, and upholstered furniture are excluded.

The test requires that, after conditioning, the surfaces be exposed to a total of nine burning cigarettes on the bare mattress or pad and the char length on the mattress surface must not be more than 2 in. in any direction from any cigarette. Tests are also conducted with nine burning cigarettes placed between two sheets. Mattress pads that are treated with flame retardant chemicals must be laundered as specified before testing.

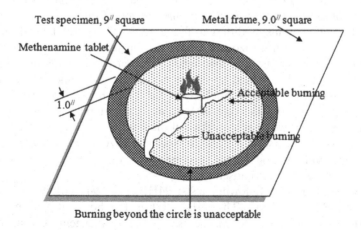

FIGURE 2.6 Surface Flammability test.

2.5.9 UL94 FLAMMABILITY TEST

There are a variety of industrial flammability tests that are widely used in industry. Each one is designed to measure different fire response characteristics. Unfortunately, most of them fail to predict the behavior of plastics under the massive effect of heat present in a large-scale fire. Different sectors of the industry adopt different flame test specifications. The standard determines the material's tendency to either extinguish or spread the flame once the specimen has been ignited.

UL testing is a method of classifying a material's tendency to either extinguish or spread a flame once it has been ignited and although originally developed by UL, it has now been incorporated into many National and International Standards (ISO 9772 and 9773).

The testing involves applying a flame to a sample in various orientations and assessing the response of the material after the flame is removed. Materials that burn slowly or self-extinguish and do not drip flaming material rank highest in the UL classification scheme.

UL 94 is a test for flammability of plastic materials for parts in devices and appliances. Underwriters Laboratories Inc. (ANSI, 1996) of USA intends to serve as a preliminary indication of acceptability of an FR plastic. This edition of UL 94, including all revisions, is approved by the American National Standards Institute. The test for flammability of plastic material used in parts for electronic devices and appliances is commonly known as UL-94 (ANSI, 1996).

The samples are tested by UL. If the results are acceptable, the device is listed. The manufacturer can attach the UL label. Follow-up procedures are established.

In this specification, the behavior of plastic when exposed to fire is expressed quantitatively. These numbers are obtained from measurements of the *after-flame* time or from the amount of material burned in a specific length of time. After-flame time refers to the length of time, in seconds, a material continues to burn after removal of the heat source. The amount of material burned refers to the length of sample that burns in a specified period of time.

The UL-94 contains 6 different flame tests. They are divided into two categories: vertical and horizontal testing. In the vertical flame test, a flame is applied to the base of the specimen held in the vertical position and the extinguishing times are determined upon removal of the test flame. In the horizontal flame tests, the flame is applied to the free end of specimens held in horizontal position and the rate of burning is determined as the flame front progresses between two bench marks. All methods described in the UL-94 specification involve the use of a standard specimen size, a controlled heat source and a conditioning period for the specimen prior to the test. These parameters vary according to the test chosen.

UL 94 contains the following tests: 94HB, 94V, 94VTM, 94-5V, 94HBF, 94HF, and Radiant Panel. The 94HB test is the horizontal burn method, while Methods 94V and 94VTM are used for more stringent vertical burning. The 94-5V test is for products that are not movable or are fixed to a conduit system. The 94HBF and HF tests are applicable for nonstructural foams, i.e., acoustical foam materials. Radiant panel test (ASTM, E162) determines the flame spread of a material exposed to fire.

2.5.9.1 Horizontal Burning (HB) Test

(ASTM D 635, D 4804, IEC 707, ISO 1210)

Three specimens are to be tested. Each specimen (12.5 x 100 mm) ($\frac{1}{2}''$x $5''$) is to be marked with two lines perpendicular to the longitudinal axis of the bar, 25 ±1 mm ($1''$) and 100 ±1 mm ($4''$) from the end that is to be ignited.

Figure 2.7 shows the experimental setup. The specimen is clamped at the end farthest from the 25 mm ($1''$) mark, with its longitudinal axis horizontal and its transverse axis inclined at 45 ±2 degrees. The wire gauze is to be clamped horizontally beneath the specimen, with a distance of 10 ±1 mm between the lowest edge of the specimen and the gauze with the free end of the specimen even with the edge of the gauze (ANSI, 1996).

The methane gas is supplied to the burner and adjusted to produce a gas flow rate of 105 ml/min with a back pressure less than 10 mm water.

The burner is placed at a remote place from the specimen and ignited. The burner is adjusted to produce a blue flame 20 ±1 mm high. The flame is to be obtained by adjusting the gas supply and the air supply of the burner until an approximate 20 ±1 mm yellow-tipped blue flame is produced. The air supply is increased until the yellow tip disappears. The height of the flame is measured again and adjusted it if necessary.

If the specimen sags at its free end during the initial set up, the support fixture is to be positioned under the specimen with the small extending portion of the support fixture at least 20 mm from the free end of the specimen. Enough clearance is to be provided at the clamped end of the specimen so that the support fixture is capable of being freely moved sidewards. As the combustion front progresses along the specimen, the support fixture is to be withdrawn at the same approximate rate.

The flame is applied to the free end at the lower edge of the specimen. The central axis of the burner tube is to be in the same vertical plane as the longitudinal bottom edge of the specimen and inclined toward the end of the specimen at an angle of approximately 45 degrees to the horizontal.

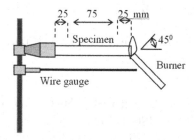

FIGURE 2.7   UL94 Horizontal flammability testing.

The burner is so positioned that the flame impinges on the free end of the specimen to a depth of 6±1 mm for 30±1 s. The burner is removed after 30 s or as soon as the combustion front of the specimen reaches the 25 mm mark. The timer is started when the combustion front of the specimen reaches the 25 mm mark. The after-flame burning time in s for the combustion front to travel from the 25 mm mark to the 100 mm mark is noted. If the combustion front passes the 25 mm mark, but does not pass the 100 mm mark, the elapsed time is recorded in seconds and the damaged length, L, in millimetres between the 25 mm mark and where the combustion front stops.

$$V = \frac{60L}{t} \qquad (2.1)$$

Where V= linear burning rate
   L is the damage length in mm
   t is time in seconds.
   e.g., if the flame front passed the 100 mm mark, L = 75 min.
A material shall be classified HB when tested by this method. A material classed HB shall meet the following conditions:

• Not have a burning rate exceeding 40 mm per minute over a 75 mm span for specimens having a thickness of 3.0–13 mm, or
• Not have a burning rate exceeding 75 mm per minute over a 75 mm for specimens having a thickness less than 3.0 mm, or
• Cease to burn before the 100 mm reference mark,
• A material classified HB in the 3.0+0.2 mm thickness shall automatically be classed HB down to a 1.5 mm minimum thickness without additional testing.

If only one specimen from a set of three specimens does not comply with the requirements, another set of three specimens is to be tested. All specimens from this second set shall comply with the requirements in order for the material in that thickness to be classified HB. H-B rated materials are considered "self-extinguishing". This is the lowest (least flame retardant) UL94 rating (Gold, 2019).
   Vertical test under UL94 may be of two types, namely (ANSI, 1996):

1. (20m) Vertical Burning Test; V-0, V-1, or V-2 (ASTM D 3801, IEC 707, or ISO 1210)
2. (125 mm) Vertical Burning Test; 5VA or 5VB (ASTM D 5048 or ISO 10,351)

2.5.9.2   (20 mm) Vertical Burning Test of Small Bars, V-0, V-1, or V-2
(ASTM D 3801, IEC 707, or ISO 1210)
   The test is conducted with 20 mm blue flame on which the specimen is placed vertically for a short period.

Some materials, due to their thinness, distort, shrink, or are consumed up to the holding clamp when subjected to this test. These materials may be tested in accordance with the test procedure in Thin Material Burning Test; VTM-0, VTM-1, VTM-2.

Standard bar specimens are to be 125 ±5 mm long by 13.0 ±0.5 mm wide, and provided in the minimum and maximum thicknesses. The maximum thickness is not to exceed 13 mm. Specimens in intermediate thicknesses are also to be provided and shall be tested if the results obtained on the minimum or maximum thickness indicate inconsistent test results. Intermediate thicknesses are not to exceed increments of 3.2 mm. Also, the edges are to be smooth, and the radius on the corners is not to exceed 1.3 mm.

The specimen is clamped from the upper 6 mm of the specimen, with the longitudinal axis vertical, so that the lower end of the specimen is 300 ±10 mm above a horizontal layer of not more than 0.08 g of absorbent 100% cotton thinned to approximately 50 x 50 mm and a maximum thickness of 6 mm (See Figure 2.8).

The methane gas is supplied to the burner and adjusted to produce a gas flow rate of 105 ml/min with a back pressure less than 10 mm of water.

The burner is adjusted to produce a blue flame 20 ±1 mm high. The flame is obtained by adjusting the gas supply and air ports of the burner until a 20 ±1 mm yellow-tipped blue flame is produced. The air supply is increased until the yellow tip just disappears. The height of the flame again is measured and readjusted it if necessary.

The flame is adjusted centrally to the middle point of the bottom edge of the specimen so that the top of the burner is 10 ±1 mm below that point of the lower end of the specimen, and it is maintained at that distance for 10 ±0.5 seconds, moving the burner as necessary in response to any changes in the length or position of the specimen. If the specimen drips molten or flaming material during the flame application, the burner is tilted at an angle of up to 45 degrees, and it is just sufficiently beneath the specimen to prevent

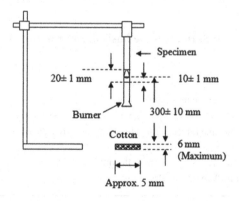

FIGURE 2.8   UL94 vertical flammability test for v= 0,1,2 classification.

material from dropping into the barrel of the burner while maintaining the 10 ±1 mm spacing between the center of the top of the burner and the remaining portion of the specimen, ignoring any strings of molten material. After the application of the flame to the specimen for 10 ±0.5 seconds, immediately the burner is withdrawn at a rate of approximately 300 mm/sec, to a distance, at least 150 mm away, from the specimen and simultaneously the after-flame time t_1 in seconds is measured.

As soon as after-flaming of the specimen ceases, even if the burner has not been withdrawn to the full 150 mm distance from the specimen, the burner is placed immediately again under the specimen and is maintained the burner at a distance of 10 ±1 mm from the remaining portion of the specimen for an additional 10 ±0.5 seconds, while moving the burner clear of dropping material as necessary. After this application of the flame to the specimen, the burner is removed immediately at a rate of approximately 300 mm/sec to a distance of at least 150 mm from the specimen and simultaneously the after-flame time, t_2, and the afterglow time, t_3 are measured.

If it is difficult to visually distinguish between flaming and glowing, a small piece of cotton, approximately 50 mm square, is to be brought into contact with the area in question by holding with tweezers. Ignition of the cotton is indicative of flaming. The following parameters are to be observed and recorded:

1. After-flame time after first flame application, t_1;
2. After-flame time after second flame application, t_2;
3. Afterglow time after second flame application, t_3;
4. Whether or not specimens burn up to the holding clamp; and
5. Whether or not specimens drip flaming particles that ignited the cotton indicator.

The material classification on the basis of 20 mm burning test is shown in Table 2.6.

TABLE 2.6

The classification of materials on the basis of their performance in 20 mm burning test

Criteria	V-0	V-1	V-2
After-flame time for each individual specimen t_1 or t_2.	≤10s	≤30s	≤30s
Total after-flame time for any condition set (t_1 plus t_2 for the 5 specimens)	≤50s	≤250s	≤250s
After-flame plus afterglow time for each individual specimen after the second flame application (t_2+t_3)	≤30s	≤60s	≤60s
After-flame or afterglow of any specimen up to the holding clamp	No	No	No
Cotton indicator ignited by flaming particles or drops	No	No	Yes

---

2.5.9.3 Thin Material Vertical Burning Test, VTM-0, VTM-1, or VTM-2
(ASTM D 4804 or ISO 9773)

This test is intended to be performed on materials that, due to their thinness, distort, shrink, or are consumed up to the holding clamp when tested using the test described in 20 mm Vertical Burning Test; V-0, V-1, or V-2. The materials shall also possess physical properties that will allow a 200 mm long by 50 mm wide specimen to be wrapped longitudinally around a 13 mm diameter mandrel.

Test specimens are to be cut from sheet material or film to a size 200 ±5 mm in length by 50 ±1 mm in width, in the minimum and maximum thicknesses that are to be tested covering the thickness range under consideration.

Test specimens are to be prepared by marking a line across the specimen width 125 mm from one end (bottom) of the cut specimen. The longitudinal axis of the specimen is to be wrapped tightly around the longitudinal axis of a 12.7 ±0.5 mm diameter mandrel to form a lapped cylinder 200 mm long with the 125 mm line exposed. The overlapping ends of the specimen are to be secured within the 75 mm portion above the 125 mm mark (upper tube section) by means of pressure sensitive tape. The mandrel is then to be removed. The specimen is then subject to vertical test as described previously.

2.5.9.4 (500 W, 125 Mm) Vertical Burning Test, 5VA or 5VB
(ASTM D 5048 or ISO 10,351)

Material shall be classified 5VA or 5VB on the basis of test results obtained on small bar and plaque specimens when tested. The experimental setups for the (125 mm) vertical burning test depends on the materials; set-ups are different for bar specimens and plaque specimens (Port Plastics, 2019).

Plaque means "an ornamental tablet, typically of metal, porcelain, or wood that is fixed to a wall or other surface in commemoration of a person or event."

Materials classified 5VA or 5VB shall also comply with the requirements for materials classified V-0, V-1 and V-2.

Material shall be classified 94-5VA or 94-5VB on the basis of test results obtained on small bar and plaque specimens.

- Specimens that do not exhibit a burn through (hole) are classified 94-5VA
- Specimens that burn through (hole) are classified 94-5VB

Specimens:
Bars: 5" x 12" x thickness
Plaques: 6" x 8" x thickness
Typically, thicknesses are 1/16",1/8", and 1/4"

Procedure for Bar Specimens (Method A):

1. Total of 10 specimens (2 sets) are tested per thickness.
2. Five of each thickness are tested after conditioning 48 hours at 23°C and 50% RH.

3. Five specimens of each thickness are tested after conditioning for 7 days at 70°C.
4. For bar specimen, the specimen is mounted with its long axis vertical (Figure 2.9).
5. A 5 in overall high Bunsen burner flame with a 1½ in. blue inner cone is applied to a lower corner of the specimen at an angle of 20° from the vertical such that the tip of the blue cone touches the specimen.
6. The flame is to be applied for 5 seconds and removed for 5 seconds. The operation is to be repeated until the specimen has been subjected to five applications of the test flame.

Classification Requirements:

- Not have any specimen that burns with flaming combustion for more than 60 seconds after the fifth flame.
- Not have any specimens that drip flaming particles that ignite the dry absorbent surgical cotton located 12in. below the specimen.

Procedure for Plaque Specimens (Method B):
Same as for bars except that three plaques (two sets) are to be mounted in the horizontal plane with the flame applied to the center of the bottom surface of the plaque (Figure 2.10).
94-5V Flame Class Requirements:

- Maximum Flame Time: 60 seconds
- Maximum Glow Time: 60 seconds
- Drip Particles That Ignite: No

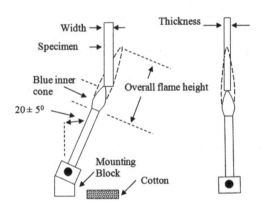

FIGURE 2.9   (125 mm) Vertical Burning Test for bar specimen.

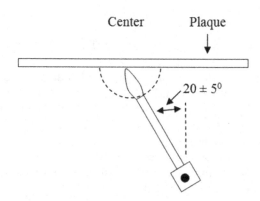

FIGURE 2.10   (125 mm) Vertical Burning Test for plaque specimen.

2.5.9.5   Radiant Panel Flame Spread Test

The Radiant Panel Flame Spread test measures the surface flammability of building products (ASTM E162) and cellular plastics (ASTM D3675) by using a gas-fired radiant heat panel. It is intended to measure and describe the properties of materials, products, or assemblies in response to heat and flame under controlled laboratory conditions and the results of this test may be used as elements of a fire risk assessment that takes into account all of the factors that are pertinent to an assessment of the fire hazard of a particular end use. An index is determined from the flame spread and heat evolution factors. This radiant panel index is a required parameter in various specifications, especially for the mass transit industry (buses and trains). The test is made on specimens of small size that are representative, to the extent possible, of the material or assembly being evaluated. The rate at which flames travel along surfaces depends upon the physical and thermal properties of the material, its method of mounting and orientation, the type and level of fire or heat exposure, the availability of air, and properties of the surrounding enclosure (FTT, 2019).

This test method of measuring surface flammability of materials employs a radiant heat source consisting of a 12 x 18″(300 x 460 mm) panel in front of which an inclined 6 x 18″(150 x 460 mm) specimen of the material is placed. The orientation of the specimen is such that ignition is forced near its upper edge and the flame front progresses downward. After ignition, the downward opposed flow flame spread rate and exhaust stack temperature rise are measured simultaneously to determine a flame spread factor and heat evolution factor, which are combined into a single radiant panel index for assessing material flammability. A factor derived from the rate of progress of the flame front and another relating to the rate of heat liberation by the material under test is combined to provide a flame spread index. The flame spread index, I_S, of a specimen as the product of the flame spread factor, Fs, and the heat evolution factor, Q, as follows:

$$I_S = F_S Q \qquad\qquad (2.2)$$

The Radiant Panel Flame Spread Apparatus (FTT, 2019) consists of:

- Porous cement and cast iron gas operated radiant panel (12" x 18") with electric spark igniter and automatic safety flame out detector,
- Stainless steel specimen holder, with observation marks every 3 in. for assessing the progress of the flame front,
- Stainless steel specimen support,
- Stainless steel pilot burner assembly,
- Pyrometer to determine the surface temperature of the radiant panel, with mounting bracket included,
- Air flow meter and gas control valve to control the mixture to the radiant panel,
- Stainless steel exhaust stack with a removable panel to enable easy cleaning of thermocouples,
- The stack is provided with eight thermocouples as required by the standards,
- Calibration burner with methane gas flow meter, and
- Safety gas controls and cut off circuitry.

Radiant Panel Flame Spread Tester of Festec International Co., Ltd (www. festec.co.kr) has the following specifications (Festec, 2019) (Figure 2.11):

- Reflection panel is composed of a cost iron frame with vertically applied porosity fire resistant material with a 305×457 mm size exposed reflection surface and is designed to function at 815°C.
- Filter-attached blower is used to supply air to Radiant panel.
- Auto ignition is enabled by Pilot Burner.

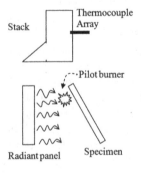

FIGURE 2.11   Radiant Panel Flame Spread Test.

- The Calibration burner should have 1.8 mm holes in 16 mm intervals, and should be manufactured along the middle line.
- Ventilation system is attached with air velocity of exhaust stack of 0.5m/s.
- In order to control the surface heat of the reflection panel, 180°C to~230°C (Black body temperature) range should be given, and a ±5°C accuracy, fit to observe a surface of 250 mm radius at 2 m distance.
- In order to measure the temperature of Stack, insulated thermocouple is attached.
- In order to measure radiation heat of radiant panel, Heat Flux Meter $(0{\sim}100kW/m^2)$ is used.

So far, UL94 testing methods are not used for testing textile materials. The flammability of textile materials are mostly tested by touching flame to small textile materials placed horizontal, vertical, or inclined position under controlled conditions. However, in real-life situations, the conditions may be very different and the flammability performances may differ from those predicted by the aforementioned tests. UL94 and other application-based tests may give further insight for more effective flammability tests of textile materials.

2.5.10 REACTION-TO-FIRE TESTS

A reaction-to-fire test assesses how easily a product can be ignited and contributes to fire growth. It relates mostly to the early stages of a fire development and is arguably mostly relevant to those products directly exposed to the fire source, i.e., wall linings, ceiling linings, and external wall surfaces. It is also relevant for assessing the performance of construction products during construction or building maintenance, e.g., welding of the building elements.

Due to advances in building technology and increased understanding of fire phenomena, there has been a worldwide move toward replacing prescriptive regulations by performance criteria and performance design procedures with respect to fire risks in buildings. One example of a performance-based test is the cone calorimeter test. Two of the methods, the cone calorimeter test and the single burning item test, are very frequently used to approximate material properties with respect to flammability, especially the cone calorimeter.

In order to determine the fire behavior of a material, we have to test various reactions of the same material to fire, i.e., how it starts, evolves, and spreads combustion. We need to determine levels of, for example, flammability, combustibility, smoke toxicity, and combustion heat; we can then classify the materials according to their performance. Manufacturers of the products that may be exposed to a possible fire should develop materials that are less vulnerable to fire to ensure the safety of their applications in transportation, buildings, or any other environment where there is a risk of fire. The applicable legislation varies according to the sector and the product's use.

2.5.11 PU Europe FIRE HANDBOOK, European FIRE Standards and National Legislation

The following tests are used to determine the Euro class for all construction products, except floor coverings and cables:

- EN ISO 1182 Noncombustibility test,
- EN ISO 1716 Determination of the heat of combustion,
- EN ISO 13823 Single Burning Item (SBI), and
- EN ISO 11925–2 Small flame ignitibility test

The first stage of testing is EN ISO 11925–2, which simulates a small flame ignition, such as a cigarette lighter, being applied for a short time (15 seconds) to the edge or surface of the product being tested. This can result in an E or F classification, or it is a prerequisite for the SBI test (30 seconds, instead of 15 seconds exposure), when the aim is to achieve classes B, C, or D.

In the SBI test method according to EN ISO 13823, a specimen is exposed to a gas flame of 30kW, simulating a single burning item in the corner (e.g., waste paper basket).

The exhaust gases are analyzed. From the amount of oxygen consumed and the amount of CO released, the heat released by the burning specimen can be calculated. The main classification is based on the criteria FIGRA (Fire Growth Rate) and THR (Total Heat Release within 10 minutes), and for the higher classes lateral flame spread is also taken into account.

In the exhaust duct, smoke obscuration is also measured, resulting in the criteria SMOGRA (Smoke Growth Rate) and TSP (Total Smoke Production) measured over 10 minutes), which form the basis for the smoke classification of the product. The third parameter for classification is based on whether burning droplets are visually observed (outside the burner area) during the test. Smoldering becomes a criterion in the reaction to fire classification at the request of some national regulators. A new test is under development. As this test is not yet available as a harmonized method, EU member states are allowed to have additional national tests and rules for CE-marked products (PU Europe, 2016).

2.5.11.1 The Cone Calorimeter Test (ISO 5660)

The cone calorimeter is a fire testing device based on the principle of oxygen consumption during combustion. The cone calorimeter is considered the most significant bench scale instrument in fire testing. This apparatus has been adopted by the International Organization for Standardization (ISO 5660–1) for measuring heat release rate (HRR) of a sample. It has been shown that most fuels generate approximately 13.1 MJ of energy per kg of oxygen consumed. Therefore, HRR is based on the fact that the oxygen consumed during combustion is proportional to the heat released. This device analyzes the combustion gases and measures the produced smoke from a test specimen that is being exposed to a certain heat flux (Lindholm et al., 2019).

In the late 1970s and early 1980s, fire researchers were in search of reliable bench-scale tools to measure material flammability based on heat release rate because this was thought to be the most reliable and accurate measure of the flammability of a material. Unlike any other apparatus, the cone calorimeter measures smoke optically and soot yield gravimetrically. It is now considered one of the most important devices for fire testing and fire protection engineering, and its usage in research has grown over the years.

Unfortunately, only a few heat release measurement tools were available at that time. They were difficult to operate and the data was very inconsistent. In 1982, the Fire Research Division at National Institute of Standards and Technology (NIST), then called the Center of Fire Research at the National Bureau of Standards, introduced the next generation instrument to measure material flammability. This instrument was awarded the prestigious American R&D 100 Award in 1988. The commercial cone calorimeters were available in the mid-1980s. At present, there are more than 300 cones in service worldwide, and the cone is the basis of more than a half-dozen fire testing standards, such as ASTM E1354 and D5485, ISO 5660–1, NFPA 271, and CAN\ULC-S135. The cone has become a reliable, accurate, and most commonplace method to access material flammability.

The cone calorimeter has long been considered the most significant bench scale instrument in the field of fire testing. This was substantiated in 2016 by the DiNenno Prize, which recognized oxygen consumption calorimetry as a significant technical achievement that has had a major impact on public safety. A cone calorimeter is a modern device used to study the fire behavior of small samples of various materials in condensed phase. It is widely used in the field of fire safety engineering. Its name comes from the conical shape of the radiant heater that produces a nearly uniform heat flux over the surface of the sample under study. It gathers data regarding the ignition time, mass loss, combustion products, heat release rate, and other parameters associated with the sample's burning properties.

Measurement of the rate of oxygen consumption provides a simple, versatile, and powerful tool for estimating the rate of heat release in fire experiments and fire tests. The method is based on the generalization that the heats of combustion per unit of oxygen consumed are approximately the same for most fuels commonly encountered in fires. Huggett's principle (Hugget, 1980) states that a measurement of the rate of oxygen consumption can then be converted to a measure of rate of heat release. The device allows a sample to be exposed to different heat fluxes over its surface.

A cone calorimeter is a useful apparatus in fire testing and research. It allows characterization of the fire properties of small samples of materials (approx. 100 mm^2). The fire characteristics of a material can be determined from several different standard models of the cone calorimeter that can be used to evaluate different aspects of the flammable materials. Research using cone calorimeters can be applied in product safety, environmental, and health services testing.

The cone calimeter is important when addressing safety issues. By using this device, it is easier to see how many different materials react with fire.

Knowing such information, safety regulations can be easily made to protect those who come in contact with, or work with the material often. It is important to know and understand the flammability, heat of combustion, ignitability, heat release, and smoke production of many materials in order to maintain a safe environment, all of which can be measured using a calorimeter. The cone calorimeter is a reduced-scale apparatus. Scale effects must be considered when using cone results to predict real-world fires.

The cone is a fire testing tool based on the principle that the amount of heat released from a burning sample is directly related to the amount of oxygen consumed during combustion. The amount of heat a material generates is directly aligned with the severity of a fire, such as fire growth rate. In order to determine a material's flammability, it is exposed to an external radiant heat source. Therefore, since this is a forced combustion test, the cone values are most often thought to reflect flammability of a second item ignited.

Figure 2.12 (AATCC, 2020) shows a schematic diagram of a cone calorimeter. A sample is placed below a *cone*-shaped radiant heater. Typically, the samples are exposed to an external flux from the heater at 35 kW/m^2. However, for more fire-resistant materials, the heater is frequently increased to 50 kW/m^2. Once enough pyrolysis products are generated, ignition occurs. The combustion product travel through the cone heater and through an exhaust hood and exhaust pipe. The

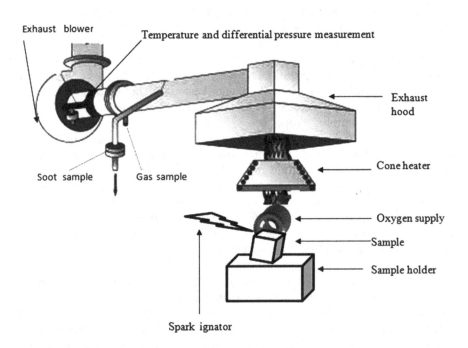

FIGURE 2.12 Cone calorimeter.

exhausted gas and soot samples are analyzed, and the temperature and differential pressure are measured. The typical values measured or calculated include, but are not limited to, the time until ignition; the mass loss rate during combustion; time until, and the value of the maximum amount of heat released during combustion; and total amount of heat released during the test.

The fire calorimeter is used by encasing a small sample in aluminum foil, wool and a retainer frame that is ignited below the exhaust hood. A conical heater is placed in between in order for materials to combust. The cone-shaped Inconel heating element provides a controllable radiant flux onto the sample, turning electricity into heat not unlike an electric toaster or oven. Inconel is a family of austenitic nickel-chromium-based superalloys or high-performance alloys. The austenitic is gamma-phase iron (γ-Fe), a metallic, nonmagnetic allotrope of iron or a solid solution of iron. The flammability of a sample can be characterized as a function of heat flux onto a sample. The conical heater is open in its center, allowing products of combustion to flow upwards into an exhaust duct.

Ventilation is also a very important part of the device, as well as the electrical power to run the conical heater. A small water supply is necessary to cool and regulate the heat in the system of the device. Since temperature and pressure are being evaluated, two different measurement tools are needed in the exhaust tube. Gas samples, smoke measurements, and soot collections are also acquired using this device.

Some of the test standards under cone calorimeter test are:

- ASTM E 1354,
- ASTM D 5485,
- CAN/ULC-S135,
- ISO 5660-1, and
- NFPA 271.

The cone calorimeter test (ISO 5660–1, 1993) is a bench-scale test for determining the reaction to fire for surface lining materials used in building construction. The test apparatus consists basically of an electric heater, an ignition source, and a gas collection.

The test specimen measures 100 mm x 100 mm and has a thickness of 6 mm and 50 mm. During the test, the specimen is mounted horizontally on a low-heat-loss insulating ceramic material. The orientation of the specimen can also be vertical, but this position is mainly used for exploratory studies.

After the test specimen has been mounted and placed in the right position, it is exposed to a heat flux from the electric heater. The output from the heater can be chosen in the range of 0–100 kW/m^2, but usually the heat output is in the range of 25–75 kW/m^2. When the mixture of gases above the test specimen is higher than the lower flammability limit, it is ignited by an electric spark source. The test is normally done for 10 minutes, but the time may be varied per the requirement for the material. Many variables are measured, but the main results of each test are:

- Time to ignition (TTI): Time at ignition was considered as the time to ignition. This was defined as the minimum exposure time required for the specimen to ignite and sustain flaming combustion.
- Mass loss rate (MLR): Loss of fuel or solid per unit time, Mass loss rate (g.s^{-1}) = dm/dt.
- Heat release rate (HRR): Heat release rate (HRR) is the rate of heat generation by fire. It is typically measured in joules per second or watts, since the output of a fire can generate more than a watt. For easier quantification, megawatts or kilowatts are used. HRR is the heat that is available in every square foot of surface absorbing heat within a particular surface. The heat release rate dQ/dt is proportional to the mass generation rate of volatile fuel or to the mass loss rate of the solid.

The results from a cone calorimeter experiment include a range of parameters and data types that can be used for several different purposes, for example (Lindholm et al., 2019):

- Fire modeling,
- Prediction of real scale fire behavior,
- Ranking of products by fire performance, and
- Pass/fail tests when developing new materials and products.

The advantages of cone calorimeters are (Efectis, 2019):

- Only very small sample sizes are required, which makes it very cost-effective in terms of e.g., material costs and transportation costs.
- It gives an indication of how to improve the tested product.
- The test is in accordance with EN 5660–1 or ISO 5660–1:2015, i.e., reaction-to-fire tests – heat release, smoke production and mass loss rate.
- The test can reasonably predict the outcome of a single burning Item test (EN 13823).
- The test can be used for flooring products that are tested with the Flooring Radiant Panel tester (EN ISO 9239).
- The test is fast, safe and accurate, with immediate results.
- The instrument can test multiple samples and assess the differences between them.
- Testing done with smaller specimen allowing testing if larger samples are not available, such as monumental buildings.
- Can compare before and after use, e.g., with exposure to weathering or frost.

Sources of errors

Heat release rate cannot be directly measured. It has to be calculated from several other measurements, and as a consequence, it includes some errors and assumptions. The error in the value for the heat of combustion can be up to 5% (Hugget, 1980). Zhao and Dembsey did an extensive study on measurement uncertainty for the cone calorimeter (Zhao et al., 2008). They concluded that the relative uncertainty of HRR decreases as HRR increases. The relative error when firing methane is "approximately 20-30% for 1 kW fires, 10% for 3 kW fires and less than 10% for 5 kW fires". The uncertainties for HRR measurements are considered reasonable (Lindholm et al., 2019).

The British national reaction to fire classes are based on the performance in the fire tests BS 476 Parts 4, 6, 7, and 11 (BSI, 1997). The European classification system for reaction-to-fire testing consists of six standards; a suite of four test standards, a classification standard and a standard covering specimen conditioning and substrate selection. The documents are as follows (DCLG, 2012):

1. BS EN 13501–1: Fire classification of construction products and building elements. Part 1: Classification using test data from reaction to fire tests.
2. BS EN ISO 1182: Reaction to fire tests for building products. Non-combustibility test.
3. BS EN ISO 1716: Reaction to fire tests for products – Determination of the heat of the heat of combustion (calorific value).
4. BS EN 13823: Reaction to fire tests for building products – Building products excluding floorings exposed to the thermal attack by a single burning Item.
5. BS EN ISO 11925–2: Reaction to fire tests–Ignitability of building products subjected to direct impingement of flame Part 2: Single-flame source test.
6. BS EN 13238: Reaction to fire tests for building products Conditioning procedures and general rules for selection of substrates.

2.5.11.2 Single Burning Item Test
BS EN 13823 (Single Burning Item (SBI) test) simulates the conditions experienced by a construction product in the corner of a room, when exposed to the thermal attack of a single burning item positioned in that corner.

The fire test methods in BS 476 Parts 6 and 7 measure different characteristics of fire performance from the European fire test methods BS EN 13823 and BS EN ISO 11925–2.

When designing a building a very important consideration is how it behaves in fire and ensure that the elements of structure do not collapse but remain standing or hold back the fire for a prescribed time. The building regulations stipulate the rules and the degree of fire resistance of the elements of structure. However the British standard 476 dictates the appropriate fire tests for these elements of structure/materials and grades the level of fire resistance (FSAC, 2019).

In particular, the BS 476 Fire tests are material tests in which the fire performance is determined by the characteristics of the surface of the material, whereas the SBI test is a test of the performance of the construction product in an arrangement representative of end use, that is, it is tested with joints, air gaps and/or fixings that are typical of its end use application and the level of thermal exposure in the test method resulting from direct flame contact means that the construction product is tested through its thickness.

The SBI test (EN 13823) was developed for the European Commission to provide the key test method for assessing the reaction to fire performance of construction products. The SBI test method was planned to assess the performance of building products in a (real scale) room corner scenario. The ISO 9705 Room corner test (ISO, 1993), which is a full-scale test method intended to evaluate the contribution to fire growth provided by a surface product applied in a room, was put forward as the reference test for this scenario.

The SBI test simulates a single burning item burning in a corner of a room. The test apparatus is presented in Figure 2.13. A triangular shaped propane diffusion gas burner running at 30 kW acts as heat and ignition source representing a burning wastepaper basket. It is placed at the base of the specimen corner. The performance of the specimen is evaluated for 20 minutes. There is a floor in the test configuration but no ceiling. The floor, specimen, and burner are installed on a trolley that can be removed from the room for easy mounting of the specimens. The combustion gases are collected in a hood and transported through a duct. The duct contains a measurement section with a differential pressure probe, thermocouples, a gas sample probe, and a smoke measurement system to measure heat and smoke production.

The Single Burning Item test is an intermediate-scale, open-corner method for measuring lateral flame spread, rate of heat release, propensity for the production of flaming drips, and rate of smoke production. The total exposed specimen surface area is 1,5 m x 1,5 m. The specimen consists of two parts

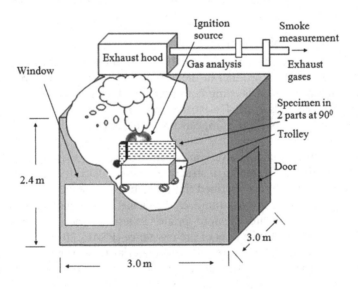

FIGURE 2.13   Single Burning Item Test.

(height 1.5 m, width 0.5 and 1.0 m) which form a right-angled corner. Eventual corner joints, as applied in end-use conditions, form part of the product under test. This provides data suitable for comparing the performances of materials, composites, or assemblies that are used primarily as the exposed surfaces of walls or ceilings. The test procedure sets out to simulate the performance of these products when fixed to the walls and ceiling of a small room under end-use conditions in which the ignition source is a nominal 30 kW single-burning item, such as a wastepaper basket in a corner of a room.

The apparatus is seated in a test room of dimensions 3 m long by 3 m wide by 2.4 m high. The room incorporates an opening of dimensions 2150 mm by 1450 mm through which a trolley holding the specimen is inserted, such that it locates within a free-standing frame, positioned adjacent to one wall of the test room. The frame supports a rectangular hood through which the fire effluent is drawn by means of a fan and connecting duct system. Two windows in the enclosure walls allow viewing of the specimen. The duct system, which has an internal diameter of 315 mm, incorporates a measurement section positioned at the end of a 3125 mm straight section of ducting that contains guide vanes to reduce air turbulence at the measurement position. Two K-type thermocouples and a bidirectional probe connected to a differential pressure transducer are incorporated in this section. The resultant temperature and differential pressure measurements are used to calculate volume flow within the ducting. The air flow rate is set at 0.6 m^3/s prior to the start of the test.

A sampling probe, which facilitates the withdrawal of a transverse sample of the fire effluent passing through the duct, is also incorporated in the duct. The effluent sample is continuously drawn through a paramagnetic oxygen analyzer, allowing rate of heat release to be continuously calculated by means of the Oxygen Consumption Method.

The obscuration of light caused by the smoke in the fire effluent passing through the exhaust duct is determined by a white light lamp and photocell system.

The trolley incorporates a triangular propane sand burner of side length 250 mm, which is positioned in the base of the corner formed by the specimen, with a horizontal separation of 40 mm between the edge of the burner and the lower edge of the specimen. This burner, known as the primary burner, has an output of 30 kW. A secondary propane sand burner is attached to the fixed frame, beneath the hood but at the furthest possible distance from the specimen when the trolley is in place. This burner is used to obtain baseline data without affecting the assembled specimen. The trolley incorporates a grill in its base and this is the sole source of ventilation for the test enclosure while a test is in progress.

The specimen is assembled as described previously, and the trolley is wheeled into position within the test enclosure. Horizontal reference lines may be drawn on the long wall specimen surface at elevations of 500 mm and 1000 mm to allow the measurement of the lateral flame spread. The volume flow rate through the hood and duct system is set at 0.6 m^3/s and logging of data from the various instruments is initiated. When 2 minutes of data have been logged, the secondary burner is supplied with sufficient propane to give a heat release rate of 30 kW. Ignition of the burner is facilitated by a small pilot flame. When 3 minutes of baseline data have been logged, the gas flow is switched to the primary burner. The 30 kW flame from this burner impinges upon the specimen for 21 minutes before the propane supply is switched off. The performance of the specimen is evaluated over a period of 20 minutes. In addition to the aforementioned heat release and smoke measurements, during and after the period of primary burner flame impingement, the following parameters are measured (Warringtonfiregent, 2019):

1. Lateral flame spread (LFS). Note is made of whether flaming between 500 mm and 1000 mm up the long wall reaches the edge of the specimen furthest from the corner.
2. Falling of flaming particles or droplets is recorded outside of a zone 300 mm from the corner line.
3. Any other relevant specimen behavior observed during the test is noted.

Measurements are made of heat release rate and smoke production rate as functions of time. From these, values of FIGRA (FIre Growth RAte index), an index representing the speed of growth in heat release rate; the total heat released over the first 10 minutes (THR600s); a simple lateral flamespread to the end of the specimen (LFS edge); an index representing the speed of growth in smoke production rate SMOGRA (a Smoke Growth RAte index); the total smoke produced over the first 10 minutes (TSP600s); and a parameter defining three levels of flaming droplets and particles (FDP) (Mierlo and Sette, 2005) are calculated. FIGRA is basically a parameter that measures the rate at which a construction product contributes heat to a fire.

The SBI test method has had a turbulent history of development and strong, both political and economic, interests have governed the debates. Over the years, the test method has also gained confidence both in test institutes and in industry. It is our believe that it is now time to have a major revision of the test method and its calibration procedures based on both new theoretical insights and on experience gained during the second SBI round robin exercise. The revision should not lead to a new test method, but to higher measurement accuracy and increased confidence (Mierlo and Sette, 2005).

Efectis (2019) combines the cone calorimeter with a sophisticated software model. This model predicts likely fire and smoke class in accordance with the SBI test (EN 13823); while a Single Burning Item test needs three specimens with an area of 2.25 m2 each, the cone calorimeter only uses a maximum of twelve specimens of 0.01 m^2. This might prove valuable in the early development stages of a new product. For instance, the cone calorimeter test makes it easier to quickly test several versions of one material and determine which one performs best before moving on to an SBI test. In some cases, it is even allowed to compare and validate the current status of a material using earlier cone calorimeter test results. Cone calorimeter vs SBI (Efectis, 2019) can be described as:

- SBI test: 3 x 1.5m x 1.5m x 0.2m = 1.35m^3
- Cone calorimeter test: 12 x 0.1m x 0.1m x 0.05m = 0.006m^3

- A factor of (maximum) 225 in volume difference

Assessing fire performance using time until ignition, heat release, spread of flame, smoke production rate, and formation of flaming droplets/debris, the products are classified under six Euro classes (compiled under EN 13 501–1) are as follows:

1. No contribution to fire
2. Very limited contribution to fire
3. Limited contribution to fire
4. Acceptable contribution to fire
5. Acceptable reaction to fire
6. No performance determined

Comparison of reaction to fire tests with current standards is shown in Table 2.7.

Euro class A is a very severe category primarily meant for noncombustible materials. The composites may fall into this category. These are the furnace test based on ISO 1182 and the calorimeter test based on ISO 1716.

At the other end of the scale, for products of appreciable combustibility, (classes E and F), the materials are tested using a simple ignitability test for vertical specimens based on an existing small burner test. Materials within classes B, C and D are subjected to the single burning item test (Fomicom, 2019).

2.5.12 APPLICATION-BASED TESTS

Protective clothing for firefighting is one of the important equipment to protect the personal safety of firefighters active in the first line of firefighting

TABLE 2.7
Comparison of reaction to fire tests with current standards in various countries

Euro class	Germany	France	UK
EN 13501-1	DIN 4102	NFP 92-501	BS 476 pts 6 & 7
A	A	M0	-
	A2	M0	Class 0
B	B1	M1	Class 0
C	B1	M1	Class 1
D	B2	M2	Class 1
E	B2	M3	Class 2
F	B3	M3	Class 3

against heat and fire. These have also been developed according to changing needs. Since 2013, protective clothing for firefighting have can be roughly divided into two types including integrated type and two-piece type.

The protective clothing protects the neck, arms, legs, and trunk and consists of (Flasa, 2019):

- A coverall
- Jacket and trousers (overlapping at least 30 cm)
- A set of over- and undergarments to be worn together

This clothing should have the following properties (Flasa, 2019):

- hindering freedom of movement as little as possible and be compatible with the other parts of the fire fighter's equipment
- as light as possible
- easy to clean and maintain
- offer protection at the wrist against penetration of flaming or hot materials
- If a combination of layers or garments is needed to achieve the required level of protection, then these layers/garments should be permanently attached to each other, or clear indications should be given on the label that adequate protection can only be guaranteed when wearing the right combination of layers/garments.
- Seams, fasteners, zippers, labels, and other accessories must not impair the performance of the garment. Reflective strips shall be applied in accordance with the user's requirements.

In spite of different wearing methods, all protective clothing for firefighting must have excellent performances in such properties as closure system, heat resistance, flame spread limitation, tensile strength, and resistance to penetration by liquid chemicals.

Here are some primary performance requirements of protective clothing for firefighting (SBS Zipper, 2019).

All materials including seams, outer garment surfaces, and the surface of the innermost lining tested according to EN ISO 15025:2002 shall achieve flame spread index 3 of prEN ISO 14116:2013. When exposed to flame, protective clothing should limit flame spread; otherwise, burning is a threat to the firefighter.

2.5.13   HEAT RESISTANCE

According to EN 469:2014, all materials used in protective clothing for firefighting shall not melt, ignite, or shrink more than 5%, and all hardware such as zippers shall remain functional when exposed to an atmosphere having a temperature of 180°C ± 5°C.

2.5.14 Tensile Strength

As specified in EN 469:2014, outer materials shall bear a breaking load more than 450N, and main seams shall have a maximum force to seam rupture more than 225N.

2.5.15 Antiwicking Barrier and Moisture Barrier

These barriers are used to prevent liquid transferring from outside to inside the garment and can provide the properties of hydrostatic pressure resistance and water vapor resistance.

1. Hydrostatic pressure resistance: for garments with a moisture barrier, the resistance should more be no less than 20 kPa.
2. Water vapor resistance: level 1 is from 30 m^2Pa/W to 45 m^2Pa/W and level 2 is less than 30 m^2Pa/W.

2.5.16 Resistance to Penetration by Liquid Chemicals

Besides fire, firefighters may be faced with other dangers such as chemicals. In this circumstance, protective clothing should completely protect their safety as well. There shall be no penetration to the innermost surface and the index of repellency shall be no less than 80.

2.5.17 Visibility Material

Visibility materials such as retroreflective/fluorescent materials should be attached at least one band encircling arms, legs, and torso to the firefighter's protective clothing to ensure visibility. The minimum area of separate retro-reflective material attached to the outermost surface should be no less than 0.13 m^2, and the minimum area of fluorescent or combined performance material applied to the outermost surface should be no less than 0.2 m^2. In addition, the photometric requirements, color and fabric should be in accordance with EN ISO 20471:2013.

2.5.18 Dimensional Change

The dimensional change of woven materials should be no more than 3% and, that of nonwoven materials should be no more than 5%.

2.5.19 Metal Accessory

For safety and comfort, metal accessories attached to protective clothing should not be exposed on the innermost surface, and should not have sharp edges, roughness, or projections.

2.5.20   CLOSURE SYSTEMS

Closure systems, fastening openings of clothing to achieve a secure closure, also play a significant role in protection. For instance, flame-resistance zippers, the tapes of which are made from fibers with flame resistance, should accord with the requirements to slow, stop, or prevent flaming.

Protective clothing for firefighters must comply with a number of requirements contained in standards, varying from country to country (ASTM, European Standards etc.). Before protective clothing, gloves, helmets and other personal protective equipment for firefighters are approved for use, they must pass demanding tests and obtain a certificate. In Europe, standards in the area of protective clothing are presented in the EN standards, published by the European Committee for Standardization, which specifies research techniques in a wide spectrum of possibilities. Standards developed in the United States are the ASTM standards, introduced by the American Society for Testing and Materials. The other standards that are applicable in the United States are those introduced by the NFPA (National Fire Protection Association). The Russian standard for clothing for firefighters was developed by The Ministry of the Russian Federation for Civil Defense, Emergencies and Elimination of Consequences of Natural Disasters (Roguski et al., 2016).

The major hazards during firefighting are radiant or convective heat, explosions, falling objects, debris, fine airborne particles, limited oxygen supply, hot liquid, molten substances, noise, toxic chemicals, smoke, and hot gases (Nayak et al., 2014).

2.5.20.1   EN 367:1992, Protective Clothing, Protection against Heat and Fire

British standard, EN 367:1992, Protective Clothing, Protection against Heat and Fire, Method for Determining Heat Transmission on Exposure to Flame was published on December 15, 1992. This standard specifies a test method for comparing the heat transmission flame ($X_{f\ (flame)}$) through materials or material assemblies used in protective clothing. A horizontally oriented test specimen is partially restrained from moving, and is subjected to an incident heat flux of 80 kW/m^2 from the flame of a gas burner placed beneath. The heat passing through the specimen is measured by means of a small copper calorimeter on top and in contact with the specimen. The time (s) for the temperature to raise 24° ± 0.2°C is recorded. The mean result of three specimens is calculated as the heat transfer index (HTI_{24}) (lowest level: ≥ 9s, highest level: ≥ 13s). The time (s) for the temperature to rise 12° ± 0.2°C may also recorded (HTI_{12}) (lowest level: ≥ 3s, highest level: ≥ 4s) (Flasa, 2019). However, the standard was withdrawn on December 31, 2016.

2.5.20.2   (EN 469:2005) Performance Requirements for Protective Clothing for Firefighting

This EU standard specifies the requirements for protective clothing for structural firefighting, mainly addressing protection against heat and flame. This standard does not cover protective clothing for special tasks or risks, e.g., cleanup of chemical spills, forest fires, firefighting at close proximity of the fire, or assistance in road emergencies.

This standard specifies the minimum performance requirements for protective clothing worn during firefighting operations and associated activities (e.g., rescue work, assistance during disasters).

The standard outlines general clothing design, the minimum performance levels of the materials used, and the test methods used to determine these performance levels. The required performance levels may be achieved by the use of one or more garments. In addition, the standard covers the event of an accidental splash of chemical or flammable liquids, but does not cover special clothing for use in other high-risk situations (e.g., reflective protective clothing) nor does it cover protection for the head, hands, and feet or protection against other hazards (e.g., chemical, biological, radiological, and electrical hazards or the hazards encountered during chemical and gas cleaning operations).

The current standard is EN 469:2005 "Performance Requirements for Protective Clothing". This is a second generation standard, the first having been EN 469:1995. Both were developed on the same basic principles with an important but controversial new concept concerning lifespan of protective clothing introduced into the 2005 edition. The basic principles adopted when developments commenced in the early 1990s were that the standard be a product specification setting out obligatory key properties of the clothing considered necessary to protect the wearer.

Because the clothing required to meet this specification (1995 or 2005) typically consists of a few layers of different textile related materials, the required tests are essentially determine the strength of the outer shell fabric, the resistance to ignition/flame spread of this outer shell and of the internal lining material (i.e., the innermost layer next to the wearer's undergarments or skin) and the resistance to penetration of heat from flames and from a radiant source through all layers of the intended clothing component materials when tested as a combination.

These basic tests are all textile and related fabric manufacturing industry tests, not tests on the complete clothing. They are therefore very small in scale compared to complete items of clothing, and can use specimens either cut from the clothing or more usually cut from the rolls of material intended to be used by the clothing manufacturer and submitted to the test house.

The 1995 and 2005 standards include methods of testing for various other properties of materials used in clothing, such as waterproofness and water vapor transmission, i.e., a measure of perspiration removal. These properties are important because fire-protective clothing should not allow water penetration/absorption via its outer shell component or retain water moisture from perspiration. This is because water has 21 times more potential to conduct/transmit heat to the wearer of clothing than the air that is *trapped* in the individual clothing materials and in any space between these materials.

Some important protection factors are (Flasa, 2019):

1. Flame spread (test method EN 532):

   a. No holes, melting or dripping
   b. After-flame time: ≤ 2 s
   c. After-glow time: ≤ 2 s.

2. Convective heat (test method EN 367 EN ISO 9151 April 2017):
   $HTI_{24}= 13$ s, $HTI_{24}- HTI_{12} = 4$ s.
3. Radiant heat (EN 366, method B, flux: 40 kW/m^2)
   $t_2 \geq 22$ s
   $t_2 - t_1 \geq 6$ s
   Heat transmission $\leq 60\%$
4. Heat resistance of outer layer (EN 366, method A, flux 10 kw
   Residual tensile strength ≥ 450 N
5. Heat resistance of all materials and accessories
   (5 minutes at$190°C$)

The material should not burn, melt, drip, or shrink more than 5%.
Additional requirements:

- Tensile strength: ≥ 450 N in all directions
- Tear strength: ≥ 50 N in all directions
- Water repellence: ≥ 4
- Dimensional stability: $\leq 3\%$ in all directions
- Resistance to chemicals EN 368: Run off $\geq 80\%$ NaOH, HCl, H_2SO_4, white spirit
- Impermeability: to be specified
- Water vapor permeability (comfort parameter): to be specified.

The only objective test, i.e., one with a quantifiable outcome that is listed in both the 1995 and 2005 editions of EN 469 for the complete clothing items is an optional one for determining the burn injury prediction (BIP) caused by heat transmission through the clothing to heat sensors on the surface of a manikin dressed in the clothing. The test method was finally published year as ISO 13506 "Protective clothing against heat and flame – Test method for complete garments – Prediction of burn injury using an instrumented manikin".

The main aspects of this method of testing are the use of a full-scale, human form, male- or female-shaped manikin, equipped with at least 100 heat sensors distributed over most of the surface. When dressed in the fire protective clothing, the manikins are subjected to a flame engulfment event via flames from an array of gas burners for 8 seconds to simulate a wearer's being caught in a flashover event.

EN 469: 2005 is already being revised, in part, to address some ambiguities and inadequately detailed text identified initially by the UK BSI committee.

Discussion has now moved on to issues such as practical performance testing, and, most recently, whether we should start with a 'clean sheet of paper'! Given that there seems to be little if any evidence of inadequate protection being provided by the current standard (or its predecessor), this proposal by certain end users and test facilities specializing in physiological and related human subject-based testing seems questionable in our view as a test house working with both manufacturers and end users.

One aspect of both EN 469 and EN 1486 that needs further work is to determine whether it is possible to set maximum burn injury prediction values for manikin fire tests.

The EN 469:2005 standard demands high performance criteria that are described in 20 most important tests (Khanna, 2019):

- Flame spread tested according to EN ISO 15025 procedure A (surface ignition)
- Heat transfer flame tested according to EN ISO 9151–2 levels
- Heat transfer flame tested according to EN ISO 6942–2 levels
- Residual tensile strength of material when exposed to radiant heat
- Heat resistance 180°C, 5 min according to EN ISO 17493
- Textile durability requirements
- Surface wetting
- Resistance to penetration by liquid chemicals tested according to EN ISO 6530
- Resistance to water penetration according to EN 20811–2 levels
- Resistance to water vapor penetration according to EN 31092–2 levels
- Ergonomic performance
- Visibility according to EN 471:2003
- Optional whole-garment testing

The standard critically addresses four key test values which are to be checked on garment labels, with levels of protections mentioned thereafter (Sioen, 2019):

1. X_f (Flame), EN 367 test method: heat transmission (flame), Heat flux of 80 kW/m^2
2. X_r (Radiation), EN 6942 test method: the transmission of heat (radiation). Heat flux of 40 kW/m^2.
3. Y (waterproofness), Waterproofness test, EN 20811
4. Z (breathability), Water vapor resistance, EN 31092

Level 1

- Lower level of protection
- Generally two-layered assembly, light weight
- High breathability, limited heat protection
- Suitable for urban search, rescue teams, and fire attack support teams

Level 2

- Higher level of protection
- Generally three-layered assembly, medium weight
- Medium breathability, high-heat protection
- Suitable for frontline/first response firefighting and proximity operations.

2.5.20.3 NFPA 1971, Standard on Protective Ensembles for Structural Fire Fighting and Proximity Fire Fighting

This standard sets the minimum requirements for design, performance, testing, and certification of the elements of the ensemble for body protection in structural firefighting and in proximity firefighting–coats, trousers, one-piece suits, hoods, helmets, gloves, footwear, and interface elements such as wresters.

As with all NFPA standards, the 2018 edition of NFPA 1971 replaced the 2013 edition, and all previous editions. The 2018 edition was approved by the American National Standards Institute on August 21, 2017, with a final completion date of August 21, 2018. The 2018 edition continues to incorporate design and performance requirements for optional CBRN requirements, and includes several new definitions and revised labeling requirements.

Changes have been made to the performance requirements for all of the ensemble elements, which are reflected in revised and/or new test methods. For garments, one of the most noteworthy new requirements is an optional particulate inward leakage test. This new requirement and associated test method is in response to the growing concern over cancer rates in fire service. The term *optional* indicates that this is not a mandated requirement, but that if a manufacturer is going to claim this type of protection, then minimum requirements must be met. The test is run on full ensembles, including the coat, pant, helmet, glove, and footwear elements, and with every SCBA specified for the ensemble by the ensemble manufacturer. The protective hood is also tested when it is not integrated into the coat (Globe, 2019).

Table 2.8 presents a summary of requirements of thermal factors in the standards NFPA 1971 (Standard on Protective Ensembles for Structural Fire Fighting and Proximity Fire Fighting, USA), EN 469 (Protective clothing for firefighters–Performance requirements for protective clothing, EU), НПБ 162–02 (Special Protect Clothing For Fire-Fighters Isolation Type. General Technical Requirements, Russia).

Standard НПБ 162–02 has the highest requirements in terms of resistance to contact with aggressive media. It also specifies requirements, depending on the temperature of the substance. On the other hand, the NFPA 1971 standard includes testing with aggressive media that are in common use. Only the NFPA 1971 standard includes a test of contact heat in the context of safety (time before second-degree burns). In EN 469, a test of contact heat in not included at all, but in НПБ 162–02 there is a test of the shrinkage of material caused by contact heat (Roguski et al., 2016).

Small-scale testing is important in assessing the level of protection provided by fabrics because it offers an inexpensive means of fabric testing. However,

TABLE 2.8

Thermal performance requirements for protective clothing (Roguski et al., 2016)

Radiant Heat Requirement standard	Parameter	Limit	Standard for testing
NFPA 1971	TPP (Thermal protection performance)	≥ 35 at heat flux 84Kw/m^2	ISO 17492
	RPP (Radiant heat performance)	intersect time ≥ 25s	ASTM F 2702
	Total Heat Loss THL)	≥ 205 W/m^2	ASTM F1868
EN 469	RHTI$_{24}$ (heat transfer - radiation)	≥ 18.0 for level 2 at: heat flux during test 40 kW/m^2	EN ISO 6942
	RHTI$_{24}$ – RHTI$_{12}$ (heat transfer – radiation)	≥ 10.0 for level 1 at: heat flux during test 40 kW/m^2	
НПБ 162-2002*		Heat flux on inner surface not more than 2.5 kW/m^2 at: heat flux during test 40 kW/m^2	НПБ 161
Flame			
NFPA 1971	Flame resistance	*After flame no more than 2 s no melting or dripping*	ASTM D 6413
EN 469	HTI$_{24}$ (heat transfer - flame)	HTI$_{24} \geq 13.0$ for level 2 HTI$_{24} \geq 9.0$ for level 1	EN ISO 9151
	HTI$_{24}$ – HTI$_{12}$ (heat transfer – flame)	HTI$_{24}$ _ HTI$_{12} \geq 4.0$ for level 2 HTI$_{24}$ – HTI$_{12} \geq 3.0$ for level 1	
	Flame spread	index 3 of prEN ISO 14116	EN ISO 15025
НПБ 162-02	Flame resistance	Flame resistance after flame no more than 2 s, shrinkage $\leq 10\%$	ISO 6941

Special protect clothing for firefighters isolation type general technical requirements. Test methods, Moscow 2003.

the drawback of small scale tests is that the level of protection offered by a garment constructed from the fabrics cannot be determined. The results from these small-scale tests can only be used to evaluate the performance of fabrics in the tested fire scenario, and no application to a real fire can be accurately extrapolated. Also, the construction of the testing apparatus and the orientations of the test fabrics are not indicative of the normal applications of the fabrics. In each of these small-scale tests, the material is tested in a static, dry fashion and the test is not an accurate depiction of a real-fire scenario. Both the fatigue of movement and the amount of perspiration in a garment can affect the performance of the garment and these factors are not accounted for in small-scale fabric tests.

All clothing apparel provides some level of protection from rain, sun, and wind. To measure the level of protection against fire and extreme heat, several performance test methods are available that give a comparative analysis of a fabric's performance.

Evaluating current test methods requires some distinctions to be made between different types of garments. It would not be appropriate to evaluate a firefighter's turn out gear against a daily-wear uniform; thus, different classifications are used to describe a garment's level of protection. These classifications are: primary, secondary, and tertiary concerns. These categories group garments by the fire scenarios to which they are likely to be exposed.

2.5.21   PRIMARY CONCERNS

The greatest level of protection is demanded from garments worn by individuals who actively fight fires. The primary concerns for these garments are their response to both the fire itself and the heat flux from the fire. The temperatures in a compartment fire can be in the range of several hundred degrees Celsius with isolated and limited heat fluxes in the flames approaching 100 kW/m^2. It is a common occurrence for a firefighter to be in these harsh fire conditions for a period of time greater than a few seconds. Therefore, garments that would fall into a primary area of concern must be able to maintain their protective properties for a longer time than those garments which fall into a secondary concern category.

2.5.22   SECONDARY CONCERNS

This category of garments does not require the heavy heat and flame resistance that firefighters' gear requires, but a high level of protection is still needed. Individuals who need this level of protection usually consist of uniformed professionals who work in an environment with a possible fire scenario.

The potentials for fire events during the jobs these people perform are greater than for most average professionals. While these people would not face the same exposure to heat conditions as firefighters, their clothing still needs to protect them from possible short fire exposures.

2.5.23   TERTIARY CONCERNS

This level of concern accounts for garments with a basic level of flame resistance and the majority of garments worn by the average person falls into this category. This basic level is meant to provide some standard that would prevent an accidental encounter with a flame source that could be hazardous (Lantz, 1994). However, these garments should not be worn in a hazardous environment. All textiles should have this basic level of protection and any garment that does not meet this level should be considered unsafe.

2.5.23.1 Thermal Exposure Tests

In thermal exposure tests, a heat source is used to measure the thermal properties of the material, rather than applying a flame to the material and allowing the material to ignite. These tests tend to be more expensive and more involved than the flaming exposure tests.

One thermal exposure test method incorporates a copper slug calorimeter as a heat flux-measuring device. Four J-type thermocouples are connected to a 1.57 in (40 mm) diameter copper disk sunken into an insulating material so that the faces of each are completely flush. Knowing the average temperature rise of the thermocouples, along with the time-change, and the thermal properties of copper and the area, the heat flux may be calculated. This heat flux is used to predict if a second-degree burn would have occurred. This is done by using tables providing correlations between second-degree skin burn data and heat flux. The difference between these thermal exposure tests lies in the heat source.

Changes in the mechanical strength, both during and after thermal exposure, are critical in assessing the high temperature performance of geopolymers, especially for materials intended for use in structural applications (Riessen et al., 2009).

2.5.23.2 ASTM D 4108-87 Standard Test Method for Thermal Protective Performance of Materials for Clothing by Open-Flame Method

The standard ASTM D 4108–87 uses an open flame to supply 2.0 cal/cm^2-s (84 kW/m^2) to the sample in a static, faced-down position. The temperature and the time are then recorded. This test method incorporates mainly convective heat transfer, along with some radiation to the material tested from the source flame (Barker et al., 1999).

This test method rates textile materials for thermal resistance and insulation when exposed to a convective energy level of about 2.0 cal/cm^2 (8.3 W/cm^2) for a short duration. It is not intended for evaluating materials exposed to any other thermal exposure, such as radiant energy or molten metal splash. This test method is not applicable to textile materials that undergo complete flaming combustion when tested vertically for flame resistance.

ASTM D4108-87 Thermal Protective Performance (TPP) test measured the amount of thermal protection a fabric would give a wearer in the event of a flash fire. The TPP is defined as the energy required to cause the onset of second-degree burn when a person is wearing the fabric. The higher the TPP, the more protective the fabric is.

NFPA published addressing a manikin test in (NFPA, 2012) Standard on Flame-Resistant Garments for Protection of Industrial Personnel against Flash Fire, 2001 Edition (NFPA, 2012). The passing criteria for NFPA 2112–2007 edition was a minimum TPP of six calories in a spaced configuration, and a minimum TPP of three calories in the contact configuration (ISHN, 2000).

TPP was first utilized in the 1980s as a method to evaluate the effectiveness of fire fighter turnout gear. When the ASTM F23 Committee on Protective Apparel reviewed this test method early in the 21st century, the Committee determined that TPP did a poor job of measuring the heat a fabric would block to prevent a burn – the protective part of TPP. As such, the Committee created HTP (Heat Transfer Performance), in which part of the transition from Thermal Protective Performance (TPP) was to remove all references to protection.

2.5.23.3   ASTM F 1060–87, Standard Test Method for Thermal Protective Performance (TPP)

The "Standard Test Method for Thermal Protective Performance (TPP) of Materials for Protective Clothing for Hot Surface Contact", also known as the TPP test, measures the performance of insulating materials in contact with a hot surface based on human tissue response as simulated by a copper calorimeter, to the conducted heat. Each material is measured under a contact pressure of 0.5 psi (3 kPa). A 4 inch by 6 inch (100 mm by 150 mm) specimen is sandwiched between a hot plate capable of maintaining 600°F (300°C) and a copper calorimeter/steel block assembly supplying the pressure and recording the measurements. Although this test method does not incorporate a flame, it is important for fire testing purposes, as the evaluation of the thermal protection is an important parameter in determining the level of protection provided by a garment (ASTM, 1993).

2.5.23.4   ASTM F2700, Standard Test Method for Unsteady-State Heat Transfer Evaluation of Flame Resistant Materials for Clothing with Continuous Heating

This test method is intended for the determination of the heat transfer performance (HTP) value of a material, a combination of materials, or a comparison of different materials used in flame resistant clothing for workers exposed to combined convective and radiant thermal hazards.

This test method evaluates a material's unsteady-state heat transfer properties when exposed to a continuous and constant heat source. In the test method, a swatch of fabric is positioned horizontally and exposed to a two cal/second flame. A sensor measures the amount of energy that passes through the fabric during the exposure. Once sufficient heat has passed through the fabric to cause second-degree burning, the flame exposure is terminated. The difference between the amount of energy the fabric is exposed to, and the amount of energy that passes through the fabric during the exposure, is essentially that fabric's HTP (ASTM, 2013).

2.5.23.5   Radiative Protection Performance Test Method

The *Radiative Protection Performance* (RPP) test method uses a quartz lamp to provide a heat flux of 0.5 cal/cm^2-s (21 kW/m^2) to the sample. Unlike the TPP test methods, the RPP test method places the sample in a horizontal fashion. This test method is designed to prevent flame impingement and thus evaluates only the radiative portion of the heat source (NFPA, 1998).

All three thermal exposure test methods create a heat transfer network to test for thermal properties of the sample material. The copper calorimeter is described in the standards; however, any number of sensors may be used. Unlike the flaming exposure tests, the thermal exposure tests usually require data acquisition systems and can correlate the heat flux data to an equivalent burn on human skin tissue.

2.5.23.6 Other Tests Methods and Requirements of Protective Clothing for Firefighters

The firefighters' protective cloths should pass several tests such as tear strength (EN ISO 4674–1 Method B), resistance to penetration by liquid chemicals (EN ISO 6530), water (EN 20811) and water vapor (EN 31092), surface wetting (EN ISO 4920), visibility (EN 469 part 6.14.1), heat transfer–flame (EN 367), heat transfer–radiation (EN ISO 6942), and residual tensile strength of material when exposed to radiant heat (EN ISO 6942+).

2.5.23.7 Large-scale Instrumental Manikin Tests

Currently the large-scale test method for clothing flammability uses a fully dressed manikin to represent a person. This manikin is then subjected to a fire environment and an evaluation of the entire garment assembly is made, rather than just a small sample of textiles from the small-scale test methods.

The instrumented manikin test was developed for garment testing in the 1960s (Leblanc, 1998).

Typically, the manikin is subjected to a flash fire scenario where eight jets of fire, in four corners of a small room, surround the manikin for anywhere from 2.5 to 10 seconds at a time. The surrounding eight propane jets expose the manikin to a heat flux of 2.0 cal/cm^2-s (84 kW/m^2) (Behnke et al., 1992).

This type of instrumented manikin test has been developed into the standard ASTM 1930. Use this test method to measure the thermal protection provided by different materials, garments, clothing ensembles, and systems when exposed to a specified fire.

2.5.23.8 ASTM 1930 (1918) Test

ASTM 1930 (1918) test is the standard test method for evaluation of flame-resistant clothing for protection against flash fire simulations using an instrumented mannequin. The test method does not simulate high radiant exposures; for example, those found in electric arc flash exposures, some types of fire exposures in which liquid or solid fuels are involved, nor exposure to nuclear explosions.

- This test method provides a measurement of garment and clothing ensemble performance on a stationary upright manikin of specified dimensions. This test method is used to provide predicted skin burn injury for a specific garment or protective clothing ensemble when exposed to a laboratory simulation of a fire. It does not establish a pass/fail for material performance.
- This test method is not intended to be a quality assurance test. The results do not constitute a material's performance specification.

- The effects of body position and movement are not addressed in this test method.
- The measurement of the thermal protection provided by clothing is complex and dependent on the apparatus and techniques used. It is not practical in a test method of this scope to establish details sufficient to cover all contingencies. Departures from the instructions in this test method have the potential to lead to significantly different test results. Technical knowledge concerning the theory of heat transfer and testing practices is needed to evaluate if and which departures from the instructions given in this test method are significant. Standardization of the test method reduces, but does not eliminate, the need for such technical knowledge.
- This test method is used to provide predicted human skin burn injury for single-layer garments or protective clothing ensembles mounted on a stationary upright instrumented manikin which are then exposed in a laboratory to a simulated fire environment having controlled heat flux, flame distribution, and duration. The average exposure heat flux is 84 kW/m^2 (2 cal/s·cm^2), with durations up to 20 s.
- The visual and physical changes to the single-layer garment or protective clothing ensemble are recorded to aid in understanding the overall performance of the garment or protective clothing ensemble and how the predicted human skin burn injury results can be interpreted.
- The skin burn injury prediction is based on a limited number of experiments where the forearms of human subjects were exposed to elevated thermal conditions. This forearm information for skin burn injury is applied uniformly to the entire body of the manikin, except the hands and feet. The hands and feet are not included in the skin burn injury prediction.
- The measurements obtained and observations noted can only apply to the particular garment(s) or ensemble(s) tested using the specified heat flux, flame distribution, and duration.
- This standard is used to measure and describe the response of materials, products, or assemblies to heat and flame under controlled conditions, but does not by itself incorporate all factors required for fire hazard or fire risk assessment of the materials, products, or assemblies under actual fire conditions.
- This method is not a fire test response test method.
- This standard does not purport to address all of the safety concerns, if any, associated with its use. It is the responsibility of the user of this standard to establish appropriate safety, health, and environmental practices and determine the applicability of regulatory limitations prior to use.
- Fire testing is inherently hazardous. Adequate safeguards for personnel and property shall be employed in conducting these tests.

This standard should be used to measure and describe the response of materials, products, or assemblies to heat and flame under controlled conditions and should not be used to describe or appraise the fire-hazard or fire-risk of materials, products, or assemblies under actual fire conditions.

However, results of this test may be used as elements of a fire-hazard assessment or a fire-risk assessment which takes into account all of the factors which are pertinent to an assessment of the fire hazard or fire risk of a particular end use.

The ASTM F1930 standard was most recently updated in 2017. Changes in the 2017 edition were primarily related to ASTM style. However, several charts showing statistical data related to percentage body burn, collected from multiple tests conducted in different test facilities were updated with the most current data.

2.5.23.9 Manikin Pit Test

The Navy Clothing and Textile Research Facility (NCTRF) modified the instrumented manikin by placing the manikin on a boom and dynamically moving the manikin through a fire. This modification is called a Manikin Pit Test (Audet and Spindola, 1986). The thermal exposure is an n-heptane pool fire that is capable of up to 2.0 cal/cm^2-s (84 kW/m^2). The manikin is first lowered and moved to an initial exposure to the fire and is then moved through the fire and away from the fire in a certain amount of time to better simulate the exposure one might face aboard a vessel. It is important to note that this modification is not considered a standard.

The NCTRF (Navy Clothing and Textile Research Facility, USA) modified the ASTM F 1930 test by placing the manikin on a boom and dynamically moving the manikin through a fire. This modification is known as the manikin pit test. In this test, heat exposure is created by burning a Heptane pool fire that is capable of up to 84 kW/m^2 . The manikin is propelled along a track at a calculated velocity that results in a desired exposure time. The data obtained during this test can be used to observe the development of skin burns as a function of time (Fay, 2002).

Clothing regulations exist that tie these test methods together. These regulations are for firefighter clothing, both turnout gear and officewear. In order to be approved by these regulations garments must pass certain flammability tests. National Fire Protection Association (NFPA), USA developed in 1998 a number of clothing standards (National Fire Codes) namely NFPA 1971, NFPA 1975, NFPA 1976, and NFPA 1977 (NFPA, 1998) .

These clothing regulations along with government standards such as CFR 16 Parts 1610, 1615, and 1616 protect the consumer from fire injury. These regulations address the fabrics that could be hazardous in a fire scenario and provide test procedures that can be used to validate a fabric's level of safety. A test method can determine the level of safety in an article of clothing; however, the government standards make meeting those criteria a law.

2.5.23.10 Flame-resistant Clothing for Protective Clothing

Since flame-resistant (FR) clothing for petrochemical and other utility workers is essential and not a choice, one may be bewildered with the wide variety of choices of fabric and garments. Before making any decisions about selection, one must know which fabrics and garments are meeting the required performance specifications, and how they are determined (ISHN, 2000).

2.5.23.11 Regulatory Requirements

The first step toward compliance is to understand which regulations must be met. With the Occupational Safety and Health Act of 1970, the US Congress created the Occupational Safety and Health Administration (OSHA) to assure safe and healthful working conditions for working men and women by setting and enforcing standards, and by providing training, outreach, education, and assistance. Three OSHA regulations (OSHA, 2019) are used as the basis of requiring flame-resistant clothing.

2.5.23.12 OSHA's 1910.269

On April 1, 2014, OSHA announced the final rule revising the 1910.269 standard for electric power generation, transmission, and distribution. This revised rule implements significant changes to utilities' requirements for protecting workers from electric arcs and using flame resistant clothing, among other areas.

The employer shall ensure that each employee exposed to hazards from electric arcs wears protective clothing and other protective equipment with an arc rating greater than or equal to the heat energy estimate whenever that estimate exceeds 2.0 cal/cm^2. "This protective equipment shall cover the employee's entire body," except for certain exemptions for hands, feet, and head protection.

FR clothing is now considered Personal Protection Equipment (PPE). Previously, FR clothing was not explicitly considered PPE. In the ruling, OSHA makes clear that it is the employer's responsibility to ensure proper care and maintenance of employees' protective clothing. OSHA is equally clear the final rule does not require employers to launder protective clothing for employees. They must train their employees in proper laundering procedures and techniques, and employers must inspect the clothing on a regular basis to ensure that it is not in need of repair or replacement.

The maintenance standard mandates that personnel who work around energized parts must not wear clothing that, if exposed to an electric arc, could contribute to the extent of burn injury. In simple terms, this means that the clothing cannot ignite, so wearing polyester, nylon rayon, or acetate (unless FR-treated) is prohibited.

2.5.23.13 OSHA 1910.132

OSHA's general duty clause requires employers to identify risks and protect employees from hazards in the workplace. OSHA 1910.132 offers employers guidelines on protective equipment for workers. The personal equipment, including personal protective equipment for eyes, face, head, and extremities, protective clothing, respiratory devices, and protective shields and barriers, shall be provided, used, and maintained in a sanitary and reliable condition wherever it is necessary by reason of hazards of processes or environment, chemical hazards, radiological hazards, or mechanical irritants encountered in a manner capable of causing injury or impairment in the function of any part of the body through absorption, inhalation, or physical contact. The rule applies to many types of personal protective equipment, and has been used to cite employers who did not require the use of flame-resistant protective apparel.

2.5.23.14   OSHA 1910.119

This section contains requirements for preventing or minimizing the consequences of catastrophic releases of toxic, reactive, flammable, or explosive chemicals. These releases may result in toxic, fire, or explosion hazards.

Process Safety Management Regulation requires employers to assess risk throughout the entire manufacturing process to ensure that the process is safe. While the standard does not specifically require FR clothing, OSHA has used this standard more frequently than the General Duty Clause as the basis of citing employers for not requiring FR clothing.

2.5.23.14.1   Performance Specifications

Once one knows the standards required to be met, it is important to know the differences between the performance specifications related to the flame-resistant clothing to be provided for employees.

ASTM F1506–98, Standard Performance Specification for Textile Materials for Wearing Apparel for Use by Electrical Workers Exposed to Momentary Electric Arc and Related Thermal Hazards.

This performance specification covers the flame resistance of textile materials to be used for wearing apparel for use by electrical workers exposed to momentary electric arc and related thermal hazards. At present, a bench scale arc test for laboratory use is not available. It is the intent of the Committee to continue the search for an acceptable laboratory test based on either an electric arc exposure or an acceptable alternative, which form the basis of a modification of this specification. Protective properties relate to thermal exposure from momentary arc and associated exposure to open flame and radiant heat.

2.5.23.15   ASTM F1891–06, Standard Specification for Arc and Flame-Resistant Rainwear

The standard applies to flame-resistant, waterproof materials used in rainwear. These garments can be made from coated or laminated fabrics. The standard is currently being revised to include a fabric flammability test more suitable to coated fabrics.

Used for rainwear compliance with OSHA 1910.269, NFPA 70E, and CSA Z462, ASTM F1891 Standard Specification for Arc and Flame Resistant Rainwear outlines minimum physical, arc flash, and thermal performance criteria for rainwear used by workers with a risk exposure to momentary electric arcs. The ASTM F1959 test method is followed; additionally, both the Arc Thermal Performance Value (ATPV) and Energy Breakopen Threshold (EBT) results are reported, and the material is exposed to twice the arc rating to ensure that no melting or dripping of the rainwear occurs.

2.5.23.16 NFPA 2112, Standard on Flame-Resistant Clothing for Protection of Industrial Personnel against Short-Duration Thermal Exposures from Fire

Essential for manufacturers and certifying agencies, this standard protects workers from flash fire exposure and injury by specifying performance requirements and test methods for flame-resistant fabric and garments.

2.5.23.17 NFPA 2113: Selection, Care, Use, and Maintenance of Flame-Resistant Garments for Protection of Industrial Personnel against Short-Duration Thermal Exposures from Fire

Protects industrial workers from thermal exposures with NFPA 2113's up-to-date provisions for PPE selection, use, and maintenance.

Diffuse fuel contributes to rapid growth and spread of the fire, and any potential for short-term thermal exposure conditions requires protecting workers from burns and other risks in exposed areas. Developed primarily for garment end users to reduce the risks caused by incorrect maintenance, contamination, or damage, NFPA 2113 provides minimum requirements for the correct selection, care, use, and maintenance of flame-resistant clothing (FRC) in compliance with the NFPA 2112: Standard on Flame-Resistant Garments for Protection of Industrial Personnel against Flash Fire for use in areas at risk.

2.6 CHANGES IN THE 2015 NFPA 2113

This new section on responsibility clarifies that both the employer and employee have specific responsibilities for selection, care, use, cleaning, and maintenance of protective garments.

This widely used standard is essential for any company or facility that handles or processes flammable gas, flammable or combustible liquids, or combustible dust that may accumulate and/or deflagrate, exposing personnel to short duration thermal exposures from fire.

2.6.1 TEST METHODS

While knowing various performance specifications has benefits, such knowledge is meaningless, unless one is familiar with the test methods used to determine whether fabrics meet these specifications.

2.6.1.1 Meeting Specifications for Clothing Worn by Electric Utility Workers

ASTM's F1506-98 fabric specification for clothing worn by electric workers relies upon the FTM 5903.1 Vertical Flammability Test, which determines whether a fabric will ignite and continue to burn after exposure to an ignition source. The test method sets criteria on how the test should be conducted (e.g., sample size, number of trials, type of flame), but does not establish performance requirements. To pass this test and be included in ASTM's Fabric

Specification for Clothing Worn by Electric Workers, the garment must have a vertical flame test maximum of two seconds after-flame and six-inch char length when it is new, as well as after 25 home launderings.

ASTM's F1959-99 Arc Thermal Performance Value Test, which measures the amount of thermal protection a fabric would give a wearer if the person was caught in the vicinity of an electric arc. The ATPV is defined as the arc energy required to cause the onset of second-degree burning when a person is wearing the fabric. The higher the ATPV, the more protective the fabric is.

This test is only conducted on flame-resistant fabrics to measure protection from the heat and flame by-products of an electric arc. It does not indicate any protection from contact with an electric arc. The standard is now superseded by ASTM F1959/F1959M, Standard Test Method for Determining the Arc Rating of Materials for Clothing.

2.6.1.2 Meeting Specifications for Arc and Flame Resistant Rainwear

ASTM's Standard for Arc and Flame-Resistant Rainwear (F1891-98) relies on the Vertical Flame Test and the Arc Thermal Performance Values described previously. To pass the test and be included in the rainwear standard, garments must have a Vertical Flame Test with a maximum two seconds after-flame and six-inch char length when it is new and after 25 washings.

2.6.1.3 Meeting Specifications for Flame-Resistant Garments for Industrial Personnel

To meet the pending NFPA 2112 standard regarding flame-resistant garments for the protection of industrial personnel against flash fires, garments must meet several tests. Aside from the aforementioned Vertical Flame Test (requirements for this specification are a vertical flame test with a maximum two seconds after-flame and four inch char length when it is new and after 100 launderings), other tests include:

ASTM D4108-87 Thermal Protective Performance (TPP) Test, which measures the amount of thermal protection a fabric, would give a wearer in the event of a flash fire. The TPP is defined as the energy required to cause the onset of second-degree burn when a person is wearing the fabric. The higher the TPP, the more protective the fabric is. To meet the criteria of this standard, the fabric must possess a minimum TPP of six-inch spaced configuration and a minimum TPP of three-inch contact configuration.

Heat resistance or thermal stability requirements mandate that the fabric would not melt and drip, separate, or ignite in a 500°F oven for 5 minutes. Thermal Shrinkage Resistance requirements mandate that fabric will have a maximum of 10% shrinkage after being in a 500°F oven for 5 minutes.

NFPA 2112 Standard on Garments for Protection of Industrial Personnel against Flash Fire, is a commonly used standard specification for the evaluation of flame-resistant clothing. In the most recent edition of NFPA 2112, gloves and shrouds, hoods, and balaclavas (a close-fitting garment covering the whole head and neck except for parts of the face, typically made of wool.)

were included; certification programs for these products are available. The standard is cited by OSHA for those drilling in the oil and gas industry.

Garments must also pass the ASTM Flash Fire Manikin Test for inclusion in the NFPA 2112 category. This test is a full-scale manikin test designed to test fabrics in completed garment form in a simulated flash fire. A manikin, with up to 122 heat sensors spaced around its body, is dressed in the test garment, and is then exposed to a flash fire for a predetermined length of time. Tests are usually conducted at heat energies of 1.8–2 cal/cm^{2}sec., and for durations of 2.5 to 4.5 sec.. Results are reported in percentage of body burn; with a maximum of 50% total body burn indicated when a standard coverall is tested for three seconds with underwear worn under the test garment.

2.6.1.4   Testing of Upholstered Furniture

British standards specify requirements for the resistance of upholstered furniture used for seating to ignition when tested in accordance with BS 5852, BS EN 1021–1, or BS EN 1021–2, as appropriate. The levels of ignition resistance have been set after careful consideration of the fire risk of the particular end-use environment involved. These levels do not necessarily reflect the behavior of the upholstered seating in a fully developed fire. Upholstered seating for domestic use and transport is not covered by this standard.

BS 5852:2006, Methods of test for assessment of the ignitability of upholstered seating by smoldering and flaming ignition sources.

BS 5852–2:1982, Fire tests for furniture – Part 2: Methods of test for the ignitability of upholstered composites for seating by flaming sources 1.

BS EN 1021–1:1994, substituted by BS EN 1021–1:2006 and subsequently by BS EN 1021–1:2014, Furniture – Assessment of the ignitability of upholstered furniture – Part 1: Ignition source smoldering cigarette.

BS EN 1021–2:1994, substituted by BS EN 1021–2:2006 and subsequently by BS EN 1021–2:2014, Furniture – Assessment of the ignitability of upholstered furniture – Part 2: Ignition source match flame equivalent.

As discussed in Chapter 1, Smoldering is the slow, low-temperature, flameless form of combustion, sustained by the heat evolved when oxygen directly attacks the surface of a condensed-phase fuel. Many solid materials can sustain a smoldering reaction, including coal, cellulose, wood, tobacco, synthetic foams, charring polymers including polyurethane foam, and some types of dust. Common examples of smoldering phenomena are the initiation of residential fires on upholstered furniture by weak heat sources and the persistent combustion of biomass behind the flaming front of wildfires. To understand the hazards posed by the upholstered fabrics, the development of the small-scale composite test BS 5852 represented a remarkable development of realistic model tests that accurately indicate the ignition behavior of full-scale products of complex structure. This is also known as Smolder Ignition Testing of Upholstery Fabrics. The test is simple to use, cost-effective, and reproducible; it may be conducted in the manufacturing environment, as well as in formal test laboratory environments.

In all flammability test procedures, the conditions should attempt to replicate real use, while atmospheric conditions are specified in terms of allowable relative humidity and temperature ranges (Zammarano et al., 2012).

2.6.1.4.1 BS 5852:2006

BS 5852:2006 is the method of test for assessment of the ignitability of upholstered seating by smoldering and flaming ignition sources. The standard is used to assess the ignitability of material combinations, such as covers and fillings used in upholstered seating, when subjected to a smoldering cigarette and a match flame equivalent as ignition source.

In United Kingdom, all items of domestic upholstered furniture must meet the "Furniture and Furnishings (Fire) (Safety) Regulations 1988" with amendments. The test is carried out with the combinations of covering and filling materials that are used in upholstered furniture. The covers or fabrics can also be tested in combination with standard fillings.

Figure 2.14 shows various components and their assembly to conduct the BS 5852:2006 composite test.

The following components are required to conduct the test (Figure 2.14).

Two pieces of covering fabric measuring:

A. (20.0×8.0) in.^{2} $((508 \times 203)$ mm$^2)$.
B. (13.0×8.0) in.^{2} $((330 \times 203)$ mm$^2)$.
C. One piece of cigarette cover sheeting (UF-400 type; Test Fabrics Inc.) measuring (5.0×5.0) in.^{2} $((127 \times 127)$ mm$^2)$.
D. Wooden sample holder (L-shaped body made of wood for seating foam pieces and other components. Two pieces of high smoldering standard polyurethane foam measuring:
E. $(8.0 \times 8.0 \times 3.0)$ in.^{3} $((203 \times 203 \times 76)$ mm$^3)$E.
F. $(8.0 \times 5.0 \times 3.0)$ in.^{3} $((203 \times 127 \times 76)$ mm$^3)$.
G. Assembly of components to conduct smoldering test.

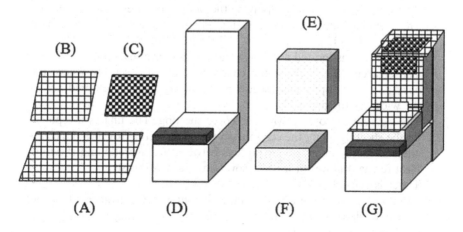

FIGURE 2.14 Components and Composite for Smoldering Test.

The components are conditioned at 55 % ± 5 % RH and 22°C ± 3°C for 24 hr prior to the tests.

The smoldering test procedure (BS 5852:2006) is according to (Zammarano et al., 2012):

- The weights of the fabrics and foams are noted separately before and after completion of the test to note mass loss of each component.
- Each piece of foam (E and F) is wrapped securely with a piece of fabric. The corners of the bigger piece of fabric (A) is pinned to the back of the larger foam (E) specimen (Back specimen) with four pins at each corner of the fabric. The edges of the fabric are arranged symmetrically on the back of the foam specimen so that a 2-in. -(51 mm) wide strip is left uncovered in the middle in the back side of the bigger foam.
- The smaller piece of fabric (B) is pinned to the top and bottom of the smaller foam specimen (F) (Front specimen) with four pins at each corner of the fabric so that one of the (8.0×3.0) in.^{2} $((203 \times 76)$ mm$^2)$ sides of the foam specimen is left uncovered.
- The dimensions of the foam pieces are measured afterwards with a ruler to make sure that the specimens have not changed shape from the pressure of handling and wrapping them. If the dimensions have changed by more than 5% of the value, the fabric is removed and reassembled again following these instructions.
- The two wrapped foam specimens are placed in the wooden sample holder (D). The Back specimen (E) is placed upright in the back of the frame while the Front specimen is laid flat on the bottom of the frame.
- The two parts of the sample holder are slid together.
- The cigarette cover sheeting (C) is placed on the top of the bigger Back specimen (E) keeping half portion on its top and the rest portion hangs vertically, fixed with pins.
- The orientation in the foam specimens (e.g., rise direction) or the sewing pattern in the fabric can affect smolder propagation. Therefore, always the assemblies are built by applying the same component configuration.
- The assembled mock-up is placed in a tray made of heavy duty aluminum foil measuring $(12.0 \times 12.0 \times 0.5)$ in.^{3} $((305 \times 305 \times 13)$ mm$^3)$ (W×L×H).
- The cigarette is ignited with the cigarette igniter apparatus.
- The cigarette is placed in the center of the test assembly where the *back* and Front specimen meet. The whole cigarette is centered across the width of the assembly.
- The cigarette is covered with the cover sheeting. The fingers are pulled on the sheeting along the cigarette so that the sheeting touches the cigarette and there remain no air pockets between the cigarette and sheeting. Heat resistant gloves or leather gloves are used for this step.
- Timer is started.
- After 45 minutes, the cover sheeting is removed from the Back specimen and place in the metal tray. Since this is the bigger specimen, it smolders more intensely.

- The fabric is removed and weighed on the balance and the fire is extinguished with a spray bottle in the metal tray. The still-smoldering foam specimen is weighted on the balance with char.
- To prevent the smoldering core from accidental separation from the specimen, lay it down on the balance so that the core is on top. A wide metal spatula may be used to keep the core in place while handling between the platform and balance and then between the balance and metal tray.
- After weighing, carve the smoldering core off into the metal tray with a spatula and extinguish it with the spray bottle and/or squirt bottle. Weigh the foam piece without the core.
- The process is repeated for the Front specimen.
- After extinguishing, wrap the charred fabrics and charred cores with the aluminium foil.
- After the test, the specimens are checked three times in 10-minute intervals for visible smoke; flip them over with a spatula to check both the bottom and top to make sure that the smoldering was extinguished properly.
- The sample holders are unassembled and checked if the smoldering cigarette butt has fallen onto the bottom of the frame. If so, drop it into the metal tray.

EN 1021–1:2014 and EN 1021–2:2014

Assessment of the ignitability of upholstered furniture – Part 1: Ignition source smouldering cigarette.

Assessment of the ignitability of upholstered furniture – Part 2: Ignition source match flame equivalent.

The test methods similar to BS 5852:2006, varies with material specifications (RISE, 2019).

The primary test metric is total mock-up mass loss after 45-min. exposure to the cigarette smoldering ignition source.

The standard is used to assess the ignitability of material combinations, such as covers and fillings used in upholstered seating, when subjected to a smoldering cigarette and a match flame equivalent as ignition source.

In Sweden, upholstered furniture for domestic use shall not ignite when subjected to a smoldering cigarette according to EN 1021–1. In public spaces, such as cinemas, hotels, restaurants, we recommend that furniture be able to resist both a smoldering cigarette and a match flame equivalent.

These test methods only evaluate the ignitability of upholstery combinations. To evaluate the burning behavior, the furniture needs to be tested in full scale with larger ignition sources.

The following materials can be tested following to this standard:

- Combinations of covering and filling materials that are used in upholstered furniture.
- Covers/fabrics that can also be tested in combination with standard fillings.

Sample dimensions:

- Complete test with 1 ignition source
- Cover/interliner/wadding, 2 pieces with the dimensions 800 x 650 mm.
- Filling, 2 pieces with the dimensions 450 x 300 x 75 mm.
- Filling, 2 pieces with the dimensions 450 x 150 x 75 mm.
- The thickness of 75 mm refers to the filling together with any wadding.

If the filling differs between the seat, back, armrests, etc. all fillings must be tested separately.

Indicative test:

- One piece each of the aforementioned material.

2.6.1.5 Pretreatment
According to the British furniture regulation, all interliners and cover fabrics that have been treated with a flame retardant shall go through a water-soaking procedure before testing.

2.6.1.6 Short Test Description
The tests are performed in a test cabinet with a calibrated air flow. The cover fabric and the filling are put in a test rig to create a small sofa with a 90-degree angle between seat and back. The ignition sources are located in the junction between seat and back.

During the cigarette test, the test assembly is not allowed to smolder after one hour from the beginning of the test. The test assembly is also not allowed to smolder to the extremities of the specimen.

The test assembly is subjected to a gas flame equivalent to a match flame for 15 seconds. No flaming is allowed to continue for more than 120 seconds after removal of the burner tube.

The cover fabric and the filling are put in a test rig to create a small sofa with a 90-degree angle between seat and back. The ignition sources are located in the junction between seat and back.

2.6.2   Scientific Methods for Mechanistic Elucidation

With the rapid penetration of polymeric materials into the building industry, flame retardancy of these materials has become very important. Products containing flame retardant additives must exhibit adequate flame resistance properties without sacrificing appearance or physical properties. Thermogravimetric Analysis (TGA), which provides information about the thermal stability and decomposition rate of materials, is an excellent technique for rapidly comparing the behavior of materials tested with different flame retardants.

Thermogravimetric Analysis (TGA), which measures weight changes in materials as a function of temperature or time, provides a simple, fast alternative to the following ASTM methods:

- D757-49: Flammability Test for Self-Extinguishing Plastics,
- D1692-57T: Test for Flammability of Plastic Foams and Sheeting, and
- D1360-58: Test for Fire Retardancy of Paints.

By monitoring the weight loss under different heating conditions, the burning characteristics of different material formulations can be compared. For the flame-retarded material the rate of oxidation is consistently about half the rate of decomposition (first stage of weight loss). In the standard material, however, the rate of oxidation is dependent on the heating rate. It is postulated that at slow heating rates (2.5°C/min or slower), a more porous carbon skeleton is formed, allowing oxidation to proceed at a higher rate. Faster heating rates apparently produce less porous carbon skeletons, resulting in slower oxidation. Repeatability of TGA decomposition studies is excellent, showing the results of three successive runs at a heating rate of 10°C/min (TA Instruments, 2019).

The rate of heat release from an unwanted fire is a major threat by the fire to life and property. A reliable measurement of a fire's heat release rate was a goal of fire researchers as early as the 1960s. Historically, heat release measurements of burning materials were based on the increased temperature of ambient air as it passed over the burning object. As the fraction of heat released by radiant emission varies with the type of material and not all the radiant energy contributes to temperature rise of the air, there were large errors in the measurements. Attempts to account for the heat that was not captured by the air required installing numerous thermal sensors about the fire to intercept and detect the additional heat. This approach proved to be tedious, expensive, and susceptible to large errors, particularly while burning large objects such as a full-sized room filled with flammable furnishings and surface finishes.

A novel alternative technique for determining heat release rate was developed at NBS during the 1970s. It had distinct advantages over the customary approach, but its widespread acceptance was hampered in the fire science community as the technique was used in less-than-ideal circumstances. Clayton Huggett, a fire scientist at NBS, published a seminal paper Hugget (1980) that convinced the fire science community that the new technique was scientifically sound and sufficiently accurate for fire research and testing. The technique is now used worldwide and forms the basis for several national and international standards.

The underlying principle of the new heat release rate technique was *discovered* in the early 1970s. Faced with the challenge of measuring the heat release of combustible wall linings during full-scale room fire tests, William Parker, Huggett's colleague at NBS, investigated an alternative approach based on a simple fact in physics: in addition to the release of heat, the combustion process consumes oxygen. As part of his work on the ASTM E 84 tunnel test, Parker (1977) explored the possibility of using a measurement of the reduction of oxygen in fire exhaust gases as an indicator of the amount of heat released by the burning test specimens. Indeed, for well-defined materials with known chemical composition, heat release and oxygen consumption can both be calculated from thermodynamic data. The problem with applying this approach to fires is that in most cases, the chemical compositions of modern materials/composites/mixes that are likely to be involved in real fires are not known. In the process of examining data for complete combustion (combustion under stoichiometric or excess air conditions) of the polymeric materials with which Parker was working, he found that, although the heat released per unit mass of material consumed (i.e., the specific heat of combustion) varied greatly, the amount of heat released per unit volume of oxygen consumed was fairly constant, i.e., within 15% of the value for methane, 16.4 MJ/m3 of oxygen consumed. This fortunate circumstance – that the heat release rate per unit volume of oxygen consumed is approximately the same for a range of materials used to construct buildings and furnishings – meant that the heat release rate of materials commonly found in fires could be estimated by capturing all of the products of combustion in an exhaust hood and measuring the flow rate of oxygen in that exhaust flow. The technique was dubbed *oxygen consumption calorimetry*, notwithstanding the absence of any actual calorimetric (heat) measurements (Steckler, 2019).

Spontaneous combustion of coal is a global problem that plagues the industry. It poses a safety risk, which can be life-threatening in extreme cases. The low temperature oxidation of coal is commonly accepted as the main cause of spontaneous combustion. A method, based on low temperature oxidation, using a thermogravimetric analyzer (TGA) as a tool to predict the spontaneous combustion propensity of coal, has recently been suggested. This method is a relatively simple and quick method, if proven repeatable enough. In this study, the accuracy of this method was tested on five different coal samples of different rank. Small samples (< 30mg) of fine coal (<212μm) were reacted with air in a TGA/DSC at five different heating rates (3°C, 5°C, 7°C, 10°C, and 20° C per minute). The TGA was used to measure (Graan and Bunt, 2016) the mass change of the coal as a function of temperature and time. The thermograms depicting mass as a function of temperature were used to predict the ignition temperature of each coal. The first derivative of mass as a function of time for the region where oxygen sorption takes place, prior to combustion was used to calculate a spontaneous combustion index. This index was used to classify the coals according to their spontaneous combustion propensity. The one drawback of the TGA method is that it requires more extensive data analysis.

A high resolution thermal gravimetric analysis (TGA) was used to determine thermal decomposition temperatures and rates of decomposition (Abu-Isa et al., 2019). TGA runs were conducted in nitrogen and air atmospheres. For the different polymers investigated, the ranges of decomposition temperatures were between 223°C and 552°C in nitrogen, and 240°C to 565°C in air. Correlation was made between the thermal properties and the flammability characteristics quantified in this study.

Ignition temperatures estimated from the Critical Heat Flux (CHF) values were about 14% higher than the decomposition temperatures from the thermal properties measurements. The experimental Thermal Response Parameter (TRP) values were about 28% higher than the TRP values calculated from thermal analysis.

High resolution thermal gravimetric analysis (TGA), and modulated differential scanning calorimetry were the two techniques employed. Thermal analysis results were compared with flammability parameters obtained using a flammability apparatus capable of measuring ignition, combustion, and heat release variables. Good agreement was observed between measured flammability parameters, such as ignition temperature, critical heat flux, and thermal response parameter, and the thermal analysis results such as specific heat, thermal conductivity, and decomposition temperature (Abu-Isa et al., 2019).

2.6.2.1 Limiting Oxygen Index Test

The limiting oxygen index test (described, for example, in ASTM D2863) is probably the most widely used method for assessing flammability. This test measures the minimum oxygen fraction in an oxygen/nitrogen mixture that would enable a slowly rising sample of the gas mixture to support combustion of a candlelike sample under specified test conditions. The top edge of the test sample is ignited, and the oxygen concentration in the flow is decreased until the flame is no longer supported. The reasons for the differences between the polymers vary, but in particular two factors may be noted:

1. The higher the hydrogen to carbon ratio in the polymer, the greater is the tendency to burning (other factors being equal).
2. Some polymers on burning emit blanketing gases that suppress burning.

While the limiting oxygen index (LOI) test is quite fundamental, it does not characterize the burning behavior of the polymer.

ASTM D2863–17a (Standard Test Method for Measuring the Minimum Oxygen Concentration to Support Candlelike Combustion of Plastics (Oxygen Index).

This test method provides for measurement of the minimum concentration of oxygen in a flowing mixture of oxygen and nitrogen that will just support flaming combustion of plastics. During the test the specimen is held vertically in different O_2: N_2 atmosphere and is ignited from the top. The correlation with burning characteristics under actual use conditions is not implied. In this test method, the specimens are subjected to one or more specific sets of laboratory test conditions. If different test conditions are substituted, or the end-use conditions are changed, it is not always possible by or from this test to predict changes in the measured fire–test–response characteristics. Therefore, the results are valid only for the fire–test–exposure conditions described in this test method.

This fire–test–response standard describes a procedure for measuring the minimum concentration of oxygen, expressed as percent volume that would just support flaming combustion in a flowing mixture of oxygen and nitrogen.

This test method provides three testing procedures.

1. Procedure A involves top surface ignition,
2. Procedure B involves propagating ignition, and
3. Procedure C is a short procedure involving the comparison with a specified minimum value of the oxygen index.

Test specimens used for this test method are prepared into one of six types of specimens:

- Test specimens of Types I, II, III, and IV are suitable for materials that are self-supporting at these dimensions.
- Test specimens of Form V and VI are suitable for materials that require support during testing.
- Test specimens of Form VI are suitable for film materials that can be rolled into a self-supporting specimen by the described procedure.

The LOI tester (Lewin and Sello, 1984; ASTM, 2019) was first applied to the assessment of the flammability of polymers and subsequently of fabrics in 1966. The apparatus (Figure 2.15) consists of a Pyrex glass chimney (A) as a flame holder. The chimney is about 38 cm x 9 cm in dimension. The inlet or gas mixture is at the bottom of the chimney, a portion of which is filled with glass beads (B). Sample (C) is placed onto a hinged U-shaped holder (D) while the free end of the holder is held with clamps. The holder is mounted vertically in the chimney on the chimney axis. Known mixtures of O_2 and N_2 in varying ratios are passed upward through the chimney at a velocity of 30–110 mm/sec. for 1–2 min. The gases from individual cylinders are first passed through separate metering orifices (E) which are fitted with separate Bourdon test gauge (F, 0–100 psi/g) and then mixed together in determined ratio and then passes through the chimney. A laboratory-type gas burner is used for the ignition of the sample at its top end at the time of flow of the gas mixture. The concentration of O_2, which is just sufficient to sustain the flame, is determined and its volume fraction in the gas mixture is calculated as LOI.

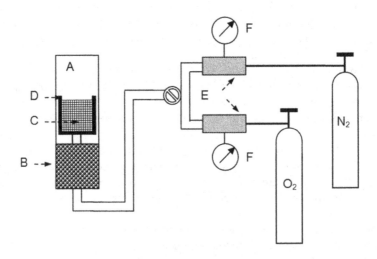

FIGURE 2.15   A schematic diagram of LOI measurement apparatus.

2.6.2.2   NASA STD 6001 Upward Flammability Test 1

Spacecraft tire safety emphasizes fire prevention, which is achieved primarily through the use of fire-resistant materials. Materials selection for spacecraft is based on conventional flammability acceptance tests, along with prescribed quantity limitations and configuration control for items that are nonpass or questionable (Friedman, 1999). NASA STD 6001 Test 1 (NASA, 1998) is the major method used to evaluate flammability of materials intended for use in the habitable environments of US spacecraft. The method is an upward flame-propagation test in a quiescent environment using a well-defined igniter flame at the bottom of a vertically mounted sample.

LOI testing procedures in flowing environments are described in ASTM D 2863. The downward flame propagation tests were standard. Upward flame propagation tests were conducted on vertical samples ignited at the bottom. A material passes this test if the vertical burn length is less than 15.2 cm and there is no evidence of transfer of burning debris (NASA, 1998). The upward flammability test is conducted in the most severe flaming combustion environment expected in spacecraft. Test 1 provides conservative results by sustaining material-flaming combustion in less severe environments than those in which extinguishment occurs in quiescent microgravity environments (Hirsch et al., 2000). For many years, this test method has provided data that have allowed the US to achieve an outstanding spacecraft fire safety record (Hirsch and Beeson, 2001).

Although reasonable from a flammability safety point of view, NASA STD 6001 Test 1 has a few drawbacks. The test may eliminate materials that may be safe for use on spacecraft. On the positive side, it is conservative, but it may be overly conservative on occasion. Its degree of conservativeness varies for different materials and cannot be estimated from the data, since it is impossible to estimate how far a material is removed from the combustion threshold conditions. The Test I pass/fail test logic does not allow a precise quantitative comparison with other ground or microgravity materials flammability test results; therefore its use is limited, and possibilities for an in-depth theoretical analysis and realistic estimates of spacecraft fire extinguishment requirements are practically eliminated. Attempts for precise quantitative correlations between the results provided by Test 1 and other ground flammability tests were generally not very successful. Previously, a version of the NASA STD 6001 Test 1 was compared with Critical Oxygen Index test results conducted with a method similar to ASTM G 125-95el, Standard Test Method for Measuring Liquid and Solid Material Fire Limits in Gaseous Oxidant, (Judd et al., 1981). The data indicated that if a material had a critical oxygen index of at least 35, it could be used in a space lab environment containing 23.8% oxygen. The empirical correlation determined based on these tests, was later shown not to always hold (Hirsch, 1989). The difficulty of quantifying NASA STD 600 results has been revealed in a study (Ohlemiller and Villa, 1991) that attempted to correlate its results with Heat Release Rate Tests, conducted according to ASTM E 1354–99 (Standard Test Method for Heat and Visible Smoke Release Rates for Materials and Products Using an Oxygen Consumption Calorimeter) and the Lateral

Ignition and Flame Spread Tests (LIFT), conducted per ASTM E 1321-97a (Standard Test Method for Determining Material Ignition and Flame Spread Properties). This study deduced that the mean upward spread velocities in the NASA tests appear to correlate inversely with the minimum heat flux for opposed flow spread and the minimum heat flux for ignition in the LIFT tests. Furthermore, the study indicates that the peak heat release determined by cone calorimeter would not predict flammability performance in the NASA test (Ohlemiller and Villa, 1991). A different result was reported for three composites in a study in which the upward flame spread rate and flame spread length were shown to increase with the peak heat release rate (Hshieh and Beeson, 1995).

2.6.2.3 Microscale Combustion Calorimetry (MCC)

The Micro Combustion Calorimeter was developed by the Federal Aviation Administration (FAA) to offer industry a research tool to assist the FAA in its mandate to dramatically improve the fire safety of aircraft materials. The tester is becoming a mainstay in research laboratories due to its ability to obtain meaningful test data with a sample size in the range of 0.5 mg to 50 mg. The instrument has been validated by ASTM International, and is now the subject of an ASTM standards publication designated D7309. Microcombustion calorimetry (MCC), sometimes referred to as pyrolysis–combustion flow calorimetry (PCFC), measures the rate at which the heat of combustion of fuel gases are released by a solid during controlled pyrolysis in an inert gas stream. The fuel gases are mixed with excess oxygen and combusted (oxidized) at high temperature, and the instantaneous heat of combustion of the flowing gas stream is measured by oxygen consumption calorimetry. Some of the commercial MCC equipment are Concept Equipment (www.concept-e.co.uk/) Control Instrument Corporation (www.controlinstruments.com/) etc.

The hazard in a fire is determined by the rate at which heat is released by a burning material, particularly in an enclosed space such as a building, a ship, or an aircraft cabin. Several different bench scale fire calorimetry methods have been developed for measuring heat release rate during flaming combustion of materials, products, and components. These methods require multiple samples on the order of 100 grams each, and the results are highly dependent on the ignition source, sample thickness, sample orientation, ventilation, and edge conditions (Kashiwagi and Cleary, 1993), all of which combine to make the test data dependent on configuration and the effect of material properties and composition on burning behavior are not clearly expressed.

The desire for a quantitative analytical laboratory test that correlates fire behavior or flame test performance with material properties has been the motivation to relate thermogravimetric analyses to flammability (Gracik and Long, 1992). To date, most thermogravimetric investigations of flammability have relied on a single thermal stability parameter (e.g., char yield or thermal decomposition temperature) to relate the chemical composition of a material to its fire or flame test performance (e.g., char yield versus limiting oxygen index). Individually, these thermal stability parameters have found limited success as material descriptors of flammability and their interrelationship in the context of flaming combustion has remained obscure until recently, when it was shown that a particular combination of thermal stability and combustion parameters could correlate fire behavior (Lyon, 2001).

Laboratory methods for the direct measurement of heat release rate of milligram-sized samples of materials were developed to understand the effect of material properties and chemical composition of materials on combustibility under controlled conditions. At first, Susott (1982) measured the heat of combustion of organic materials using transient heating and oxygen consumption. Milligram-sized samples of forest products (foliage, wood, stems, and bark) were pyrolyzed in an inert gas stream at a constant rate of temperature rise, and the pyrolysis gases were combined with oxygen prior to entering a catalytic reactor. In order to prevent oxidation of the char residue pyrolysis was conducted under inert gas flow. The rate of oxygen consumption was determined from the electrolytic oxygen generation rate using a null-balance, closed-loop technique. However, only qualitative information was obtained because the oxygen consumption signal was distorted by the instrument and, therefore, was not synchronized with the mass loss history of the sample. Pyrolysis residue/ char fraction was measured by weighing the sample before and after the test and the heat of combustion of the char was determined separately. Sample heating rates were limited to less than 16 K/min by the dynamic capability of the oxygen generator. The rate of heat released by complete oxidation of the pyrolysis gases during thermal decomposition of the sample was calculated from the measured oxygen consumption rate using an average gross heat of combustion for forest products of 14.0 ±0.5 kJ/g-O_2. The relevant heat of combustion is the net value obtained by subtracting the heat of vaporization of water (produced on combustion) from the gross heat of combustion. Converting Susott's gross heat of combustion to a net value gives13.3 ±0.5 kJ/g-O_2, which is equivalent to the currently accepted value 13.1 ±0.7 kJ/g-O_2 used in oxygen consumption fire calorimetry (ASTM (1990) as determined from data on a wide range of organic solids and polymers (Hugget, 1980), as well as liquids and gases. Susott (1982) did not measure mass loss during the pyrolysis–combustion experiment but normalized the total heat of combustion of the volatiles to the starting sample mass on an ash-free, dry fuel basis. The gross heat of combustion of the char determined in separate oxygen bomb calorimetry experiments (ASTM, 1985) was found to be relatively independent of the fuel type at 32.0 ±0.9 kJ/g for the 43 typical forest fuels tested.

Parker (1988) claimed an improved method by using a step-change in temperature to pyrolyze the sample at a heating rate considered to be more typical of the surface conditions in a fire. The samples of cellulose were inserted into a preheated, nitrogen-purged furnace resulting in a rapid uncontrolled temperature change and subsequent pyrolysis. The oxygen is mixed with the inert gas-pyrolyzate stream in a catalytic reactor. Combustion gas concentrations (O_2, CO_2, H_2O) are measured and the heat release rate of volatile fuel combustion is calculated from the oxygen consumed. Although mass loss is not measured directly during the test, transient oxygen depletion and combustion gas generation were used to calculate the mass loss rate of cellulose from its known chemical structure.

In the study by Reshetnikov et al. (1999) milligram polymer samples were decomposed isothermally in an inert gas stream and excess oxygen is added to the volatile fuel stream at various temperatures to effect oxidation. A gas chromatograph with a flame ionization detector was used to sample the fuel stream prior to mixing with oxygen in order to separate and identify the individual products of decomposition. Gas phase oxidation kinetics of the fuel species are computed from the measured oxygen consumption history using isothermal methods of analysis. Reshetnikov's device measures the rate of combustion (oxidation) of the fuel gases but not the rate at which these gases are produced by the decomposing solid as it is heated, the latter being the rate limiting process in a fire.

Lasers have Lasers have been used to pyrolyze polymers for flammability studies by Price (1999) and Angel (1998). The pyrolysis gases were analyzed by mass spectrometry and laser-induced fluorescence of hydroxyl radicals, respectively. In combination with thermochemical calculations of the heat of combustion of the fuel species, laser pyrolysis methods can provide an estimate of the total heat released by the polymer.

Coupling the commercial thermogravimetric analyzers (TGA) to an evolved gas analysis (EGA) detector, which is synchronized with the mass loss signal, provides dynamic (rate) capability during the test and allows for interpretation of the transient evolved gas data in terms of the decomposition kinetics of the solid using established methods of non-isothermal analysis (Lyon, 1997) and thermal degradation models for charring and noncharring polymers (Lyon, 1998).

Gracik et al. (1995) used flaming combustion in a TGA in combination with evolved gas (CO, CO_2) analysis to study the flammability of fiber-reinforced polymer composites. A 50 mg sample was heated at a rate of 20 K/min in air in a commercial thermogravimetric analyzer until ignition occurred. Flaming combustion of the sample in the TGA produced carbon dioxide (CO_2) and some carbon monoxide (CO) which are measured and used to calculate the heat release rate. An advantage of CO_2/CO generation (CDG) calorimetry compared to oxygen consumption calorimetry as a measure of combustion heat is the higher sensitivity and lower cost of the CO_2/CO detector(s), but the method is more sensitive to fuel type.

The plastic certification program of Underwriters Laboratories Inc. (UL) requires Follow-Up Services to conduct surveillance of certified products. MCC or PCFC do not replace the requirement for UL 94 flame ratings in end-product standards; it serves as an additional analytical test to verify material combustion consistency in follow-up service.

PCFC measures the rate at which the heat of combustion of the fuel gases is released by a solid during controlled pyrolysis in an inert gas stream. The fuel gases are mixed with excess oxygen and combusted (oxidized) at high temperature, and the instantaneous heat of combustion of the flowing gas stream is measured by oxygen consumption calorimetry.

Figure 2.16 is a schematic diagram showing the means of reproduction of flaming combustion component processes in pyrolysis–combustion flow calorimetry. The apparatus is based an original concept of Susott (1982) of linear programmed heating of milligram samples in an inert (nonoxidizing) atmosphere to separate the processes of char formation and gas phase combustion which normally occur in a fire. In this device, the sample is heated using a linear temperature program, and the volatile thermal degradation products are swept from the pyrolysis chamber by an inert gas and combined with excess oxygen in a tubular furnace at flame temperatures to force complete, nonflaming combustion (oxidation) of the fuel. Combustion products CO_2, H_2O, and acid gases are scrubbed from the gas stream and the transient heat release rate is calculated from the measured flow rate and oxygen concentration after correcting for flow dispersion. The maximum (peak) value of the PCFC heat release rate normalized for the initial sample mass and heating rate is a material flammability parameter with units of heat release capacity (J/g-K) that depends only on chemical composition of the sample and is proportional to the burning rate of the material in a fire. Time integration of the PCFC heat release rate yields the heat of complete combustion of the pyrolysis gases, and the char yield is measured by weighing the sample before and after the test. If the pyrolysis is conducted in air so that there is no possibility of char remaining after the test, time integration of the oxygen consumption signal gives the net heat of complete combustion of the solid, as would be determined in a high-pressure oxygen bomb calorimeter (ASTM-D238, 1988).

The maximum heating rate capability of the TGA (100–200 K/min) was well below the heating rates in fires, which can be as high as several hundred degrees per minute. For these reasons, a temperature-controlled pyrolysis chamber was

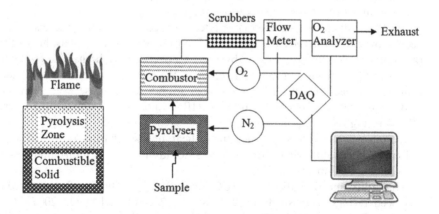

Flaming Combustion   Pyrolysis Combustion Flow Calorimeter

FIGURE 2.16 Schematic diagram of Pyrolysis Combustion Flow Colrimeter and normal flamming combustion.

designed that could be continuously purged with gas and coupled directly to the combustion furnace, and accept a commercial probe pyrolyzer (Pyroprobe 1000/2000, CDS Analytical) containing 25-mm-long platinum resistance coil, which heats the sample at a constant rate to gasify the sample. This arrangement provided consistent temperature and minimum dead volume with the probe in place for the experiment. The coiled combustion tube is contained in a ceramic furnace capable of maintaining a maximum temperature of 1200°C. Published studies of the oxidation of the products of flaming combustion showed that a residence time of 60 seconds at 1000°C is required to completely oxidize the largest size soot particles observed in real fires (Babrauskas et al., 1994).

The repeatability of heat release rate measurements for a single operator is estimated to be ±3% and accuracy is ±7%, with the majority of the error associated with moisture pickup occurring during the weighing and handling of small (1 mg) samples and in the noise in the deconvoluted signal. Repeatability and accuracy of total heats of combustion obtained by time integration of the heat release rate is ±5% (Lyon and Walters, 2002).

Small-scale MCC was developed to assess heat release properties of pure polymers, and is utilized as a research tool for screening new materials (Wilkie et al., 2006). The method simulates flaming combustion by pyrolyzing the pure polymer in a N_2 environment to produce fuel gases which are subsequently oxidized in an O_2 rich environment.

The peak specific heat release rate (PSHRR, W/g) of a material is calculated on the basis of maximum oxygen consumption rate of fuel gases from pyrolysis of a polymer. The maximum heat release potential of a polymer, i.e., the heat release capacity (HRC, J/g-K) is calculated by dividing PSHRR by sample heating rate. The temperature at PSHRR(T@PSHRR) can also be measured and has been related to ignition temperature of a polymer (Lyon, 2000).

The total amount of heat released (THR, kJ/g) can be calculated based on time evolution of oxygen consumption in MCC.

MCC parameters have been correlated with conventional flame tests. It has been shown that for pure polymers HRC correlates well with peak heat release rate (PHRR, kW/m^2) measured by cone calorimetry at high heat flux where pyrolysis is relatively complete (Lyon and Walters, 2004; Lyon et al., 2006). Limiting Oxygen Index (LOI) was also found to decrease with increasing HRC (Lyon et al., 2006). A range of HRC values was established to screen UL 94 rating of pure polymers.

Despite its usefulness for measuring heat release characteristics of a pure polymer, the use of MCC for polymers containing flame retardants has not been widely demonstrated.

The utility and limitations of MCC for polymers containing flame retardants has been recently published (Lyon et al., 2007a; Schartel et al., 2007). It has been found that MCC correlations with UL 94, LOI, and cone calorimetry are limited for polymers containing flame retardants because the condensed phase and gas phase processes of flaming combustion are necessarily incomplete, particularly in the presence of these compounds at low heat flux and at flame extinction such as UL 94 and LOI (Lyon et al., 2007b).

MCC is a thermal analysis technique that establishes a procedure for determining flammability characteristics of combustible materials such as plastics. UL uses ASTM D7309, Method B to perform this test. This test is conducted in a laboratory environment using controlled heating of a milligram sized specimen (~2.5 mg) and complete thermal oxidation of the specimen gases. The rate of heat released by combustion of the specimen gases produced during controlled thermo-oxidative decomposition of the specimen is measured from the rate of oxygen consumption. The specimen temperatures over which combustion heat is released are also measured. This method no longer requires the flame bars, since only a small amount of material in any form, including pellets, is needed to conduct the test (UL, 2019).

The benefits of MCC test are (UL, 2019):

- A more quantitative comparison of combustion characteristics of polymeric materials can be obtained.
- A quantitative information related to flammability behavior of new formulations is available that might be useful in new product development.
- The specimen conditioning inherent in the UL 94 test method is not required; a result cycle-time to obtain results is faster.
- The need for molding is eliminated, since MCC and ID tests can be done on pellets.
- Waste associated with molding bars is eliminated, i.e., no more compounding of extra material to obtain only a limited number of bars.
- The need to discard molded unburned bars and unburned portions of bars is with no more purging of compounding and molding equipment.
- Combustion by-products from UL 94 flammability testing is significantly reduced.

MCC has recently become commercially available for assessing flammability of polymers using only milligrams of test specimen. Although significant data and correlations with conventional flammability tests such as UL 94, Limiting Oxygen Index, and cone calorimetry are available in the literature, correlation data for polymers containing flame retardants are still limited (Lin et al., 2007).

MCC was used to analyze the flammability of 15 commercially available halogenated and nonhalogenated FR wire and cable compounds (Lin et al., 2007). Conventional flammability tests such as UL 94, LOI, and cone calorimetry were also conducted on the same FR compounds. The correlations between MCC and conventional flammability tests were analyzed. Key correlation results are summarized next.

UL 94 ratings of FR compounds were found to have sharp transition at about 30% LOI, above which the FR compounds have V-0 rating and below which the FR compounds have HB rating.

In MCC, T@PSHRR has strong correlation with HRC (R=0.86) and moderate correlation with THR (R=0.73). The positive correlations of T@PSHRR with HRC and THR are unexpected. It is possible that the very broad range of compositions and FR chemistries in the study by Lin et al. (2007) results in some formulations dominating the correlations leading to this surprising result. THR correlated poorly with char yield for halogenated FR compounds, but had moderate correlation with char yield for nonhalogenated FR compounds (R=-0.83).

LOI was found inversely proportional to $HRC^{0.25}$ with a correlation coefficient of 0.29 for FR compounds. This power law dependence on HRC for FR compounds is similar to that of pure polymers reported in the literature. The poor correlation between LOI and HRC for FR compounds could be attributed to the incomplete combustion in the LOI flame test. LOI has strong correlations with THR and char yield in MCC for nonhalogenated FR compounds, but is independent of THR and char yield for halogenated FR compounds. For each MCC parameter, there is a transition region where FR compounds could have either UL 94 V-0 or HB ratings. In general, FR compounds with T@PSHRR below 499°C, THR below 21 kJ/g, and HRC below 380 J/g-K are expected to have high probability of achieving UL 94 V-0 rating.

In cone calorimetry, PHRR was found to have the strongest correlation with FIGRA (R=0.93) and FPI (R=-0.89). THR has strong correlation with char yield (R=-0.96) for nonhalogenated FR compounds, but has no correlation with char yield for halogenated FR compounds. For nonhalogenated FR compounds, LOI was found to have a strong correlation with THR (R=-0.92) and char yield (R=0.91), similar to what was observed in MCC. UL 94 rating does not correlate with TTI . On the other hand, FR compounds are expected to have a high probability of achieving UL 94 V-0 rating with TTPHRR above 210 seconds, PHRR below 280 kW/m^2, THR below 12 kJ/g, FIGRA below 1.4 kW/m^2-s, and FPI above 0.5 $s-m^2/kW$. For the relationship between MCC and cone calorimetry, T@PSHRR has marginal to moderate correlations with PHRR, TTPHRR, FIGRA, and FPI. THR (MCC) has strongest correlations with FIGRA (R=0.91) and FPI (R=-0.89). On the other hand, HRC has poor correlation with cone calorimetry parameters. Its correlation coefficient with PHRR is 0.43, compared to the correlation coefficient of 0.93 for pure polymers reported in the literature. This HRC-PHRR correlation difference between systems of FR compounds and pure polymers can be attributed to significant interactions of incomplete gas-phase combustion, gas-phase flame inhibition reactions, char formation, and heat/mass transfer of burning FR compounds in cone calorimetry.

Good correlation is observed between HRC and PHRR in forced flaming combustion at high heat flux because burning is relatively complete under these conditions, at least for the condensed phase.

In contrast, burning is incomplete at flame extinction and such extrinsic factors associated with flame retardancy as flame inhibition and reduced heat transfer due to charring/swelling become important. These extrinsic factors are comparable in magnitude and effect to the intrinsic thermal combustion properties at flame extinction, so MCC alone cannot correlate flame resistance or fire behavior over a broad range of flame retardant chemistry, material composition, and test conditions.

Despite many challenges in correlating MCC with other conventional flammability tests, MCC remains an appropriate small-scale apparatus for rapid flammability screening of FR compounds (Lin et al., 2007).

2.7 FLAMMABILITY OF COMPOSITE MATERIALS

A composite material (also called composite) is a material made from two or more constituent materials with significantly different physical or chemical properties that, when combined, produce a material with characteristics different from the individual components. The individual components remain separate and distinct within the finished structure, differentiating composites from mixtures and solid solutions. The new material may be preferred for many reasons; common examples include materials which are stronger, lighter, or less expensive when compared to traditional materials. Textile composite materials consist of a polymer matrix (thermoplastic or thermoset) combined with a textile reinforcement. Textile-reinforced composites are increasingly used in various industries such as aerospace, construction, automotive, medicine, and sports due to their distinctive advantages over traditional materials such as metals and ceramics. Fiber-reinforced composite materials are lightweight, stiff, and strong. They have good fatigue and impact resistance. Their directional and overall properties can be tailored to fulfill specific needs of different end uses by changing constituent material types and fabrication parameters such as fiber volume fraction and fiber architecture.

Flammability of composites is mainly determined by the resin matrix, since the contributions of glass and other high performance fibers to the fire are much smaller than the resin matrix. In general, natural fiber composites (NFCs) are more flammable than advanced composites, because the natural fibers are far more flammable than glass or carbon fibers with the same matrix resin. The burning of composites and plastics is characterized by the following properties: ease of ignition, flame spread, heat release rate, ease of extinguish, smoke emission/toxicity, and burning characteristics. Strong pointed out that the purpose of flammability testing can be divided into three groups (Kim, 2012):

1. Official tests for official requirements,
2. Laboratory tests for product development/improvement, and
3. Full scale tests for simulating actual use condition.

Official tests are mandated by various market sectors. The most important official test is the cone calorimeter (ISO 5660/ASTM E1354), which provides most of the fundamental combustion characteristics (ease of ignition, hear release rate, weight of sample as it burns, temperature of sample as it burns, rate of weight loss, rate of smoke release, and yield of smoke) under a wide range of heat and ignition conditions. This vast amount of data can be used to develop a combustion model of the composite material. Other official flammability tests are the flame spread index (ASTM E162) by radiant heat panel test, and the smoke density generating rate under both flaming and smoldering condition

(ASTM E662) by smoke chamber test. For example, automotive seats, panels, walls, partitions, and ceiling made of composites are tested with ASTM E162 and ASTM E662.

Laboratory tests are used to adjust formula or product design, when the official tests are not feasible in terms of cost and time. However, laboratory tests are not suitable for ranking the combustion characteristics of widely different materials. The most common laboratory tests are LOI test (ASTM D2863), the vertical burn test (ASTM D568 and D3801), and the horizontal burn test (ASTM D635). The LOI test is the most accurate lab test, which determines the minimum oxygen content required to sustain burning. A higher LOI index is associated with decreased flammability.

Full-scale tests are used to determine performance under the actual combustion situation, which may be speculated based on the results from the official and/or laboratory tests. Flammability research plays an important role in new product development and improving existing products to meet many agencies' goals of protecting the public and limiting liability through testing (Kim, 2012).

2.8   FUTURE TRENDS

It happens with some regularity–news of a fire or explosion in an industrial facility that causes considerable property damage or, even worse, worker casualties or deaths. Textiles, wooden furniture, and house decorating materials are frequently responsible for fire outbreaks. The frequent occurrence of fires and explosions in the process industries that use flammable materials is typically as a result of several factors, such as the presence of an explosive mixture in the vapor space, lack of knowledge of the properties of the chemical's inherent safety implications, or inadequate safety procedures. Flammability testing of raw materials, as well final products is therefore, very important.

In order to minimize the risk of fire or explosion, it is important to evaluate the flammability characteristics of the material to understand key characteristics, such as the lower flammability limit (the lowest concentration at which a mixture of flammable vapor or gas and air is flammable), and upper flammability limit (the highest concentration at which a mixture of flammable vapor or gas and air is flammable), A variety of flammability tests may be performed to allow for determination of these characteristics, and the understanding of these conditions is essential while implementing proper safety practices.

When conducting flammability testing, it is important that the customers communicate what data is being sought so that testing can be properly designed in order to determine the necessary flammability property of a chemical mixture.

Nearly every country has its own set of textile fire testing standard methods that are claimed to relate to the special social and technical factors peculiar to each of them. In addition, test methods are defined by a number of national and international bodies, particularly governmental departments relating to industry, transport, defense, and health.

A variety of standard flammability tests have been developed over time on specific activities, many of them are very much different, while many of them have very little variation. Standardization bodies tried to harmonize these standard test methods to facilitate innovation, better market accessibility for the manufacturers, and cost reduction for the end-users.

The complexity of the burning process for any material such as textile proves difficult to quantify and hence rank in terms of its ignition and post-ignition behavior, because it is *thermally thin*, and also has a high specific volume and oxygen accessibility relative to other polymeric materials. Most common textile flammability tests are currently based on ease of ignition and/or burning rate behavior, which can be easily quantified for fabrics and composites in varying geometries. Few of these tests, however, yield quantitative and fire-science-related data, unlike the often-maligned oxygen index methods. LOI, while it proves to be a very effective indicator of ease of ignition, has not achieved the status of an official test within the textile arena.

A good flammability testing regime takes into consideration the many different variables that affect the flammability of a specific chemical; these variables include oxidizing environment, temperature, pressure, ignition energy, size, and geometry of the vessel, and gas composition. There are a variety of pressure vessels varying in size and geometry that can be used for flammability testing purposes, depending on particular need. The choice (e.g., spherical, cylindrical, large, small, glass, and steel) depends on the particular test design. A well-defined ignition source is also necessary, as is a good data acquisition system for monitoring pressure and temperature.

Accounting for these variables can result in test data that is much more applicable to specific processes than information taken from literature. Experts are happy to discuss flammable hazard concerns and work to design tests that provide needed information. The goal is to provide specific data – not just data.

Data acquired from this testing can be used to implement proper safety procedures and design to help minimize the chance of an explosive event (Peters, 2014).

Fire tests are a critical component used in the process of designing a fire-safe environment. Fire tests must be carefully developed and monitored to make sure they correspond to reality. Fire experimentalists are always concerned with real-world fire parameters in the development, execution, and application of fire tests and fire test data. While materials properties and the effects of fire are well understood, the circumstances and environment specifically contributing to ignition scenarios are more complex. Fire tests tell us about the behavior of a material(s) under the very specific conditions of the test(s). The test method and data generated need to be researched and assessed to determine if the specific conditions of the test are applicable to the given fire scenario(s) or the enduse of the test subject. If the test method is applicable to the fire scenario, the test can be used to determine if a material, product or assembly meets expected requirements and/or specifications. The limitations of standard test method(s) are that the test conditions may not capture known or potentially important fire parameters, and that the methods may not be sufficiently versatile for different environments and new technologies. Modeling large scale fires is more cost effective but lacks the *reality factor* – a necessary compromise.

A single test cannot capture all properties of all materials. Existing test methods should be reviewed and compared with respect to the properties and geometry of the test subject and the environment of its enduse or application. One test method may be sufficient, but test data from multiple tests provide more information. Test data from multiple tests should be assessed individually and in combination (Elizabeth, 2019).

In terms of equipment and test procedures, new technologies tend to reduce specimen size and facilitate interpretation of the burning behavior of textile materials. For example, MCC(ASTM D 7309) provides better understanding of the specific burning of textiles, while the specimen size is much smaller than in conventional flammability test.

A major disadvantage of current test methods for protective clothing is that they do not take into account actual fire conditions. The tests are carried out in a static way on small specimens and often do not take into account the seams, reflective trim, reinforcements, pockets, etc. (Roguski et al., 2016).

While the need for FR fabrics for military textiles has always been prevalent, with the rise of the use of Improvised Explosive Devices (IEDs), the need for innovative and protective FR fabrics has become more urgent for militaries around the world.

Protection from fire is not the only challenge facing designers of military textiles. One very important heat-related factor to be considered in FR design choices is the effect of the fabric on the wearer's thermal signature, which refers to how targetable the wearer would be through near-infrared detection technologies. Some ways of advancing flame-resistance may simultaneously, positively or negatively, affect infrared detectability.

Heat and flame as well as ballistics–protection are not the only considerations in designing and specifying military uniforms and fabrics. Some of the key parameters for all military textiles identified by a researcher named Aravin Periyasamy, of the Technical University of Liberec, are "damage resistance, comfort, sweat management ... and the integration of high-tech materials into uniforms".

Chemical treatment of natural fibers may fail in military uniforms due to laundering or exposure to chemicals in a soldier's working environment. However, processes have been developed to overcome some of this fragility, and militaries do use some of these treated natural fibers, such as in FR-treated cotton undergarments and T-shirts.

For military textiles, many countries may specify ISO (International Organization for Standardization), ASTM (American Society for Testing and Materials), or ANSI (American National Standards Institute) standards for FR testing of their fabrics. For example, ASTM D6413/D6413-13b, a vertical flame test for flame resistance of textiles, is cited in many military specifications (AATCC, 2020).

Flammability testing critically ensures the safety and trust of consumer products. It plays an important role in quality assessment in industries related to textiles and consumer goods, aerospace and transportation, bedding, and furniture materials. Consumer products must satisfy safety requirements and regulations before marketing, including those based on rate of burning, heat, and smoke release criteria.

REFERENCES

AATCC (2020). *AATCC: Flammability Requirements for Military Textiles*, May 12, 2015, http://textilesupdate.com/aatcc-flammability-requirements-for-military-textiles, accessed on 2.5.2020.

Abu-Isa I.A., Cummings D.R., LaDue D.E., and Tewarson A. (2019). Thermal properties and flamability behavior of automotive polymers, https://www.google.com/search?q=Abu-IsaI.A.%2C+CummingsD.R.%2C+LaDueD.E.%2C+and+TewarsonA.+(2019).+Thermal+properties+and+flamability+behavior+of+automotive+polymers%2C+wwwnrd.nhtsa.dot.gov%2Fpdf%2Fesv%2F.%2F98s4p17.pdf%2C+accessed+on+2.4.2019.&oq=Abu-IsaI.A.%2C+CummingsD.R.%2C+LaDueD.E.%2C+and+TewarsonA.+(2019).+Thermal+properties+and+flamability+behavior+of+automotive+polymers%2C+wwwnrd.nhtsa.dot.gov%2Fpdf%2Fesv%2F.%2F98s4p17.pdf%2C+accessed+on+2.4.2019.&aqs=chrome..69i57.793j0j7&sourceid=chrome&ie=UTF-8, accessed on 2.4.2019.

Angel S.M. (1998). *Situ Flame Chemistry by Remote Spectroscopy," in Fire Resistant Materials: Progress Report*. R.E. Lyon, Ed., DOT/FAA/AR-97/100, p. 229, November. https://www.researchgate.net/publication/235192604_Fire-Resistant_Materials_Progress_Report.

ANSI (1996). *UL-94 - Test for Flammability of Plastic Materials for Parts in Devices and Appliances*, 5th edition (October 96) Underwriters Laboratories Inc. (UL), Northbrook, IL. http://www.xn--lyddmpning-g6a.dk/documents/pictu5509.pdf.

ASTM (1990). *Standard Test Method for Heat and Visible Smoke Release Rates for Materials and Products Using an Oxygen Consumption Calorimeter*. ASTM E 1354-90, ASTM Fire Test Standards, 3rd edition, American Society for Testing of Materials, Philadelphia, PA, p. 803, 817.

ASTM (2013). ASTM F2700-08 (2013) standard test method for unsteady-state heat transfer evaluation of flame resistant materials for clothing with continuous heating, www.astm.org/Standards/F2700.htm.

ASTM (2019). *Annual Book of ASTM Standards*. ASTM International, West Conshohocken, PA, www.astm.org/BOOKSTORE/BOS/index.html, accessed on 8.3.2019.

ASTM (2019). ASTM D2863: Standard test method for measuring the minimum oxygen concentration to support candle-like combustion of plastics (oxygen index), in: *Annual Book of ASTM Standards*. ASTM International, West Conshohocken, PA.

ASTM D2015 (1985). *Standard Test Method for Gross Calorific Value of Coal and Coke by the Adiabatic Bomb Calorimeter*. ASTM D 2015-85, ASTM Fire Test Standards, 3rd edition American Society for Testing of Materials, Philadelphia, PA, pp. 222–229.

ASTM designation: D1230-17 standard test method for flamability of apparel textiles, ASTM Volume 07.01 Textiles (I).

ASTM-D238 (1988) Standard Test Method for Heat of Combustion of Hydrocarbon Fuels by Bomb Calorimeter (High Precision Method), American Society for Testing of Materials.

Audet N.F. and Spindola K.J. (1986) U. S. navy protective clothing program, Performance of Protective Clothing, ASTM STP 900.

Babrauskas V. (1981). Combustion of mattresses exposed to flaming ignition sources, part II. Bench scale tests and recommended standard tests, NBSIR, 80-2186.

Babrauskas V., Parker W.J., Mulholland G., and Twilley H. (1994). The phi meter: A simple, fuel-independent instrument for monitoring combustion equivalence ratio, *The Review of Scientific Instruments*, 65(7), 2367–2375.

Barker R.L., Hamouda H., Shalev I., and Johnson J. (1999). *Review and Evaluation of Thermal Sensors for Use in Testing Firefighters Protective Clothing*. Annual Report. North Carolina State Univ., Raleigh, NIST GCR, 99-772. National Institute of Standards and Technology, Gaithersburg, MD, p. 72.

Belshaw R.L. and Jerram D.L. (1986). Garments designed to reduce fire hazard, *Fire Safety Journal*, 10, 19–28.

Bennett R.D. (1973). Flammability and consumer, *Canadian Textile Journal*, February.

Bhatnagar V. (1974), Flammability in Apparel: Progress in Fire Retardancy Series Vol.7, Technomic Publishing Co. Inc., Westport, CT.

BSI (1997). BS 476-7:1997 Fire tests on building materials and structures. Method of test to determine the classification of the surface spread of flame of products. Also BS 476-4:1970, BS 476-6:1989 and BS 476-11:1982.

Buc E.C. (2019). Fire testing and fire reality: What do fire tests really tell us about materials?, *Fire & Building Safety in the Single European Market*, https://www.fireseat.eng.ed.ac.uk/sites/fireseat.eng.ed.ac.uk/files/images/06-Buc.pdf. accessed on May 5, 2020.

Carter W.H., Finley E.L.A., and Farthing B.R. (1973). *Journal of Fire and Flammability*, 4, 106–111.

CPSC (1975). Guide to Fabric Flammability', U.S. Consumer Product Safety Commission. W.D.C.- 20207, April.

CPSC (2019). Laboratory Test Manual for 16 CFR Part 1610: Standard for the flammability of clothing textiles United States Consumer Product safety Commission, October 2008, https://www.cpsc.gov/PageFiles/115435/testapparel.pdf accessed on 8.3.2019.

Davis Rick D. (2019). Cone Calorimeter, NIST Created June 05, 2014, updated August 21, 2018, www.nist.gov/laboratories/tools-instruments/cone-calorimeter, accessed on 6.3.2019.

DCLG (Department for Communities and Local Government) (2012). The impact of European fire test and classification standards on wallpaper and similar decorative linings, https://assets.publishing.service.gov.uk/March 2012, ISBN: 978-1-4098-3385-7.

Efectis (2019). The cone calorimeter, https://efectis.com/wp-content/uploads/2017/08/Leaflet_ConeCalori.pdf, accessed on 6.3.2019.

Fay T.S. (2002). Development of an Improved Fabric Flammability Test, Master of Science Thesis Submitted to Worcester Polytechnic Institute, MA, USA in Fire Protection Engineering, May. https://web.wpi.edu/Pubs/ETD/Available/etd-0625102-153152/unrestricted/Fay.pdf, accessed on 19.2.19.

Festec (2019). www.gtrade.or.kr/accessed on 17.3.2019.

Flasa (2019). www.flasa.ch/de/normes/367.pdf, accessed on 23.2.2019.

Fomicom (2019). The Single Burning Item (SBI) test, www.fomicom.com, accessed on 7.3.2019.

Friedman R. (1999). *Fire Safety in the Low-Gravity Spacecraft Environment*, SAE Technical Paper 1999-01-1937, (Also NASNTM-209285).

FSAC (Fire Sefety Advise Centre) (2019). British Standard 476, *Fire Tests*. www.firesafe.org.uk/british-standard-476-fire-tests/accessed on 19.3.2019.

FTT (2019). ASTM E162, ASTM D3675 radiant panel flame spread apparatus, *Fire Testing Technology*. www.fire-testing.comaccessed on 16.3.2019.

Globe (2019). NFPA 1971, http://globeturnoutgear.com/, accessed on 21.3.2019.

Gold C. (2019). UL-94 - Test for flammability of plastic materials for parts in devices and appliances, *Laird Technologies*, http://cdn.lairdtech.com. accessed on 27.2.2019.

Graan M.V. and Bunt J.R. (2016). Evaluation of A TGA method to predict the ignition temperature and spontaneous combustion propensity of coals of different rank, *International Conference on Advances in Science, Engineering, Technology and Natural Resources (ICASETNR-16)*, Nov. 24-25, 2016, Parys (South Africa). DOI: 10.15242/IAE.IAE1116404.

Gracik T.D. and Long G.L. (1992). Prediction of thermoplastic flammability by thermogravimetry, *Thermochimica Acta*, 212, 163–170.

Gracik T.D. and Long G.L. (1995). Heat release and flammability of a small specimen using thermoanalytical techniques," in *Fire Calorimetry*, Proceedings from the 50th Calorimetry Conference, R.E. Lyon and M.M. Hirschler, Eds., Gaithersburg, MD, DOT/FAA/CT-95/46, p. 101.

Hirsch D. (1989). Comparison of Results of the European space agency oxygen index test and the NASA upward propagation test, NASA JSC WSTF TR-581-001, March.

Hirsch D., Beeson H., and Friedman R. (2000). Microgravip Effects on Combustion of Polymers NASA/TM-2000-209900.

Hirsch D.B. and Beeson H.D. (2001). Improved test method to determine Flammability of aerospace materials, Halon Options Technical Working Conference, 24-26.

Horrocks A.R. (1989). Developments in flame-retarding polyester/cotton blends, *JSDC*, 105, 346–349, October.

Horrocks A.R. (2011). Flame retardant challenges for textiles and fibres: New chemistry versus innovatory solutions, *Polymer Degradation and Stability*, 96, 377–392.

Horrocks A.R. and Price D. (2001). Fire resistance material, *Woodhead Publishing*, 2001.

Horrocks A.R., Price D., and Tunc D. (1989). The burning behaviour of textiles and its assessment by oxygen index methods, *Textile Progress*, 18(1–3), 1–205.

Hshieh F.Y. and Beeson H. (1995). Cone calorimeter testing of epoxy/fiberglass composites in normal oxygen and oxygen-enriched environments, in: D. Janoff, T. Royals, and M. Gunaji, Eds., *Flammability and Sensitivity of Materials in Oxygen-Enriched Atmospheres. 7b Volume.* ASTM STP 1267, ASTM.

Hugget C. (1980). Estimation of rate of heat release by means of oxygen consumption measurements, *Fire and Materials*, 4(2), 61–65.

Instruments T.A. (2019). *Thermal Analysis Application: Brief Use of TGA to Distinguish Flame-Retarded Polymers from Standard Polymers.* www.tainstruments.com/pdf/literature/TA135.pdf accessed on April 4, 2019.

ISHN (2000). What you need to know about flame-resistant clothing, Industrial Safety and Hygene News, September 29, 2000, www.ishn.com/

ISO (1993). ISO Room Corner; International Standard – Fire Tests – Full-scale Room Test for Surface Products; ISO 9705:1993; International Organisation for Standardisation, Geneva.

Judd M.D. et al. (1981). The critical oxygen index test as a potential screening test for aerospace materials, *Fire andMaterials*, 5(4), 175–176.

Kashiwagi T. and Cleary T.G. (1993). Effects of sample mounting on flammability properties of intumescent polymers, *Fire Safety Journal*, 20, 203–225.

Khanna V. (2019). multilayer fire proximity suit, http://www.foremostsafety.com/images/Fire%20Proximity%20Suit%20-%20Technical%20&%20Testing.pdf accessed on 30.4.2020.

Kim Y.K. (2012). Natural Fibre Composites (Nfcs) for Construction and Automotive Industries in Handbook of Natural Fibres: Processing and Applications, vol. 2, Woodhead Publishing.

Kolhatkar A.W. (2006). Investigation of factors influencing flammability and prediction of hazard potential of saree, Ph.D. thesis under P C. Patel (Guide), M.S. University, Baroda, 27/n09/2006.

Lantz R.V. (1994). Ignition Index – An engineering based methodology For *Fabric Flammability Testing*, (M.S. Thesis, Worcester Polytechnic Institute).

Leblanc D. (1998). *Fire environments typical of navy ships, draft No. 3*, (M.S. Thesis, Worcester Polytechnic Institute.

Lewin M., Atlas S.M., and Pearce E.M., (Eds.) (1982). *Flame Retardant Polymeric Materials*. Plenum Publishers, New York.

Lewin M. and Sello S.B. (1984). *Chemical Processing of Fibers and Fabrics—Functional Finishes, Part B*, M. Dekker, Eds., New York.

Lin T.S., Cogen J.M., and Lyon R.E. (2007). Correlations between Microscale Combustion Calorimetry and Conventional Flammability Tests for Flame Retardant Wire and Cable Compounds. International Wire & Cable Symposium176 Proceedings, USA, http://www.iwcs.org/archives/2007

Lindholm J., Brink A., and Hupa M. (2019). Cone calorimeter – A tool for measuring heat release rate, www.ffrc.fi/FlameDays_2009/4B/LindholmPaper.pdf, accessed on 3.4.2019.

Lyon R.E. (1997). An Integral Method of Nonisothermal Kinetic Analysis, *Thermochimica Acta*, 297, 117–124.

Lyon R.E. (1998). Pyrolysis Kinetics of Char Forming Polymers, *Polymer Degradation and Stability*, 61(2), 201–210.

Lyon R.E. (2000). Solid State Thermochemisty of Flaming Combustion, in: A.F. Grand and C.A. Wilkie, Eds., *Fire Retardancy of Polymeric Materials*. Mercel Dekker, Inc., New York, pp. 391–447.

Lyon R.E. (2001). "Heat Release Capacity," Proceedings of the Fire & Materials Conference, San Francisco, CA, January 22–24, p. 11.

Lyon R.E. and Walters R. (2002). A microscale combustion calorimeter, Report no. DOT/FAA/AR-01/117 Office of Aviation Research (US), Washington, February.

Lyon R.E. and Walters R.N. (2004). Pyrolysis combustion flow calorimetry, *Journal of Analytical and Applied Pyrolysis*, 71, 27–46.

Lyon R.E., Walters R.N., Beach M., and Schall F.P. (2007a). "Flammability screening of plastics containing flame retardant additives," Proceedings of the 16th International Conference ADDITIVES.

Lyon R.E., Walters R.N., and Stoliarov S.I. (2006). Thermal analysis of flammability, *Flame Retardant*, 111–122.

Lyon R.E., Walters R.N., and Stoliarov S.I. (2007b). Screening flame retardants for plastics using microscale combustion calorimetry, *Polymer Engineering & Science*, 47, 1501–1510.

JP M. and NW H. (1992). *Performance of Protective Clothing: Fourth Volume*, ASTM, STP1133, ASTM International, West Conshohocken, PA.

Mehta R.D. and Martin J.R. (1975). *Journal of Fire and Flammability*, 6, 105–115, April.

Mierlo R.V. and Sette B. (2005). The Single Burning Item (SBI) test method – A decade of development and plans for the near future, *HERON*, 50(4), 191–207.

Miller B. and Goswami B.C. (1971). Effects of constructional factors on the burning rates of textile structures: Part I: Woven thermoplastic fabrics, *Textile Research Journal*, 41, 949.

Miller B. and Martin J.R. (1975). *Journal of Fire and Flammability*, 6, 105–115, April.

Miller B., Martin J.R., Meiser C.H., and Gargiulio M. (1976). The flammability of polyeter-cotton mixtures, *Textile Research Journal*, 46, 530–538.

Miller B., Martin J.R., and Turner R. (1980). Fabric ignition by flame impingement, *Textile Research Journal*, 50(4), 256–260, April.

Moussa A., Toong T.Y., and Backer S. (1973). An experimental investigation of flame-spreading mechanisms over textile materials, *Combustion Science and Technology*, 8(4), 165–175. DOI: 10.1080/00102207308946640.

NASA (1998). Flammability, odor, offgassing, and compatibility requirements and test procedures for materials in environments that support combustion, NASA STD 6001, Test 1, Upward FlamePropagation, February 9, 1998 (formerly NHB 8060.1C).

Nayak R., Houshyar S., and Padhye R. (2014). Recent trends and future scope in the protection and comfort of fire-fighters' personal protective clothing, *Fire Science Reviews*, December, 3: 4.

NFPA (1998). NFPA 1977, *1998 National Fire Codes.* National Fire Protection Association, Quincy, MA. www.nfpa.org/. accessed on 23.2.2019.

NFPA (2112). *Standard on Flame-Resistant Garments for Protection of Industrial Personnel against Flash Fire 2001 Edition.* National Fire Protection Agency, Quincy, MA.

Ohlemiller T.J. and Villa K.M. (1991). *Material Flammability Test Assessment for Space Station Freedom.* NISTIR 4591, NASA CR-187115National Institute of Standards and Technology, Gaithersburg, MD.

OHSA (2006). A guide to the globally harmonized system of classification and labeling of chemicals (GHS), Occupational Health and Safety Administration, www.osha. gov/dsg/hazcom/ghs.html, accessed on July 13, 2007.

OSHA (2019). Standard Interpretations, Occupational Safety and Health Administration, United States, Department of Labor, www.osha.gov/, accessed on 22.32019.

Pailthorpe M. (2000). Flameproofing of textiles, unpublished report 2000.

Parker W.J. (1977). *An Investigation of the Fire Environment in the ASTM E 84 Tunnel Test.* NBS Technical Note 945, National Bureau of Standards, Washington, DC, August.

Parker W.J. (1988). "Prediction of the heat release rate of wood," Ph.D. thesis, The George Washington University, April.

Peters S. (2014). Why flammability testing in industry is essential, posted on 12. 11.14, www.fauske.com/

Port Plastics (2019). Specifications – UL test procedures. https://media.digikey.com/pdf/ Other%20Related%20Documents/GC%20Elect/UL_Test_Procedure.pdf, accessed on 16.3.19.

Price D., Gao F., Milnes G.J., Eling B., Lindsay C.I., and McGrail P.T. (1999). Laser pyrolysis/time-of-flight mass spectrometry studies pertinent to the behavior of flame retarded polymers in real fire situations, *Polymer Degradation and Stability*, 64, 403–410.

PU Europe (2016). European fire standards and national legislation, PU Europe FIRE HANDBOOK. http://highperformanceinsulation.eu/wp-content/uploads/2016/08/ PU-Europe-Fire-Handbook-European-fire-standards-and-national-legislation.pdf.

Reeves W.A. and Drake G.L., Jr., (1971). *Flame Resistant Cotton*, Merrow Publishing Co. Ltd., Watford, England.

Reeves W.A., Drake G.L., Jr., and Perkins R.M. (1974). *Fire Resistant Textiles Handbook.* Technomic Pub, Westport, CA.

Reshetnikov S.M.A. and Reshetnikov I.S. (1999). *Polymer Degradation and Stability*, 64, 379–385.

Riessen A.V., Rickard W., and Sanjayan J. (2009). *Thermal Properties of Geopolymers in Geopolymers.* Woodhead Publishing, pp. 315–342.

RISE (2019). Information about EN 1021-1:2014 and EN 1021-2:2014, Testing of ignitability for upholstered furniture. Research Institutes of Sweden (RISE). www.sp.se/ en/index/services/firetest_furniture/ss_en_1021/Sidor/default.aspx, accessed on 25. 3.2019.

Roguski J., Stegienko K., Kubis D., and Błogowski M. (2016). Comparison of requirements and directions of development of methods for testing protective clothing for firefighting, *Fibres & Textiles in Eastern Europe 24*, 5(119), 132–136. DOI: 10.5604/ 12303666.1215538.

Schartel B., Pawlowski K.H., and Lyon R.E. (2007). Pyrolysis combustion flow calorimeter: A tool to assess flame retardated PC/ABS materials, *Thermochimica Acta*, 462, 1–14.

Schindler W.D. and Hauser P.J. (2004). *Chemical Finishing of Textiles*. Woodhead, Cambridge, England.

Sioen N.V. (2019). EN 469 – Protection clothing for firemen, www.vidal-protection.com/strengths/standards/en-469, accessed on 19.3.2019.

Steckler K.D. (2019). Estimation of rate of heat release by means of oxygen consumption measurements, https://nvlpubs.nist.gov/nistpubs/sp958-lide/280-282.pdf, accessed on 3.4.2019.

Susott R.A. (1982). Characterization of the thermal properties of forest fuels by combustible gas analysis, *Forest Sciences*, 28(2), 404–420.

Trevera (1997). *Standards for Testing of Interior Textiles*. Frankfurt, Trevira (formerly Hoechst) GmbH.

UL (2019). Microscale Combustion Calorimetry (MCC) testing for polymeric materials, https://industries.ul.com/wp-content/uploads/accessed on 30.3.2019.

Warringtonfiregent (2019). EN 13823, single burning item test, www.wfrgent.com/, accessed on 3.3.2019.

Wilkie C.A., Chigwada G., Gilman J.W., and Lyon R.E. (2006). High throughput techniques for the evaluation of fire retardancy, *Journal of Materials Chemistry*, 16, 2023–2029.

Zammarano M., Krämer M.S., and Davis R.D. (2012). NIST Technical Note 1775, Standard Operating Procedures for Smolder Ignition Testing of Upholstery Fabrics, https://ws680.nist.gov/publication/get_pdf.cfm?pub_id=912584

Zhao L. and Dembsey N.A. (2008). Measurement uncertainty analysis for calorimetry apparatuses, *Fire and Materials*, 32(1), 1–26.

Zipper S.B.S. (2019). Performance requirements of protective clothing for firefighting flame spread, www.sbs-zipper.com/blog/performance-requirements-of-protective-clothing-for-firefighting/, accessed on 20.3.19.

3 Inherent FR Fibers

3.1 INTRODUCTION

Fire-safe polymers are resistant to degradation at high temperatures. There is need for fire-resistant polymers in the construction of small, enclosed spaces such as skyscrapers, boats, and airplane cabins. In these tight spaces, ability to escape in the event of a fire is compromised, increasing fire risk. In fact, some studies report that about 20% of victims of airplane crashes are killed not by the crash itself but by ensuing fires. Fire-safe polymers also find application as adhesives in aerospace materials, insulation for electronics, and in military materials, such as canvas tenting.

For making apparel and home textiles, cotton is mostly preferred due to its soft feel, good moisture regain, and adequate thermal insulation. Because it is cellulosic in nature with a low limiting oxygen index (LOI) of 18%, cotton catches flame readily and burns vigorously in an open atmosphere, which is difficult to extinguish, and also sometimes causes accidental death. It may be noted that textiles with an LOI $\leq$ 21% catch flame readily and burn rapidly in an open atmosphere. Samples with an LOI $\geq$ 21% to $\leq$ 27% also catch flame, but burn slowly in an open atmosphere. On the other hand, samples with an LOI $\geq$ 27% are generally considered to be flame retardant. For a textile material to be considered flame retardant, it should have an LOI of more than 27%.

When LOI values rise above approximately 26–28%, fibers and textiles may be considered to be flame retardant and pass most small flame fabric ignition tests in the horizontal and vertical orientations. Nearly all flammability tests for textiles, whether based on simple fabric strip tests, composite tests (e.g., BS5852:1979, ISO 8191/2, EN1021-1/2, and EN597-1/2) and more product/hazard related tests (e.g., BS6307 for carpets, BS6341 for tents, and BS6357 for molten metal splash) are essentially ignition-resistance tests (Horrocks and Price, 2001).

The situation is slightly better with respect to lignocellulosic textiles, such as jute with LOI of 21% or more, making them suitable for packaging of agricultural crops, food products, upholstery, and home-furnishing applications. For wool and silk, being protein fibers, the situation is still better due to their higher combustion temperatures (Samanta et al., 2015). Among natural fibers, wool is a fiber of low flammability. Its LOI is 25.2, which is marginally above the atmospheric oxygen content. It burns in air under favorable conditions only. It is non-thermoplastic and has a very high T_g of 600^0C. Its low heat of combustion (4.9.kcal/g) and high ignition temperature (570°C–600°C) are due to its higher moisture regain (8%–16%, depending on relative humidity), high nitrogen (15%–16%) and sulfur (3%–4%) content, and low hydrogen (6%–7%) content (% by weight). While organo-sulfur compounds are generally flame retardant to some degree, the disulphide-containing cystine links are easily oxidizable, and this can offset some of the anticipated natural flame retardancy. Preoxidation of wool, and hence cystine, to cysteic acid residues restores this expected retardant activity and oxidized wools can have greater inherent flame retardancy as a consequence. When high densities of structure and horizontal orientation (e.g., carpets) are required in a product, wool fabrics often pass the required flame retardancy tests untreated. Wool drapes, furnishings, and carpets do not create major fire hazards. Any fires slowly, if at all.

Among thermoplastic fibers, both polyamide (LOI 23%) and polyethylene terephthalate (LOI 23%), support combustion in air. Thermoplasticity, however, has an important effect on burning behavior. Synthetic fibers shrink considerably below the melting point. This causes materials to shrink away from ignition sources, making established combustion less likely. Moreover, molten polymers fall away and remove the heat likely to spread combustion. Thus, in free hanging garments such as nightdresses, brushed nylon provides an acceptable risk and good aesthetic properties (Oulton, 1995).

Most man-made polymers are flammable, unless specifically modified to make them self-extinguishing, by physical or chemically incorporating fire-retardant chemicals.

3.2 CLASSIFICATION OF FR MATERIALS

Flame retardant textiles can be manufactured via the following two routes (Gaan et al., 2011):

1) Using inherently flame retardant fibers, or
2) Finishing or coating of flammable materials with flame retardants. Flame retardants are applied at the fabric stage by a finishing process and subsequently fixed by various techniques such as thermal curing, ammonia curing, UV and plasma polymerization.

Inherently flame retardant fibers can be further manufactured in the following three ways:

1) Fibers from inherently flame retardant (IFR) polymers. Aramids, polyimide, melamine, glass, basalt, halogen-containing olefins, oxidized polyacrylonitrile, and polyphenylene, sulfide-based fibers are some examples of inherently flame retardant polymers.
2) Fibers from copolymers containing special flame retardant comonomers, e.g., modified polyesters such as Trevira CS and modacrylics.
3) Fibers containing special nonreactive flame retardant additives. Viscose FR and Visil are FR cellulosic fibers containing very high concentration of FR additives (~30%).

3.3   IFR AND FR-TREATED POLYMERS

Flame retardant polymers can be divided into two distinct classes:

1. Inherently flame retardant (IFR) polymers, and
2. Flame retardant treated (FRT).

The inherently flame-resistant property is permanent and lasts for the complete life cycle of the product. The process of additional fiber-retardant or flame resistant after treatment can therefore be avoided for such fabrics and garments (CIRFS, 2019).

Another approach is to treat conventional flammable polymers (e.g., cotton) with flame retardants in fabric form by finishing process. The additives or flame retardants are various organic and/or inorganic compounds that are added in the finishing bath in conjunction with or without a synergist. The additives belong to various chemical classes; the two major classes are halogens (chlorine and bromine) and organo-phosphorous compounds. During a fire, fire retardants rely on certain chemical reactions to extinguish the flame or to prevent spreading of fire. This reaction is triggered by the heat of the fire while the fabric is exposed to the fire. Certain fire retardants also depend on physical phenomena to prevent the spreading of fire.

A finish may affect the handle, texture, and moisture management properties of the fabric and hence, the comfort of the user. Finishes may also be removed by washing or abrasion, creating environmental pollution. The applied chemicals may adversely affect human skin and the body as a whole.

Treated fabrics usually weigh more than the inherent FR fabrics, which adds to their performance in many workspace conditions by putting more mass between the wearer and the hazard. However, this may result in a compromise for the comfort of the fabric or even add to heat strains.

It is very difficult to determine whether protection has been compromised with an FRT fabric. There are ways to test it. Unfortunately, all of these test methods are destructive–there really is no way to test flame resistance or arc protection value without destroying agarment in a flame or arc flash test. In a carpet, generated heat rises away from the fuel and the ignition source is lost. In a curtain, heat rises towards the fresh fuel; the burning behavior is therefore quite different. According to the desired nature of flame retardancy, there are a large number of test methods. A particular FR test can be linked with a particular flame or firing environment–there is no universal FR test.

FR finishes make up the majority of the flame retardant market due to lower costs for processing these commodity materials into flame retardant polymers. Finishes work well for most flame retardant applications, but they also have a few drawbacks. Flame retardant finishes leach out of the polymers over time, making the polymer less flame-resistant and polluting the environment. Another drawback to using flame retardant finishes is their tendency to degrade physical properties such as tensile strength or impact resistance of the native polymer.

The US National Fire Protection Agency developed a set of standards for determining the fire safety of a textile or fabric, known as NFPA 701: Standard Methods of Fire Tests for Flame Propagation of Textiles and Films. Although NFPA 701 itself is not a law, many local and state governments do require that textiles used in public spaces comply with it.

IFR fibers have flame resistance built into their chemical structures. DuPont created IFR aramid fibers such as poly (para-aramid) fiber namely Kevlar (DuPont) and poly (meta-aramid) fiber, Nomex (DuPont). The actual structure of the fiber itself is not flammable. For IFR fibers, the protection is built into the fiber itself and can never be worn away or washed out. An essential thing to keep in mind when assessing flame-resistant technologies is that inherent flame-resistant properties cannot be washed out or damaged through exposure to chemicals in the workplace or laundering practices, whether at-home or commercial. This means that the flame-resistant properties of garments made of inherent fibers cannot be compromised. It is crucial for the wearer to know that flame-resistant protection is always there.

FR-treated garments, however, may be damaged by chlorine bleach, the combination of hydrogen peroxide ("oxygen bleach") with hard water, or exposure to oxidizing chemicals in the workplace.

IFR polymers are more complicated since these polymers have flame-resistant moieties incorporated either in the backbone of the polymer or as a pendant group. They represent a smaller class of compounds usually referred to as high-resistant engineering polymers. They tend to be expensive due to the high cost of processing these materials and their smaller sales volumes. These polymers are usually found in specialty applications in which cost is not as important a factor.

Substantial research efforts have been invested in the manufacture of IFR textile fibers. All the major polymer types with the exception of poly-olefins have fire-retardant versions. The chemical aspects of modifications are not generally revealed.

The following features of IFR fibers may be noted (CIRFS, 2019):

1. They are capable of being used alone or in blend with other fibers to make a fabric or other article flame retardant or flame-resistant.
2. They can be used to produce fabrics that have greatly reduced, or zero rate of flame spread.
3. They can be used to produce fabrics that provide protection from flame and heat.
4. Most relevant standards specify laundering and/or other care procedures to ensure the permanence of the flame retardance or flame resistance for the intended use.
5. Fibers treated with FR agents cannot be described as IFR.
6. Methods exist for the safe disposal or recycling of fabrics made with IFR fibers at the end of their life.

3.4   IFR FIBERS

IFR polymers are mostly linear, single-stranded polymers with cyclic aromatic components. Most IFR polymers are made by incorporation of aromatic cycles or heterocycles, which lend rigidity and stability to the polymers. Polyimides, polybenzoxazoles (PBOs), polybenzimidazoles, and polybenzthiazoles (PBTs) are examples of polymers made with aromatic heterocycles.

Polymers made with aromatic monomers have a tendency to condense into chars upon combustion, decreasing the amount of flammable gas that is released. Syntheses of these types of polymers generally employ prepolymers, which are further reacted to form the IFR polymers.

It is evident that in the case of those polymers having highly aromatic structures that have associated high second order transition temperatures, T_g of 275°C or greater and ill-defined melting temperatures that are even higher, these fibers may be used at service temperatures of at least 150°C or higher since they show neither significant thermoplastic properties nor tendencies to thermally degrade below about 350°C. Consequently, the inherently FR fibers based on conventional doping method, may be used in textile structures having both heat and fire resistance. The melamine-formaldehyde fibers are highly cross-linked and extremely char-forming in character, even though they are not aromatic in structure. They also may resist temperatures of 150°C or higher during service life because they start to cross-link and then form char at temperatures above this level (Horrocks, 2013).

Inorganic and semi-organic polymers often employ silicon-nitrogen, boron-nitrogen, and phosphorus-nitrogen monomers. The nonburning characteristics of the inorganic components of these polymers contribute to their controlled flammability. For example, instead of forming toxic, flammable gases in abundance, polymers prepared with incorporation of cyclotri-phosphazene rings give a high char yield upon combustion. Polysialates (polymers containing frameworks of aluminum, oxygen, and silicon) are another type of inorganic polymer that can be thermally stable up to temperatures of 1,300°C–1,400°C. (Barbosa et al., 2000). Some of the IFR synthetic fibers are presented in Table 3.1.

The limiting oxygen indices (LOI) of the respective polymer are as reported by Gaan et al. (2011) and Horrocks (2013).

Several of the previously-mentioned polymers can be said to be IFR through having below-average flammabilities. IFR polymers tend to have relatively high thermal stabilities by virtue of significant aromatic content (e.g., the aromatic polyesters, polyethers and polyamides; the polyarylates and polycarbonates; and the phenolic resins), to decompose to give gas-phase inhibitors of combustion (e.g., PVC), or to contain significant quantities of nitrogen or similar heteroatoms (e.g., polyamides and aminoresins). In addition to these materials there are, however, a few other important groups that include the poly(aryl ketone)s and ether ketones, the poly(aryl sulfide)s and sulfones, the polyimides, and the polybenzimidazoles, benzoxazoles and benzthiazoles,

TABLE 3.1

Details of Inherent FR polymers

Name of polymer	Chemical name	Trade name and manufacturer	LOI%
Meta aramid	Poly(m-phenylene isophthalamide	Nomex (Du Pont)Teijinconex (Teijin Aramid)	30
Para aramid	Poly (p-phenylene terephthalamide	Kevlar (Du Pont), Twaron (Acordis, formerly Enka)	25–28
Melamine-formaldehyde fiber		. Basofil (BASF)	32
Phenol-formaldehyde	Novoloid	Kynol (Kynol GmbH)	30–34
Polyphenylene sulfide (PPS)		Torelina (Toray), Ryton(Solvay), DIC.PPS (DIC Corp).	34
Polybenzazole group	Polybenzimidazole (PBI)	Celazole	41
	polybenzoxazole (PBO)	Zylon	68
Oxidized PAN fibers		Sigrafil O, Panox	40–52
Polyimide fibers		P84 (Evonik)	38
Polyvinyl chloride			23–43
Polyvinylidene chloride			60
Polytetrafluoroethylene (PTFE)			. 90
Inorganic and ceramic fibers			

all of which show above-average thermal and thermo-oxidative stability, both of which are important to flame retardance.

3.5 ARAMID FIBERS

The most commonly used thermally resistant aramids are based on a metachain structure as typified by the original Nomex (DuPont) fiber. Meta-aramid or poly (m-phenylene isophthalamide) is produced in the Netherlands and Japan by Teijin Aramid under the trade name Teijinconex; in Korea by Toray under the trade name Arawin; in China by Yantai Tayho under the trade name New Star and by SRO Group (China) under the trade name X-Fiper; and a variant of meta-aramid in France by Kermel under the trade name Kermel.

Based on earlier research by Monsanto Company and Bayer, para-aramid fiber with much higher tenacity and elastic modulus was also developed in the 1960s and 1970s by DuPont and AkzoNobel, both profiting from their knowledge of rayon, polyester, and nylon processing.

In 1973, DuPont was the first company to introduce a para-aramid (poly (p-phenylene isophthalamide) fiber, which iscalled Kevlar, to the market. Aramid fibers, such as Kevlar (DuPont), are synthetic fibers characterized by high strength (five times stronger than steel on an equal-weight basis) and high heat-resistance (some more than 500°C). Aramids have wide applications, including those in composites, ballistics, aerospace, the automotive industry, protective clothing against heat/radiation/chemicals, substitutions for asbestos, and optical fiber cables in telecommunications.

The word aramid comes from the words *aromatic* and *polyamide*. It is a general term for a manufactured fiber made of a long-chain synthetic polyamide, in which at least 85% consists of amide linkages (-CO-NH-) attached directly to two aromatic rings, (as defined by the US Federal Trade Commission). The worldwide aramid market is shared by three aramid fiber manufacturers, namely DuPont (US), Teijin (Japan), and Kolon Industries (South Korea). The brand names are Kevlar and Nomex (DuPont), Twaron and Technora (Teijin) and Heracron (Kolon).

Aramid fiber is produced by spinning a solid fiber from a liquid chemical blend. This causes the polymer chains to orientate in the direction of the fiber length, increasing strength.

Kevlar is expensive and dangerous to manufacture, partly because it is dissolved in concentrated sulfuric acid. This is necessary to keep the highly insoluble polymer in solution during synthesis and spinning.

Unlike high molecular weight polyethylene (HDPE), the Kevlar molecule is polar. This allows other substances including water to attach them to the Aramid. This allows it to be more active chemically than UHMWPE (Dyneema, Spectra). Ultrahigh-molecular-weight polyethylene (UHMWPE, UHMW) is a subset of the thermoplastic polyethylene. Also known as high-modulus polyethylene, (HMPE), it has extremely long chains, with a molecular mass usually between 3.5 and 7.5 million amu (atomic mass unit). It also means that it can be bonded for example to epoxy and it is wettable.

The development of aromatic polyamides changed dramatically in good direction with the discovery of lyotropic liquid crystalline aramid. The first commercial applications of aramid fiber appeared in the early 1960s. Stephanie Kwolek did great work while working at DuPont in 1961. It took a long time to figure out how to make any useful product out of aramid because it is insoluble in most of the solvents. The company introduced the para-aramid fiber with the brand name *Kevlar*. For her great invention of Kevlar, Stephanie Kwolek was inducted in July 1995 into the United States Inventor's Hall of Fame.

In the case of Nomex, when exposed to flame, the aramid fiber swells and becomes thicker, forming a protective barrier between the heat source and the skin. This protective barrier stays supple until it cools, giving the wearer vital extra seconds of protection to escape.

Nomex generally comes in three kinds. It's either used by itself (as 100% Nomex), blended with up to 60% Kevlar, or blended with Kevlar and some antistatic fibers. In this last form, it is known as Nomex III. This blend is now well-established in fire fighters' clothing both in the United States and in the UK for outer-shells as well as underwear, hoods, socks, and gloves. Other uses include industrial workwear, aluminized proximity clothing, military protective clothing and fire barrier/blocker applications.

3.5.1 CHARACTERISTICS OF ARAMID

Aramid fibers consists a series of synthetic polymers in which repeating units containing large phenyl rings are linked together by amide groups. Amide groups (CO-NH) form strong bonds that are resistant to solvents and heat. Phenyl rings (or aromatic rings) are bulky six-sided groups of carbon and hydrogen atoms that prevent polymer chains from rotating and twisting around their chemical bonds. Structures 3.1 (a) and 3.1 (b) show the chemical structures of meta-aramid and para-aramid respectively.

STRUCTURE 3.1 A   Chemical structure of meta-aramid.

STRUCTURE 3.1 B   Chemical structure of para-aramid.

3.5.2 CHEMICAL PROPERTIES

All aramids contain hydrophilic amide links. Not all aramid products absorb moisture to the same extent. The PPD-T (poly-phenylene terephthalamide) fiber has very good resistance to many organic solvents and salt, but strong acids can cause substantial loss of strength. Aramid fibers are difficult to dye due to their high glass transition temperatures. Moreover, the aromatic nature of para-aramid is responsible for oxidative reactions when exposed to UV light that leads to a change in color and loss of some strength.

3.5.3 THERMAL PROPERTIES

Aramid fibers do not melt but decompose simultaneously. The maximum service temperature of m-aramide is 210°C (410°F). They burn only with difficulty because of high LOI values. Some aramid types can retain about 50% of their strength at 300°C. Aramids shrink negligibly at high temperature due to high crystallinity.

3.5.4 MECHANICAL PROPERTIES

Aramid is more than five times stronger than steel under water and twice as strong as glass fiber or nylon. High strengths of aramids result from the presence of aromatic and amide groups and high crystallinity. Aramid fibers retain strength and modulus at high temperatures up to 300°C. Under tension, aramids behave like elastic materials. On bending severely, they show nonlinear plastic deformation. On applying tension, no failure is observed, even at impressively high loads and a large number of cyclic load. Creep strain for aramid is only 0.3%.

3.5.5 USES

Nomex is best known as a barrier to fire and heat. Apart from race-car drivers, Nomex is worn by astronauts, fire-fighters, and military personnel. It is also widely used in more mundane ways, such as in household oven gloves. In sheet form, heatproof Nomex finds many uses in automobiles, including high-temperature hoses and insulation for spark plugs.

Nevertheless, Nomex is not only useful for protective clothing. The molecular structure that stops heat from passing through also stops electricity flowing through it; this means Nomex is an extremely poor conductor–almost a perfect insulator. Nomex, made into the form of a paper sheet or board, is a superb insulating material for all kinds of electrical equipment, from motors and generators to transformers and other electrical equipment.

Aramid fibers are used for the manufacture of the following:

- Various composite materials;
- Sail cloth;
- Snowboards;
- Protective gloves, helmets, and body armor;
- For winding around pressure vessels;
- Flame and cut resistant clothing;
- Substitute for asbestos;
- Ropes and cables;
- Optical fiber cable systems;
- In jet engine enclosures;
- For tennis strings and hockey sticks;
- The reeds of wind instrument;
- Tyres and rubber Reinforcement; and
- Circuit board reinforcement.

3.5.6 BENEFITS

The main advantages of aramid fibers are high strength and low weight. Like graphite, they have a slightly negative axial coefficient of thermal expansion, which means aramid laminates can be made thermally stable in dimensions.

Unlike graphite, they are very resistant to impact and abrasion damage. It can be made waterproof when combined with other materials such as epoxy. They can be used as composites with rubber to retain flexibility. High tensile modulus and low breakage elongation, combined with very good resistance to chemicals make aramid fibers the right choice for different composite structural parts in various applications.

3.5.7 LIMITATIONS

Aramid also has some disadvantages. The fibers absorb moisture, so aramid composites are more sensitive to the environment than glass or graphite composites. For this reason, aramid must be combined with moisture-resistant materials such as epoxy systems. Compressive properties are relatively poor as well. Consequently, aramid is not used in the construction of bridges or wherever moisture resistance is needed. Also, aramid fibers are difficult to cut and to grind without special equipment (e.g., special scissors for cutting, special drill bits). Finally, they undergo some corrosion and are degraded by UV light. For this reason, they must be properly coated.

3.5.8 ARAMID IN COMPOSITES

In a world where lightweight and durable composites are increasingly replacing conventional materials, aramid and para-aramid fibers play an important role. They are essential for reinforcing composites in which weight reduction and excellent damage tolerance are required. Many different kinds of composite goods are reinforced with aramid because of the strength, stiffness, and dimensional stability of laminates that contain it.

Aramid and Kevlar fibers are compatible or can be used with many types of resin systems. The best choice of resin system is epoxy because it adheres best to the fiber surface. Vinyl ester and isophthalic polyester may also be used. Ortho-phthalic polyester should be avoided as it does not provide sufficient adhesion to the fiber.

3.6 THERMOSET POLYMERIC FIBERS

Thermoset Polymeric Fibers are typified by the melamine-formaldehyde fiber, Basofil (BASF) and the phenol-formaldehyde (or novoloid) fiber, Kynol (Kynol GmbH). Melamine, if combined with formaldehyde, produces melamine resin (Structure 3.2), which is a very durable thermosetting plastic, with melamine rings terminated with multiple hydroxyl groups that are derived from formaldehyde. Basofil is the only commercial member of melamine fiber, developed by BASF AG in the early 1990s. Basofil is a fiber made from a condensation polymer of melamine, melamine derivatives, and formaldehyde-supplying products. In the condensation reaction, methylol compounds are formed which then react with one another to form a three-dimensional structure of methylene ($-CH_2-$) and dimethylene ether ($-CH_2-O-CH_2-$)

bridges. Faster, stronger, lighter, safer, are some of the demands which constantly make melamine interesting to today's researchers and manufacturers. Melamine fiber has recently entered the high temperature fiber market, one of the newest fibers, and has had a rapid impact. Melamine fiber is an advanced synthetic fiber having superior heat and flame resistance with decomposition temperature above 350°C (Maity and Singha, 2012).

STRUCTURE 3.2   Melamine formaldehyde resin (R–H, alkyl).

Basofil fiber has the same characteristics typical of other common melamine-formaldehyde-based materials: heat stability, low flammability, high wear performance, solvent resistance, and ultraviolet resistance. According to its manufacturer, Basofil fiber meets all environmental regulations regarding processing and use (Wakelyn, 2008).

Kynol novoloid fibers are cured phenol-aldehyde fibers made by acid-catalyzed cross-linking of melt-spun novolak resin to form a fully cross-linked, three-dimensional, amorphous network polymer structure similar to that of thermosetting phenolic resins. Because of their basic chemical structure, Kynol fibers are infusible (nonmelting) and insoluble, and possess physical and chemical properties that clearly distinguish them from all other man-made and natural fibers. The uniqueness of this fiber structure is implicit in the generic term *novoloid*, officially recognized by the US Federal Trade Commission, designating a manufactured fiber containing at least 85% of a crosslinked novolak. Kynol is a registered trademark and the only commercially available novoloid fiber. When actually exposed to flames, Kynol materials do not melt, but gradually char until completely carbonized, without losing their original fiber structure. The stable surface char radiates heat away from the material, presents a minimum reactive surface to the flame, and retards further production of volatiles. The low shrinkage and absence of melting allows the charred material to retain its integrity as a barrier to keep heat and oxygen away from the interior of the fiber structure, and the low thermal conductivity of the uncharred interior material further limits penetration of the heat. Kynol fibers and materials are highly flame resistant; in addition, they are excellent thermal insulators. However, they are not high temperature materials in the usual context (Kynol Europa GmbH, 2019).

Phenolic fibers (Novoloid fibers) are highly flame retardant with ignition temperature above 2,500°C. These fibers are obtained by spinning and post-curing phenol-formaldehyde resin precondensate. The fiber is soft and golden colored, with moisture regain of 6%. When strongly heated or placed in an open flame, the phenolic fabric is slowly carbonized, with a resulting loss of strength at high temperature. A new phenolic fiber, Philene was developed in France. This new fiber was said to have outstanding flame resistance and had been recommended as a precursor to general purpose carbon fibers. Philene is a highly cross-linked phenolic resin (resit), and is an aromatic glassy polymer with a high carbon content of 72% by weight. The moisture regain of the fiber is 7.3% and it is said to be nonflammable and self-extinguishing with an LOI of 39%. It does not show any change in tensile properties after being heated for 24h at 140°C or for 6h at 200°C (Bajaj, 1992).

When heated, phenolic polymers continue to polymerize, cross-link, and thermally degrade to coherent char replicas. Derived chars have especially high flame and heat resistance as a consequence of their high carbon contents, although their relatively low strengths prevent their being processed easily into yarns, and they are therefore more often incorporated into nonwoven fabric structures. In addition, their inherent color (pink for Basofil and gold for Kynol) ensures that they are used as barrier fabrics and not in face fabrics, although the melamine-formaldehyde structure in Basofil does allow the fiber to be dyed with small molecular disperse dyes.

The inherent fire retardant behavior of the phenolic resins (Structure 3.3) is due to the char-forming tendency of the cross-linked chemical structure. Owing to the large proportion of aromatic structures in the cross-linked cured state, phenolic resins carbonize in a fire, and hence extinguish once the source of fire is removed. They may thus be said to encapsulate themselves in char, and therefore do not produce much smoke. Epoxy and unsaturated polyesters, on the other hand, carbonize less than do phenolics and continue to burn in a fire, and structures based on these aromatic compounds produce more smoke. Although phenolics have inherent flame-retardant properties, their mechanical properties are inferior to other thermoset polymers, such as polyester, vinyl ester, and epoxies. For this reason, they are less favorable for use in load-bearing structures. Epoxies, on the other hand, because of their very high mechanical strength, are the more popular choice.

STRUCTURE 3.3   Phenol-formaldehyde resin.

Typical end-use applications of both Basofil and Kynol in thermal protection include fire-blocking and heat-insulating barriers, as well as heat- and flame-protective apparel. Typically, fibers may be blended with meta- and para-aramid fibers to improve tensile properties, including strength and abrasion resistance in both nonwoven felts and fleeces for fire blocking, aircraft seat fabrics, and fire-fighters' clothing. Such fabrics may be aluminized to improve heat reflection, and hence, fire performance (Horrocks et al., 2013).

3.7   POLYPHENYLENE SULFIDE (PPS) FIBERS

Polyphenylene sulfide (PPS) (Structure 3.4) is an organic polymer consisting of aromatic rings linked by sulfides. Synthetic fiber and textiles derived from this polymer resist chemical and thermal attack. PPS is used in filter fabric for coal boilers, papermaking felts, electrical insulation, film capacitors, specialty membranes, gaskets, and packing. PPS is the precursor of a conductive polymer of the semiflexible rod polymer family. The PPS, which is otherwise insulating, can be converted to the semiconducting form by oxidation or use of dopants.

Polyphenylene sulfide is an engineering plastic, commonly used today as a high-performance thermoplastic. PPS can be molded, extruded, or machined to tight tolerances. In its pure solid form, it may be opaque white to light tan in color. Maximum service temperature is 218°C (424°F). It has not been found to dissolve in any solvent below this temperature. An easy way to identify the compound is by the metallic sound it makes when struck.

STRUCTURE 3.4   Polyphenylene sulfide (PPS) Fiber.

PPS is one of the most important high temperature thermoplastic polymers because it exhibits a number of desirable properties such as resistance to heat, acids, alkalis, mildew, bleaches, aging, sunlight, and abrasion. It absorbs only small amounts of solvents and resists dyeing.

The US Federal Trade Commission definition for sulfur fiber is "A manufactured fiber in which the fiber-forming substance is a long chain synthetic polysulfide in which at least 85% of the sulfide (–S–) linkages are attached directly to two (2) aromatic rings."

The PPS (polyphenylene sulfide) polymer is formed by reaction of sodium sulfide with p-dichlorobenzene in the presence of polar solvent at 250°C (Equation 3.1).

$$ClC_6H_4Cl + Na_2S \rightarrow 1/n[C_6H_4S]_n + 2NaCl \tag{3.1}$$

The process for commercially producing PPS (Ryton) was initially developed by Drs. H. Wayne Hill Jr. and James T. Edmonds at Phillips Petroleum Company and the first US commercial sulfur fiber was produced in 1983. N-Methyl-2-pyrrolidone (NMP) was used as the reaction solvent because it is stable at the high temperatures required for the synthesis, and it dissolves both the sulfiding agent and the oligomeric intermediates. Linear, high-molecular-weight PPS that is capable of being extruded into film and melt spun into fiber was invented by Robert W. Campbell.

Polyphenylene sulfide (PPS) fiber has become one of the most promising special protective clothing fiber materials by virtue of its excellent thermal stability, corrosion resistance and inherent flame retardancy.

Some remarkable properties of PPS are (Omnexus, 2019):

- Exceptionally high mechanical strength,
- High temperature resistance,
- Dimensional stability,
- Inherent flame resistance,
- Good chemical resistance, and
- Electrical insulation properties.

A shortcoming of PPS fiber is that it suffers from brittleness due to highly crystalline structure (overcome by filling with glass up to 40%).

The promotion of PPS fiber in the field of application is restricted by factors such as low moisture absorption rate and poor dyeability (Mao et al., 2013). In order to give full play to the good performance of PPS fiber, it is necessary to balance the relationship between protective effectiveness and wearing comfort (Saf et al., 2015). There is commitment to developing protective clothing more in line with ergonomic design. Current research on the modification of PPS mainly focuses on three aspects; chemical reactions are triggered by ultraviolet light, radiation, and other irradiation to produce free radicals on the surface of the fabric on which hydrophilic monomers are grafted and polymerized (Pervin et al., 2014). However, the polymerization process is difficult, and the product properties are unstable. Many surface modification techniques have been used to develop PPS fibers that provide physical and/or chemical composition alterations on the PPS surface, including grafting, plasma, and laser (Anagreh et al., 2008). However, its durability is poor; the hydrophilic group gradually disappears in the use process. On the other hand, several studies propose that adding modifiers to PPS can improve the hygroscopicity of PPS, and the operation of melt blending is simple. However, it is important to consider the compatibility of the modifier with PPS and the spinnability of the composite fiber (Mousa et al., 2018). In comparison to the aforementioned methods, this study aims to prepare a kind of hygroscopic and dyeable PPS composite fiber by melt spinning for obtaining long-lasting modification effects. Due to the particularly high processing temperature of PPS, modifiers must also have good thermal stability. Sodium polyacrylate, a kind of polyelectrolyte, has been chemically cross-linked and the macro molecular chains have been intertwined to form a lightly crosslinked spatial network (Ito et al., 2019).

3.7.1 APPLICATIONS

- Automotive (under hood)
- Electrical (lamp sockets, fuse holders)
- Semiconductor manufacturing
- Oil and gas (valves, seals, pump housing)
- Nuclear

Some of the key producers of PPS include:

- Toray Resin Company–TORELINA, TORAYCA,
- RTP Company–RTP 1,300 series,
- Solvay–Ryton, PrimoSpire, Tribocomp Celanese–FORTRON, Cool-Poly, Celstran,
- Polyplastics–DURAFIDE,
- SABIC–LNP LUBRICOMP, LNP STAT-KON, LNP THERMO-COMP, and
- Lehman & Voss–LUVOCOM.

3.8 POLYBENZAZOLE (PBI AND PBO) FIBERS

Imidazole derivatives are known to be stable compounds. Many of them are resistant to the most drastic treatments with acids and bases and are not easily oxidized. The high decomposition temperature and high stability at over 400°C suggest a polymer with benzimidazole as the repeating unit may also show high heat stability. These fiber-forming polymers are called *ladder polymers* and are essentially wholly aromatic polymer chains. In chemistry, a ladder polymer is a type of double stranded polymer with the connectivity of a ladder. In a ladder polymer, the monomers are interconnected by four bonds. Inorganic ladder polymers are found in synthetic and natural settings. Ladder polymers are a special case of cross-linked polymers because the cross-links exist only with pairs of chains.

The two common examples available commercially are the polybenzimidazole, PBI (Celanese) (Structure 3.5) with the full chemical name poly (2,2′ (m-phenylene)-5,5′-bibenzimidazole), and the polybenzoxazole, Zylon (Toyobo) with the full chemical name poly(para-phenylene benzobisoxazole) and generic acronym, PBO (Structure 3.6). Their similarity in polymer chain structures and high degree of chain rigidities gives both of these fibers their superior thermal properties with thermal degradation temperatures well in excess of 400°C and superior LOI values well over 40%.

STRUCTURE 3.5   Chemical structure of polybenzimidazole (PBI).

STRUCTURE 3.6   Chemical-structure-of-poly-p-phenylenebenzobisoxazole (PBO).

PBI is a synthetic fiber with a very high decomposition temperature and does not exhibit a melting point. It has exceptional thermal and chemical stability and does not readily ignite. PBI has been introduced to the commercial markets only during the last 20 years or so in spite of its development during the early 1960s. The current PBI fiber is a sulfonated version and this improves shrinkage resistance at high temperature. Like many highly aromatic polymers it has an inherent color, bronze, and cannot be dyed. The fiber is more often used as a blend and one well-known blend is PBI Gold in which a yarn is spun with both PBI and Kevlar in a 40%–60% blend. This gives rise to gold-colored fabrics with fire protective properties that have been claimed to be superior even to those made from Nomex III. They are claimed to give better protection against fire than aramid fiber fabrics, and, in addition, remain flexible, maintain their integrity, and exhibit no afterglow (Jeffries, 1988). PBI is several times more expensive than the meta-aramids; thus, the superior performance comes at a price. Due to its high stability, polybenzimidazole is used to fabricate high-performance protective apparel such as firefighters' gear, astronauts' space suits, high-temperature protective gloves, welders' apparel, and aircraft wall fabrics. Polybenzimidazole has been applied as a membrane in fuel cells. PBI's moisture regain is useful in protective clothing; this makes the clothing comfortable to wear, in sharp contrast to other synthetic polymers. The moisture regain of PBI (13%) compares favorably with that of cotton.

A series of subjective wearers' evaluations has shown that PBI fiber exhibits comfort ratings equivalent to those of 100% cotton. In high temperature applications, PBI fiber has also been found suitable as an asbestos replacement.

Zylon or Poly-*p*-phenylenebenzobisoxazole (PBO) (Structure 3.6) forms the strongest synthetic polymer fiber known so far. It provides excellent mechanical properties paired with extreme thermal stability, making PBO the optimum material for such applications as lightweight bulletproof vests and fire-resistant suits. PBO is spun from a lyotropic melt in polyphosphoric acid (PPA) under stretching by an adjustable spin–draw ratio (SDR). PPA is removed by coagulation in a water bath, resulting in the as-spun (AS) fiber, which is subsequently heat-treated to form the final high-modulus (HM) fiber (Ran et al., 2011).

While there are at least two variants of fiber, Zylon-AS and Zylon-HM, of which the latter has the higher modulus, both have the same thermal and burning parameter values. Principal examples of thermally protective textiles include heat-protective clothing and aircraft fragment/heat barriers in which its cost, similar to that of PBI, restricts its use to applications in which strength, modulus and fire resistance are at a premium.

3.9 OXIDIZED PAN FIBERS

Inidex is a polyacrylate fiber with a high degree of polymerization that is non-melting and has a very high limiting oxygen index (43%). Smoke release is very low from Courtaulds PLC. These non-thermoplastic crosslinked polyacrylate fibers have considerable inertness to heat and the action of chemicals. The fibers are stable when ignited to dry heat up to about 160°C, burn with difficulty (releasing extremely low levels of smoke and gases), and have very good resistance to ultraviolet light. Suggested applications are in nonwoven fabrics for use as upholstery, nonflammable barrier fabrics, acoustic insulation in aircraft and other vehicles, fire and welding blankets, wall coverings, and hot-gas filters.

The Oxidized PAN fibers, also known as semicarbon fibers, are essentially carbon fibers, but, unlike true carbon fibers, retain acceptable textile properties. Within the group, the oxidized acrylics represent the sole commercial group and are produced following controlled, high-temperature oxidation of acrylic fibers during the first stages of carbon fiber production. Since the early 1980s, a number of commercial products have come into the market, including Celiox (Celanese), Grafil O (Courtaulds), Pyron (Stackpole), Sigrafil O (Sigri Elektrographit, now SGL), and Panox (SGL UK Ltd., formerly R K Textiles); many of theseare now obsolete. Current examples include Panox (SGL Carbon Group), Pyromex (Toho Rayon), and Lastan (Asahi). Their low tenacity results in difficult in processing of these weak fibers, although they can be spun into yarns by the woolen system. Thus, they are produced as a continuous tow that is stretch-broken by conventional means for eventual conversion into coarse woolen-type yarns. The limiting oxygen index is typically about 55, so fabrics are extremely thermally resistant, giving off negligible smoke and toxic gases when subjected to even the most intense of flames. Unfortunately, the fibers are black and are therefore rarely used alone, except in military and law enforcement cover all clothing when the color is an advantage. Therefore, oxidized acrylic fibers are usually blended with other fibers, typically wool and aramid, in order to dilute the color and introduce other desirable textile properties. Because of their extreme fire resistance and lower cost compared to PBI and PBO, they have applications as blends in antiriot suits, tank suits, FR underwear, fire blockers for aircraft seats, as well as heat-resistant felts (insulation), hoods, and gloves. When aluminized, they are very effective in fire entry/fire proximity suits (Horrocks, 2013).

Oxidized polyacrylonitrile (OPAN) fibers, such as Zoltek OX fiber (Zoltek Corporation, Toray Group), is designed for cost-effective, flame-resistant, and heat-resistant solutions in textile, industrial, aircraft, and automotive markets. OPAN fibers do not melt or burn and most importantly, they do not shrink, even after a 30-second, 1,250°C flame or molten metal drip exposure.

In addition to providing outstanding protection against direct flame, OPAN fiber products exhibit low thermal conductivity and make excellent thermal insulators.

The performance features of oxidized polyacrylonitrile fiber include:

- Limiting Oxygen Index (LOI) values between 45 and 55,
- Unsurpassed flame and heat dimensional stability,
- Easy processability into yarns, woven, knits, and nonwoven products,
- Soft, comfortable fabrics,
- Electrically nonconductive and excellent chemical resistance, and
- No halogens and very low toxic gas emissions, upon flame exposure.

The manufacturing process begins with polyacrylonitrile precursor fiber (PAN). The PAN precursor fiber is solution-spun and then processed through a high-temperature oven, in air, to stabilize its molecular structure. (Handermann, 2017).

PANOX (SGL) is an oxidized polyacrylonitrile (PAN) fiber that does not burn, melt, soften, or drip. It is made by thermal stabilization of PAN at 300°C. This results in an oxidized textile fiber with carbon contain of approximately 62%. Its special properties include high LOI (>45%), depending on the density of oxidation. PANOX is optimized for textile processing and demonstrates ideal blending and processing behavior.

3.10 ARIMID OR POLYIMIDE FIBERS

Among a number of reported arimid or polyimide fibers, only the P84 fiber has been commercially exploited. It was initially introduced by Lenzing during the mid-1980s, and was then sold to Inspec Fibres until the group became Evonik Fibres GmbH (Austria) in 2007.

This fiber (Structure 3.7) has a T_g value of 315°C and a decomposition temperature of approximately 500°C. It has an LOI of 36%–38%. These fibers do not melt and can withstand constant use without any essential changes in mechanical properties at temperatures up to 260°C in air (Bajaj, 1992). Polyimides are known to be used in a wide range of operating temperatures, starting from cryogenic applications and ending with high temperature applications at the limits of polymer-based materials. Thermal stability is based on the aromatic backbone of the polymer. The fibers do not melt. Despite their halogen free structure, they exhibit a high LOI of 38%, and are classified as nonflammable. The maximum service temperature is 260°C (500°F). The key advantage of P84 fiber is its unique multilobal cross section, which offers up to 90% more surface area compared to conventional round fibers. This increased surface area results in the highest filtration efficiency of conventional fibers, even for submicron particles (Evonik, 2019).

where R = $-C_6H_4.CH_2-$, $-C_6H_4.CH_2.C_6H_4-$

STRUCTURE 3.7   Polyimide fiber.

The P84 fibers have a very rigid polymer backbone with a very low H–C ratio (< 1) and hence, these factors give the fibers superior thermal and flame-resistant properties compared to aramid. The fibers are slightly weaker than meta-aramid fibers such as Nomex, and they can be processed using normal textile equipment. Thus, P84 finds use in protective outerwear, underwear, and gloves, either as 100% or blended with lower cost fibers such as flame-retardant viscose and meta-aramid. For instance, a 50%/50% P84/Viscose FR (Lenzing) blend is available for knitted underwear with high moisture absorbency. Spun-dyeing of P84 fibers enables their natural gold color to be replaced by those often demanded by customers who may require more appropriate and bright safety colors.

The other major member of this grouping is the poly (aramid-aramid) fiber Kernel, (Structure 3.8), which was produced initially by Rhone-Poulenc of France in 1971, and since 2007, by an independent company, Kermel (France) (Horrocks, 2016).

Kermel fibers have very similar properties to those of the meta-aramids, but having better dyeing properties. They have poor UV stability, and must therefore be protected from intense radiation sources. They compete in protective clothing markets where again, they are used as 100% or as blends with other fibers, including FR viscose and wool. Composite yarns with high modulus aromatic fibers such asthe para-aramids have yielded the modified Kermel HTA, a yarn with a para-aramid core (35%) and a Kermel fiber wrapping (65%) that have improved abrasion resistance.

STRUCTURE 3.8 Poly (aramid-arimid) fiber Kermel.

IFR polyimides (and ether imides) can be further improved by the incorporation of reactive fluorine-containing (Rogers et al., 1993) and phosphorus-containing groups (Tan et al., 1998). Polyimides have also been made more flame retardant by phosphorylation with phosphorus-containing reagents such as diethylchlorophosphate (Liu et al., 1997). The resulting materials are slightly less thermally stable than the unmodified precursors, but give high char yields and have LOIs in excess of 48. Flame retardant, randomly segmented copolymers with phosphorus-containing poly(arylimide) hard segments and poly(dimethyl siloxane) soft segments have also been described (Wescott et al., 1994). Char yields in these copolymers were found to depend principally upon the siloxane content.

Polyamide-imides are either thermosetting or thermoplastic, amorphous polymers that have exceptional mechanical, thermal, and chemical resistant properties. Polyamide-imides are used extensively as wire coatings in making magnet wire. They are prepared from isocyanates and TMA (trimellic acid-anhydride) in N-methyl-2-pyrrolidone (NMP). Polyamide-imides display a combination of properties from both polyamides and polyimides, such as high strength, melt processability, exceptional high-heat capability, and broad chemical resistance.

Poly(amide–imide) (PAI) brings together both superior mechanical properties of polyamide and the high thermal stability and solvent resistance characteristics associated with polyimide. Some fillers can strongly influence the combustion characteristics of a polymer, such as its resistance to ignition. In the synthesis of nanocomposites by Mallakpour and Khadema (2015), a kind of halogenated modifier was used to obtain a flame retardant and dispersed system of NPs. Tetrabromo phthalic anhydride had flame retardant characterization. p-Aminophenol was applied for opening phthalic anhydride cycle and constructing functional groups that could be grafted onto the surface of NPs.

Chirality is a geometric property of some molecules and ions. A chiral molecule/ ion is non-superimposable on its mirror image. The presence of an asymmetric carbon center is one of several structural features that induce chirality in organic and inorganic molecules. The term *chirality* is derived from the Greek word for hand (Fox and Whitesel, 2004).

Thermally stable chiral PAI was successfully synthesized by a simple efficient and green approach using a mechanical stirrer with TBAB/TPP as the condensing agent without the need for any additional promoters. The aim of the study was to use modified ZrO_2 NPs for manufacturing NCs by flame retardant properties. The ZrO_2 NPs were treated with 2,3,4,5-tetrabromo- 6-[(4-hydroxyphenyl) carbamoyl]benzoic acid (HCBA) (Structure 3.9) as the flame retardant coupling agent that had two functional groups,–COOH and–OH, for grafting to–OH groups on the surface of NPs. The chiral PAI was synthesized during polycondensation reaction of chiral and biocompatible diacid and aromatic diamine under green conditions using molten Tetrabutylammonium bromide (TBAB) as a molten IL for PAI synthesis. The results showed that the NPs were uniformly dispersed in the obtained PAI matrix, and all the samples were self-extinguishing.

STRUCTURE 3.9 Chemical structure of HCBA.

3.11 POLYHALOALKENES

PVC fibers are relatively nonflammable. They do not burn, nor do they emit flames or release molten incandescent drops capable of spreading fire on combustible materials. When subjected to an intense flame, PVC fibers disintegrate, but the residue can be touched, since it is not hot, and thus, there is no risk of burning the skin. Polyvinylidene chloride fibers have behavior similar to that of PVC fibers. Their flameretardance can further be improved by the use of additives, copolymerization with monomers containing phosphorus, and coatings with FR finishes. Rhovyl and Polyhaloalkenehavebeen used for thermalwear. Polyhaloalkenes are claimed to be totally unaffected by water, i.e., they neither shrink nor swell in water. Nonflammable curtains made from them are in use, and polyhaloalkenes have also been used in athletic clothing for its thermal comfort (Bajaj, 1992).

The major fibers of importance here are those based on poly(vinylidene choride) or PVDC. The chemical structure of the polymeric repeat unit–$CCl_2.CCl_2$–creates a polymer with a high degree of chemical resistance and a high degree of order. This latter creates a polymer thatis difficult to process, with the result that in commercially useful forms, copolymers with other vinyl and acrylic comonomers, such as vinyl chloride, acrylonitrile, and methyl acrylate (usually present at <15% w/w), are present. Dow Chemical Company developed the Saran fiber based on this polymer in the 1940s, and the name Saran is a registered trademark with respect to the polymer which is used in coatings, films, monofilaments, and other extrusions. Currently, the resin Saran 510 is that recommended for monofil applications and it is in fact a copolymer of vinylidene chloride and vinyl chloride. Fibers have been marketed under the names Permalon and Velon. Similar fibers, such as Fugafil, produced in Germany by Saran GmbH & Co., are also available.

The aforementioned polymer melts over the range 160°C–170°C, and is melt spun at about 180°C by conventional melt spinning methods to yield both multifilaments and monofilaments. It softens over the range 115°C–160°C, depending upon its copolymeric character and this limits its service temperature limit. In addition to chemical resistance, it possesses an inherently low flammability and an LOI value of 60%. When in a fire, however, it gives off hydrogen chloride which is both toxic and corrosive.

The fluorocarbons constitute another class. They have excellent fire-resistant properties and decompose at high temperatures and it is difficult but not impossible to ignite them. The toxic potency of fluoro-polymers is very complex and has been reviewed by Purser (1992). Most work has been done on PTFE, but other per-fluorinated polymers appear to behave similarly to PTFE. For per-fluorinated polymers, the toxic potency under flaming conditions is around 200 $gm^{-3}min$, which is not dissimilar to that of PVC. Under nonflaming decomposition conditions, the toxic potency depends very much upon the exact decomposition conditions, varying over a very wide range from 0.5–87 $gm^{-3}min$, depending upon the extent to which extreme toxic potency particulates are formed.

Poly(vinyl chloride) or chlorofiber is homo–polymer, while polyvinylidene chloride, e.g., SaranTM (Asahi Kasei) is a copolymer.

3.12 INORGANIC AND CERAMIC FIBERS

Inorganic and Ceramic Fibers have no tendency to burn (no known LOI values), and may therefore be used in applications in which high temperature stability and poor chemical reactivity determine their durability. This group includes the various forms of glass, silica, and alumina, although more exotic forms such as stainless steel, boron nitride, and silicon carbide are also available for specialist end users. Such extreme thermal protection is useful in applications such as furnace linings, hot component insulation in car exhaust catalysts, or around combustion chambers in jet engines in which working temperatures and occasional flash temperatures are in excess of 500°C and even 1,000°C in extreme circumstances. However, while fire resistance is an intrinsic feature of these inorganic fibers and textiles, their poor aesthetics limits their use to these extreme technical applications, although glass or ceramic-cored, organic fiber-wrapped yarns may be used.

The most well-established family of glass fibers has been exhaustively reviewed by Jones (2001), while ceramic fibers have been reviewed recently by Price and Horrocks (2013). These fibers are used when heat and fire resistance are essential. The prime use has usually focused on their reinforcing quality and so the fibers, because of an inherent brittleness and poor textile general character, are used as reinforcing elements in flexible textile as well as in rigid composite structures. The glass fiber assemblies are used in nonreinforcing applications, filter media for high temperature gas and liquid filtration, as battery separators, and as fire and acoustic insulation in aircraft and other transport systems When heated above 850°C, devitrification and partial formation of polycrystalline material occurs as the former glass fibers become more similar in character to ceramic materials. This devitrified form melts at 1,225°–1,360°C, which is high enough. Vitrify means "to convert (something) into glass or a glasslike substance", typically by exposure to heat. The Olympic Stadium in Berlin with a surface area of 42,000 m^2 is a recent example in which the upper roof is made up of a highly tear-resistant fiberglass fabric of coated with PTFE (polytetrafluoroethylene), resulting in a lifespan of at least thirty years, excellent fire resistance, and a self-cleaning surface (Koch, 2004).

On the other hand, ceramic fibers have even poorer textile properties, are often more expensive, and are mostly used as refractory fibers in insulating and fire barrier materials for applications requiring resistance to temperatures of at least 1,000°C for prolonged periods. They tend to have polycrystalline structures; hence, their exceptional high temperature characteristics are not often produced in appropriate fibrous dimensions for normal textile process-ing, and are more usually available as nonwoven or wet-laid webs. This group may be classified into four major groups, namely, alumina-based, silica-based, alumina-silica, and silicon carbide fibers. Polycrystalline fibers such as Saffil (Saffil Ltd., UK), are available as a lofty, nonwoven, wet-laid web or *blanket* and may be used in refractory and fire barrier applications at temperatures as high as 1,600°C (Horrocks, 2013).

Finally, the silica-based fibers, such as Quartzel (Saint-Gobain, France), while having slightly inferior fire and heat performance, are available as con-tinuous filament yarns, filament-based nonwoven products, and wet-laid papers. Thus, they may be used in furnace insulation, combustion chamber insulation in aircraft, ablative composites for military and other markets, and hot corrosive gas and liquid filtration.

Falling between the two aforementioned extremes are the alumina-silica fibers exemplified by the Nextel range of products (3M, USA), which are determined by the alumina–silica ratio, and rarely contain less than 60% alu-mina. Varying the amount of silica leads to various forms of alumina-silica fibers with a range of high temperature behavior. For example, Nextel 610 comprises 99% (similar to Saffil), Nextel 720 comprises 85%, and Nextel 312 comprises 62% Al_2O_3 yielding a maximum user temperature range of 1,260°–1,370°C. These are also available in yarn, fabric, and nonwoven forms for similar applications (Horrocks, 2013).

3.13   COMPARISON OF PERFORMANCES

Bourbigot and Flambard (2002a) have produced one of the few recent comparisons of the fire performance of a number of high temperature and fire resistant polymer-based fibers based on cone calorimetric data. The parameter rate of heat release (RHR) per second is known as the fire growth index (FIGRA) and this may be plotted against time to give a better measure of relative fire-propagating behavior.

The following inherently flame retardant polymers can be arranged in the increasing order of FIGRA (Equations 3.2 and 3.3)

$$\text{PBO (Polybenzoxazole)} < \text{Kynol (phenol} - \text{formaldehyde)} \sim \text{para–aramid}$$
$$< \text{Technora (copolymer of para–aramid)} < \text{Oxidized acrylic}$$

$$(3.2)$$

The decreasing order in terms of LOI as a fire measure is:

$$\text{PBO} > \text{Oxidized acrylic} > \text{para} - \text{aramid} \sim \text{Kynol (phenol–formaldehyde)}$$
$$> \text{Technora (copolymer of para–aramid)}$$

$$(3.3)$$

Where

Para-aramid–poly (para-phenylene terephthalamide), e.g.,Kevlar (DuPont) and Twaron (Teijin).

Technora fiber is based on the 1:1 copoly(terephthalamide) of 3,4'-diaminodiphenyl ether and p-phenylenediamine – a copolymeric derivative of the para-aramid fibers introduced by Teijin in 1985.

Oxidized acrylic–controlled, high temperature oxidation of acrylic fibers during the first stages of carbon fibre production, e.g., Celiox (Celanese), Grafil O (Courtaulds), Panox (SGL Carbon Group), Pyromex (Toho Rayon), and Lastan (Asahi). Many of these are now obsolete.

3.14   BLENDS OF INHERENT FLAME RETARDANTS

Blends of the aforementioned fibers have been commercially exploited for a number of years, with aramid-FR viscose being perhaps one of the first attempts to reduce the cost of the final fabric while maintaining a high level of fire protection. The blends of meta-aramids with para-aramid or PPTA (e.g., Nomex III, Du Pont; Kermel HTA, Rhodia) and PBI and para-aramid (PBI Gold, Celanese) have been developed to give a balance between the fire properties of both with the higher modulus and strength of the para-aramid. Very few, if any, blends have been produced to generate synergies of fire resistance and other properties while, perhaps, improving other desirable features, including cost. Bourbigot et al. (2002b) have recently reported a number of possibly synergistic blends made by mixing yarn by yarn of wool with PPTA, improving flame retardancy and general thermal stability of the whole fabric. In the suggested mechanism, the char of wool mixes with polymer melt, which coats adjacent para-aramid fibers, hindering the diffusion of oxygen to them and thereby negating their sensitivity to oxygen from the air and consequent thermal oxidative degradation. Subsequent research (Bourbigot et al., 2007) has indicated that intimate blends of wool/PPTA show synergy when only 30% or greater PPTA is present as opposed to 70% or more in the previously blended yarn results. Furthermore, synergy was noted in wool–Technora blends which can show reduced peaks of heat release rates with respect to 100% Technora; some wool–PBO blends show similar encouraging results, suggesting the possibility of enhanced fire performance at reduced cost and improved aesthetics is feasible commercially using these interesting blends.

3.15   FLAME RETARDANCY BY DOPING OR COPOLYMERISATION

Conventional manmade fibers may be rendered inherently flame retardant by either incorporating a flame retardant additive in the polymer melt or solution prior to polymer extrusion or by copolymerization before, during, or immediately after processing into filaments or staple fibers.

The additives should be thermally stable up to 300°C under mechanical force for considerable time, should dispense easily and have a suitable melting point and a high degree of dispersion, without degrading or reacting detrimentally with the polymer. The problems of compatibility, especially at the high temperatures used to extrude melt-extruded fibers such aspolyamide, polyester, and polypropylene and in reactive polymer solutions such as viscose dope and acrylic solutions have ensured that only a few such IFR fibers are commercially available. A major problem in developing successful IFR fibers based on conventional fiber chemistries is that any modification, if present at a concentration much above 10% by weight (whether as additive or comonomer), may seriously reduce tensile properties, as well as other desirable textile properties, such as dyeability, luster and appearance, and handle.

Table 3.2 shows some of the conventional regenerated and synthetic fibers and a few corresponding IFR counterparts with respective LOI values under bracket (Horrocks, 2013). The table clearly shows much higher LOI values of the IFR counterparts (fibers in Column 2 of the table). No significant changes in tensile properties were observed between the original fiber and the corresponding modified (FR) counterpart. In other words, the mechanical properties of the modified fibers are comparable to the respective unmodified analogues. Other properties such as dyeability, are also little affected by the respective flame retardant modifications present.

Various additives are added in the melt or solution of polymers before spinning of polymers or copolymerization of conventional polymers in order to impart flame retardancy.

The most successful flame retardant additives on the market today are halogenated flame retardant additives. These compounds are used in conjunction

TABLE 3.2

The conventional manmade fibers and the corresponding IFR fibers (LOI % under bracket)

Conventional Flammable Fiber	Corresponding Modified IFR(s)
Viscose (18–19%)	Lenzing FR (28%), Visil AP (28–30%)
Polyester (20%)	Trevira CS (28–30%), Toyobo GH/Heim (30%)
Nylon 6,6 (21–22%)	Nexylon FR (28%)
Acrylic (18%)	Modacrylic (29–32%)
Polypropylene (18%)	FR Polypropylene (Est. 23%)
Polyethylene (18–19%)	Poly(vinyl chloride) (37%), Poly(vinylidene chloride) (60%)

with a synergist to aid increasing flame retardancy with lower loadings. The halogenated flame retardant retards the flame efficiently by acting quickly as a radical trap. Halogen-containing flame retardants costing about one billion dollars are consumed every year to keep plastics safe from combustion. The majority of these halogenated additives are similar. Poly-brominated-biphenyls (PBB), have long been used in commercial polymers to impart flame retardancy. Halogenated compounds can have negative environmental and toxicological impacts that have deterred many countries from using them in commercial products. The European Union is trying to remove halogenated compounds from all plastics due to environmental concerns. Leaching additives can end up in drinking water. Additionally, halogenated organic compounds are considered Persistent Organic Pollutants (POPs) which are not easily broken down or oxidized by the environment. Another problem is the use of antimony oxide, which carries with it heavy metal concerns and has a possible link to Sudden Infant Death Syndrome (SIDS).

3.15.1   FR VISCOSE

The development of flame retardants for viscose is very challenging as the flame retardant has to withstand severe alkaline and acidic conditions during the fiber manufacturing process. Flame retardant viscose fiber has better skin affinity and dyeability than flame retardant polyester fiber, and moisture regain is about 13%, which belongs to the category of moisture regain of comfortable human bodies. In addition, flame retardant viscose fiber is a kind of regenerated cellulose produced from natural resources. The waste after use is decomposed by micro-organisms present in the soil, water, and air, and is degraded into carbon dioxide and water, so its raw material is inexhaustible, and the waste is harmless. The fiber raw material belongs to green environmental protection and sustainable development.

Additives for viscose are organophosphorus and nitrogen/sulfur-containingspecies e.g.,10–15 wt% 2, 2-oxybis 5,5-dimethyl-1,2,3-dioxaphosphorinane-2,2–disulphide, $C_{10}H_{20}O_5P_2S_2$ (Clariant 5060), is used in Lenzing

FR (Lenzing AG); 30 wt% polysilicic acid and complexes are used in Visil AP (Sateri). Silicon-based polymers, unlike carbon-based materials, are generally incombustible. VISIL 33 A.P. (Kemira, Finland) is a viscose fiber with a silicic acid backbone. The smoke emission is low and is free from toxic fumes (Oulton, 1995). The polysilicic acid present in Visil viscose fiber is particularly interesting in that it is not only largely phosphorus-free, but on heating, both a carbonaceous char and a siliceous char are formed. The aluminum salt after treatment (probably aluminum phosphate; hence the "AP" in Visil AP) raises the LOI from an otherwise value of 26–27 to about 30%, increases sensitivity to alkalis in the pH range 7–9, and also increases wash durability to acceptable commercial levels. The presence of the silica in the residue ensures that the thermally-exposed fabrics revert to a ceramic char, thus affording high levels of protection to temperatures as high as 1,000°C. Both flame retardant viscose types have acceptable tensile properties as well as general textile properties.

These FR fibers usually have flame retardant additives incorporated into the spinning dopes during their manufacture, which therefore yield durability and reduced levels of environmental hazard, eliminating the need for a chemical flame retardant finishing process. Additives such as Sandoflam 5060 are phosphorus-based and are therefore similar to the majority of FR cotton finishes in terms of their mechanisms of activity (condensed phase), performance, and cost-effectiveness. Environmental desirability may be questionable and this issue has been minimized by Sateri (formerly Kemira) Fibers, Finland with their polysilicic acid-containing Visil flame retardant viscose fiber (Horrocks, 1996). This fiber has not only eliminated the need for phosphorus, but also the need for chars to form a carbonaceous and silica-containing mixed residue which offers continued fire barrier properties above the usual 500°C where carbon chars quickly oxidize in air.

3.15.2   FR POLYESTER

Polyester flame retardant fibers belongs to petrochemical products, which are prepared from the scarce resource of petroleum. The energy consumption ratio of petroleum fiber to fiber is much higher than that of viscose fiber, and the fiber waste causes pollution.

Flame retardant polyester fibers were produced in laboratory scale using various organo-phosphorous compounds as melt additives. To ensure proper blending at the molecular level and stability at the spinning temperatures, four organo-phosphorous compounds with melting point lower than that of poly-ethylene terephthalate (PET) and boiling point higher than the processing temperature were selected. These compounds are triphenylphosphine oxide (TPPO), diphenylchloro phosphate (DPCP), phenyl phosphonic acid (PPA), and 2,4-diterbutylphenyl phosphite (DTBPP). The flame retardant polyester fibers were melt spun at 80 m/min on a laboratory melt-spinning unit and subsequently drawn-heatset. The limiting oxygen index (LOI) has been found to be in the range 25%–27% at 2.5%–10.0% loading. The effect of additives on mechanical properties of fibers was investigated and correlated with the structure of the fibers. TPPO was found to be the most appropriate additive among the selected compounds for providing adequate flame retardancy with very small loss in mechanical properties. The, flame retardant PET fibers containing 10% TPPO showed tenacity comparable to the control sample. The study suggests the possibility of developing mechanically strong flame retardant polyester fibers using a more economical dope additive route (Singh, 2015).

IFR polyesters include:

- Trevira CS (Trevira GmbH, formerly Hoechst): Organophosphorus species;
- Fidion FR (Montefiber), phosphorus-containing additive, phosphinic acidic comonomer;
- Fidion FR (Toyobo) polyester is believed to be based on the sulfone-phosphonate copolymer; the permanent flame retardant fiber can be prepared by incorporating organic phosphorus compounds into low-degree polymer chains without affecting basic fiber properties.

Various organophosphorus additives are used, namely phosphinic acidic comonomer in Trevira CS, (Trevira GmbH); phosphorus-containing additive in Avora CS (KoSa), sulphonyl bis phenol phenylphosphinate oligomer in Heim/Toyobo GH (Toyobo, Japan), phosphorus-containing additive in Fidion FR (Montefibre) and Brilén FR (Brilén, Spain).

IFR polyesters are commercially available from various manufacturers such as Trevira GmbH (Trevira CS), Invista (Avora) and Toyobo (Heim). These kinds of fibers contain reactive organophosphorus additives based on phosphonic acid derivatives. A phosphorus content of 0.7%–1% is sufficient to achieve satisfactory levels of flame retardancy. The mechanism of flame-retardant action of these phosphorus compounds may be due to reduction in melt viscosity caused by acid hydrolysis. The phosphorus atom in these polyesters could be an integral part of the main backbone, or it could be present as a pendant group to the main chain. The positioning of phosphorus atom in the polyester does not have any effect on the flammability of the fiber, but it does affect the hydrolysis stability of the polymer.

One successful FR polyester group is typified by the well-established Trevira CS, which contains the phosphinic acid comonomer. Some FR additives for polyester are as follows:

- Structure 3.10 (a) shows additive for FR polyester-phosphinic comonomer for Trevira CS;
- Structure 3.10 (b) Bisphenol-S Additive (comonomer) for Toyobo GH; and
- Structure 3.10 (c) Cyclic dimeric additives for FR polyester (Amgard or Antiblaze P45, Rhodia (formerly Albright & Wilson); it is the former Mobil Chemical Antiblaze 19 compound that is available in dimeric form as a melt additive and in monomeric form as a polyester textile finish (Antiblaze CU).

All of these FR polyester variants do not promote char, but function mainly by reducing the flaming propensity of molten drips normally associated with unmodified polyester. As yet, no char-promoting flame retardants exist for any of the conventional synthetic fibers and this must constitute the real challenge for the next generation of acceptable inherently FR synthetic fibers.

Trevira CS can be used to prepare drapes, beddings, linen, curtains, room dividers and more, and hence, can protect us from fires in homes, hotels, public buildings, hospitals and other care facilities, maritime, air travel, road rail travel and in the private market. It is made from environmentally friendly manufactured materials. It is certified to Oekotex 100 Standard and little toxic fumes are developed in the event of a fire.

IFR poly(lactic acid) was produced by chemically incorporating an effective organophophorus-type flame retardant (FR) into the PLA backbone via the chain extension of the dihydroxyl-terminated prepolymer with 1, 6-hexamethylene diisocyanate (HDI). The thermal analysis revealed that the char yield of IFR-PLA and PLA-FR blend above 400°C was greatly enhanced compared to that of pure PLA. The LOI value was significantly improved from 19 for pure PLA to 29 when 1 wt% of phosphorus content was introduced and all IFR-PLA samples achieved V-0 rating in the UL-94 tests. PLA-FR blends had an LOI value of 25–26 and UL-94 V-2 rating at 20 wt% of IFR-PLA content (Yuan et al., 2011).

$$\underset{\underset{X}{|}}{HO - P - Y - COOH} \quad \overset{\overset{O}{\parallel}}{}$$

X = H or alkyl
Y = alkylene

STRUCTURE 3.10 A copolymer and additives for FR polyester- Phosphinic Comonomer for Trevira CS.

$$-O-\langle\bigcirc\rangle-SO_2-\langle\bigcirc\rangle-O-\overset{\overset{O}{\parallel}}{\underset{\underset{Ph}{|}}{P}}-$$

STRUCTURE 3.10 B copolymer and additives for FR polyester–Bisphenol-S additive for Toyobo GH.

STRUCTURE 3.10 C copolymer and additives for FR polyester–Cyclic dimeric for Amgard or Antiblaze P45.

3.15.3   FR POLYAMIDE

Polyamides are now widely found in a variety of high performance plastics applications in automotive, electrical and electronic, and other industrial markets, for instance. Flame retardants play an increasingly important role in extending the reach of polyamides in these areas.

Nexylon FR (Ems-Chemie AG) is an IFR polyamide fiber (FR-PA66), which is rated as hardly combustible (BI 4102–1) and has a LOI of 28. The FR additive is wash resistant and free of halogens and toxic compounds. The additive does not influence the textile properties of the PA fiber; rather, the abrasion resistance of the fiber is improved via a special polymer recipe. They are easily dyeable and improve the abrasion resistance of garments.

For Nexylon FR, a phosphorus based FR additive is used which acts permanently and is wash resistant (100 launderings at 75°C were simulated on lab scale). During burning, the additive sets free a radical scavenger thereby stopping the flame. No halogens are used and smoke development is low with no toxic substances in the smoke. Fiber blends with an amount of 30–45% Nexylon FR can significantly improve abrasion resistance and lifetime service performance (Bender et al., 2013).

The paucity of FR polyamides reflects their high melt reactivities and hence, their poor potential flame retardant additive compatibilities. This problem has been further discussed by Weil and Levchik (2009).

Some of the FR additives produced by M/s ICL that are particularly suitable for IFR-PA passing class V-0 according to the UL 94 standard (ICL IP, 2019) are described next.

FR-803P 1 (brominated polystyrene- 67% bromine), owing to its polymeric nature, contributes to high comparative tracking index (CTI, the electrical breakdown (tracking) properties of an insulating material) and excellent thermal stability. The advantages of FR-803P 1 in PA are:

- It is nonblooming.
- High-temperature processing stability up to 310°C contributes to excellent color stability.
- It has excellent electrical properties.

FR-1025 (Brominated Polyacrylate–71% bromine), is a proprietary polymeric flame retardant offered by ICL IP. The main advantages of FR-1025 in PA 6 applications are:

- It has effective flame retardancy,
- It is nonblooming,
- It provides good temperature stability and longterm heat–aging stability.

It also finds application in PA 6,6 provided temperature in the melt is maintained strictly below 290°C. This processing condition is easy to achieve taking into account the processing aid effect of FR-1025.

F-2400 (highest molecular weight brominated epoxy polymers–53% bromine in ICL-IP BEs product line), are melt blendable and are UV stable efficient polymeric FR additives. They are nonblooming.

F-3100, recently introduced modified brominated epoxy polymer (53% bromine) has excellent cost performance with well-balanced properties in PA.

The manufacturer ICL-IP claimed that all these FR additives have undergone extensive toxicological and environmental testing and have been proven to pose no risk to health and the environment.

The halogen-free, IFR polyamide 6 (PA6), fabrics may be produced in which component fibers still have acceptable tensile properties and low levels (preferably ≤10 wt %) of additives by incorporating a nanoclay along with two types of flame retardant formulations. The latter include (i) aluminum diethyl phosphinate (AlPi) at 10 wt %, known to work principally in the vapor phase and (ii) ammonium sulfamate (AS)/dipentaerythritol (DP) system present at 2.5 and 1 wt % respectively, believed to be condense phase active. The nanoclay chosen is organically modified montmorillonite clay. The AS/DP-containing formulations with total flame retardant levels of 5.5 wt % or less showed far superior properties and with nanoclay, showed fabric extinction times ≤ 39 s and reduced melt dripping (Horrocks et al., 2016).

3.15.4   FR ACRYLIC

Acrylic fibers bear a LOI value of 18%, one of the lowest among all textile fibers, which have been widely used for clothing and home textiles. It needs to be fireproofed because most fire accidents happened at home according to the reports.

Poly(methylmethacrylate) (PMMA) shows high strength and transparency but is a flammable material. In a study, the surface of aluminum hydroxide was modified with methacrylate containing phosphoric acid moieties before dispersion in MMA, and organic–inorganic-nanohybrid materials were obtained by bulk polymerization in the presence of the surface-modified aluminum hydroxide. The resulting hybrid materials retained the high transparency of PMMA, with transparency values similar to that of pure PMMA. Moreover, the flame resistance of the hybrid materials was improved in comparison with that of pure PMMA, with depression of the horizontal burning rate becoming a maximum at an inorganic content of 3 wt% (Daimatsua et al., 2007).

Flame retardant acrylics are usually so highly modified in terms of comonomer content that they are termed modacrylics. This latter group has been commercially available for 40 years or so, but at present, few manufacturers continue to produce them. This is largely because of the success of backcoatings applied to normal acrylic fabrics, which create high levels of flame retardancy more cost-effectively.

Polyacrylics can cross-link via their pendant nitrile substituents to give a carbonized structure as evidenced by their carbon fiber precursor suitability. The high flammability of acrylic fibers is associated with the rapid heating rate associated with the burning process, which favors the volatilization (probably by unzipping) of the polyacrylonitrile chains. Slow heating rates favor the oligomerization and cross-linking reactions that are more usually associated with carbon fiber production conditions. Any effective char-promoting flame retardant should, therefore, reduce the former volatilization tendency at high heating rates and enhance oligomerisation in the first instance (Horrocks and Price, 2001). Velicren (Montefiber) and Kanecaron (Kaneka Corp.) FR acrylic fibers are made by copolymerization with halogenated comonomers (35–50% w/w) plus antimony compounds. Halo-organic compounds usually are brominated derivatives, e.g., Sandoflam 5072 (Clariant, formerly Sandoz).

Novel flame retardant chemical additives and polymers were synthesized and their flammability was measured. Self-extinguishing compositions were obtained for poly- (acrylonitrile butadienestyrene) and high-impact polystyrene by adding as little as 10 weight % of boronic acid derivatives (Structure 3.11) or halogen-containing bisphenylethenes (BPH). Self-extinguishing compositions were obtained for polyethylene by adding as little as 10 weight % BPH. The efficacy of BPH additives as flame retardants suggested incorporating these moieties directly into the polymer to further reduce flammability. Consequently, polymers and copolymers were synthesized, having BPH backbone and pendant groups, including backbone copolymers containing acetylene and phosphineoxide.

$$R-B \big\langle \begin{matrix} OH \\ OH \end{matrix}$$

STRUCTURE 3.11   Boronicacid.

3.15.5　FR Polypropylene

FR polypropylene includes brominated organic compounds e.g., tris-(tribromopentyl) phosphate (FR-370, ICL); hindered amine stabilizer, e.g., Ciba Flamstab NOR 116, plus bromo-organic species.

Polypropylene has a low melting point (~165°C). It is challenging to make this fiber flame retardant because of its tendency to undergo random scission to highly flammable smaller hydrocarbons and its lack of tendency to form char (Zhang and Horrocks, 2003). Compared to other synthetic fibers, polypropylene is easier to produce. Hence, the manufacturers prefer to produce their own flame retardant versions by traditional additive formulations based on bromo-organic species in the presence of a synergist i.e., tris(tribromopentyl) phosphate (FR-370, ICL) and pentabromoacrylate (FR-1025, ICL) both of which may be used without the need for an antimony III oxide synergist. Weil and Levchik (2009) reviewed alternative synergists to include free radical generators such as 2,3-dimethyl-2,3-diphenylbu-tane (Perkadox 30, Akzo) andthe hindered amine radical stabiliser Flamstab NOR 116, which is an N-alkoxy-2,2,6,6-tetramethyl-4-substituted morpholine as also reviewed by Zhang and Horrocks (2003). The latter enables lower than expected levels of bromo-organic species to be used (Kaprinidis et al., 2002).

3. 15.6　Polyurethane

Rigid polyurethane foams (RPUFs) have become commercially important as building thermal insulation materials. A novel reactive flame retardant triol (TDHTPP) based on a triazine and phosphate structure was designed and used as additive FR for rigid polyurethane foams (RPUFs). This triol was chemically incorporated in the main chains of RPUFs as a chain-extender to prepare inherently flame retardant RPUFs. 2,4,6-triphosphoric acid diethyl ester hydroxy-methyl phenoxy-phosphonate (TDHTPP) showed good solubility in the polyols (polyol 4110 and PEG 400) of RPUF, making it convenient for industrial fabrication, such as in spray foaming. Excellent compatibility of TDHTPP with the polymer matrix endowed flame retardant RPUFs (IFR-RPUFs) much higher compressive strength than that of the neat RPUF, and a low thermal conductivity of about 0.03 W/(m·K). Notably, with only 5 wt % of TDHTPP incorporated, the resultant RPUF displayed a UL-94 V-0 rating and more importantly, exhibited great persistent flame retardancy during thermal accelerated aging tests at 140°C for 96 h. Further characterization revealed that TDHTPP possessed both vapor-phase and condensed-phase flame retardant behaviors in which vapor-phase action was dominant. This work provides a facile route to synthesize RPUFs with persistent flame retardancy, excellent thermal insulation, and mechanical properties (Wang et al., 2018).

3.16　FUTURE TRENDS

It also seems likely that the use of IFR polymers will increase, especially as new ways are found of processing such polymers, many of which are also rather troublesome owing to high softening temperatures and/or melting points or high degrees of crosslinking.

"Instead of adding new fire retardant chemicals that ultimately may be shown to cause health problems, we should be asking whether we need to use these chemicals or if there are other ways to achieve equivalent fire safety," contends Arlene Blum, a biophysical chemist and visiting scholar at the University of California, Berkeley (Pipe and Drape Online, 2019).

One promising approach is to incorporate flame retardants into the materials themselves. The aerospace industry currently uses some inherently nonflammable plastics, but they are too expensive for commodity-type applications such as electronics housings, given the industry's profit margins. More recently, scientists have begun trying to develop plastic polymers that are inherently nontoxic and nonflammable.

A number of man-made IFR fibers are already in the market. However, most of these are costly, and hence, cannot substitute FR-treated counterparts. Therefore the future researches will be focussed to find cheaper IFR polymers. No char-promoting flame retardants exist for any of the conventional synthetic fibers and this must constitute the real challenge for the next generation of acceptable IFR synthetic fibers.

REFERENCES

Anagreh N., Dorn L., and Bilke-Krause C. (2008). Low-pressure plasma pretreatment of polyphenylene sulfide (PPS) surfaces for adhesive bonding, *International Journal of Adhesion and Adhesives*, 28(1–2), 16–22.

Bajaj P. (1992). Fire retardant materials, *Bulletin of Materials Science*, 15(1), 67–76.

Barbosa V.F.F., MacKenzie K.J.D., and Thaumaturgo C. (2000). Synthesis and characterization of materials based on inorganic polymers of alumina and silica: Sodium polysialate polymers, *International Journal of Inorganic Materials*, 2(4), 309–317. DOI: 10.1016/S1466-6049(00)00041-6.

Bender K., Schach G., and Rosa S. (2013). New flame-retardant polyamide fiber, *Melliand International*, 1, 18–20.

Bourbigot S., Duquesne S., Bellayer S., Flambard X., Rochery M., and Devaux E. (2007). *Advances in the Flame Retardancy of Polymeric Materials–Current Perspectives Presented at FRPM'05*. B. Schartel, Ed. Herstellung und Verlag, Norderstedt, pp. 159–180.

Bourbigot S. and Flambard X. (2002a). Heat resistance and flammability of high performance fibres: A review, *Fire and Materials*, 26(4–5), 155–168.

Bourbigot S., Flambard X., Ferreira M., and Poutch F. (2002b). Blends of Wool with High Performance Fibers as Heat and Fire Resistant Fabric, *Journal of Fire Sciences*, 20(1), 3.

CIRFS. (2019). *Inherently Flame Retardant and Resistant Fibers*. European Man Made Fibers Association, Brussels, www.cirfs.org/, accessed on 1.6.2019.

Daimatsua K., Sugimotoa H., Katoa Y., Nakanishi E., Inomata K., Amekawa Y., and Takemura K. (2007). Preparation and physical properties of flame retardant acrylic resin containing nano-sized aluminum hydroxide, *Science Direct*. DOI: 10.1016/j.polymdegradstab.2007.05.012.

Evonik. (2019). P84polyimide fibres. www.p84.com/, accessed on 10.6.19.

Fox M.A. and Whitesel J.K. (2004). *Organic Chemistry*, 3rd edition. Jones & Bartlett Publishers, Sudbury, MA. ISBN 0763721972.

Gaan S., Salimova V., Rupper P., Ritter A., and Schmid H. (2011). Chapter 5. Flame retardant functional textiles, in: N. Pan and G. Sun, Eds., *Functional Textiles for Improved Performance, Protection and Health*. Woodhead, Cambridge, UK, pp. 98–130, eBook ISBN: 9780857092878.

Handermann A. (2017). Oxidized polyacrylonitrile fiber properties, products and applications, www.researchgate.net, September 2017.

Horrocks A.R. (1996). Developments in flame retardants for heat and fire resistant textiles–The role of char formation and intumescence, *Polymer Degradation and Stability*, 54, 143–154.

Horrocks A.R. (2013). Inherently flame resistant fibres, in: J. Alongi, A.R. Horrocks, F. Carosio, and G. Malucelli, Eds., *Update on Flame Retardant Textiles: State of the Art, Environmental Issues and Innovative Solutions*. Smithers Rapra, Shawbury, UK, pp. 178–206.

Horrocks A.R. (2016). Technical fibres for heat and flame protection, in: *Handbook of Technical Textiles*, 2nd edition, Vol. 2. Technical Textile Applications, Woodhead, Cambridge, UK, p. 243.

Horrocks A.R. and Price D. (2001). *Fire Retardant Materials*. Woodhead and CRC, Cambridge, UK.

Horrocks A.R., Sitpalan A., Zhou C., and Kadola B. (2016). Flame retardant polyamide fibres: The challenge of minimising flame retardant additive with added nanoclays, *Polymers*, 8(8), 288. DOI: 10.3390/polym8080288.

ICL IP. (2019). Flame-retardants for polyamides, http://icl-ip.com/, accessed on 15.6.19.

Ito F., Nishiyama Y., and Duan S. (2019). Development of high-performance polymer membranes for CO2 separation by combining functionalities of polyvinyl alcohol (PVA) and sodium polyacrylate (PAANa), *Journal of Polymer Research*, 26(5), 106. DOI: 10.1007/s10965-019-1769-6.

Jeffries R. (1988). Clothing for work and protection, *Textile Asia*, 19(11), 72–82.

Jones F.R. (2001). Glass fibres, in: J.W.S. Hearle, Ed., *High Performance Fibres*, Woodhead Publishing, Cambridge, UK, pp. 191–238.

Kaprinidis N., Shields P., and Leslie G. (2002). *Flame Retardants 2002*. Interscience Communications, London, pp. 95–106.

Koch K. (2004). New roof of the Olympic stadium in Berlin, *TUT Textiles a Usages Techniques*, 3(53), 10.

Kynol Europa GmbH. (2019). Kynol novoloid fibers, www.kynol.de/, accessed on 6.6.2019.

Liu Y.L., Hsiue G.H., Lan C.W., Kuo J.K., Jeng R.J., and Chiu Y.S. (1997). Synthesis, thermal properties, and flame retardancy of phosphorus containing polyimides, *Journal of Applied Polymer Science*, 63(7), 875–882.

Maity S. and Singha K. (2012). Melamine fiber–Synthesis, features and applications, *Chemical Fibers International*, 62(4), 183–186.

Mallakpour S. and Khadema E. (2015). Recent development in the synthesis of polymer nanocomposites based on nano-alumina, *Progress in Polymer Science*, 51, 74–93.

Mao Y.H., Guan Y., and Zheng Q.K. (2013). Carrier dyeing of polyphenylene sulphide fabric with disperse dye, *Co loration Technology*, 129(1), 39–48.

Mousa A.A., Youssef Y.A., Mohamed W.S., Farouk R., Giebel E., and Buchmeiser M.R. (2018). Organoclays assisted vat and disperse dyeing of poly(ethylene terephthalate) nanocomposite fabrics via melt spinning, *Coloration Technology*, 134(2), 126–134.

Omnexus. (2019). Polyphenylene Sulfide (PPS): A comprehensive guide on high heat plastic, https://omnexus.specialchem.com/, accessed on 6.6.19.

Oulton D.P. (1995). Fire-retardant textiles, in: C.M. Carr, Ed., *Chemistry of the Textiles Industry*. Springer-Verlag, New York, pp. 103–124.

Parker D., Bussink J., van de Grampel H.T., Wheatley G.W., Dorf E.-U., Ostlinning E., Reinking K., Schubert F., Jünger O., and Wagener R. (2002). Polymers, high-temperature, in: *Ullmann's Encyclopedia of Industrial Chemistry*. Wiley-VCH, Weinheim. DOI: 10.1002/14356007.a21_449.

Pervin S., Prabu A.A., and Kim K.J. (2014). Dyeing behavior of chemically modified poly(1,4-phenylene sulfide) fiber towards disperse, anionic, and cationic dyes. *Fibers and Polymers*, 15, 1168–1174. DOI: 10.1007/s12221-014-1168-x.

Pipe and Drape Online. (2019). Inherently flame retardant, www.pipeanddrapeonline.com/, accessed on 25.6.2019.

Purser D.A. (1992). Recent developments in understanding the toxicity of PTFE thermal decomposition products, *Fire Mat*, 16, 67–75.

Ran S., Burger C., Fang D., Cookson D., Yabuki K., Teramoto Y., Cunniff P.M., Viccaro P. J., Hsiao B.S., and Chu B. (2011). Structure formation in high-performance PBO fibers, Science Highlights, 147–152, National synchrotron light source, activity report.

Rogers M.E., Brink M.H., McGrath J.E., and Brenan A. (1993). Semicrystalline and amorphous fluorine-containing polyimides, *Polymer*, 34(4), 849–855.

Saf A.O., Akin I., Zor E., and Bingol H. (2015). Preparation of a novel PSf membrane containing rGO/PTh and its physical properties and membrane performance, *RSC Advances*, 5(53), 42422–42429, DOI: 10.1039/C5RA06371J.

Samanta K.K., Basak S., and Chattopadhyay S.K. (2015). *Sustainable Flame-Retardant Finishing of Textiles Advancement in Technology From: Handbook of Sustainable Apparel Production*, CRC Press, www.routledgehandbooks.com/doi/10.1201/b18428-5, accessed on 24.5.2019.

Singh M.K. (2015). Inherent flame retardant polyester fiber using organophosporous compounds as dope additives, *IOSR Journal of Polymer and Textile Engineering (IOSR-JPTE)*, 2(2), 39–47. DOI: 10.9790/019X-0223947. e-ISSN: 2348-019X, p-ISSN: 2348-0181.

Tan B., Tchatchoua C.N., Dong L., and McGrath J.E. (1998). Synthesis and characterization of arylene ether imide reactive oligomers and polymers containing diarylalkylphosphine oxide groups, *Polymers for Advanced Technologies*, 9(1), 84–93.

Wakelyn P.J. (2008). Environmentally friendly flame resistant textiles, in: *Advances in Fire Retardant Materials*, A.R. Horrocks and S.C. Anand, Eds.

Wang S.-X., Zhao H.-B., Rao W.-H., Huang S.-C., Wang T., Wang L., and Wang Y.-Z. (2018). Inherently flame-retardant rigid polyurethane foams with excellent thermal insulation and mechanical properties, *Polymer*, 153, 616–625.

Weil E.D. and Levchik S.V. (2009). *Flame Retardant Plastics and Textiles*. Hanser, Munich, p. 197.

Wescott J.M., Yoon T.H., Rodrigues D., Kiefer L.A., Wilkes G.L., and McGrath J.E. (1994). Synthesis and characterization of triphenylphosphine oxide-containing poly (aryl imide)-poly(dimethyl siloxane) randomly segmented copolymers, *Journal of Macromolecular Science Part A Pure and Applied Chemistry*, A31(8), 1071–1085.

Yuan X.-Y., Wang D.-Y., Chen L., and Wang Y.-Z. (2011). Inherent flame retardation of bio-based poly(lactic acid) by incorporating phosphorus linked pendent group into the backbone, *Polymer Degradation and Stability*, 96(9), 1669–1675. DOI: 10.1016/j.polymdegradstab.2011.06.012.

Zhang S. and Horrocks A.R. (2003). *Progress in Polymer Science*, 28(11), 1517.

4 Flame Retardants

4.1 INTRODUCTION

The worldwide consumption of flame retardants amounts to more than 2.25 million tons per year. Aluminum hydroxide is the largest single flame retardant at 38% share. The second largest categories consumed are the brominated and chlorinated flame retardant products, which are commonly used together with the synergist antimony trioxide, in total 31%. Organophosphorus and other flame retardants, such as inorganic phosphorus compounds, nitrogen, and zinc based flame retardants make up the remainder at 31%. Over the last decade there has been a trend toward substituting legacy halogenated flame retardants with more sustainable nonhalogenated products.

According to a 2017 market study of IHS Consulting, the consumption of flame retardants had grown substantially in the past 4 years, notably in electronics, and will continue to grow at a global annualized rate of 3.1% upto 2021. Flame retardants are mainly consumed by the plastics/resin industry. Textiles and rubber products account for most of the rest. Asia consumed the largest volume of flame retardants in 2016 with a 50% share, with China being the largest single consumer at 26% (IHS, 2019).

Globally, flame retardant fabrics are mostly relevant to the defense forces, law enforcement agencies, firefighting services and transport. They are also growing in popularity for home and institutional furnishings. Industrial protective clothing accounted for about 53% of global demand for flame retardant fabric in 2013.

Flame retardants are a diverse class of compounds used primarily as additives in materials. These additives reduce the chance of ignition of these materials and/or decrease the rate of combustion, thereby leading to greater safety.

The term flame retardant (FR) fabric is used to describe fabrics which do not support combustion and are self-extinguishing. In case of accidental fire, this type of fabrics do not contribute to the spreading of flame. Other terms such as flame-resistant, flameproof, and fireproof are often meaningless or misleading. Nearly all fabrics are combustible to some degree. The rate of burning ranges from that of nitrocellulose, which burns so rapidly that it produces explosion similar to the extent produced by asbestos, which is virtually unaffected by fire.

Flame retardant finishes are chemicals that are added to combustible materials to render them resistant to ignition. They are designed to minimize the risk of a fire in case of contact with small heat sources such as cigarettes, candles or electrical faults. On ignition of flame-retarded material, the flame retardant slows down combustion and preventsfire from spreading to other items.

The flame retardant must have a decomposition temperature that is sufficiently lower than that of the polymer in order to ensure its efficiency. For example, on polypropylene: a flame retardant based on aliphatic bromines shows better efficiency than that based on aromatic bromines, which are thermally more stable.

4.2 CLASSIFICATION OF FRs

Several types of compounds and polymers are used as FR treatments of textile materials, namely inorganic acids, acid salts and hydrates, organo-phosphorus and organo-bromine compounds, and antimony salts/halogen systems. They may be classified into three categories:

1. Primary FR based on phosphorous and/or halogen – the phosphorous derivatives act usually in solid or condensed phase, while halogen (chlorine or bromine) are active in the gaseous or vapor phase.
2. FR using synergists such as nitrogen and antimony for phosphorous and halogen-based FRs respectively. Synergists themselves are not flame retardants. FRs exhibiting P/N and/or Sb/X synergisms are durable in nature.
3. Adjunctive or physical FRs include alumina trihydrate, boron compounds, silicates, and carbonates. Their activity is mainly physical, although recently some evidence of a chemical effect has been cited. They are nondurable and used only if durability to laundering is not important.

From an application point of view, FR systems may be classified as follows (Lewin and Sello, 1984):

1. Nonreactive systems
   a. Topical deposition
   b. Two-stage precipitation
   c. Exhaustion from bath
   d. Incorporation in fiber before spinning.

2. Reactive systems
   a. The reaction of FR with fiber
   b. Graft copolymerization with fiber
   c. In situ polymerization of FR compounds
   d. Copolymerization of FR compounds with fiber-forming monomers
   e. Inherently FR fibers.

Flame retardants (FRs) are mainly of two types:

1. Additive flame retardants, and
2. Reactive flame retardants.

Mineral flame retardants are typically additive while organo-halogen and organo-phosphorus compounds can be either reactive or additive.

FRs may be classified in two ways:

1. On the basis of durability, or
2. On the basis of chemical constitution.

On the basis of durability, FRs may be classified as:

1. Nondurable, or
2. Durable.

FRs are usually classified according to their durability to laundering. Indeed, a nondurable FR is washed off immediately when soaked in water, but may resist dry-cleaning. Conversely, semidurable FRs are able to resist water-soaking and possibly a few washes, while durable FRs endure some 50 or 100 washing cycles.

4.3   NONDURABLE FRs

Nondurable FRs are generally water-insoluble inorganic compounds that are easily removed by water, rain, or perspiration.

Some nondurable FRs are described next:

4.3.1   BORIC ACID AND BORAX

Boron-containing compounds act by the stepwise release of water and by the formation of a glassy coating protecting the surface. The mixtures of boric acid and borax are used as FR for cellulose.

4.3.2   PHOSPHORIC ACID DERIVATIVES

Ammonium salts decompose thermally into phosphoric acid by the loss of ammonia (Equation 4.1); the calcium salts do not. Presumably, the calcium salts are not volatile and buffer the acidity of phosphoric acid so the generation of char is diminished.

$$(NH_4)_2HPO_4 \rightarrow 2NH_3 + H_3PO_4 \qquad (4.1)$$

The water-insoluble ammonium polyphosphate is an effective flame retardant and is added to coatings and binder systems, for example, for pigment printing.

4.3.3   AMMONIUM SALTS

Ammonium salts of strong acids such as phosphoric acid are very much useful as nondurable FRs for cellulose. Three commercially important products are diammonium phosphate, ammonium sulfamate, and ammonium bromide. These salts liberate strong acids upon heating. Ammonium bromide is applied as ~ 10% solids add-on and is effective in the gas phase (Equation 4.2).

$$NH_4Br \rightarrow NH_3 + HB_r \tag{4.2}$$

Diammonium phosphate and ammonium sulfamate are used at less than 15% solids add-on;

4.3.4 SULFAMIC ACID AND AMMONIUM SULFAMATE

Combinations of these compounds also function as nondurable flame retardants (Equation 4.3).

$$NH_4OSO_2NH_2 \rightarrow 2NH_3 + H_2SO_4 \tag{4.3}$$

4.4 DURABLE FRs

For many end-uses of textiles and other products, fast flame retardancy fast to washing. Durable FR finishes can be obtained by one or more of the following chemical compounds.

1. Minerals
2. Halogens (Bromine and Chlorine)
3. Phosphorus
4. Nitrogen
5. Intumescent
6. Nanomaterial-based FRs e.g. nanocomposites.
7. Reactive FRs

Flame retardants are, at present, organic and inorganic, halogen-based, and non-halogen. Brominated, nitrogen, and phosphorus-based organic compounds; inorganic compounds, mainly antimony trioxide, magnesium hydroxide, aluminum hydroxide, silicon, and other flame retardant systems are in use. In general, organic FRs have very good affinity for plastics and textiles. The brominated organic FRs are in position of absolute dominance among the flame retardant systems. Many of them are criticized from environmental point of view, but they are difficult to replace by other existing FRs.

4.4.1 MINERAL FRs

Inorganic FRs include metal oxides, zeolites, hydroxides, borates, stannates, inorganic phosphorus compounds (red phosphorus and ammonium polyphosphate), and graphite. They are usually combined with halogen FRs, as well as with phosphorus – and/or nitrogen-based ones. Mineral or inorganic FRs include aluminum hydroxide (ATH), magnesium hydroxide (MDH), Chalk (calcium carbonate), huntite and hydromagnesite, various hydrates, and boron compounds, mostly borates.

They are inexpensive, but require loadings in polymers of about 60% or more. These materials thermally decompose through strongly endothermic reactions. If enough energy is absorbed by these reactions, the pyrolysis temperature of a fiber is not reached and combustion cannot take place. An example of this method is the use of alumina trihydrate and calcium carbonate as fillers in polymers and coatings. Alumina trihydrate in amounts of 40%–60% has been found to be much more effective than the anhydrous alumina Equations 4.4–4.6).

$$Al_2O_3 \cdot 3H_2O \quad \rightarrow \quad Al_2O_3 + 3H_2O$$
$$200\,^{\circ}C$$
$$(4.4)$$

$$2\,Al(OH)_3 \quad \rightarrow \quad 3\,H_2O + Al_2O_3$$
$$+ 1050\,kJ\,/\,kg$$
$$300\,^{\circ}C$$
$$(4.5)$$

$$Mg(OH)_2 \quad \rightarrow \quad H_2O + MgO$$
$$+ 1300\,kJ\,/\,kg$$
$$(4.6)$$

Fine precipitated ATH and MDH (< 2 µm) are used in melt compounding and extrusion of thermoplastics such as cable PVC or polyolefins for cables. For use in cable, ATH and more often MDH are coated with organic materials to improve their compatibility with the polymer. Coarser ground and air separated grades can be used in liquid resin compounding of thermosets for electrical applications, seats, panels and vehicle parts.

Semidurable FRs are obtained using insoluble salts of amphoteric cations and anions – stannates, tungstates, aluminates, borates, and phosphates of Zn, Sn, Al, and easily reducible metallic oxides – Sn, Fe, Ti, Cr, Zr, Ce, Bi, W, As, and Si.

Inorganic flame retardants are metal hydroxides (such as aluminum hydroxide and magnesium hydroxide), ammonium polyphosphate, boron salts, inorganic antimony, tin, zinc, and molybdenum compounds, as well as elemental red phosphorus. Both aluminum hydroxide, also sometimes called aluminum trihydrate (ATH), and magnesium hydroxide are used as halogen-free alternatives to brominated flame retardants, and they also function as smoke suppressants. Inorganic phosphorus compounds are widely used as substitutes to brominated flame retardants. Inorganic flame retardants are added as fillers into the polymer and are considered immobile in contrast to the organic additive flame retardants. The whole group of inorganic flame retardants represents around 50% by volume of the global flame retardant production, mainly as aluminum trihydrate (ATH), which, in terms of volume, is the biggest flame retardant category in use on the market.

4.4.1.1 Aluminum Hydroxide (ATH)

ATH has been used as a flame retardant and smoke suppressant since the 1960s, and it is available in a variety of particle sizes as commercial products.

Flame retardation by ATH has been shown to be partly due to the *heat sink* effect, and partly due to the dilution of combustible gases by the water formed as a result of dehydroxylation. Alumina which is formed as a result of thermal degradation of ATH slightly above 200°C has been shown to form a heat-insulating barrier on the surface that prevents further fire propagation of the matrix material. The major concern with ATH is the required high loading levels in order to obtain equivalent flame retardant properties as by other additives. These loads can be reduced with a correct choice of particle size, surface modification, and proper dispersion in the matrix material (Swaraj, 2001). Furthermore, recently developed coated filler products (e.g., ZHS-coated ATH) offer the possibility of equivalent or better flame retardancy and smoke suppression at significantly reduced incorporation levels.

4.4.1.2   Magnesium Hydroxide

Magnesium hydroxide acts, in general, the same way as ATH, but it thermally decomposes at slightly higher temperatures, around 325°C. Combinations of ATH and magnesium hydroxide function as very efficient smoke suppressants in PVC.

4.4.2   HALOGEN-BASED FLAME RETARDANTS (HFRS)

Halogen-based flame retardants are almost always organic products containing halogen.These FR types are currently of the largest value in the FR marketplace. They are classic and are widely used – there are many halogen-based products. Among halogenated flame retardant additives for plastics, only bromine, chlorine, and fluorine are in use.

Organohalogen class includes organo-chlorines such as chlorendic acid derivatives and chlorinated paraffins, organobromines such as decabromodiphenyl ether (decaBDE), decabromodiphenyl ethane (a replacement for decaBDE); and polymeric brominated compounds such as brominated polystyrenes, brominated carbonate oligomers (BCOs), brominated epoxy oligomers (BEOs), tetrabromophthalic anyhydride, tetrabromobisphenol A (TBBPA), and hexabromocyclododecane (HBCD). Most, but not all, halogenated flame retardants are used in conjunction with a synergist to enhance their efficiency. Antimony trioxide is widely used, but other forms of antimony, such as the pentoxide and sodium antimonate, are also used.

These elements are included in many flame retardant products. The largest halogen product today by far is polyvinyl chloride (PVC) – something most people have heard of but probably do not think of or know as a flame retardant. PVC is widely used, is inherently flame retardant, and is found in automotive interiors, plastic pipe, wiring in homes and businesses, packaging, and a host of other common products.

When they were allowed to be used, brominated FRs were very popular as one of the most important additives in different plastics and resins, electrical equipment, electronics parts, polyurethane foams and synthetic textile fibers.

Commercial organo-halogen flame retardants include aliphatic, alicyclic, and aromatic chlorine, as well as bromine compounds. Aliphatic compounds are the most effective, and the aromatic compounds are the least effective, with the alicyclic compounds in between. They are classic and most widely used because of their low cost and high efficiency.

Halogenated phosphorous FRs works both in condensed phase and in vapor phase. Tris (1,3-dichloro-2-propyl) phosphate (TDCPP) and its blends are used as flame retardants in the manufacture of polyurethane foam.

4.4.3 SUBSTITUTE FOR HALOGEN-BASED FRs

Halogen-based FRs are very popular globally because of their low cost and high efficiency. Unfortunately, due to toxicity, the majority of halogen-based FRs are banned globally and intensive searches for halogen-free FRs (HFFRs) have started a fresh.

Since the 1990s, flame retardants have begun to raise environmental concerns, because some brominated flame retardants (e.g., polybrominated diphenylethers, PBDE) were found to form halogenated dioxins and furans in uncontrolled combustion, and also because flame retardants were found in various environmental compartments and biota, including humans. The producers of flame retardants have responded to these concerns, and are developing more environmentally compatible products, mainly nonhalogenated alternatives.

The challenges of finding substitutes for banned halogen-containing flame-retardants include:

- Lower flame retardancy: Only 10 wt% of BFRs is equivalent to about 30%–50% by wt of inorganic FRs.
- High cost: Inorganic FRs are low-cost but require high loading. New HFFRs, on the other hand, are mostly costlier. The cost ratios (£) of BF-FR with phosphorous-based and non-phosphorous-based HFFR are about 1:6 and 1:2, respectively.
- Higher thermal stability: Many HFFRs decompose at a higher temperature (about 400°C) than their halogenated counterparts (about 330°C).
- Melt-dripping: Melt dripping occurs with most polymers, e.g., polyethylene, polypropylene, PET, and ABS. Flammable drips act as secondary ignition sources. FRs should make drips nonflammable.
- Deterioration in mechanical properties: The deterioration of mechanical properties increase proportionately with the amount of FR. Therefore, the deterioration is greater with less efficient FRs (David Suzuki Foundation, 2016).

4.4.4 PHOSPHOROUS BASED FRs (PFRs)

PERs are easy to use, are compatible with many textile chemicals, and are favored from both environmental and toxicological perspectives. Phosphorus flame retardants (PFRs) – (organic and inorganic– are more environmentally friendly, have good thermal stability, and have superior performance, resulting from the synergistic effect of P–N. They do not tend to form toxic gases because phosphorus is mostly locked into the char. Phosphorous FRs may be reactive (chemically bound) or additive (physical mixing) (Schartel, 2010).

A large number of FRs are based on phosphorous – both inorganic and organic derivatives. Inorganic phosphorous derivatives, mostly nondurable or semidurable, entail primarily phosphoric acid and its ammonium salts, e.g., diammonium phosphate (DAP); $(NH_4)_2HPO_4$) are highly effective flame retardants and smoldering inhibitors.

The organo-phosphorus class includes organophosphates such as triphenyl phosphate (TPP), resorcinol bis (diphenylphosphate) (RDP), bisphenol A diphenyl phosphate (BADP), and tricresyl phosphate (TCP); phosphonates such as dimethylmethylphosphonate (DMMP); and phosphinates such as aluminum. In one important class of flame retardants, compounds contain both phosphorus and a halogen. Such compounds include tris (2,3-dibromopropyl) phosphate (brominated tris), and chlorinated organophosphates such as TRIS (1,3-dichloro-2-propyl)phosphate (chlorinated TRIS or TDCPP) and tetrakis (2-chlorethyl) dichloroisopentyldiphosphate. These are most effective when they are in the highest oxidation state (P^{+5}). Various phosphazene derivatives can be incorporated in the spinning bath of rayon filaments.

4.4.5   Nitrogen FRs(NFRs)

Nitrogen compounds are a small but rapidly growing group of flame retardants (FR) that are the focus of public interest concerning environmentally friendly flame retardants. Today, their main applications are melamine for polyurethane flexible foams, melamine cyanurate in nylons, melamine phosphates in polyolefin, melamine and melamine phosphates or dicyandiamide in intumescent paints, guanidine phosphates for textiles, and guanidine sulfamate for wallpapers. The main common advantages of nitrogen FRs (NFRs) are their low toxicity, their solid state and, in case of fire, their absence of dioxin and halogen acids as well as their low evolution of smoke. Their efficiency lies between that of halogen compounds and that of aluminum trihydrate and magnesium hydroxide. The metallic hydroxides split off water and are environmentally friendly, but their low activity requires high concentrations which change the mechanical properties of the matrix to which they are applied. In contrast to many halogen compounds, flame retardants based on nitrogen do not interfere with the set of stabilizers added to every plastic material. Recyclability has become important because many plastics are recycled. Flame-retarded materials based on nitrogen compounds are suitable for recycling because the nitrogen flame retardants have high decomposition temperatures. In the field of cable jacketing, less corrosive gases evolve during combustion and do not damage electrical installations. With regard to waste disposal, NFRs are comparable with fertilizers because they possess the same elements of importance, namely, nitrogen and phosphorous. In comparison with metallic hydroxides they are more efficient and deteriorate the mechanical properties of the plastic material less.

NFRs act by several mechanisms. In the condensed phase, melamine gets cross-linked, promoting char formation. Gaseous ammonia is released as a by-product. The nitrogen enhances the attachment of the phosphorus to the polymer. Molecular nitrogen may also release in the gas phase, diluting the volatile polymer decomposition products.

Melamine is mainly used in polyurethane foams, whereas melamine cyanurate is used in polyamides or in polypropylene intumescent formulations in conjunction with ammonium polyphosphate. The phosphate, polyphosphates, and pyrophosphates of melamine contain both nitrogen and phosphorus, and are used in polyamides. In some specific formulations, triazines, isocyanurates, urea, guanidine, and cyanuric acid derivatives are used as reactive compounds. (Pinfa, 2018).

4.4.6 INTUMESCENT FRS

The French verb tumere means "to swell". The Latin equivalent tumescere can be translated as "to swell up". Therefore tumid or tumescent means swollen or bulging, and the process of getting to a swollen state is intumescence. In flame retardant terms, exposure to heat initiates a series of chemical and physical processes, leading to a tumescent condition. This state is characterized by fire-resistant insulating foam. The foam serves to isolate heat and oxygen from the fuel source, extinguishing the fire. Intumescent char is a carbonaceous char formed by a large number of small bubbles that act as an insulating layer to protect the substrate (Figure 4.1).

The last decade has witnessed the development of so-called intumescent FRs (IFRs), intended primarily for the protection of wood, plastics, and metals. These are applied to the material surface and expand under the influence of high temperature, forming an insulatory fire-resistant layer at the surface of the material, protecting it from further pyrolysis and burning. The char residue has a characteristic foamy appearance. These systems act in the condensed phase and involve some highly complex, mutually dependent components. Phosphorous compounds (phosphoric acid and their derivatives, polyphosphate) in the intumescent system cause the polymer to phosphorilate (with a C-O bond), stimulating its dehydration and carbonization. Liquid polymer foams during carbonization as it releases nonflammable gases (NO, NO_2) generated by the decomposition of nitrous compounds (e.g., melamine, urea, dicyandiamine) (Drobny, 2007).

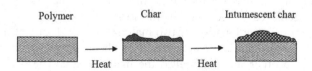

FIGURE 4.1 Schematic diagram of working principle of Intumescent FRs.

IFRs consist of three components, namely:

1. Acid source (e.g., inorganic acids, organic acids and salts, phosphates of amine or amide, organo-phosphorus compounds),
2. Carbonization agent (e.g., starch, dextrins, pentaerythritol, phenol-formaldehyde resins, methylol melamine), and
3. Blowing agents (e.g., urea, urea-formaldehyde resins, dicyandiamide, melamine).

Intumescence is the result of a combination of charring and foaming at the surface of the burning polymer, which protects the underlying material from the action of the heat or flame. IFRs are often used for applications requiring high levels of flame retardancy. They are highly efficient and low toxic. They provide very robust fire safety and flame resistance performance. The carbon agent forms multicellular charred layers, and the char may be soft or hard (Rakotomalala et al., 2010).

- Soft char IFRs are composed of a carbon source pentaerythritol (PER), acid source (ammonium polyphosphate) and a gas-blowing additive (melamine).
- Harder char IFRs are composed of sodium silicates and graphite. These are suitable for use in plastic pipe fire-stops as well as exterior steel fireproofing.

PER is quite costly. A possible substitute is green carbon agent is chitosan (CS), obtained by the alkaline deacetylation of abundantly naturally occurring chitin. A good synergistic effect observed when chitosan/urea compound nano–based phosphonic acid melamine salt (HUMCS), was added to an IFR system for poly-propylene (PP) (Rakotomalala et al., 2010).

4.4.7 Nanocomposite FRs

Polymer nanocomposites are a new class of FR additives that work only in the condensed phase. Organically treated layered silicates (clays), carbon nano-tubes/nanofibers, or other submicron particles at low loadings (1–10 wt%) are used for their manufacture.

Clay exists naturally in the form of platelets around a nanometer thick, stacked up to form particles on the order of a few microns. Most of the bene-ficial properties of clay as a filler material are only realized when the particles are dispersed and exfoliated to separate the individual nanoscale layers – this is most often achieved by treatment with ultrasound.

Composites are a combination of two materials in which one of the materials, called the reinforcing phase, is in the form of fibers, sheets, or particles, and is embedded in the other material, called the matrix phase. The ductile or tough matrix material is reinforced into strong and low-density reinforcing materials. In the finished form, these two phases are sometimes not visibly distinguishable.

There is, however, a difference from an engineered structure containing more than one material, such as alloys, in the sense that although the composite material may seem to be homogeneous, there are still different constituent phases. The term *composite* originated in engineering science when two or more materials were combined together in order to rectify some shortcomings of a particularly useful component. A composite comprises a large number of strong stiff fibers called the reinforcement, embedded in a continuous phase of a second material known as the matrix. The resulting product has the advantage of being lower in weight, greater in strength, and higher in stiffness than individual constituents.

Nanocomposite is a multiphase solid material in which one of the phases has one, two, or three dimensions of less than 100 nanometers, or structures with nanoscale repeat distances between the different phases that make up the material (Wang et al., 2010; Shen et al., 2013; Zhang et al., 2013). A nanocomposite is a matrix to which nanoparticles have been added to improve a particular property of the material. Nanocomposites or polymers and exfoliated clays have been shown to exhibit excellent flame retardant properties. The spread of fire through the material is slowed, and dripping is reduced, which helps prevent the fire from spreading to other objects nearby. The nanoclay also promotes formation of chars, which are thermally insulating and act as a barrier to flame propagation.

The most attractive property of clays, however, is that they are completely natural materials. This makes preprocessing much more straightforward, and there are very few environmental issues associated with disposal.

Nanocomposites have been gaining increasing attention since the late 1990s as potential new flame retardants. Some nanocomposites are made by polymer-layered aluminosilicate clay minerals like montmorillonite, composed of layers with gaps (gallery spaces) in between. These silicates have the ability to incorporate polymers. Research with nanocomposites has focused on plastics such as polymethyl-methacrylate (PMMA), polypropylene, polystyrene, and polyamides. Nanocomposites particularly prevent dripping and promote char formation. Therefore, nanocomposites are used as synergists in some polymer – flame retardant combinations. However, they require special processing and, for the time being, are not considered to become viable stand-alone flame retardants (www.pinfa.eu, accessed on 14.3.18).

If a structure is multilayered, it burns in distinct stages as the heat penetrates subsequent layers and degradation products move to the burning zone through the fibrous layers. In general, the thickness of a structure can affect the surface flammability characteristics up to a certain limiting value, after which the full depth of the material is not involved in the early stages of burning and the material is said to be "thermally thick" (Mikkola and Wichman, 1989).

The structures and properties of the composite materials are greatly influenced by the component phase morphologies and interfacial properties. Nanocomposites are based on the same principle and are formed when phase mixing occurs at a nanometer-dimensional scale. As a result, nanocomposites show superior properties over their micro counterparts or conventionally filled polymers.

The flame retardant polymer nanocomposites have gained popularity and have made tremendous progress (Morgan and Wilkie, 2007). The detailed understanding of the char-forming mechanism is the key to develop a new type of flame retardant. The combustion behavior of polymer nanocomposites depends on two-fold mechanisms brought by nanofillers, namely:

- A physical barrier effect, and
- A chemical charring catalytic action occurring in the condensed phase.

Flame retardant nanofillers such as clays, metal oxide nanoparticles, and carbon nanotubes, when added to the polymers, typically operate in the condensed phase by forming a surface layer that limits the supply of the polymer to the fire (Wang, 2013).

By themselves, polymer nanocomposites greatly lower the base flammability of a material, making it easier to flame-retard the polymer containing a nanocomposite structure. They are effective when combined with just about all types of FR additives. They work best when combined with other FR additives. In effect, nanocomposites are a class of nearly universal FR synergists; exceptions do exist, of course.

4.4.8 Reactive FRs

Reactive flame retardants, that is, those that are covalently attached to polymer chains, offer several advantages over those that are merely additives. For example, reactive flame retardants are inherently immobile within the polymer matrix, and therefore are not susceptible to lose during service through migration to the polymer surface (blooming) or solvent leaching. Reactive flame retardants incorporated at the time of the synthesis of the polymer can also be homogeneously dispersed throughout the polymer at the molecular level. Hence, to achieve the desired level of flame retardance, reactive FRs may require lesser quantities compared to comparable additives. The lower levels of incorporation of flame retardant groups may affect the overall properties of the polymer (chemical, physical, and mechanical) to a lesser extent when compared with those of the non-flame-retarded counterpart. Moreover, reactive flame retardants are prevented from forming a separate phase within the polymer matrix; this is particularly important in the use of polymers as fibers in which a heterogeneous phase structure is likely to bring with it problems during fiber-spinning and reductions in the modulus of the resultant fibres.

However, the reactive incorporation of flame-retarding groups can also bring problems, not least of which is the relative difficulty and expense of reactively modifying polymers for which commercially well-established methods for manufacturing the unmodified variants already exist.

Furthermore, the extensive reactive modification of a partly crystalline polymer is likely to lead to a significant loss of crystallinity, whereas if an additive is introduced to a partly crystalline polymer, it would most probably end up in the amorphous phase and have little impact upon crystallinity (although it may

plasticize the amorphous regions). Moreover, in general, the reactive modification of chain reaction polymers is less readily accomplished than for step
reaction polymers, unless the reactive modification is applied after the manufacture of the primary chain, for example, through a postpolymerization grafting
reaction. Grafting reactions may also be useful for covalently attaching flame-
retardant groups to the surfaces of polymer-based plastics mouldings, films, and
fibers, where they will be particularly effective at the point of first impingement
of any flame.

Crosslinking by introducing N-hydroxymethyl (3-dimethyl phosphono) pro-
pionamide and trimethylolmelamine (TMM) or hexamethylol melamine (HMM)
onto silk fabric can improve laundering durability to~ 30–50 hand wash cycles,
and the LOI of treated silk fabric were above 30%. The treated silk fabrics would
become stiff and yellow and had the problem of formaldehyde release, however
(Guan and Chen, 2006).

Developing a durable flame retardancy and formaldehyde-free process for
silk fabric is still a challenge. Guan and Chen (2010) successfully synthesized
a series of phosphorus-based flame retardants and applied them to silk fabrics
to produce durable and formaldehyde free effects.

Three kinds of silk fabrics with different weaving styles for apparel uses were
treated with a vinyl phosphorus monomer dimethyl-2-(methacryloyloxyethyl)
phosphate (DMMEP) by a graft copolymerization technique using potassium
persulfate as an initiator. The treated silk fabric can pass the vertical flammability test and can self-extinguish once ignited by candlelike fire. The permeability of silk fabrics dropped slightly. The flame retardant silk fabrics became
a little bit stiff, which was helpful for flat seam appearance, and were more difficult to distort in fabric laying up panel cutting, and making garments. The
formability and sizing stability during sewing, wearing, and pressing process
was improved. The effect of flame retardancy treatment on the drape property
of silk fabrics depends on their weaving styles. The seam strength decreased
for different weaving styles after being treated with flame retardants. Adhering
to interlining improved seam strength to levels that were even better than that
of the original silk fabrics (Guan et al., 2013).

The most obvious way of reactively flame-retarding a chain-reaction polymer
is by incorporating a P-containing comonomer during polymerization. However,
there have been no reports, to date, of significant commercial progress on this
front, although a variety of chain reaction polymers have been reactively modified with P-containing groups on the laboratory scale and have been shown to
be significantly flame-retarded when compared with their unmodified counterparts (Joseph and Ebdon, 2010).

4.4.9   HYBRID ORGANIC–INORGANIC FRs

Hybrid materials are composites consisting of two constituents at the nanometer
or molecular level. Commonly, one of these compounds is inorganic and the
other one organic in nature. Thus, they differ from traditional composites, where
the constituents are at the macroscopic (micrometer to millimeter) level. Mixing

at the microscopic scale leads to a more homogeneous material that either shows characteristics in between the two original phases, or even new properties.

The term *nanocomposite* is used if the combination of organic and inorganic structural units yields a material with composite properties. This means that the original properties of the separate organic and inorganic components are still present in the composite and are unchanged by mixing these materials. However, if a new property emerges from the intimate mixture, then the material becomes a hybrid.

Advantages of hybrid materials over traditional composites in organic polymer matrices are:

- Inorganic clusters or nanoparticles with specific optical, electronic, or magnetic properties can be incorporated in organic polymer matrices.
- Contrary to pure solid state inorganic materials that often require a high temperature treatment for their processing, hybrid materials show a more polymerlike handling, either because of their large organic content, or because of the formation of cross-linked inorganic networks from small molecular precursors, just as in polymerization reactions.
- Light scattering in homogeneous hybrid material can be avoided and therefore optical transparency of the resulting hybrid materials and nanocomposites can be achieved.

Organic–inorganic hybrids offer the possibility of integrating organic and inorganic characteristics within a single "molecular composite" (Day and Ledsham, 1982). Organic materials are generally lightweight, mechanically flexible, chemically versatile (e.g., molecules of different length, width, conformation), and can provide useful optical and electrical properties. Inorganic materials offer mechanical robustness, good thermal stability, high electrical and thermal conductivity, and interesting magnetic or ferroelectric transitions. Nature provides many examples of organic–inorganic composites that exhibit superior properties compared with those of the constituent organic or inorganic components (e.g., seashell, bone, tooth enamel) (Day and Ledsham, 1982).

Organic–inorganic hybrid materials do not only represent a creative alternative to design new materials and compounds for academic research; their improved or unusual features also allow the development of innovative industrial applications. Nowadays, most of the hybrid materials that have already entered the market are synthesized and processed by using conventional soft chemistry based routes developed in the 1980s. These processes are based on:

- The copolymerization of functional organosilanes, macromonomers, and metal alkoxides,
- The encapsulation of organic components within sol–gel derived silica or metallic oxides,
- The organic functionalization of nanofillers, nanoclays, or other compounds with lamellar structures, etc.

The chemical strategies [self-assembly, nanobuilding block approaches, hybrid Metal Organic Frameworks (MOF), integrative synthesis, coupled processes, bioinspired strategies, etc.] offered nowadays by academic research allow, through an intelligent tuned coding, the development of a new vectoral chemistry that is able to direct the assembling of a large variety of structurally well-defined nanoobjects into complex hybrid architectures hierarchically organised in terms of structure and functions (Sanchez et al., 2005).

4.4.10 SILICONE FRs

Silicone is eco-friendly, widely available in nature. It is easy to prepare flame retardants from silicone. The silica ash layer can also prevent oxygen from reaching the matrix. Like other silicone rubber products, the flame retardant series also possess high temperature resistance, chemical resistance, ozone and UV resistance, and easy fabrication. This material is also nontoxic and fungus resistant. Silicone is so widely used due a variety of useful properties, including flexibility, adhesion, insulation, and low toxicity. However, one of the most important characteristics of silicone is its heat resistance, allowing silicone products to maintain their properties when exposed to both high and low temperatures.

The dominant polymer in the silicone industry is polydimethylsiloxane (PDMS). The main structural elements of polysiloxanes have direct or indirect influence on their stability at elevated temperatures, including: inherent strength of the siloxane (Si–O) bond, pronounced flexibility of the – [Si–O]x – chain segments, and entropically higher stability of low-molecular-weight cyclic siloxanes compared to their high molecular weight linear counterparts against thermal degradation (Dvornic et al., 2000). Silicones have comparatively low-heat release rates (HRR), minimal sensitivity to external heat flux and low yields of carbon monoxide release Silicones also show a slow burning rate without a flaming drip and when pure, no emissions of toxic smokes. Based on these fire properties, silicones offer significant advantages for flame retardant applications. It is not surprising that due to their properties against flames, PDMS has been put in the top list of polymers for applications at high temperature such as in electrical wires and cables. Unlike organic polymers, silicones exposed to elevated temperatures under oxygen leave behind an inorganic silica residue. The shielding effects provide some of the fundamentals for the development of silicone-based fire retardants. Silica residue serves as an *insulating blanket*, which acts as a mass transport barrier delaying the volatilization of decomposition products. Therefore, it reduces the number of volatiles available for burning in the gas phase, and thus, the amount of heat that feeds back to the polymer surface. The silica residue also serves to insulate the underlying polymer surface from incoming external heat flux.

Silicones also show a slow burning rate without a flaming drip and when pure, no emissions of toxic smokes. Based on these fire properties, silicones offer significant advantages for flame retardant applications. It is not surprising that due to their properties against flames, PDMS has been put in the top list of polymers for applications at high temperature such as in electrical wires and cables. Unlike organic polymers, silicones exposed to elevated temperatures under oxygen leave behind an inorganic silica residue. The shielding effects provide some of the fundamentals for the development of silicone-based fire retardants. Silica residue serves as an *insulating blanket*, which acts as a mass transport barrier delaying the volatilization of decomposition products. Therefore, it reduces the number of volatiles available for burning in the gas phase, and thus, the amount of heat that feeds back to the polymer surface. The silica residue also serves to insulate the underlying polymer surface from incoming external heat flux.

Various methods have therefore been proposed for forming inorganic protective layers on the surface of burning polymers, since inorganic filler particles act through a dilution effect and reduce heat feedback due to large amount of ash (Marosi et al., 2002). The physical network formed by such additives of high surface area (e.g., aerosol) in the polymer melt also reduces dripping but, on the other hand, significantly restricts the processability of such systems. Similar problems occur when porous fillers such as zeolites are used in high concentrations, especially if the average pore size (>10 nm) allows the polymer chains to penetrate (Wen and Mark, 1994).

Silicones are greatly acknowledged for their better thermal and thermo-oxidative stabilities compared to most carbon-based polymers. This acute resistance against flame has put PDMS in the top list of polymers for applications at high temperature where flame retarding behavior is required. As a flame retardant, PDMS offers outstanding advantages such as being a halogen-free flame retardant with very low or almost zero emission of toxic smokes, and thus being considered as among the "environmentally friendly" category of additives.

Silicones can be used as flame retardant agents through direct blending within the polymer matrix, incorporation into porous fillers, or by synthesizing block/graft copolymers including silicone segments.

Fabrics with high flame retardancy have been extensively applied for numerous applications including textile, garments, automobile industries, pants, shirts, suits bed sheets, and indoor decorations.

Fine white powder, Laponite, is a synthetic clay which swells to produce a clear, colorless thixotropic gel when dispersed in water. It is used in the conservation of stone, metals, organic materials, ceramics, and paintings. Coatings consisting of ammonium hexametaphosphate (NH_4-HMP), laponite (LAP), and hexadecyltrimethoxysilane were synthesized through sol–gel method; they were then employed on the cotton fabrics to enhance their hydrophobicity and flame retardancy. The influences of LAP concentration on fire-retardancy of the samples were evaluated. The combustion behavior, morphological structures, thermal stability, and hydrophobic properties of the cotton fabrics were studied. Results indicated the excellent flame retardant property of the treated cotton fabrics as they immediately extinguished upon removal of the flame source. The limiting oxygen index of the treated cotton was enhanced to 29% in comparison to that of the pure one (19.5%). The findings also indicated that a higher concentration of LAP is useful for improving flame retardancy of the coated substrate (Nabipour et al., 2020).

4.5 MECHANISM OF FLAME RETARDANCY

Thermal degradation of cotton and other cellulosic materials, at a temperature above 300°C produces gaseous, liquid, tarry, and solid products. The flammable components in gaseous products burn to produce additional heat to convert the liquid and tarry products into flammable vapors that further propagate the flaming combustion. The process continues till a carbonized residue is left. As the residue does not support flaming, the first phase, i.e., flaming combustion comes to an end. The very first step in heat degradation of cotton (cellulose) is the formation of water and 1,2-anhydroglucose (Equation 4.7) which rearranges to 1,6-anhydroglucose or levoglucosan (Equation 4.8) as heating continues.

$$(C_6H_{10}O_5)_n \quad \rightarrow \quad n\,(1,2-\text{anhydroglucose}) + n\,H_2O \qquad (4.7)$$
$$\text{(Cellulose)} \quad \text{(Heat)}$$

$$[4.8]$$

(1, 2- anhydroglucose)   (1, 6- anhydroglucose)

As the temperature soars this compound breaks up into various products (Equation 4.9) (Nair, 2001).

$$1,6-\text{anhydroglucose} \quad \rightarrow \quad \text{Gases} + \text{Liquids} + \text{Tar} + \text{Char}$$
$$\text{(Heat)} \qquad\qquad\qquad\qquad\qquad\qquad (4.9)$$

The combustible gases burn and generate more heat while the liquids and tar break up releasing further gases and char.

The degree of protection during combustion provided by a char depends on both its chemical and physical structure. Whereas pure graphite is highly stable to heat and oxygen, chars from polymer combustion do not have this property. Although chars are richer in carbon than the original polymer, they are rarely all carbon. The fire retardant property of char depends on its physical structure.

The char may be:

1. Ideal or
2. Nonideal.

The ideal char for fire retardant properties is an intact structure of closed cells containing pockets of gas. For this to happen the bubbles of gas must become frozen into the expanding and thickening polymer melt, which ultimately solidifies to produce the honeycombed structure. This prevents the flow of volatile liquids or vapors into the flame and provides sufficient thermal gradient to keep the remaining polymer or polymer melt below its decomposition temperature.

The nonideal or poor char structure does not contain closed cells, but contains channels or fissures through which gaseous decomposition products or polymer melt can escape. Of these two effects, the more important is the movement of liquid products, which can be drawn by capillary action into hotter regions where they are more likely to decompose. This negates any heat insulating effects that the char may have on the virgin polymer beneath.

Factors which influence the type of char formed are still not properly understood but include melt viscosity, the surface tension of the melt, gas interface, and the kinetics of gasification and polymer cross-linking (Price et al., 2001).

The first step, i.e., formation of 1,6-anhydroglucose (levoglucosan), is significant from flammability point of view. Levoglucosan, a precursor of flammable volatiles, is the main pyrolytic product of cellulose. It is cyclic-acetal-created when the α-1, 4-glucosidic linkage is split and a molecule of water is lost (Equation 4.7).

Some other volatile pyrolytic products are shown in Figure 4.2, namely:

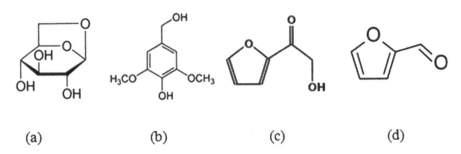

(a)   (b)   (c)   (d)

FIGURE 4.2   Various volatile pyrolytic products of cellulose

Figure 2(a): 1,6-anhydro-β–glycofuranose
Figure 2(b): 5-hydroxymethyl-2-furfural
Figure 2(c): 2-furylhydroxymethyl ketone
Figure 2(d): furfural

Levoglucosan and the aforementioned volatile substances mix with oxygen to act as fuel and to propagate combustion process.

Theoretically, cellulose can be made to decompose into carbon and water when there is no flammable gas (Equation 4.10).

$$(C_6H_{10}O_2)_n \rightarrow 6nC + 5nH_2O$$
$$\text{(Heat)}$$

(4.10)

The generation of carbon and water are ideal decomposition products to prevent flaming of cotton. Least amounts of flammable gases are released during burning when flame retardant finishes are deposited on textile materials. A study (Nair, 2001) showed that the ratio of volatile nonflammable to volatile flammable matters increased 8.6 and 4.5 times when ferric oxide (33%) and stannic oxide (28%), respectively, are deposited on the textile materials. In contrast, the ratio increases only by 1.3 times when silica (34%) is deposited which is not effective as FR. Metallic oxides are known for their catalytic action in the oxidation of organic compounds, as well as in decomposition of the intermediate products. Thus, the ferric oxide can completely oxidize methyl alcohol to carbon dioxide or in case of oxygen deficiency to carbon. The precipitated ferric oxide decomposes acetaldehyde at $400°C$ to carbon monoxide and methanol. Both stannic oxide and ferric oxide act as oxidation catalysts when present in fabrics to decompose tar partially to carbon dioxide and water which are noncombustible.

The simplest way to reduce flammability is by enhancement of decomposition temperature of the material. Usually, the mechanism has not been exploited by flame retardants. This mechanism is more usual in inherently flame- and heat-resistant fibers.

Flame retardant systems for synthetic or organic polymers act in five basic ways: (1) gas dilution; (2) thermal quenching; (3) protective coating; (4) physical dilution; (5) chemical interaction (Pettigrew, 1993); or through a combination of these mechanisms.

Flame retardants fulfill their purpose primarily by either physical or chemical action.

Physical action can be subdivided into three modes;

1. Cooling: An endothermic process is triggered by additives cooling the substrate to a temperature below that required for sustaining the combustion process.
2. Formation of protective layer: The combustible layer is shielded from the gaseous phase with a solid or gaseous protective layer. The oxygen required for the combustion process is excluded and heat transfer is impeded.
3. Dilution: Fillers are incorporated that evolve inert gases on decomposition diluting the fuel in the solid and gaseous phase so that the lower ignition limit of the gas mixture is not exceeded.

Chemical action can be subdivided into two modes namely:

1. Reaction in the solid phase: The flame retardant causes a layer of carbon to form on the polymer surface. This can occur through dehydration of the flame retardant forming a carbonaceous layer by cross linking. The carbonaceous layer acts as an insulation layer, preventing further decomposition of the material.
2. Reaction in the gas phase: The free radical mechanism of the combustion process that takes place in the gas phase is interrupted. The exothermic processes are thus stopped, the system cools down and the supply of flammable gases is suppressed.

Pyrolysis reveals various mechanisms namely:

1. Heat sink or endothermic degradation
2. Dilution of fuel
3. Thermal shielding by melting and swelling
4. Reduced release of flammable volatiles
5. Formation of radicals.

Each mechanism is discussed in further detail in the following sections.

4.5.1 ENDOTHERMIC DEGRADATION

Under the action of heat at comparable temperatures, the flame retardant chemical and the fiber are decomposed. Thermal breakdown of the flame-retardant chemical extracts energy from the combustion process which has an endothermic effect.

Certain compounds break down endothermally when they are subjected to high temperatures. Magnesium and aluminum hydroxides, various carbonates, and hydrates such as mixtures of huntite (a carbonate mineral with the chemical formula $Mg_3Ca(CO_3)_4$ and hydromagnesite (a hydrated magnesium-carbonate mineral with the formula $Mg_5(CO_3)_4(OH)_2 \cdot 4H_2O$) are examples of endothermic degradation. Heat is removed by the high heat of fusion and/or degradation and/or dehydration (e.g., inorganic and organic phosphorus-containing agents, aluminum hydroxide, or "alumina hydrate" in back-coatings).The use of hydroxides and hydrates is limited by their relatively low decomposition temperature, which limits the maximum processing temperature of the polymers (typically used in polyolefins for wire and cable applications).

These FR materials thermally decompose through strongly endothermic reactions thereby disrupting the combustion cycle for textile substrates by sinking heat in the fiber. If enough energy is absorbed by these reactions, the pyrolysis temperature of the fiber is not reached and combustion cannot take place.

4.5.2 DILUTION OF FUEL

Flammable gases and liquids are fuels for combustion. Dilution of fuel and thereby reduction of combustion may occur when the release of noncombustible gases is increased and/or the release of combustible gases is decreased.

Substances, which evolve inert gases on decomposition, dilute the fuel in the solid and gaseous phases and causes reduced access to oxygen. Inert fillers (e.g., talc powder or calcium carbonate) act as diluents, lower the combustible part of the material, and lower the extent of heat produced during burning for a unit volume of material. Thus, the concentrations of combustible gases fall under the ignition limit. For example, the release of carbon dioxide by decomposition of calcium carbonate or similar compounds is as in Equation 4.11.

$$CaCO_3 \quad \rightarrow \quad CaO + CO_2 \uparrow$$
$$(Heat)$$

(4.11)

The decomposed noncombustible gases dilute combustible gases and oxygen in the flame. Inert gases or vapor, mostly carbon dioxide and water, act as diluents of the combustible gases, lower their partial pressures and the partial pressure of oxygen, thereby slowing the reaction rate. These gases are produced by thermal degradation of some materials. Hydrated retardants and some char-promoting retardants release water; halogen-containing retardants release hydrogen halide.

Flame retardant chemicals intervene in the pyrolysis reaction with a dehydrating effect. This promotes the formation of carbonization residues and, at the same time, reduces the production of combustible gases.

4.5.3 THERMAL SHIELDING

Under the action of heat and the expenditure of energy, flame retardant chemicals are converted into a melted state that impedes the access of air and inhibits the evolution of combustible gases.

Sometimes a thermal insulation barrier is created between the burning and the yet-to-burn parts. The physical effect of certain FRs is the formation of an insulating and protective layer of glassy matter or char hindering the passage of the combustible gases and transfer of heat from pyrolyzing polymer to its surface. Boric acid and its hydrated salts have lower melting points and dehydrate in two stages into metaboric acid and boric oxide as shown in Equation 4.12 at 130°–200°C and 260°–270°C, respectively.

$$2H_3BO_3 \rightarrow 2HBO_2 \rightarrow B_2O_3$$
$$\quad - 2H_2O \quad\quad - H_2O$$
$$\quad (Heat) \quad\quad (Heat)$$

(4.12)

On heating, borax dissolves in its own water of hydration and then changes to a clear melt. A mixture (sodium pentaborate) of borax and boric acid (7:3) is an effective flame retardant for cellulose – only 5% add-on is required for retardancy, but 20% for a substantial decrease of glow resistance. Addition of diammonium phosphate enhances flame retardancy and anti smouldering activity. A clear glassy layer (stable up to 500°C) is formed that adheres uniformly to the cellulose fibers. Borate esters form dehydrate cellulose, enhancing char formation.

Sodium tetrafluoroborate decomposes into sodium fluoride and gaseous BF_3, the latter catalyzes the formation of ketone groups and carbon double bonds, which cross-link with cellulose enhancing formation of char.

Intumescent material swells on exposure to heat, with a subsequent increase in volume and decrease in density. An intumescent is typically used in passive fire protection. Intumescent additives are sometimes applied for turning the polymer into carbonized foam, separating the flame from the material and slowing down the heat transfer to the unburned fuel.

The role of a char on the surface of a polymer is to protect the underlying polymer from the heat of the fire. Ideally, a char layer should be an excellent insulator. In addition, it should be difficult to combust, must remain in place so that it can be effective (i.e., be adherent), must have structural integrity so that it cannot be easily ruptured. The typical char is a carbonaceous material that is formed during pyrolysis. The best example of this type of char may be produced in the degradation of polyacrylonitrile and yield elemental carbon if appropriately treated (Mathur et al., 1992).

The process envisaged was to devise a way by which a polymeric precursor of a char could be attached to a polymer in such a way that the precursor would thermally degrade and offer protection to the polymer. The challenges which are faced in this endeavor are:

1. To identify suitable char-formers, and
2. To develop processes by which these could be delivered to the surface of the relevant polymer. It must be remembered that this should be of utility with a wide variety of polymers, so the process of delivery of the char former must be general.

4.5.4 REDUCED RELEASE OF FLAMMABLE VOLATILES

One way to achieve flame retardancy is to influence the pyrolysis reaction to produce less flammable volatiles and more residual chars. Phosphorous-containing flame retardants follow this condensed phase mechanism. Phosphoric acid produced by thermal decomposition cross-links with hydroxyl-containing polymers thereby alters the pyrolysis to yield less flammable by-products and prevents the formation of undesirable levoglucosan, the precursor of flammable volatiles. Most phosphorus and nitrogen-containing flame retardants (PNFRs) in cellulose and wool, as well as heavy metal complexes in wool decrease formation of flammable volatiles and increase char formation.

The retardancy mechanism can largely be divided into two major types namely:

1. Condensed phase mechanism and
2. Gas phase radical quenching.

4.5.5   CONDENSED PHASE MECHANISM

The condensed-phase mechanism postulates a chemical interaction between the flame retardant agent and the polymer. This interaction occurs at temperatures lower than those of the pyrolytic decomposition.

The pyrolysis reaction is modified in order to produce less flammable volatiles and more residual chars. Phosphorous-containing flame retardants follow this condensed phase mechanism. Phosphoric acid produced by thermal decomposition cross-links with hydroxyl-containing polymers (Equations 4.13 and 4.14) thereby altering the pyrolysis to yield less flammable by-products and prevent the formation of undesirable levoglucosan, the precursor of flammable volatiles.

$$
\text{Cell-CH}_2\text{OH} \xrightarrow{\text{P,N, heat}} \text{Cell-CH}_2\text{-O-}\overset{\overset{\displaystyle O}{\|}}{\underset{\underset{\displaystyle O}{\,}}{P}}\text{-O} \; + \; \text{H}_2\text{O} \qquad [4.13]
$$

$$
\text{Cell-CH}_2\text{-O-}\overset{\overset{\displaystyle O}{\|}}{\underset{\underset{\displaystyle O}{\,}}{P}}\text{-O} \; + \; \text{Cell-CH}_2\text{OH} \xrightarrow{\text{H+}} \text{Cell-CH}_2\text{-O-}\overset{\overset{\displaystyle O}{\|}}{\underset{\underset{\displaystyle O}{\,}}{P}}\text{-O-CH}_2\text{-Cell} \qquad [4.14]
$$

However, there are other explanations for this initial dehydration, including single esterification of the primary hydroxyl group without cross-linking; these phosphorous esters catalyze dehydration and prevent the formation of undesirable levoglucosan, the precursor of flammable volatiles.

The cross-linking and the single type of esterification of cellulose polymer chains by phosphoric acid reduces levoglucosan generation, catalyzes dehydration and carbonization, and thus functions as an effective flame retardant mechanism. This carbonization of cellulose is similar to the well-known carbonization process of treatment of wool with sulfuric acid. In an idealized equation, flame-retardant-finished cellulose $(C_6H_{10}O_5)_n$ would be decomposed to $6n$ C and $5n$ H_2O. The resulting char is much less flammable than the volatile organic pyrolysis products of untreated cellulose. The first step of this reaction is shown in Equation 13 (Schindler and Hauser, 2004).

As the aforementioned reactions occur in the solid phase, the mechanism is known as a condensed phase. In condensed phase mode, the reduction of volatile, flammable products and the increase of residual carbonaceous chars are caused by two mechanisms namely:

1) Dehydration and/or
2) Cross-linking.

These processes are well established for cellulosics and operate probably to some extent in other polymers as well. Weil (1978) suggested that the varying efficiency of phosphorous in different polymers is directly related to the possibility of dehydrative char formation – the efficiency decreases with decreasing polymer oxygen content. 2% phosphorous is adequate to make cellulosic flame retardant, whereas 5%–15% is required for polyolefins.

Cross-linking promotes the stabilisation of the structure of cellulose by providing additional covalent bonds between the chains, which are stronger than the hydrogen bonds, and which have to be broken before the stepwise degradation of the chain that occurs on pyrolysis. However, low degrees of cross-linking can decrease the thermal stability by increasing the distance between the individual chains and consequently weakening and breaking the hydrogen bonds. Thus, although the LOI of cotton increases marginally with increasing formaldehyde cross-linking, that of rayon decreases markedly (Roderig et al., 1975).

The formation of char in celluloses is initiated by rapid auto-cross-linking due to the formation of ether oxygen bridges formed from hydroxyl groups on adjacent chains. The auto-cross-linking is evidenced by a rapid initial weight loss, owing to the evolution of water, in the first stage of pyrolysis at 251°C, and is linearly related to the amount of char. Formaldehyde cross-linking of rayon interferes with the auto-cross-linking reaction, decreases the initial weight loss, and reduces char formation (Roderig et al., 1975).

In highly crystalline cotton, auto-cross-linking occurs to a small extent and the dominant effect of cross-linking is stabilization and reduced flammability.

As the first stage of combustion is pyrolytic decomposition, fine structural features of polymers such as intermolecular forces, chain rigidity, degree of crystallinity affecting the energy required for melting, and degradation of polymers are all critical factors in flammability. The studies on cotton and rayon showed that degradation mostly occurs in the amorphous region. The increase in the orientation of the cellulosic chains showed a decrease of the rate of pyrolysis in the air at 250°C.

Organic nitrogen is thought to control the pH during cross-linking reactions of phosphoric acid. The nitrogen may be protonated, reducing the amount of acid available. At low pH, cellulose undergoes acid hydrolysis rather than cross-linking. Again at high pH, acid catalyzed cross-linking may not take place. Organic nitrogen may be converted into phosphorous acid amides that also catalyze dehydration and carbonization of cellulose.

When external heat is applied, the substrate undergoes thermal decomposition or pyrolysis with the generation of combustible fuel measured in terms of heat generated, ΔH_2 cal/g polymer. Only a part of the fuel, thus generated, is combusted into flame by atmospheric oxygen releasing heat of ΔH_1 cal/g. The residual part ΔH_3 ($\Delta H_2 - \Delta H_1$) can also be combusted with excess oxygen in the presence of a catalyst. A portion of the heat of the flame causes continued pyrolysis of the substrate, perpetuating the cycle; the remaining portion is dissipated and lost to the environment.

The heat balance of polymer combustion can be represented as in Equation 4.15.

$$\left(\Delta H^0{}_C\right)_P = \Delta H_2 + R\left(\Delta H^0{}_C\right)_{char} \qquad (4.15)$$

where $\left(\Delta H^\circ{}_C\right)_P$ is the standard heat of combustion of the polymer in cal/g polymer and $\left(\Delta H^\circ{}_C\right)_{char}$ is the heat of combustion of the char, R in grams, from burning in air, cal/g residue. Both can be determined by combustion with oxygen under pressure. ΔH_1 can be determined in the Isoperibol calorimeter (Barker and Drews, 1976).

In the condensed-phase mechanism, the amount of fuel produced (ΔH_2) during pyrolysis is reduced, i.e., ΔH_2 decreases with increase in applied FR, but the ratio $\Delta H_1 - \Delta H_2$ remains constant. In the gas phase mechanism, ΔH_2 remains constant, both ΔH_1 and the ratio $\Delta H_1/\Delta H_2$ decrease. In other words, the pyrolytic processes remain essentially same with or without FR – the mode of combustion in the flame changes with the use of FR. The amount of fuel consumed in the flame (ΔH_1), and consequently, the heat generated ($\Delta H_1/\Delta H_c$) are decreased with increase in the amount of FR. The amount of heat returned to the polymer surface diminishes, and due to the reduction of surface temperature, pyrolysis is slowed down or halted. As the retarding effect is in the gas phase, the FR moiety itself or its decomposition products should be volatile to reach the gas phase. The residual FR in the char after combustion is much less than that in case of condensed-phase FR. Hence, the mode of action of FR chemical can be determined by the chemical analysis of char. Flammability in these cases should be independent of polymer structure and the pyrolytic process remains unchanged.

The flammability of a substrate is determined by the ease of pyrolysis, i.e., the minimum temperature at which pyrolysis starts and the character and quantity of gaseous substances are generated. An FR agent that acts via a condensed phase mechanism substantially reduces the amounts of gaseous substances produced (ΔH_2) during pyrolysis by favoring formation of carbonaceous char, carbon dioxide, and water. The demand for substantial amounts of FR agent in condensed-phase to impart flame retardancy suggests that it is a co-reactant with polymers and not a catalyst as suggested by Little (1947). Lower amounts (catalytic) of FR increase, rather than decrease, flammability. Basch and Lewin (1973) showed that at low add-on levels, acidic FRs catalyze thermal degradation with an increase in the concentration of the fuel levoglucosan, thereby enhancing flammability, and larger quantities of FRs are required to overcome catalytic actions. For better condensed-phase flame retardancy, FRs should react with polymers at a temperature lower than the pyrolysis temperature.

A decrease in $\left(\Delta H^\circ{}_C\right)_{char}$ values with increase in FR indicate a more thorough pyrolysis and combustion i.e., an activity in the condensed phase. This has been shown in the case of a polyester fabric treated with Tris-isobutylated triphenyl phosphate (TBPP), but not with tri-(phenyl) Phosphine (TPP) which behaves according to the gas-phase mechanism (Barker et al., 1979).

A simple method of assessment of the mode of action of a FR system is char analysis. The condensed phase FRs are nonvolatile and therefore, their basic elements (P or S) can be traced and determined in the char.

The condensed phase FRs are most effective as they promote char formation by converting the organic fiber structure to a carbonaceous residue or char and reduce volatile fuel formation. Indirectly, these flame retardants, which require absorption of heat for them to operate, offer the additional mode by releasing nonflammable molecules such as CO_2, NH_3, and H_2O during char formation. In addition, the char behaves as a carbonized replica of the original fabric, which continues to function as a thermal barrier, unlike flame retardant thermoplastic fibers.

Char-forming flame retardants, therefore, offer both flame and heat resistance to a textile fiber and so can compete with many of the so-called high-performance flame and heat resistant fibres such as the aramids and similar fibers.

For char formation to be most effective, the polymer backbone must comprise side-groups, which, on removal, lead to unsaturated carbon bond formation and eventually a carbonaceous char following the elimination of most of the noncarbon atoms present. Most phosphorus- and nitrogen-containing retardants, when present in cellulose, reduce the volatile formation and catalyse char formation. While this is a considerably oversimplified view of the actual chemistry involved, a brief overview of some of the essential features of the mechanism provide a model for char formation in general (Kandola et al., 1996).

Most phosphorus-containing retardants act in this double capacity because, on heating, they first release polyphosphoric acid, which phosphorylates the C_6 hydroxyl group in the anhydroglucopyranose moiety, and simultaneously acts as an acidic catalyst for dehydration of these same repeat units. The first reaction prevents the formation of levoglucosan, the precursor of flammable volatile formation and this ensures that the competing char-forming reaction is now the favored pyrolysis route and the rate of this route is increased further by the acidic catalytic effect of the released polyacid. While considerable research has been undertaken into char formation of flame retarded cellulose, the actual mechanisms of both unretarded and retarded cellulose charring are not well understood (Kandola and Horrocks, 2001).

Phosphorous- and sulfur-containing FRs act in cellulose by a condensed phase mechanism, for other polymers the mechanism is less certain. The phosphorous derivatives are most effective for polymers containing oxygen by forming acids during pyrolysis and combustion. For polypropylene and other hydrocarbon-based polymer, polyphosphoric acid is generated and it forms a glassy layer on the burning mass. Element red phosphorous, an efficient FR for polyester, acts in both condensed and gas-phase mechanisms. It decreases pyrolysis rate, but does not change the composition of pyrolysis gases. Tris-isobutylated triphenyl phosphate (TBPP) causes thermal depolymerization and a lowering of the melting point of polyester. As a result, the polymer melts and drips away from the flame, thereby decreasing flammability. However, a recent study shows that it inhibits free-radical combustion reactions. Phosphorous-containing FRs, except highly stable and volatile triphenylphosphine oxide (TPPO), reduce the flammability of polyester by affecting both pyrolytic and vapor-phase processes.

Volatile phosphorous compounds, especially TPPO, break down in the flame to small molecular species and radicals such as $PO\cdot$, $P\cdot$, and P_2. These radicals scavenge $H\cdot$ radical as in case of halogen.

Because most of the phosphorous–containing compounds exhibit the same flame retardancy on an equivalent phosphorous basis, a common esterification followed by ester decomposition mechanism is proposed (Equation 4.16).

$$\begin{aligned} R_2CH - CH_2OH + HOR(acid) \;&\rightarrow\; R_2CH - CH_2OR + H_2O \\ &\rightarrow\; R_2C = CH_2 + HOR(acid) \end{aligned} \tag{4.16}$$

The lower FR activity on rayon, as compared to cotton, is due to lower thermal stability of rayon;it decomposes before all of the phosphate esters can decompose and liberate free acid to direct the pyrolysis to lessflammable end products.

DTA analysis suggests that sulfur compounds act on cellulose via carbonium– ion catalysis (Equation 4.17).

$$\begin{aligned} R_2CH - CH_2OH \;&\rightarrow\; R_2CH - CH_2OH^+{}_2 \xrightarrow{H^+} R_2CH - CH^+{}_2 + H_2O \\ &\rightarrow\; R_2C = CH_2 + H^+ \end{aligned}$$

$$\tag{4.17}$$

Esterification-ester-decomposition mechanism is affected by the fine structure of cellulose, while carbonium ion dehydration scheme is less affected by fine-structural parameters.

Synergism of a FR may occur in the presence of a second FR or a non-FR additive such as nitrogen. Urea and similar nitrogenous (N) compounds facilitate phosphorylation of cellulose. The synergistic effect is not a function of any specific N-P ratio, but increases with increasing amounts of N compounds at fixed phosphorous ratios. Moreover, the effect is not general to all N compounds. Reeves et al. (1970) generalized that amine and amide-type nitrogen compounds are synergistic with phosphorous, whereas nitrile compounds are antagonistic. The antagonism may be due to volatization of phosphorous during pyrolysis. Polymer-bound nitrogen (e.g., acrylamide polymer) cannot interact with phosphorous and are therefore inactive. Two mechanisms are proposed for N compounds. Such compounds as urea swell cellulosic fibers thereby increase the accessibility of the phosphorylating reagent. In other cases, P-N bonds are formed that tare more reactive to cellulose than are the corresponding P-O type compounds. P-N bonds also form a cross-linked network within cellulose that inhibits release of volatile combustible fragments and promotes char formation.

4.5.6   GAS-PHASE RADICAL QUENCHING

Some chlorinated and brominated compounds release hydrogen chloride and hydrogen bromide during thermal degradation. These react with the highly reactive H· and OH· radicals in the flame, resulting in an inactive molecule and a Cl· or Br· radical. The halogen radical is of much lower energy than H· or OH· and hence, it is unable to propagate the radical oxidation reactions of combustion. Antimony compounds tend to act in synergy with halogenated flame retardants. The hydrogen chloride and hydrogen bromide released during burning are highly corrosive.

The gas-phase activity of FR is attributed to its interference in the combustion reaction in the flame. Like other fuels, polymers, upon pyrolysis, produce species capable of reacting with atmospheric oxygen. The fuel combustion propagates by branching or scavenging due to the formation of free radicals by the H_2-O_2 reaction scheme in the presence of oxygen (Shtern, 1964) (Equation 4.18).

$$H+O_2 \leftrightarrow \cdot OH + \cdot O \quad (+8 \text{ kcal/mole}) \qquad (4.18)$$

The most exothermic reaction in the flame that provides most of the energy maintaining the combustion is as follows (Equation 4.19):

$$\cdot OH + CO \leftrightarrow CO_2 + \cdot H \quad (-81 \text{ kcal/mole}) \qquad (4.19)$$

To prevent or to slowdown combustion, the chain branching reactions (Equations 4.18 and 4.19) must be hindered.

The inhibiting effect of halogen derivatives, which are considered to operate via the gas-phase mechanism, occurs by first releasing either a halogen atom if the compound is devoid of hydrogen (Equation 4.20).

$$MX \leftrightarrow M^/ + \cdot X \qquad (4.20)$$

or by releasing a hydrogen halide (Equation 4.21).

$$MX \leftrightarrow M^/ + HX \qquad (4.21)$$

The hydrogen halide acts as a flame inhibitor by effecting chain branching (Equations 4.22 and 4.23).

$$\cdot H + HX \leftrightarrow H_2 + \cdot X \qquad (4.22)$$

(X = -1 and − 17 kcal/mole for Cl and Br, respectively)

$$\cdot OH + HX \leftrightarrow H_2O + \cdot X \tag{4.23}$$

(X = -8 and – 24 kcal/mole for Cl and Br, respectively)

The X radical is much less reactive than H and OH radicals and cannot effectively propagate chain reaction. Antimony reacts with X radicals to form SbOX and SbX, both of which thermally decompose to yield halogen radicals.

The former reaction is twice faster and hence, the high ratio of $H_2/\cdot OH$ in the flame front indicates that Equation 4.23 is the main inhibiting reaction. The inhibition effect is determined by the consumption of active hydrogen atoms due to competition between reactions in Equations 4.18 and 4.22. Two free radicals are produced for each hydrogen atom by the reaction (Equation 4.18), whereas one relatively unreactive halogen atom is produced by the reaction (Equation 4.23) (not active in the H_2–O_2 reaction scheme) (Shtern, 1964).

The relative effectiveness of halogen compounds as a flame retardant is as follows (Equation 24):

$$I > Br > Cl > F \tag{4.24}$$

The activity of the halogens is also strongly affected by the strength of the respective carbon-halogen bonds. The low bond strength of I–C, and consequently, the low stability of the iodine compounds virtually exclude their use. The high stability of the fluorine derivatives and the high reactivity of the fluorine atoms in reactions prevents the radical quenching processes in the flame. The lower bond strength and stability of the aliphatic compounds, their greater ease of dissociation, as well as the lower temperature and earlier formation of the HBr molecules are responsible for their higher affectivity, compared to the aromatic halogen compounds. The higher stability of the latter, along with its higher volatility, allows these compounds to evaporate before they can decompose and furnish the halogen to the flame.

The radical trap activity is not the only activity of the halogenated flame retardants. Physical factors, such as the density and mass of the halogen and its heat capacity, have a profound influence on the flame-retarding activity of the agent. In addition, its dilution of the flame, which thus decreases the mass concentration of combustible gases present, is effective.

In FR polymer systems, the halogens appear to work by reducing the heat evolved in the combustion of the gases given off by the decomposing polymer (low or zero fuel value plus action as a heat sink), such that to sustain burning the mass rate of gasification must be increased by the application of an increased external heat flux (Larsen, 1974).

A physical effect, often mentioned but rarely demonstrated or evaluated, is the *blanketing* effect of excluding oxygen from the surface of the pyrolyzing polymer. Ignition generally takes place in the vapor phase adjacent to the condensed phase, when an ignitable fuel–air mixture is reached. There is, however, evidence that the rate of pyrolysis may be affected by the oxygen's getting to or into the condensed phase, and that in polyolefins, surface oxidation may provide energy for pyrolysis (Stuetz et al., 1975).

Schindler and Hauser (2004) compared the condensed and gas phase mechanisms which are summarized in Table 4.1.

4.6 APPLICATION METHODS

For successful flame retardancy, flame retardants should combine additively or reactively with textile materials to an acceptable level at an affordable cost; they are applicable to textile fabrics using conventional textile finishing and coating equipment. The finish is applied on open width form by one of the following methods:

1. The simple pad-dry technique is applicable with most nondurable and water-soluble finishes such as the ammonium phosphates. This sequence is typical of those used to apply crease-resistant and other heat-curable textile finishes. In the case of flame retardant finishes, best use has been found for application of the phosphonamide systems such as Pyrovatex

---

TABLE 4.1

Comparison of condensed and gas phase mechanism for flame retardancy

Type of mechanism	Condensed phase	Gas phase
Type of chemistry involved	Pyrolysis chemistry	Flame chemistry
A typical type of synergism	P/N	Sb/Br or Sb/Cl
Effective for fiber type	Mainly cellulose, also wool, catalyzing their dehydration to char	All kinds of fibers, because their flame chemistry is similar (radical transfer reactions)
Particularities	Very effective because dehydration and carbonization decrease the formation of burnable volatiles	Fixation with binder changes textiles properties such as handle and drape, preferably for back coating of, for example, furnishing fabrics and carpets
Application process	For durable flame retardancy, multistep process	Relatively simple, standard methods of coating, using additives with controlled viscosity They can affect handle and drape.
Environment toxicity	With durable flame retardancy, formaldehyde emission during curing and after finishing, phosphorous compounds in the wastewater	Antimony oxide and some organic halogen donators (DBDPO and HCBC) generate toxic dioxins and furans, and are banned

(Ciba), which are applied with resin components such as the methylolated melamines. Because the process requires the presence of acidic catalysts (e.g., phosphoric acid), the wash-off stage includes an initial alkaline neutralization stage.

2. This same sequence without the washing-off stage may be used to apply semidurable finishes where a curing stage allows a degree of interaction to occur between the finish and the cellulose fiber; a typical example is given by the ammonium phosphates, which during curing at about 160° C, give rise to phosphorylation of the cellulose. Thus, the finish develops a degree of resistance to water-soak and gentle laundering treatments.

3. The third method requires an ammonia-gas-curing process in order to polymerize the applied finish into the internal fiber voids. In this process the Proban CC condensate of tetrakis (hydroxy methyl) phosphonium chloride and urea after padding and drying onto the fabric is passed through a patented ammonia reactor which cross-links the condensate to give an insoluble polymeric finish. In order to increase the stability, and hence durability, of the finish, a subsequent oxidative *fixation* stage is required before finally washing off and drying.

4. In back-coating methods the flame retardant formulation is applied in a bonding resin to the reverse surface of an otherwise flammable fabric. In this way, the aesthetic quality of the face of the fabric is maintained while the flame retardant property is present on the back or reverse face. Flame retardants must have an element of transferability from the back into the whole fabric, so they almost always are based on the so-called vapor-phase active antimony–bromine (or other halogen) formulations as typified by decabromodiphenyloxide or hexabromocyclododecane and antimony III oxide. Application methods include doctor blade or knife-coating methods and the formulation is as a paste or foam. These processes and finishes are used on fabrics in which aesthetics of the front face are of paramount importance, such as for furnishing fabrics and drapes.

The relatively high application levels required for inorganic FRs may adversely influence fabric handle, drape, and appearance; these effects are minimized by ensuring that minimal finish remains on the fiber and fabric surfaces and on the back side of the fabric during the coating process. In addition, softening agents may be included within the formulations during application; careful selection of these is essential to ensure compatibility in the formulation and on the resulting flame retardant property.

The selection of intumescent composition to be applied on flexible textile fabrics as a coating should be done carefully so that the aesthetic properties are not hampered to a large extent.

4.7 FUTURE TRENDS

More than 60 years ago, the 1950–1960 period witnessed the development of the chemistry underlying most of today's successful and durable flame retardant treatments for fibers and textiles. In today's more critical markets in terms of environmental sustainability, chemical toxicological acceptability, performance and cost, many of these are now being questioned. "Are there potential replacements for established, durable formaldehyde-based flame retardants such as those based on tetrakis (hydroxylmethyl) phosphonium salt and alkyl-substituted, N-methylol phosphonopropionamide chemistries for cellulosic textiles?" is a frequently asked question. "Can we produce char-forming polyester flame retardants?" and "Can we really produce effective halogen-free replacements for coatings and back-coated textiles?" are others.

Current flame retardant polymer solutions are tailored to the regulatory test the polymer must pass to be sold. The molecules are therefore designed and applied to solve the following problems:

- Ignition resistance,
- Flame spread,
- Heat output,
- Structural integrity under fire and heat,
- Smoke/toxic gas output, and
- Combinations of one or more of the aforementioned.

What works for one test may not (and often does not) work for another test! The flame retardant chemist designs to test flame retardancy that is not universal. The chemist can only design by using the criteria given (e.g., fire, cost, performance, lifetime). It is impossible to design for the unforeseen criteria that may occur 10–20 years later (Morgan, 2009).

Future FRs for cellulosic material should:

1. Be nontoxic,
2. Be economically feasible,
3. Not change the appearance, color, or shade of the fabric treated,
4. Ensure pleasant feel (less roughness), adequate strength (breaking strength, elongation at break) and be wear-resistant,
5. Be waterproof for at least 50 washing cycles in alkaline medium, at high temperatures, independent of water hardness,
6. Not release free formaldehyde during processing or after it, and
7. Have high air-permeability after the treatment, regardless of the high amount of FR coating (Harrocks, 2011).

REFERENCES

Barker R.H., Drake G.L., Hendrix Y.E., and Bostic J.E. (1979). *Flame Retardancy of Polymeric Materials*, Vol 5. Marc el Dekker, New York.
Barker R.H. and Drews M.J. (1976). *Development of Flame Retardants for Polyester-Cotton Blends*, Final report, NBS-GCR-ETIP 76-22. National Bureau of Standards, Washington, DC.

Basch A. and Lewin M. (1973). Low add-on levels of chemicals on cotton and flame retardancy, *Textile Res. J.*, 43, 693–694.

David Suzuki Foundation. (2016). Toxic flame retardants are a burning issue, www.david suzuki.org/, accessed on 23.3.16.

Day P. and Ledsham R.D. (1982). Organic-inorganic molecular composites as possible low-dimensional conductors: Photo-polymerization of organic moieties intercalated in inorganic layer compounds, *Journal Molecular Crystals and Liquid Crystals*, 86(1), 163–174.

Drobny J.G. (2007). *Handbook of Thermoplastic Elastomers*. William Andrew Inc., Oxford, UK. ISBN 978-0-323-22136-8.

Dvornic P.R. (2000). Thermal properties of polysiloxanes, in: R.G. Jones, W. Ando, and J. Chojnowski, Eds., *Thermal Stability of Polysiloxanes: Silicone–Containing Polymers*. Kluwer Academic Publisher, Dordrecht, the Netherlands, pp. 185–212.

Guan J.P. and Chen G.Q. (2006). Flame retardancy finish with an organophosphorus retardant on silk fabrics, *Fire and Materials*, 6, 415–424.

Guan J.P. and Chen G.Q. (2010). Graft copolymerization modification of silk fabric with an organophosphorus flame retardant, *Fire and Materials*, 34(5), 261–270.

Guan J.P., Lu H., and Chen Y. (2013). Apparel performance of flame retardant silk fabrics, *Journal of Engineered Fibers and Fabrics*, 8(4), 30–35.

Harrocks A.R. (2011). Flame retardant challenger for textiles and fibers: New chemistry versus innovatory solutions, *Polym. Degrad. Stab.*, 96(3), 377–392.

HIS. (2019). Consulting 2017, www.flameretardants-online.com/flame-retardants/market, accessed on 12.7.19.

Joseph P. and Ebdon J.R. (2010). Phosphorus-based flame retardants, in: C.A. Wilkie and A.B. Morgan, Eds., *Fire Retardancy of Polymeric Materials*, 2nd edition. CRC Press, USA.

Kandola B.K., Horrocks A.R., Price D., and Coleman G. (1996). Flame retardant treatments of cellulose and their influence on the mechanism of cellulose pyrolysis, *Revs. Macromol. Chem. Phys.*, C36, 721–794.

Kandola B.K. and Horrocks A.R. (2001). Nanocomposites, in: R. Horrocks and D. Price, Eds., *Fire-Retardant Materials*, 1st edition. CRC Press, USA, pp. 204–219, (March 16).

Larsen E.R. (1974). Mechanism of flame inhibition. I: The role of halogens, *Journal of Fire and Flammability/Fire Retardant Chemistry*, 1, 4–12.

Lewin M. and Sello S.B., Ed. (1984). *Functional Finishes, Handbook of Fibre Science and Technology: Volume II, Part B*. Marcel Dekker, New York.

Little R.W., Ed. (1947). *Flameproofing Textile Fabrics*. Rheinhold, New York.

Marosi G., Marton A., Anna P., Bertalan G., Marosfoi B., and Sze′p A. (2002). Ceramic precursor in flame retardant systems, *Polym Degrad Stab*, 77, 259–265.

Mathur R.B., Bahl O.P., and Sivram P. (1992). Thermal degradation of polyacrylonitrile fibres, *Curr Sci*, 62, 662–669.

Mikkola E. and Wichman I.S. (1989). On the thermal ignition of combustible materials, *Fire Mater.*, 14, 87–96.

Morgan A.B. (2009). Polymer flame retardant chemistry, ABM, F.R. Chemistry, 30.9.2009. www.nist.gov/system/files/documents/el/fire_research/2-Morgan.pdf, accessed on 12.12.2019.

Morgan A.B. and Wilkie C.A. (2007). *Flame Retardant Polymer Nano-Composites*. John Wiley & Sons, Inc., Hoboken, NJ.

Nabipour H., Wang X., Song L., and Hu Y. (2020). Hydrophobic and flame-retardant finishing of cotton fabrics for water–oil separation, *Cellulose*. DOI: 10.1007/s10570-020-03057-1.

Nair G.P. (2001). Flammability in textiles and routes to flame retardant textiles – IX, Colourage, May, 41–48.

Pettigrew A. (1993). Halogenated flame retardants, in: *Kirk-Othmer Encyclopedia of Chemical Technology*, 4th edition, Vol 10. John Wiley & Sons, New York, pp. 954–976.

Pinfa. (2018). www.pinfa.eu, accessed on 14.3.18.

Price D., Anthony G., and Carty P. (2001). Introduction: Polymer combustion, condensed phase pyrolysis and smoke formation, in: A.R. Harrocks and D. Price, Eds., *Fire-Retardant Materials*. CRC Press, USA.

Rakotomalala M., Wagner S., and Döring M. (2010). Recent developments in halogen free flame retardants for epoxy resins for electrical and electronic applications, *Materials*, 3, 4300.

Reeves W.A., Perkins R.M., Piccolo B., and Drake G.L., Jr. (1970). Some chemical and physical factors influencing flame retardancy, *Textile Res. J*, 40, 223.

Roderig C., Basch A., and Lewin M. (1975). Cross-linking and pyrolytic behavior of natural and man-made cellulosic fibre, *J. Polym. Sci., Polym. Chem. Ed.*, 15, 1921–1932.

Sanchez C., Julián B., Bellevilleb P., and Popallc M. (2005). Applications of hybrid organic–inorganic nanocomposites, *J. Mater. Chem.*, 15, 3559–3592. DOI: 10.1039/b509097k.

Schartel B. (2010). Phosphorus-based flame retardancy mechanisms – Old hat or a starting point for future development?, *Materials*, 3, 4710–4745. DOI: 10.3390/ma3104710. ISSN 1996-1944.

Schindler W.D. and Hauser P.J. (2004). *Chemical Finishing of Textiles*. Woodhead, Cambridge, England.

Shen M.-Y., Chang T.-Y., Hsieh T.-H., Li Y.-L., Chiang C.-L., Yang H., and Yip M.-C. (2013). Mechanical properties and tensile fatigue of graphene nanoplatelets reinforced polymer nanocomposites, *Journal of Nanomaterials*, 2013, 1–9, https://www.hindawi.com/journals/jnm/2013/565401/.

Shtern V.Y. (1964). *The Gas-Phase Oxidation of Hydrocarbons*. Macmillan, New York.

Stuetz D.E., DiEdwardo A.H., Zitomer F., and Barnes B.P. (1975). Polymer combustion, *J. Polym. Sci., Polym. Chem. Ed.*, 13, 585–621.

Swaraj P.A.B. (2001). State of the art study for the flame retardancy of polymeric materials with some experimental results, PP Polymer.

Wang P., Jiang T., Zhu C., Zhai Y., Wang D., and Dong S. (2010). One-step, solvothermal synthesis of graphene-CdS and graphene-ZnS quantum dot nanocomposites and their interesting photovoltaic properties, *Nano Research*, 1–6.

Wang Q. (2013). Polymer nanocomposite: A promising flame retardant, *Journal of Materials Science & Nanotechnology*, 1(2)ISSN: 2348-9812, Open access, publication date: October 30, 2013.

Wen J. and Mark J. (1994). Mechanical properties and structural characterization of poly(dimethylsiloxane) elastomers reinforced with zeolite fillers, *J Mater Sci*, 29(2), 499–503.

Zhang T., Yan H.Q., Peng M., Wang L.L., Ding H.L., and Fang Z.P. (2013). Construction of flame retardant nanocoating on ramie fabric via layer-by-layer assembly of carbon nanotube and ammonium polyphosphate, *Nanoscale*, 5, 3013–3021. DOI: 10.1039/c3nr34020a.

5 Halogen-Based FRs

5.1 INTRODUCTION

Bromine, chlorine, fluorine, and iodine are the elements in the chemical group known as halogens. Halogenated flame retardants act directly on the flame, the core of the fire. They are said to act "in the vapor phase", meaning that they actually interfere with the chemistry of the flame.

This category actually covers bromine- and chlorine-containing flame retardants. Iodine would probably work but is too expensive and its compounds tend to be relatively unstable with respect to heat, and especially, to light.

Fluorine plays an important role in flame retardancy. Two important FR fluorine-based FR compounds are polytetrafluoroethylene (PTFE) and fluorinated ethylene propylene (FEP). Polytetrafluoroethylene (PTFE) is a synthetic fluoropolymer of tetrafluoroethylene that has numerous applications. The well-known brand name of PTFE-based formulas is Teflon (Teflon, 2019).

Fluorinated ethylene propylene (FEP), a copolymer of hexafluoropropylene and tetra-fluoroethylene, is used prominently in wire insulation. It differs from the polytetrafluoroethylene (PTFE) resins in that it is melt-processable using conventional injection molding and screw extrusion techniques (Peter and Schmiegel, 2000).

FEP is very similar in composition to fluro-polymer PTFE (polytetrafluoro-ethylene) and PFA (perfluoroalkoxy polymer resin). FEP and PFA both share PTFE's useful properties of low friction and nonreactivity, but are more easily formable. FEP is softer than PTFE and melts at 260°C; it is highly transparent and resistant to sunlight.

Fluorine compounds are ineffective flame inhibitors because these compounds and their possible breakdown product, hydrogen fluoride, are too stable to interrupt oxidative chain reactions in the flame. The only role of fluorine in the flame-retardant additive field is the use of small quantities (typically, less than 0.1%) of powdered poly(tetrafluoroethylene) (Teflon) as an additive to prevent dripping of burning polymers, probably by a rheological effect.

Chlorine has a long history in flame retardant additives, and is now represented by the polychloroparaffins and by Dechlorane Plus®, a polychlorinated flame retardant produced by Oxychem as a result of the Diels-Alder reaction of two equivalents of hexachlorocyclopentadiene with one equivalent of cyclooctadiene.

Among halogen compounds, iodine and fluorine compounds are virtually precluded from use as FRs. 13% bromine has been found to be as effective as 22% chlorine when comparing the tetrahalophthalic anhydrides for FR polyester. Aliphatic bromine compounds were found to be 1.5 times more effective than the aromatic compounds. Hexachlorohexane was found to be three times more active than hexachlorobenzene for treating polyurethane foams.

Ammonium bromide was found to be three times more effective than ammonium chloride. The reason for the high activity of the former may be relative ease of formation of hydrogen bromide during pyrolysis. The simultaneous release of ammonia swells crystalline regions of cellulose rapidly, and thus decreases the formation of levoglucosan with a consequent decrease in flammability.

Halogenated flame retardants are widely used in flame-retarded plastic applications in consumer electronics. The environmental issues concerning brominated flame retardants cloud their future. Consequently, the manufacturers of computers and business machines are re-evaluating the use of halogenated flame retardants.

This is particularly true for the European marketplace, where eco-friendly labels are becoming a major factor in marketing computers and other electronic equipment. To qualify for such labels as the White Swan of Sweden, a product must comply with certain environmental ground rules that include not using any halogenated flame retardants. Computer manufacturers are therefore considering switching from halogenated flame retardant systems, to a PC/ABS (polycarbonate/acrylonitrile butadiene styrene) polymers flame retarded with a phosphorous compound (Despinasse, 2016).

The main construction materials affected are extruded polystyrene (XPS) and expanded polystyrene (EPS) insulation. The main halogenated retardant used in these materials until recently was a brominated retardant called HBCD or Hexabromocyclododecane. This has been replaced or is soon to be replaced in most products by compounds that have been claimed to be safer.

Typically, a brominated or chlorinated organic compound is added to the polymer or, in suitable cases, halogenated structures are introduced into the polymer chain by copolymerization to prepare fire-retardant polymer materials. Metal compounds, such as antimony trioxide, which do not, by themselves, impart significant fire-retardant properties to polymers can strongly enhance the fire-retardant effectiveness of halogenated compounds (synergistic effect). On heating, these fire retardants evolve volatile metal halides, which are well-known flame inhibitors of much greater effectiveness as compared to hydrogen halides evolved in the absence of the metal compound. The detailed mechanism of the reactions that produce the volatile flame inhibitors is, however, not fully understood, although these systems have been used for several decades.

Antimony oxide–halogen combination represents one of the most important synergistic FR combinations. The ability of antimony to enhance the effectiveness of halogen-based flame retardants was first demonstrated for cellulosic fabrics treated with chlorinated paraffin and antimony trioxide (Little, 1947). Such synergism has also been shown in polyester resins (Learmonth and Thwaite, 1969), polystyrene resins (Lindemann, 1969), and polyolefins (Fenimore and Martin, 1966). The use of antimony compounds in conjunction with halogenated flame retardants is also documented for polyurethanes, polyacrylonitrile, and polyamides (Lyons, 1970). Zinc borate (used mainly in PVC), zinc hydroxystannate (ZHS), and zinc stannate (ZS) have primarily found use as alternative non-toxic synergists to antimony trioxide in PVC and other halogen-containing polymer systems. However, growing applications in halogen-free formulations have recently been found, and are particularly effective as partial replacements for hydrated fillers such as ATH and magnesium hydroxide, either in the form of powdered mixtures, or as coated fillers (Cusack, 2005).

The first studies (Little, 1947) conducted on the effect of antimony oxides and various chlorinated compounds on the flammability of cotton demonstrated that the flame retardance imparted increased with the ease of elimination of hydrogen halide from the chlorinated additive. Therefore, the active species inhibiting combustion was thought to be antimony oxychloride liberated within the polymer during heating; most experiments were thus carried out at an atomic ratio of antimony chloride of 1:1. Attempts to determine the optimum ratio led to the discovery that higher chlorine levels improved the efficiency of additives up to atomic ratios of about 1:2. The discrepancies were attributed to incomplete volatilization of hydrogen chloride. Coppick (1947) found for chlorinated alkane an optimal atomic ratio of about 1:3.

The effectiveness of halogen-based FRs depends largely on the strength and volatility of the C-Br bonds. The aromatic bromine compounds are less effective because they evaporate before decomposition of the strong C-Br bonds, whereas aliphatic compounds are more effective because of their poor stability.

Though halogen FRs are largely active in the gas phase, condensed or solid phase activity was observed in the case of chlorinated polyethylene. FRs containing both phosphorous and bromine clearly show condensed phase activity. Measurement of the limiting oxygen and nitrous oxide index of chlorinated polyethylene (Fenimore and Jones, 1966) suggests that, for this polymer, the flame-retardant influence of chlorine primarily takes place in the condensed phases. This suggestion is confirmed by the fact that the addition of quite large amounts of gaseous chlorine or hydrogen chloride to the nitrogen/oxygen atmosphere above the burning polyethylene does not affect the oxygen index.

Antimony oxide, Sb_2O_3 or Sb_4O_6, has no flame-retarding activity, but its presence greatly enhances the activity of halogenated compounds. The molar ratios of antimony to chlorine ranging from 1:1.7 to 1:2.9 provide the highest FR effect.

5.2 MECHANISM OF HALOGEN FLAME RETARDANTS

It is generally accepted that the combustion of gaseous fuel proceeds via a free radical mechanism. A number of propagating and chain branching mechanisms are illustrated in Equations 5.1–5.8.

$$CH_4 + O_2 \rightarrow \cdot CH_3 + H \cdot + O_2 \tag{5.1}$$

$$H \cdot + O_2 \leftrightarrow HO \cdot + O \cdot \tag{5.2}$$

$$O \cdot + H_2 \leftrightarrow OH \cdot + H \cdot \tag{5.3}$$

$$CH_4 + HO \cdot \rightarrow \cdot CH_3 + H_2O \tag{5.4}$$

$$\cdot CH_3 + O\cdot \rightarrow H_2CO + H\cdot \tag{5.5}$$

$$H_2CO + HO\cdot \rightarrow CHO\cdot + H_2O \tag{5.6}$$

$$CHO\cdot + O_2 \rightarrow H\cdot + CO + O_2 \tag{5.7}$$

$$CO + HO\cdot \rightarrow CO_2 + H\cdot \tag{5.8}$$

Here $H\cdot$, $HO\cdot$ and $O\cdot$ are radicals and chain carriers. The reaction of the $H\cdot$ radical and the O_2 molecule is an example of chain branching in which the number of carriers is increased.

The gas-phase activity of the active flame retardant consists of its interference in the combustion train of the polymer. Polymers, like other fuels, produce upon pyrolysis species capable of reaction with atmospheric oxygen, and produce the H_2–O_2 scheme that propagates the fuel combustion by the branching reactions shown in Equations 5.2 and 5.3 (Minkoff and Tipper, 1962).

In the radical trap theory of flame inhibition, it is believed that HBr competes for the radical species $HO\cdot$ and $H\cdot$ (Equation 5.9) that are critical for flame propagation (Green, 1996):

$$H\cdot + HBr \rightarrow H_2 + Br\cdot \tag{5.9}$$

The active chain carriers are replaced with the much less active $Br\cdot$ radical. This slows the rate of energy production resulting in flame-extinguishing. It also has been suggested that halogens simply alter the density and mass heat capacity of the gaseous fuel-oxidant mixture so that flame propagation is effectively prevented. This physical theory is equivalent to the way in which inert gases such as carbon dioxide and nitrogen may influence combustion (Larsen, 1974).

Suggestions have been made that the flame retardant mechanism of some bromine compounds acts mainly in the condensed phase, and also depends on the type of polymer being treated. Reaction of the flame retardant or its decomposition products with the polymer can inhibit the decomposition of the polymer, thereby influencing the flame retardancy (Weil and Levchik, 2009).

5.2.1 HALOGEN SYNERGISM

Antimony oxide itself usually renders no flame inhibition properties to polymers, but it is known as a synergist for halogen compounds. Antimony oxide is not volatile, but antimony oxyhalide (SbOX) and antimony trihalide (SbX_3) formed in the condensed phase, by reaction with the halogenated flame retardant, are volatile. They facilitate the transfer of halogen and antimony into the gas phase where they function. Antimony oxide flame retardants are therefore usually used indirectly in the form of antimony trichloride ($SbCl_3$) or antimony tribromide ($SbBr_3$). These forms are very effective retardants at typical flame temperatures.

Laboratory flammability tests indicate that the optimum halogen–antimony atom ratio in many polymers is about 2:1 to 3:1. It has been suggested that antimony halides are also highly active radical traps. Although the antimony halides appear to act exclusively in the vapor phase, some effect in the condensed phase cannot be ruled out (Pettigrew, 1993).

The impurities such as iron and aluminum, which came from catalyst residues, limited the use of the flame retardants, much like the fouling of a catalyst. The residues tend to bind with the flame retardant, making it less volatile and more prone to condense or precipitate on the surface. The levels of these impurities have since been reduced. Some aromatic bromine compounds, e.g., decabromo diphenyloxide, are thermally stable up to very high temperatures (Pettigrew, 1993).

Interference with the antimony – halogen reaction affects the flame retardancy of the polymer. For example, metal cations from color pigments and inert fillers such as calcium carbonate may lead to the formation of stable metal halides. These metal halides can render the halogen unavailable for reaction with the antimony. The result is that neither the halogen nor the antimony is transported into the vapor phase, where they provide flame retardancy. Silicones have also been shown to interfere with the flame retardant mechanism. Consequently, the total plastic composition must be considered in developing a new flame retardant product.

Other members of Group V of the periodic table, such as arsenic and bismuth, also function as synergists for halogens. Little work has been done with these compounds for toxicity reasons. The Diels-Alder adduct of hexachlorocyclopentadiene with 1,5-cyclooctadiene (Dechlorane Plus) can be used to flame-retard nylons, epoxies, and polybutylene terephthalate using synergists other than antimony oxide. These compounds include zinc compounds such as borates, oxides and phosphates, as well as iron oxides such as Fe_2O_3 and Fe_3O_4. The use of mixed synergists is also reported to lower the level of the total flame retardant required (Green, 1996).

At a temperature above $250°C$, the halogen compounds release hydrogen halide, which reacts with antimony oxide in the following ways (Equations 5.10–5.12):

$$R.HX \rightarrow R + HX \qquad (5.10)$$

$$Sb_2O_3 + 6HX \rightarrow 2SbX_3 + 3H_2O \qquad (5.11)$$

$$Sb_2O_3 + 2HX \rightarrow 2SbOX + H_2O \qquad (5.12)$$

The chemical analysis of volatiles shows the presence of $SbCl_3$. However, Brauman (1979) pointed out that at a later stage, the weight loss pattern during pyrolysis requires the formation of a less volatile Sb-containing moiety, such as SbOCl, and its decomposition products. $SbCl_3$ is released in the gas phase, whereas SbOCl, a strong Lewis acid, operates in the condensed phase by facilitating dissociation of C-HX bonds. In the case of polyester, the formation of higher molecular weight moieties and cross-links increases the amount of char. In the oxygen-containing atmosphere, a protective layer on the polymer surface is also produced.

Antimony mainly acts in the gas phase. After reaching the flame, antimony halides undergo a series of reactions releasing hydrochloric or hydrobromic acid. The antimony halides provide additional avenues for the hydrogen atoms to recombine, which are considerably faster than the usual recombination reactions in their absence. A fine dispersion of solid SbO and Sb is also produced in the flame, and catalyzes the H· combination. The lowering of H· concentration enhances FR effect. However, high cost and limited availability of the antimony trioxide have resulted in a search for a suitable alternate synergist. Zinc borate has claimed to be capable of replacing 50% of antimony oxide. Antimony tetraoxide and pentaoxide are less effective than trioxide.

A series of back-coated cotton fabrics comprising varying molar ratios of chlorine, bromine, and antimony have been studied using a simulated match ignition test, LOI, and thermal analysis (TGA and DTA). The results show that the presence of chlorine alone and in combination with bromine increases LOI and carbonaceous char values at optimum halogen–antimony molar ratios of 2.5–3.6:1 and bromine: chlorine molar ratios within the range 0.31–0.42. Only back-coated samples containing bromine pass the simulated match test, although at Br–Sb molar ratios above 4:1, LOI and char residual values decrease. It is evident that condensed and vapor phase flame retardant mechanisms are operating with efficiencies determined by halogen: Sb and Br/Cl molar ratios (Wang et al., 2000).

5.3   CLASSIFICATION OF HALOGEN FRs

Halogenated flame retardants can be divided into three classes: aromatic, aliphatic, and cycloaliphatic. Bromine and chlorine compounds are the only halogen compounds having commercial significance as flame retardant chemicals. Fluorine compounds are expensive and, except in special cases, are ineffective because the C-F bond is too strong. Iodine compounds, although effective, are expensive and too unstable to be useful (Cullis, 1987; Pettigrew, 1993). The brominated flame retardants are much more numerous than the chlorinated types because of their higher efficacy (Cullis, 1987).

With respect to processability, halogenated flame retardants vary in their thermal stability. In general, aromatic brominated flame retardants are more thermally stable than chlorinated aliphatics, which are more thermally stable than brominated aliphatics. Brominated aromatic compounds can be used in thermoplastic polymers at fairly high temperatures without the use of stabilizers, and at very high temperatures with stabilizers. The thermal stability of the chlorinated and brominated aliphatics is such that, with few exceptions, they must be used with thermal stabilizers, such as tin compounds.

Bromine-based flame retardants are highly brominated organic compounds with a relative molecular mass ranging from 200 to that of large molecule polymers. They usually contain 50% to 85% (by weight) of bromine (Cullis, 1987).

There are four main groups of commercially produced brominated FRs (BFRs):

1. Hexabromocyclododecane (HBCD) (Structure 5.1(a));
2. Tetrabromobisphenol A (TBBPA) (Structure 5.1(b));
3. Polybrominated biphenyl (PBB) (Structure 5.1(c)); and
4. Three commercial mixtures of polybrominated diphenyl ethers (PBDEs) (Structure 5.1(d)): decabromo-diphenyl ether (deca-BDE), octabromodipheylether (octa-BDE), and pentabromo-diphenyl ether (Penta- BDEs) (Shaw et al., 2010).

The highest volume brominated flame retardant in use istetrabromobisphenol A (TBBPA) (Structure 5.1(b)) (IPCS, 1995), followed by decabromodiphenyl ether (DeBDE) (Structure 5.1(d)) (IPCS, 1994). Both of these flame retardants are aromatic compounds. The primary use of TBBPA is as a reactive intermediate in the production of flame-retarded epoxy resins used in printed circuit boards (IPCS, 1995). A secondary use for TBBPA is as an additive flame retardant in ABS systems. DeBDE is the second largest volume brominated flame retardant and is the largest volume brominated flame retardant used solely as an additive. The greatest use (by volume) of DeBDE is in high-impact polystyrene, which is primarily used to produce television cabinets. Secondary uses include ABS, engineering thermoplastics, polyolefins, thermosets, PVC, and elastomers. DeBDE is also widely used in textile applications as the flame retardant in latex-based back coatings (Pettigrew, 1993). Hexabromocyclododecane (HBCD), a major brominated cyclo-aliphatic flame retardant, is primarily used in polystyrene foam. It is also used to flame-retard textiles.

Chlorine-containing flame retardants belong to three chemical groups: aliphatic compounds, cycloaliphatic compounds, and aromatic compounds. Chlorinated paraffins are by far the most widely used aliphatic chlorine-containing flame retardants. They have applications in plastics, fabrics, paints, and coatings. Bis(hexachlorocyclopentadieno) cyclo-octane is a flame retardant having unusually good thermal stability for a chlorinated cycloaliphatic. In fact, this compound is comparable in thermal stability to brominated aromatics in some applications. It is used in several polymers, especially polyamides and polyolefins for wire and cable applications. Its principal drawback is the relatively high use levels required, compared to some brominated flame retardants

(a)   (b)   (c)   (d)

STRUCTURE 5.1   Types of brominated FRs.

(Pettigrew, 1993). Aromatic chlorinated flame retardants are not used for flame-retarding polymers.

There are about 75 different commercial brominated FRs, each with specific properties and toxicological behavior. They all contain bromines and act in the vapor phase by a radical trap mechanism.

5.3.1 POLYBROMINATED BIPHENYLS (PBB) AND POLYCHLORINATED BIPHENYLS (PCB)

PBB (Structure 5.1(c)) consists of two benzene rings, but they are directly linked, i.e., not linked via an oxygen atom as is the case for PBDE (Structure 5.1(d)). As for PBDE, the maximum number of bromine atoms is 10. From a chemical point of view, PBB resembles PBC. PBB, however, cannot be compared to PCB in that bromine atoms replace chlorine atoms in PCB. Also, the atomic weight of bromine is 80, while the weight of chlorine is 35.5. Finally, at room temperature, PCB is a partly volatile liquid, while the PBB flame retardants are solid with very low volatility.

PCBs ($C_{12}H_{10-n}Cl_n$), have been widely used as fire retardants before their very high toxicity was discovered in the 1960s (Jensen, 1966). They are a class of organic compounds with 1–10 chlorine atoms attached to biphenyl.

Unfortunately, in an unwise decision, PCBs were replaced by chemically analogous PBBs as fire retardants without any prior toxicity evaluation. PBBs were substituted, with "safer" polybromo diphenyl (or biphenyl) ethers (PBDEs) (Structure 5.1(d)) or oxides (PBDOs), after a very thorough toxicity study that determined that PBDEs are biodegradable owing to replacement of the biphenyl link of bioaccumulating polychlorinated biphenyls (PCBs) and polybrominated biphenyls (PBBs) with the ether bond.

5.3.2 POLYBROMINATED DIPHENYLETHERS (PBDE)

Chemically, PBDE (Structure 5.1(d)) consists of two benzene rings linked with an oxygen atom. Instead of hydrogen, a different number of bromine atoms may be attached to the benzene rings. Up to 10 bromine atoms can attach to the diphenyl ether molecule (e.g., Decabromo diphenyl oxide (DBDPO), Structure 5.2). Usually, at least half of the possible maximum must be brominated to obtain a sufficient effect of flame protection. Normally the congener

STRUCTURE 5.2 Chemical structure of DBDPO.

may be deca-, octa-, hepta-, hexa-, or penta-Bromo DPO, depending on the number of bromine atoms attached are used.

5.3.3 The Congener

The *congener* refers to one of many variants or configurations of a common chemical structure. For example, polychlorinated biphenyls (PCBs) occur in 209 different forms, or congeners. Each congener has two or more chlorine atoms located at specific sites on the PCB molecule. Since the size of a bromine atom is comparatively large, the characteristics of the different congeners vary.

Low-brominated PBDEs, which are not usable as flame retardants, cannot be compared to high-brominated PBDEs. The latter have a high molecular weight and high thermal stability. The major applications are in styrene, polyolefins, polyesters, polyamides, and textile materials.

Polybrominated diphenyl ethers (PBDEs) are ubiquitous chemicals used as flame retardants in a wide range of consumer products, including televisions, computers, electronics, motor vehicles, carpets, and furniture. Health effects of PBDE exposure include damage to the neurological, reproductive, immune, and hormonal systems. The most widely used congener in this group, decaBDE, is also a suspected carcinogen.

5.3.4 Tetrabromobisphenol–A (TBBPA)

TBBPA is mainly used in epoxy resins, where it reacts into the polymer backbone and is bound as part of the polymer.

5.3.4.1 Brominated Polystyrene

Brominated polystyrene is commonly used in polyesters and polyamides. It is itself a polymer, and is, therefore, quite immobile in the matrix.

5.3.4.2 Brominated Phenols

Tribromophenol, a reactive FR, is often used as intermediate in the manufacture of polymeric brominated FRs.

5.3.5 Tetrabromophthalic Anhydride (TBPA)

TBPA is often used as a reactive FR in unsaturated polyesters used to manufacture of circuit boards and cellular phones (Structure 5.3).

Halogenated flame retardants are either added to, or reacted with, the base polymer. Additive flame retardants are those that do not react with the polymer. There are a few compounds that can be used as an additive in one application and as a reactive in another; TBPA is the most notable example. Reactive flame retardants become a part of the polymer either by becoming a part of the backbone or by grafting onto the backbone. The choice of a reactive flame retardant is more complex than the choice of an additive type. The development of systems based on reactive flame retardants is more expensive

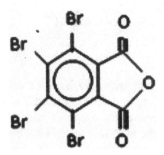

STRUCTURE 5.3   Tetrabromophthalic anhydride (TBPA).

for the manufacturer, who in effect has to develop novel copolymers with the desired chemical, physical, and mechanical properties, as well as the appropriate degree of flame retardance (Cullis, 1987; Pettigrew, 1993).

5.4 FLAME RETARDANCY OF WOOL

Titanium chloride is an effective flame retardant agent for wool. In the presence of an alpha hydroxy carboxylic acid (citric acid), a stable complex is formed that almost completely exhausts when applied at a pH < 3.5 at the boil. It is important that the titanium should not be allowed to hydrolyze to TiO_2 because as TiO_2, titanium is ineffective. This process is good for wool used in protective clothing, carpets, and upholstery. Titanium complexes make wool yellower.

An effective FR finish can be applied on wool at pH 3 (with citric acid) with potassium (hexa) fluoro zirconate or (hexa) fluoro titanate at 50°–100°C, by exhaustion (after or during dyeing with levelling or 1:1 metal-complex dyes) or by the padding method. At least 0.8% Zirconium is necessary to meet stringent FR specifications. The anions are strongly retained by the amino groups of wool by ionic and polar bonds, similar to dye anions. The flame retardancy takes place in the condensed phase by formation of ions/ compounds that enhance or catalyze char formation. The finish is durable to dry cleaning and washing up to 40°C at pH <6. At higher pH, ineffective zirconium oxide forms. The physical properties of wool fiber and its hand are not significantly affected. Titanate is more effective and cheaper, but a yellow tint is imparted that darkens on exposure to light.

5.4.1 ZIRPRO PROCESS

The need to boil wool for exhaustion of titanium and zirconium chelates leads to felting and lower dye fastness, and the process is energy intensive. To overcome these deficiencies, International Wool Secretariat (IWS) developed a process based on titanium and zirconium hexafluoride, known as Zirpro Process. Hexafluorotitanates and hexafluorozirconates are extremely stable in acid solutions and exhaust onto wool well below the boil. The titanates turn wool yellower. Hence, the zirconates are preferred commercially. K_2ZrF_6 at a pH < 3 gives 77% exhaustion at 50°C for 30 minutes and good levelling is achieved at 70°C. A bath containing 3% K_2ZrF_6, 10% HCl (37%) for 30 minutes at 75°C, gives rise to a washfast, lightfast, and improved heat- and flame-resistant fabric. Pad-batch and pad-dry application processes may also be used. It is accepted that the hexafluorozirconate ion is bound to the cationic wool in the same manner as acid dyes. Flame resistance on wool durable to dry laundering can be obtained with a number of halogen-containing compounds.

Chlorendic acid, or 1,4,5,6,7,7-hexachlorobicyclo[2.2.1]-hept-5-ene-2,3-dicarboxylic acid, is a chlorinated hydrocarbon used in the synthesis of some flame retardants and polymers. It is a common breakdown product of several organochlorine insecticides. It is also called HET acid, hexachloro-endo-methylenetetrahydronaphthalic acid, when reacted with nonhalogenated glycols, it forms halogenated polyols used as flame retardants in polyurethane foams. It is also used for production of dibutyl chlorendate and dimethyl chlorendate, which are used as reactive flame retardants in plastics.

Tetrabromophthalic anhydrides (TBPA) (Structure 5.3) and chlorendic acid (Structure 5.4) exhaust into wool fibers. Add-ons of 6%–18% give good flame retardancy, dye compatibility. They are reasonably durable to dry cleaning.

5.5 HALOGANATED FRS AND ENVIRONMENT

Halogenated FRs (HFRs) are widely used in consumer products because of their low impact on other material properties and the low loading levels necessary to meet the required flame retardancy. However, they have raised concerns due to their persistency, bioaccumulation on living organisms, and their potential toxic effects on human health.

The use of brominated FRs results in long-term carcinogenic and mutagenic effects, regardless of their concentration. During thermal stress, brominated FRs are converted into dioxins and furans, which are highly toxic,

STRUCTURE 5.4 Chemical structure of Chlorendic acid.

and can cause reproductive problems and damage the immune system. The cause of dioxin formation is often connected with the use of brominated FRs in combination with antimony oxides as synergist additives. Furthermore, brominated FRs are highly resistant to biological, chemical, and physical degradation. However, if their degradation occurs under certain environmental conditions, they are converted into even more persistent and toxic lower-brominated compounds (Horrocks, 2013). The most up-to-date and thorough reviews of the toxicology of BFRs including PBDEs, TBBPA, and HBCD has been discussed by Darnerud (2003) and many others (Janssen, 2005). In general, HBCD, TBBPA, and PBDEs are absorbed from the gastrointestinal tract and accumulate in fatty tissues. None of the BFRs discussed in this report appear to cause immediate symptoms from acute toxicity at average doses. Rather, like PCBs, health effects from chronic exposure, particularly in developing infants and wildlife, are of more concern.

Overall, the available literature on BFR toxicology is incomplete. Based on the available data, however, we know that BFRs are associated with several health effects in animal studies, including neurobehavioral toxicity, thyroid hormone disruption, and (for some PBDE congeners) possibly cancer. Additionally, there are data gaps, but some evidence that BFRs can cause developmental defects, endocrine disruption, immunotoxicity, and reproductive and long term effects, including second generation effects. Some evidence is available for estrogenic activity of PBDEs and TBBPA, but more studies are needed to determine if low-dose exposures have estrogenic activity in humans or other species (Darnerud, 2003).

5.5.1 ENDOCRINE DISRUPTION: THYROID FUNCTION

Man-made chemicals that interfere with the normal action of hormones are known as endocrine disruptors. One of the most sensitive end points of PBDE and TBBPA toxicity observed in animal bioassays appears to be effects on thyroid function. PBDEs and TBBPA bear a strong structural similarity to thyroid hormone and have been extensively studied for effects on thyroid function.

There are several ways that BFRs can interfere with thyroid hormone activity and/or function. Studies illustrate that PBDEs, TBBPA, and PCBs lower thyroid hormone levels to varying degrees and with varying consistency. PCBs also interfere with thyroid hormone activated gene transcription (Zoeller et al., 2002). A recent laboratory study has shown that HBCD, but not TBBPA or deca-BDE, can stimulate genes normally triggered by thyroid hormone (Yamada-Okabe et al., 2005). BFRs also can bind to the thyroid hormone receptor with varying affinities (as noted subsequently). Moreover, BFRs can bind to thyroid transport proteins with varying affinities (as noted subsequently), where the relevance to humans is uncertain. Lastly, some human studies of occupational exposure indicate that workers show hypothyroidism.

In laboratory models, PBDEs, hydroxylated metabolites of the lower brominated PBDEs, and TBBPA can bind to the thyroid hormone receptor, as well as to the PBDEs, hydroxylated metabolites of the lower brominated PBDEs, and TBBPA can bind to the thyroid hormone receptor and to transport protein, transthyretin. Transthyretin is responsible for binding and transporting thyroid hormone in the blood and interference with its function can result in lower thyroid hormone levels (Darnerud, 2003).

An increase in thyroid cell numbers or hyperplasia, a potentially precancerous condition, has also been observed in studies of mice and rats exposed to PBDEs (Legler and Brouwer, 2003).

5.5.2 CARCINOGENICITY

There is general agreement within the scientific community that there are insufficient data to evaluate the carcinogenicity of BFRs. Only one form of PBDEs, deca-BDE, has been tested for carcinogenicity in animals (Darnerud, 2003). It has caused statistically significant increase in hepatocellular (liver) carcinomas and marginal increase in thyroid follicular cell carcinomas in mice. The International Agency for Research on Cancer (IARC) has classified deca-BDE as having limited evidence for a carcinogenic effect in animals, but stopped short of classifying the substance as carcinogenic in humans (IARC, 1990).

It is worrisome, however, that deca-BDE, considered to be poorly absorbed and of relatively low toxicity among the PBDEs, showed evidence of carcinogenicity in rodent studies. The entire group of chemicals may pose a cancer concern. Because the highest environmental and tissue levels of PBDEs are Penta-BDE congeners, it is critical that these PBDEs be tested for their potential to cause cancer.

HBCD was tested for carcinogenicity in mice and found to be associated with increased incidence of liver tumors, but there were inconsistencies in the study, which bears repeating (Darnerud, 2003). TBBPA has not been tested for carcinogenicity in whole animal studies.

Both HBCD and lower brominated PBDEs are capable of inducing genetic recombination in mammalian cell lines, a possible indicator of carcinogenicity, similar to other environmental contaminants such as DDT and PCB (deWit, 2002). Conversely, TBBPA did not induce genetic recombination in two cellular assays using mammalian cells (Darnerud, 2003).

5.5.3 OTHER ENDOCRINE DISRUPTING CONCERNS

In addition to interfering with thyroid hormone function, some studies have examined the capacity of BFRs to interfere with estrogen hormone action. Depending on the PBDE congener being studied, some PBDEs have estrogenic activity whereas others are anti-estrogenic.

5.5.4 DIOXIN-LIKE EFFECTS

Dioxin-like activity, as displayed by the prototypical dioxin, TCDD (2,3,7,8-tetrachlorodibenzo-p-dioxin), leads to a wide variety of toxic endpoints including mortality, carcinogenicity, teratogenicity, and immunotoxicity. These effects all mediate through a cellular protein known as the Ah (aromatic hydrocarbon) receptor. Many structurally similar chemicals have been found to bind to this receptor with very different affinities. As some PCBs bind to the Ah receptor, researchers have studied BFRs for Ah-binding and dioxin-like activity (Janssen, 2005).

Brominated flame retardants are not only found in numerous household, health care, and consumer products, but they are now everywhere in our environment. In the last decade, scientists have detected BFRs in both human and wildlife tissues, as well as in house dust, sediments, sewage sludge, air, soil, and water samples. PBDEs and HBCD have been found in air samples of remote regions such as the Arctic, as well as in marine mammals from the deep seas, indicating long range transport of BFRs (Alaee et al., 2003).

The entire life cycle of BFRs likely contributes to their distribution in the environment. Industrial facilities that produce BFRs, as well as manufacturing facilities that incorporate BFRs into consumer products, release these chemicals during polymer formulation, processing, or manufacturing practices. Disintegration of foam products, volatilization (especially under conditions of high temperature), and leaching from products during laundering or use, results in the release of BFRs from products in homes and businesses. Finally, disposal of products, including combustion and recycling of waste products, as well as leaching from landfills, is the final route of entry for BFRs into the environment (deWit, 2002).

Many BFRs are highly lipophilic (fat-soluble), rather than water soluble. BFRs also have a high affinity for binding to particles, which is reflected in low measurements in water samples and higher measurements in sediment, sewage sludge, and particulate samples such as dust particles (Birnbaum and Staskal, 2004).

BFRs are generally very stable and resistant to degradation. The tendency of an environmental contaminant to resist physical, biological, and chemical degradation is called *persistence*. The persistence of a substance in a given medium is scientifically defined by its overall half-life in a medium such as soil, water, or sediment (Environmental Protection Agency, 1999).

Although new production of Penta-BDE and Octa-BDE is being phased out voluntarily in the United States and the substances have been banned in the European Union, a large number of products containing these flame retardants are still in use. This means that their release into the environment will continue throughout product lifecycles, potentially for several more decades. In addition, the increasing use of Deca-BDE makes it important to understand its degradation in relation to the occurrence of lower-brominated PBDEs in environmental samples.

Several studies have shown that lower-brominated congeners are the most toxic and are accumulating at the highest rates and levels in wildlife and human tissue samples (Birnbaum and Staskal, 2004).

Studies have shown the higher-brominated PBDEs, such as deca-BDE, undergo degradation that removes bromine atoms, resulting in the formation of the more persistent and toxic lower-brominated compounds. Several studies have found that deca-BDE breaks down to lower-brominated congeners (nona to hexa- BDEs) in sand, sediment, and soils under laboratory conditions of both artificial and natural sunlight. The breakdown of deca-BDE occurs much more quickly in UV light (half-life < 30 min), compared to natural sunlight where the estimated half-life is 53 hours on sediment and 150–200 hours on soil (Soederstroem et al., 2004). The degradation products can be more toxic than the original compounds and include lower brominated PBDEs, brominated bisphenols, and polybrominated dibenzodioxins (PBDDs)/polybrominated dibenzofurans (PBDFs).

Given the ubiquity and persistence of BFRs in our environment, it is not surprising that these chemicals find their way into tissues of both wildlife and humans. Similar to PCBs, concentrations of BFRs increase up each step of the food chain, indicating that these chemicals are readily absorbed by the body where they accumulate in fatty tissues. The lower-brominated compounds accumulate in the highest concentrations, indicating that these compounds are either preferentially absorbed, or that metabolic breakdown of higher brominated compounds occurs ((Birnbaum and Staskal, 2004).

In the past, higher-brominated BDEs such as deca-BDE were thought to be too large for absorption, and therefore, not bioavailable to organisms (Darnerud et al., 2001). Industry has used this argument to suggest that the use and accumulation of deca-BDE in the environment is not harmful. However, recent evidence has proven otherwise. Deca-BDE was detected in Swedish peregrine falcon eggs (<20–430 ng/g lipid weight) and freshwater fish (median 48 ng/g lipid weight), indicating that deca-BDE is bioavailable ((Lindberg et al., 2004).

There is also evidence for the metabolism of Octa-BDE and Deca-BDE formulations to penta-octa BDE congeners in fish (Stapleton et al., 2004). These studies indicate that higher- brominated PBDEs are absorbed and metabolized to lower-brominated PBDEs, which are highly bioaccumulative and toxic.

When chlorinated and brominated flame retardants burn, they release not only smoke or carbon monoxide, but also highly toxic substances such as dioxins and furans. These toxic chemicals are released into the environment during manufacturing and when products containing them are discarded. As products degrade, PBDE also ends up in household dust, where they can be inhaled and ingested. The finding that PBDEs are rapidly accumulating in humans and in the environment has raised serious concerns. Norway, the European Union, and several US states have banned these chemicals for health and environmental reasons (David Suzuki Foundation, 2016).

PBDEs and polybrominated biphenyls (PBBs) have been restricted in the European Union since 2004 (European Parliament and Council, 2011), and the production of penta- and octa-BDE has been phased out in the United States since 2004 (US-EPA, 2008). Penta-BDE and octa-BDE have been identified as persistent organic pollutants (POP) (SFT, 2009), and Penta-BDE is already listed as a hazardous substance under the Water Framework Directive (WFD), while octa-BDE and deca-BDE are listed among the substances to be monitored (European Parliament and Council, 2000). Deca-BDE and HBCD were included by the European Chemicals Agency (ECHA) in the list of candidates for Author-isation under REACH (The Regulation on Registration, Evaluation, Authorisa-tion and Restriction of Chemicals) Article 57d as substances of very high concern (SVHC) based on PBT (persistent, bioaccumulating, toxic) properties. TBBPA is registered under REACH and is currently not subject to any REACH restriction processes. Tris(2-chloroethyl) phosphate (TCEP) is a representative of the chlorin-ated FRs and has been included in Annex XIV of REACH ("Authorisation List") because of concerns over its reproductive toxicity (Cat 1B).

The ban, adopted by the European Commission on October 1, means that from March 1, 2021 the use of halogenated flame retardants in enclosures and stands of electronic displays will be prohibited. Meanwhile, outside the EU, the US Consumer Product Safety Commission voted in 2017 to grant an NGO petition to ban nonpolymeric, additive organohalogen flame retardants (OFRs) from plastic electronic casings and three other product categories (https://chemicalwatch.com/).

REACH restrictions on flame retardants (Annex 17) include:(www.chemsa fetypro.com/Topics/EU/REACH_annex_xvii_REACH_restricted_substance_ list.html) the following FRs:

- Pentabromodiphenyl ether* (PentaBDE, 0,1% w/w)
- Octabromodiphenyl ether* (OctaBDE, 0,1% w/w)

Not allowed in articles for skin contact (e.g. textiles):

- Tris(aziridinyl)phosphinoxide
- Tris (2,3 dibromopropyl) phosphate (TRIS)
- Polybromobiphenyls (PBB)

Annex 14 (Candidate) List of Substances of Very High Concern for Authorization:

- Hexabromocyclododecane (HBCD) – PBT substance
- Tris(chloroethyl)phosphate (TCEP) – Reprotox Cat. 1b
- Alkanes, C10-13, chloro (Short Chain Chlorinated Paraffins) – PBT and vPvB
- Boric Acid – Reprotox
- Trixylylphosphate (TXP) – Reprotox Cat. 1b

* as commercial formulations, i.e., including other congeners

Because electronics is a major application area for flame retardants, nongo-vernmental organizations (NGOs), environmental scientists, and authorities have focused their attention on this industry to date. Over the last decade, the fate of electronic waste and hazardous materials contained therein met with increasing political attention and led to the WEEE (Waste Electrical and Electronic Equip-ment Directive, 2012/19/EU on waste electrical and electronic equipment) and RoHS Directives (Restriction of Hazardous Substances Directive2002/95/EC,

February 2003) in Europe. The aim of these regulations is that electronic waste shall be recovered and recycled properly, and that new equipment shall not contain any problematic substances. Flame retardants are affected, because according to the WEEE directive, plastics containing brominated flame retardants have to be separated before further treatment of the waste.

5.6 HALOGEN-FREE FRs

With the worldwide use of brominated FRs in the past and their persistence, different brominated substances have been bioaccumulated in the food chain, which represents a serious health risk for the general population. Recent environmental and human monitoring showed that brominated compounds are present in indoor and outdoor air, dust, water, different foods, and in the human body despite the fact that the use of brominated FRs is almost completely prohibited (Frommea et al., 2016).

Scientists are developing eco-friendly alternatives to the commonly used halogenated agents that are subject to ever stricter regulation due to potential toxicity. Humans may be exposed to HFRs in various ways such as emissions during production, use of HFR-based products, leaching from landfills, combustion, or recycling at the end of the products' life.

Recently, due to perceived environmental issues, researchers are stimulated to design effective, but more environmentally-friendly products. As a result, in the last 30 years, most of the high performing halogen- or formaldehyde-based–flame retardants for fabrics have been banned or limited from commercial use, thus favoring the use of phosphorus-containing products.

According to a recent report from the Swiss science and business consultancy firm Acon AG, the worldwide market for halogen-free FRs is set to increase strongly from US$1.62 billion in 2005 to $2.72 billion in 2010, representing a global compound annual growth rate of about 11% (Additives for Polymers, December 10, 2006). In Western Europe, the United States, and Japan, public consciousness of potential hazardous halogenated products, industrial end-user initiatives, and environmental legislation are together driving the market trend toward halogen-free products, which is an opportunity for a growing demand of intumescent products. These last-mentioned products are phosphorus-based compounds which are expected, along with mineral FRs, to show the fastest growth. This study also reports that nanotechnology will play a key role in improving fire-retardant performance and reducing production costs. It makes sense because the benefit of combining nanoparticles with conventional FRs is known. Some advantages include the decrease of the total loading, the increase of the mechanical properties, the multifunctionality, and the strong increase of the flame retardancy (Bourbigot and Duquesne, 2010).

In recent years, driven by the urgent need for environmental protection, research on the environment-friendly halogen-free flame retardant (HFFR), the so-called green flame retardant, has received considerable attention.

A variety of halogen-free FRs is available on the market, such as organic (phosphorus and nitrogen based chemicals) and inorganic (metals) materials.

The Stockholm Convention (2001) listed 23 organo-halogen chemicals (including all BFRs) to be banned globally (Leonard and Fenge, 2003). The challenges of finding substitutes for banned halogen-containing flame retardants include:

1. Lower flame retardancy: Only 10 wt% of BFRs is equivalent to about 30%–50% by wt of inorganic FRs.
2. High cost: Inorganic FRs is low-,but requires high loading. New HFFRs, on the other hand, are mostly costlier. The cost ratios (£) of BF-FR with phosphorous-based and non-phosphorous-based HFFR are about 1:6 and 1:2, respectively.
3. Higher thermal stability: Many HFFRs decompose at higher temperature (about 400°C) than their halogenated counterparts (about 330°C).
4. Melt-dripping: Melt dripping occurs with most polymers, e.g., polyethylene, polypropylene, PET, and ABS. Flammable drips act as secondary ignition sources. FRs should make drips nonflammable.
5. Deterioration in mechanical properties: The deterioration of mechanical properties increase proportionately with the amount of FR. Therefore, the deterioration is more with less efficient FRs (Roy Choudhury, 2018).

In 2009, the nonhalogenated phosphorus, inorganic and nitrogen flame retardants association (PINFA) was founded as a sector group of CEFIC, the European Chemical Industry Council. PINFA (2019) represents the manufacturers and users of the three major technologies of nonhalogenated flame retardants, namely, phosphorus, inorganic FRs, and nitrogen FRs. Most of fire deaths are due to smoke toxicity, which are claimed to be reduced by PINFA FRs. The members of PINFA share the common vision of continuously improving the environmental and health profile of their flame retardant products and offering innovative solutions for sustainable fire safety.

Halogen-free flame retardants (Birnbaum and Staskal, 2004) cover a variety of chemicals which are commonly classified (PIN) as:

1) Phosphorus-based flame retardants include organic and inorganic phosphates, phosphonates, and phosphinates, as well as red phosphorus, thus covering a wide range of phosphorus compounds with different oxidation states.
2) Inorganic category comprises mainly metal hydroxides such as aluminum hydroxide and magnesium hydroxide. Other compounds such as zinc borate are used to a much smaller extent.
3) Nitrogen-based flame retardants are typically melamine and melamine derivatives (e.g., melamine cyanurate, melamine polyphosphate, melem, melon). They are often used in combination with phosphorus-based flame retardants.

Intumescent flame retardants are an example of a typical mechanism of halogen-free flame retardants. The combustible material is separated from the fire or heat source by an insulating foam forming at the surface. Intumescent flame retardant systems can be applied to decrease flammability of thermoplastic polymers, such as polyethylene, polypropylene, polyurethane, and polyester and epoxy resins.

PIN FR's used in transportation textiles are (PINFA, 2010):

• Aluminum-tri-hydroxide (ATH)	• Aluminum phosphinate
• Amino-Ether-HALS derivatives	• Ammonium phosphate
• Ammonium polyphosphate	• Ammonium sulfamate
• Ammonium sulfate	• Cyclic phosphonate
• Dicresyl phosphate	• Diethyl phosphinic acid
• Guanidine phosphate	• Isopropyl phosphate ester
• Melamine powder	• Melamine cyanurate
• Melamine phosphate	• Melamine polyphosphate
• Methyl phosphonic acid	• Amidino-urea compound
• Oxaphosphorinane powder	• Oxy-bis-dimethyl sulfide
• Potassium hexafluoro titanate	• Urea
• Zinc borate	• Zirconium acetate

In the transport sector, it is mainly the automotive industry that has to deal with specific environmental legislation and concerns related to chemicals. In Europe, the End-of-Life Vehicles Directive was published in 2000 (ELV, 2000/53/EC). The directive aims at making vehicle dismantling and recycling more environmentally friendly. It defines clear, quantified targets for reuse, recycling, and recovery of vehicles and components and encourages producers to manufacture new vehicles which are easy to recycle.

5.7 FUTURE TRENDS

Flame retardants have become a class of chemicals that are increasingly attracting scientific and public attention due to their environmental and health effects. Halogen-free technologies already assume a major share of the market, with a strong dominance in North America and Europe. Asia is the leading consumer of brominated and chlorinated flame retardants, because they are mainly applied in electronic devices, and manufacture of the latter has moved from Europe and North America to Asia in recent decades.

Brominated flame retardants (BFRs) are used in a variety of consumer products and several of those are produced in large quantities. The concentration of BFRs in products ranges from 5%–30%. The total world demand for the highest production volume BFRs in 2001 was estimated by the bromine industry at 203,740 metric tons, with TBBPA and Deca-BDE being the most widely used BFRs (BSEF, 2001). Deca-BDE is used in virtually every type of plastic polymer that requires a flame retardant. BFRs are generally very stable and resistant to degradation. Based on the available data, however, we know that BFRs are associated with several health effects in animal studies, including neurobehavioral toxicity, thyroid hormone disruption, and (for some PBDE congeners) possibly cancer. One method to reduce flammability is to replace highly flammable plastics that release toxicants when burned with materials that are more inherently flame-resistant. Finally, when chemical-free alternative materials or designs are not feasible, nonhalogenated flame retardants can be used to meet fire safety standards.

High flow during molding, good compromise between impact and stiffness, and tolerance to high temperature when in use are some of the requirements for the plastic parts used in the production of low cost component plastics for the electronics and automobile industries. Due to miniaturization and the consequent increase in operating temperature, more stringent flame retardancy properties are needed. Moreover, under the pressure from environmentalists, the European market has been looking for diphenyl oxide-free flame retardant systems. As a result, a new generation of multipurpose, environmentally friendly brominated flame retardants has been recently introduced that offers additional benefits which widen the usage of the host polymeric systems. Flame retardants with appropriate softening temperatures such as brominated indan (Reyes, 1995), tris(tribromophenyl) cyanurate (Reyes, 1996) and tris(tribromoneopentyl) phosphate (Squires, 1996). provide processing aid effects and better flow properties. Reduced cycle times during injection moulding are possible with these flame retardants and they enable production of parts with thinner walls. According to recent testing, FR-1808 (brominated indan DSBG), is not considered to pose any risk to the health of the general population or to the environment (Horrocks and Price, 2001).

Halogen-free flame retardants cover a variety of chemicals which are commonly classified as: inorganic (mainly metal hydroxides such as aluminum hydroxide and magnesium hydroxide), phosphorus-based flame retardants (include organic and inorganic phosphates, phosphonates, and phosphinates), and nitrogen-based flame retardants (typically melamine and melamine derivatives). Intumescent flame retardants are an example of a typical mechanism of halogen-free flame retardants. The combustible material is separated from the fire or heat source by insulating foam forming at the surface. Intumescent flame retardant systems can be applied to decrease flammability of thermoplastic polymers, such as polyethylene, polypropylene, polyurethane, polyester resins, and epoxy resins.

REFERENCES

Alaee M., Arias P., Sjodin A., and Bergman A. (2003). An overview of commercially used brominated flame retardants, their applications, their use patterns in different countries/regions and possible modes of release, *Environment International*, 29, 683–689.

Birnbaum L.S. and Staskal D.F. (2004). Brominated flame retardants: Cause for concern?, *Environmental Health Perspectives*, 112, 9–17.

Bourbigot S. and Duquesne S. (2010). *Intumescence-Based Fire Retardants in Fire Retardancy of Polymeric Materials*, 2nd edition. C.A. Wilkie and A.B. Morgan, Eds., CRC Press, USA.

Brauman S.K. (1979). Friedel–crafts reagents as charring agents in impact polystyrene, *Journal of Polymer Science: Polymer Chemistry Edition*, 17(4), 1129–1144. DOI: 10.1002/pol.1979.170170417.

Bromine Science and Environmental Forum (BSEF) (2001). Brominated flame retardants in consumer and commercial products, www.bsef.com, accessed in 2004.

Coppick S. (1947). Metallic oxide-chlorinated: Fundamentals of process, in: R.W. Little, Eds., *Flameproofing Textile Fabrics*, Monograph No. 104, Reinhold Publishing Company, New York, pp. 239–248.

Cullis C.F. (1987) Bromine compounds as flame retardants, *Proceedings of the International Conference on Fire Safety*, 12, 307–323.

Cusack P.A. (2005). *Proceedings of High Performance Fillers*. Rapra Technology, Cologne, Germany, Paper 6.

Darnerud P.O. (2003). Toxic effects of brominated flame retardants in man and wildlife, *Environment International*, 29, 841–853.

Darnerud P.O., Eriksen G.S., Johannesson T., Larsen P.B., and Viluksela M. (2001). Polybrominated diphenyl ethers: Occurrence, dietary exposure, and toxicology, *Environmental Health Perspectives*, 109, 49–68.

David Suzuki Foundation (2016). Toxic flame retardants are a burning issue, www.david suzuki.org/, accessed on 23.3.16.

Despinasse M.C. (2016). Influence of flame retardant structures and combinations on the fire properties of bisphenol a polycarbonate/acrylonitrile-butadiene-styrene, Ph. D thesis, Department of Biology, Chemistry and Pharmacy of Freie Universität, Berlin.

deWit C.A. (2002). An overview of brominated flame retardants in the environment, *Chemosphere*, 46, 583–624.

European Parliament and Council (2011). Directive 2011/65/EU of the European Parliamentand of the Council of 8 June 2011 on the restriction of the use of certain hazardous substances in electrical and electronic equipment, *Official Journal of the European Union*, 174(88), 1–23.

Environmental Protection Agency (1999). Persistent Bioaccumulative Toxic (PBT) chemicals; Lowering of reporting thresholds for certain PBT chemicals; addition of certain PBT chemicals; community right-to-know toxic chemical reporting, *Federal Register*, 64(209), 58666–58753.

Fenimore C.P. and Jones G.W. (1966). Modes of inhibiting polymer flammability, *Combustion and Flame*, 10(3), 295–301.

Fenimore C.P. and Martin F.J. (1966). Flammability of polymers, *Combustion and Flame*, 10(2), 135–139.

Frommea H., Becher G., Hilger B., and Völke W. (2016). Brominated flame retardants – Exposure and risk assessment for the general population, *International Journal of Hygiene and Environmental Health*, 219, 1–23. DOI: 10. 1016/j.ijheh.2015.08.004.

Green J. (1996). Mechanisms for flame retardancy and smoke suppression – A review, *Journal of Fire Sciences*, 14, 426–442.

Horrocks A.R. (2013). A flame retardant and environmental issues, in: J. Alongi, A. R. Horrocks, F. Carosio, and G. Malucelli, Eds., *Update on % flame retardant textiles: State of the art, environmental issues and innovative solutions*. Smithers Rapra Technology Ltd, Shawbury, pp. 207–239.

Horrocks A.R. and Price D. (2001). *Fire Retardant Materials*, 1st edition. Woodhead Publishing, Cambridge, UK.

IARC (1990). *Monographs: Some Flame Retardants and Textile Chemicals, and Exposure in the Textile Manufacturing Industry. Decabromodiphenyl Oxide.*International Agency for Research on Cancer, Lyon, France, pp. 73–84.

IPCS (1995). *Environmental Health Criteria 172: Tetrabromobisphenol-A and Derivatives.* World Health Organization, International Programme on Chemical Safety, Geneva, p. 139.

Janssen S. (2005). Brominated Flame Retardants: Rising Levels of Concern, Health Care without Harm, www.noharm.org.

Jensen S. (1966). Report of a new chemical hazard, *New Scientist*, 32, 612.

Larsen E.R. (1974). Mechanism of flame inhibition. I: The role of halogens, *Journal of Fire and Flammability/Fire Retardant Chemistry*, 1, 4–12.

Learmonth G.S. and Thwaite D.G. (1969). Flammability of plastics II.† Effect of additives on the flame, *British Polymer Journal*, 1(4), 154–160.

Legler J. and Brouwer A. (2003). Are brominated flame retardants endocrine disruptors?, *Environment International*, 29, 879–885.

Leonard D.D. and Fenge T. (Eds.) (2003). *Northern Lights against POPs: Combating Toxic Threats in the Arctic.* McGill-Queens University Press, Montreal and Kingston, p. 354.

Lindberg P., Sellstrom U., Haggberg L., and de Wit C.A. (2004). Higher brominated diphenyl ethers and hexabromocyclododecane found in eggs of peregrine falcons (Falco peregrinus) breeding in Sweden, *Environmental Science & Technology*, 38, 93–96.

Lindemann R.F. (1969). Flame-retardants for polystyrenes, *Industrial & Engineering Chemistry*, 61(5), 70–75. DOI: 10.1021/ie50713a007.

Little R.W. (1947). *Flameproofing Textile Fabrics.* Reinhold, London.

Lyons J.W. (1970). *The Chemistry and Uses of Fire Retardants.* Wiley-Interscience, New York.

Minkoff G.I. and Tipper C.F.H. (1962). *Chemistry of Combustion Reactions.* Butterworth, London.

Peter C.D. and Schmiegel W. (2000). Fluoropolymers, organic, in: *Ullmann's Encyclopedia of Industrial Chemistry 2000.* Walter Wiley-VCH, Weinheim. DOI: 10.1002/14356007.a11_393.

Pettigrew A. (1993). Halogenated flame retardants, in: *Kirk-Othmer Encyclopedia of Chemical Technology*, 4th edition, Vol 10. John Wiley & Sons, New York, pp. 954–997.

PINFA (2010). Innovative and sustainable flame retardants in transportation, nonhalogenated phosphorus, inorganic and nitrogen flame retardants, September, Brussels.

PINFA (2019). Phosphorus, Inorganic and Nitrogen Flame Retardants Association, Brussels, www.pinfa.eu., accessed on 26.11.19.

Reyes J. (1995). FR-1808, A novel flame retardant for environmentally friendly applications, 6th annual conf Recent Advances in Flame Retardancy of Polymeric Materials, Business Communications Co, Norwalk, CT.

Reyes J. (1996). A new brominated cyanurate as flame retardant for application in styrenics, 7th annual conf Recent Advances in Flame Retardancy of Polymeric Materials, Business Communications Co, Norwalk, CT.

Roy Choudhury A.K. (2018). Advances in halogen-free flame retardants, *Trends in Textile & Fashion Design*, 1(4), 70–74. LTTFD MS.ID.000117.

SFT (2009). *Guidance on Alternative Flame Retardants to the Use of Commercial Pentabromodiphenylether (c-PentaBDE).* Oslo, http://chm.pops.int/Portals/0/docs/POPRC4/intersession/Substitution/pentaBDE_revised_Stefan_Posner_final%20version.pdf.

Shaw S.D., Blum A., Weber R., Kurunthachalam K., Rich D., Lucas D., Koshland C.P., Dobraca D., Hanson S., and Birnbaum L.S. (2010). Halogenated flame retardants:

do the fire safety benefits justify the risks? *Reviews on Environmental Health*, 25(5), 261–305.

Soederstroem G., Sellstroem U., de Wit C.A., and Tysklind M. (2004). Photolytic debromination of decabromodiphenyl ether (BDE 209), *Environmental Science & Technology*, 38, 127–132.

Squires G.E. (1996). Flame retardant polypropylene – A new approach that enhances form, function and processing, Flame Retardants Conference, Interscience Communications Ltd, London.

Stapleton H.M., Alaee M., Letcher R.J., and Baker J.E. (2004). Debromination of the flame retardant decabromodiphenyl ether by Juvenile Carp (cyprinus carpio) following dietary exposure, *Environmental Science & Technology*, 38, 112–119.

Teflon (2019). The Teflon™ brand: A higher standard of performance and quality, www.teflon.com/, Retrieved 8.11.19.

US-EPA (2008). *Tracking Progress on U.S. EPA's Polybrominated Diphenyl Ethers (PBDEs) Project Plan: Status Report on Key Activitiest*. Washington, DC, http://www.epa.gov/sites/production/files/2015-09/documents/pbdestatus1208.pdf.

Wang M.Y., Horrocks A.R., Horrocks S., and Hall M.E. (2000). Flame retardant textile back-coatings. part 1: Antimony-halogen system interactions and the effect of rePlacement By Phosphorus-Containing agents, *Journal of Fire Sciences*, 18, 265. DOI: 10.1177/073490410001800402.

Weil E.D. and Levchik S.V. (2009). *Flame Retardant Plastics and Textiles*. Hanser Publishers, Munich.

WHO (1994). *Environmental Health Criteria, Publication no. 162: Brominated Diphenyl Ethers*. IPCS Publications, Geneva.

Yamada-Okabe T., Sakai H., Kashima Y., and Yamada-Okabe H. (2005). Modulation at a cellular level of the thyroid hormone receptor-mediated gene expression by 1,2,5,6,9,10-hexabromocyclododecane (HBCD), 4,4′-diiodobiphenyl (DIB), and nitrofen (NIP), *Toxicology Letters*, 155, 127–133.

Zoeller T.R., Dowling A.L., Herzig C.T., Iannacone E.A., Gauger K.J., and Bansal R. (2002). Thyroid hormone, brain development, and the environment, *Environmental Health Perspectives*, 110, 355–361.

6 Phosphorous-Based FRs

6.1 INTRODUCTION

Halogen-based flame retardants, particularly those derived from bromine, have played and continue to play a large role in flame retardancy. However, mainly because of environmental concerns and end-of-life issues, there has been a growing interest in halogen-free solutions. The predominant literature on nonhalogen flame retardants focuses on phosphorus-based products. Whether the halogens need replacing or will largely be replaced is still an open question and cannot entirely be resolved on scientific grounds (Weil, 2001). Thus, research on halogen alternatives such as phosphorus is very active–some phosphorus alternatives to bromine are finding market acceptance, and patents in this area are becoming prolific (Weil, 2005).

Compounds of phosphorus were recognized as flame retardants at the beginning of the 19th century when the French chemist Gay-Lussac recommended ammonium phosphate to prevent the burning of theater curtains. The advent of organic phosphorus compounds as important flame retardants for plastics occurred during the 1910–1920 period, when the extreme flammability of cellulose nitrate was brought under some degree of control by the use of tricresyl phosphate. Tricresyl phosphate was the first major commercial organophosphorus compound. Today, PFRs are among the *workhorse* products in the flame retardant field, and research is proceeding at an accelerating pace in this area in view of increasingly stringent flame retardant requirements for plastics and textiles. A large number of phosphorus-based flame retardants (PFRs) are actually in commercial use or appear to be at a serious stage of commercial development for plastics and textiles. In addition to these compounds, there have been many thousands of phosphorus compounds suggested in patents and publications as having flame retardant utility; a survey has been published by Lyons (1970). A particular aspect of phosphorus flame retardancy on which a great deal of work has been done (with only a few commercial products being arrived at) has been the synthesis of addition and condensation polymers from phosphorus monomers. An excellent review of this area was published by Sander and Steininger (1967).

The requirements for polyesters and nylons are stringent because of high processing temperatures, sensitivity to degradation caused by possible acids, and the need for long-term dimensional stability and avoidance of exudation (*blooming*). These various requirements have eliminated most of the known phosphorus-based flame retardants apart from thermally very stable materials (Levchik and Weil, 2006).

The flame retardancy of PFRs is mainly due to increased char formation, although volatile phosphorus compounds also have some vapor phase free-radical inhibiting properties. Phosphorus is often used in combination with other free-radical fire retardants such as halogens and melamine. Phosphorus is used in many forms, including elemental red phosphorus, as an inorganic (such as ammonium polyphosphate) or organic compound (such as phosphate esters) (Zhang et al., 1994).

The structure–toxicity relationships of organophosphorus compounds have been extensively studied and are relatively well understood. Generally, phosphorus-containing flame retardants, as a class, exhibit only low to moderate toxicity, as gauged by their lethal dose (LD) values. However, there are exceptions to the aforementioned norms. These include the neurotoxicity of tricresyl phosphate (mainly owing to the presence of the o-isomer) and the mutagenic/carcinogenic effects of tris (2,3-dibromopropyl) phosphate, a flame retardant used for polyester fabric in the 1970s. The main effect of phosphorus fire retardants on the overall combustion toxicity of materials seems to stem from the increased yields of toxic gases such as CO and, in some instances, HCN (Joseph and Ebdon, 2010). However, the results from various studies proved inconclusive owing to the variables involved in these experiments, such as the fuel–air ratios, burning configurations, equipment design, heat input, and the particular toxicity criterion in question (Hasegawa, 1990).

6.2 CLASSIFICATION OF PFRs

PFRs have been in use for more than 150 years, the first patent was granted in 1735. They are easy to use, compatible with many textile chemicals, and are favored from environmental and toxicological perspectives.

PFRs are the second most widely used class of flame retardants. Due to toxicity of highly efficient halogen FRs, the development of new flame retardants has been shifted strongly toward phosphorus and other halogen-free systems. The PFRs can be broadly classified into three groups namely (Green, 1996; Levchik, 2007):

1) Inorganic flame retardants, including phosphates and elemental red phosphorus;
2) Organic phosphorus-based products; and
3) Chlororganophosphates.

PFR additives cover a wide range of chemical structures and can be both gas-phase and condensed-phase FR additives. Aromatic/aliphatic structures are used for polymer compatibility purposes. Oligomers are used for cost-effectiveness or ease of manufacture. Inorganic PFRs are used when more condensed phase effects are desired. Elemental phosphorus (red) can be a very effective flame retardant for some systems. PFR additives do not typically need synergists, but sometimes they are more effective when combined with other types of flame retardants or elements, such as halogenated FR (phosphorus-halogen vapor phase synergy) and nitrogen compounds (phosphorus-nitrogen condensed phase synergy). PFRs may be additive or reactive.

The benefits of additive PFRs are (Morgan, 2009):

- They can be effective in both vapor phase and condensed phase flame retardants.
- They can be very effective at lowering heat release rate at low loadings of additive.

The drawbacks of additive phosphorus FR are:

- They tend to generate more smoke and carbon monoxide during burning.
- They are not effective in all polymers.
- They have started to come under regulatory scrutiny.

From a practical point of view, adequate flame retardancy is achieved either by the mechanical blending of a flame retardant compound with the polymeric substrate (i.e., by introducing an additive) or by the chemical incorporation of the flame retardant into the polymer molecule by simple copolymerization or by chemical modification of the preformed polymer, that is, using a reactive component. Currently, synthetic polymers are usually made more flame-retardant by adding additives. Such additives often have to be used at high loadings to achieve a significant effect, for example, 30% by weight or more, which occasionally can have a more detrimental effect on the physical and mechanical properties of polymers than that produced by reactive flame retardants. Nevertheless, additives are more generally used as they are usually cheaper and more widely applicable (Ebdon et al., 2001).

The flexible polyurethane foams used in automotive and building constructions are made of flame retardants using reactive FRs made from phosphates, phosphonates, and phosphinates. Elemental phosphorus and its various compounds have been used to retard flame on a wide variety of polymer-based materials for several decades. Environmental considerations, especially concerning the use of halogen-based systems, have paved the way in recent years, for the increased use of PFRs as alternatives to the halogen-containing compounds. Furthermore, this has generated active research in identifying novel flame retardants based on phosphorus, as well as synergistic combinations with compounds of other flame retardant elements (such as nitrogen and the halogens) and with several inorganic nanofillers (e.g., phyllosilicates and carbon nanotubes) (Joseph and Ebdon, 2010).

Ammonium phosphates were first recommended for flame-retarding theatre curtains by Gay Lussac in 1821. Monoammonium and diammonium phosphates, or mixtures of the two, are widely used to impart flame resistance to a wide variety of cellulosic materials such as paper, cotton, and wood (Moniruzzaman and Winey, 2006).

Selected reactive PFRs are used in polyester fibers and for wash-resistant. flame retardant textile finishes. Other reactive organophosphorus compounds can be used in epoxy resins in printed circuit boards (Mcwebmaster (2017). In spite of highly effective flame retardancy, phosphorus compounds are not effective for some major classes of polymers such as styrene resins and polyolefins

(Lewin and Weil, 2001). Furthermore, the basic mode of the intervention of phosphorus compounds, regardless of the phase in which they are active, is to suppress the efficiency of combustion reactions occurring in the gas phase that are mostly radical in nature. This invariably involves lesser oxidation of a carbonaceous substrate to carbon dioxide and thus leads to the production of more soot/smoke and to a comparatively higher value of the CO–CO_2 ratio (Price et al., 2005). Needless to say, the increased smoke production and toxic vapors (mainly CO) are a major concern, especially in real-fire scenarios, in which materials passively protected with PFRs are involved. However, from an ignitability point of view, albeit only a low-level indicator of the overall flammability hazard associated with a material, the use of PFRs has proved to be of great value in that they generally increase the ignition resistance of materials very significantly (Ebdon et al., 2000).

6.3   MODES OF ACTION

For polymers (cellulose, wool), with relatively high concentration of hydroxyl or amino groups, the phosphorus compounds primarily work in the condensed phase. In the case of synthetic polymers containing oxygen, and nitrogen atoms, catalytic hydrolysis of the ester or amide groups by phosphorus acids promotes an enhanced melt dripping and fast shrinkage from the flame. As far as olefin-based polymers are considered, the phosphorus compounds mainly act in the gas phase by recombining the key fuel species such as H and OH radicals and preventing their oxidation. Some minor physical effects due to volatilization of phosphorus compounds and dilution of the fuel can also occur (Salmeia et al., 2016).

Although many PFRs exhibit general modes of action, there are specifics for each of the aforementioned classes. It is generally accepted that PFRs are significantly more effective in oxygen- or nitrogen-containing polymers, which could be either hetero-chain polymers or polymers with these elements in pendant groups. Effective PFRs are more specific than halogen-based products for certain polymers. This relates to the condensed-phase mechanism of action, where the PFR reacts with the polymer and is involved in its charring (Aseeva and Zaikov, 1986).

PFRs namely, phosphate esters, ammonium orthophosphates, ammonium polyphosphates, and red phosphorus, are oxidized during combustion to phosphorus oxide, which turns into a phosphoric acid on its interaction with water. This acid stimulates the take-up of water out of the bottom layer of the material that has decomposed thermally, leading to char, thus increasing the carbonate waste and as reducing the emission of combustible gases. The phosphorous compounds work in the solid state, but can also operate in a gaseous state when they contain halogenated compounds. This group represents 20% of the world's flame retardant production (Al-Mosawi, 2016).

There is very convincing evidence, especially in oxygen-containing polymers such as cellulose and rigid polyurethane foam, that phosphorus compounds can increase the char yield. Formation of char means that less material is actually burned. Secondly, char formation is often accompanied by water release, which dilutes the combustible vapors. Moreover, the char can often protect the underlying polymer and the char-forming reactions are sometimes endothermic.

The flame retardancy of cellulose has been studied in great detail, which gave good insight for understanding the interaction of PFRs with polymers containing hydroxyl groups (Lewin and Weil, 2001). PFRs, in the form of either acid derived from decomposition of ammonium phosphate salts or of phosphate esters, react (esterify or trans-esterify) with the hydroxyl groups of the cellulose (Kandola et al., 1996). Upon further heating, phosphorylated cellulose undergoes thermal decomposition and a significant amount of char is formed at the expense of combustible volatile products that would be produced by virgin cellulose.

When cellulose is heated to its pyrolysis temperature, it normally depolymerises to a tarry carbohydrate product (mainly levoglucosan) which further breaks down to smaller combustible organic fragments. A phosphorus-containing flame retardant, present in cellulose, breaks down to phosphorus acid or anhydrides upon fire exposure. These reactive phosphorus species then phosphorylate the cellulose, generally with release of water. Phosphorylated cellulose then breaks down and forms char. A flame retardant effect results from the formation of a non-combustible, outward-flowing vapor (water), the reduction in fuel, and in some cases the protective effect of the char. A greater degree of flame retardancy seems likely if the char resists oxidation, although even a transitory char may have some inhibitory effect. Even if the char does undergo oxidation (usually by smouldering), the presence of a phosphorus compound tends to inhibit complete oxidation of the carbon to carbon dioxide, and thus, the heat evolution is lessened. Besides its effect in enhancing the amount of char, the PFR may coat the char and thus help prevent burning and smouldering by obstruction of the surface.

Two alternative mechanisms (Equations 6.1 and 6.2) have been proposed by Basch and Lewin (1973) for the condensed phase in cellulose: dehydration of cellulosics with acids and acid-forming agents of phosphorus and sulfur derivatives.

Scheme 1: Esterification and subsequent pyrolytic ester decomposition

$$R_2CH - CHR'OH + ZOH(acid) \rightarrow R_2CH - CHR'OZ + H_2O$$
$$\rightarrow R_2C = CHR' + ZOH \tag{6.1}$$

(Where Z = acyl radical of the acid) and
Scheme 2. Carbonium ion catalysis

$$R_2CH - CHR'OH \rightarrow R_2CH - CHR'OH^{2+} \rightarrow H_2O + R_2CH-C^+HR'$$
$$\tag{6.2}$$

Differential thermal analysis (DTA) and limited oxygen index (LOI) data indicated that phosphorus compounds reduce the flammability of cellulosics primarily by the Scheme 2.1 mechanism, which, being relatively slow, is affected by the fine structure of the polymer. Less-ordered regions (LOR) pyrolyze at a lower temperature than the crystalline regions and decompose before all of the phosphate ester can decompose, which decreases the efficiency of flame-retarding and necessitates a higher amount of phosphorus.

Sulfated celluloses, obtained by sulfation with ammonium sulfamate, are dehydrated by carbonium ion disproportionment (Equation 6.3) and show a strong acid activity which rapidly decrystallizes and hydrolyzes the crystalline regions. The fire-retardant activity was accordingly found not to be greatly influenced by the fine structural parameters, and the same amount of sulfur was needed to flame-retard celluloses of different crystallinities (Basch and Lewin, 1973).

$$R_2CH - CHR'OH \rightarrow R_2CH - CHR'OH_2 \rightarrow H_2O + R_2CH-C^+HR'$$

$$(6.3)$$

6.3.1 Cross-Linking and Char Formation

It was early recognized that cross-linking promotes char formation in pyrolysis of celluloses (Back, 1967) Cross-linking has been assumed to be operative in P–N synergism (Hendrix et al., 1972). Cross-linking reduces in many cases, albeit not always, the flammability of polymers. Although it increases the OI of phenolics, it does not markedly alter the flammability of epoxides (Economy, 1978). A drastic increase in char formation is observed when comparing cross-linked polystyrene (PS), obtained by copolymerizing it with vinylbenzyl chloride, to un-cross-linked PS. PS pyrolyzes predominantly to monomer and dimer units that are almost without char. Cross-linked PS yielded 47% of the char. Cross-linking and char formation were recently obtained by an oxidative addition of organometallics to polyester (Sirdesai and Wilkie, 1987).

Some nitrogen-containing compounds, such as urea, dicyandiamide, and melamine, accelerates phosphorylation of cellulose through formation of a phosphorus–nitrogen intermediate, and thus synergize the flame retardant action of phosphorus (Weil, 1999). Phosphorus–nitrogen synergism is not a general phenomenon but depends on the structure of the phosphorus and nitrogen flame retardants, as well as the polymer structure (Lewin and Weil, 2001).

Similar to cellulose, phosphate esters can also transesterify other polymers. For example, polycarbonates can undergo rearrangement during thermal decomposition, in which phenolic OH groups are formed that then become the target for attack by aromatic phosphate esters (Murashko et al., 1999). Thus, phosphorus is grafted on the polymer chain. Char is formed upon thermal decomposition of this grafted polymer. Similar phosphorylation chemistry was found for polyphenylene ether (PPE; a component of a PPE–HIPS blend), which also tends to rearrange upon heating and form phenolic OH groups (Murashko et al., 1998).

If the polymer cannot be involved in the charring because of the absence of reactive groups, a highly charring coadditive is used in combination with the phosphorus flame retardant (PFR). The coadditive is usually a polyol, which can undergo phosphorylation similar to that of cellulose; pentaerythritol is a typical example of such a polyol. Melamine can be used in conjunction with this system as well. These combinations of flame retardants are called intumescent systems because they form a viscous swollen char on the surface of the burning polymer. The char impedes the heat flux to the polymer surface and retards diffusion of volatile pyrolysis products to the flame. This mechanism of action is mostly physical because the polymer itself is not necessarily involved in the charring process, but its volatilization is retarded significantly.

PFRs can remain in the solid phase and promote charring or volatilize into the gas phase, where they act as potent scavengers of H· or OH· radicals. Volatile phosphorus compounds are among the most effective inhibitors of combustion. A recent study showed (Babushok and Tsang, 2000) that phosphorus at the same molar concentration is, on average, five times more effective than bromine and 10 times more effective than chlorine. The mechanism of radical scavenging by phosphorus was suggested by Hastie and Bonnell. The most abundant phosphorus radicals in the flame are $HPO_2·$, $PO·$, $PO_2·$, and $HPO·$, in decreasing order of significance. Some examples of radical scavenging with participation of $HPO_2·$ and $PO·$ radicals are shown in Equations 6.4–6.7. A third body is required in the reactions involving PO· radicals.

$$HPO_2 · + H· \rightarrow PO + H_2O \rightarrow PO_2 + H_2 \qquad (6.4)$$

$$HPO_2 · + OH· \rightarrow PO_2 + H_2O \qquad (6.5)$$

$$PO · + H · + M \rightarrow HPO + M \qquad (6.6)$$

$$PO · + OH · + M \rightarrow HPO_2 + M \qquad (6.7)$$

If conditions are right, phosphorus-based molecules can volatilize and are oxidized, producing active radicals in the flame. On the other hand, PFRs tend to react with the polymer or to oxidize to phosphoric acid in the condensed phase. This favors mostly condensed-phase mechanisms. It is challenging to design a PFR that will volatilize into the flame at relatively low temperatures, but will not be lost during polymer processing.

6.4 INORGANIC PHOSPHORUS FRs

Inorganic phosphorous derivatives, mostly nondurable or semidurable, entail primarily phosphoric acid and its ammonium salts, e.g., diammonium phosphate (DAP, $(NH_4)_2 HPO_4$) are highly effective FRs and smoldering inhibitors. While only 15% of either of them is required for flame retardancy, 50–70% of its sodium salt is required to have the same effect.

6.4.1 AMMONIUM POLYPHOSPHATE (APP)

Ammonium Polyphosphate (APP) is considered a high-performance inorganic flame retardant with white powder, decomposition temperature greater than 256°C, degree of polymerization between 10 and 20, and water solubility. APP is poorly soluble in water when the degree of polymerization is larger than 20. Itis cheaper than organic flame retardants, and has low toxicity and good thermal stability. It can be used individually with other flame retardant compounds for flame retardancy. The use of APP is wide; among its most important uses areas a source of acid, and in connection with carbon sources and gas sources consisting of IFRs. Also, it can be used for other fire-retardant plastics, fibers, rubber, paper, wood, and firefighting in large areas of forests and coalfields. Use of APP alone as a flame retardant has been found effective in polyamides and similar polymers.

Other PFRs include disodium hydrogen phosphate, lithium iron phosphate, magnesium phosphate, sodium phosphate, diammonium phosphate. Diammonium hydrogen phosphate is generally used in synthetic fibers and rubber, rigid and flexible foams, forest fires, paper, and wood, but disodium hydrogen phosphate is used in textile, paper, and wood fire-retardants.

6.4.2 RED PHOSPHOROUS

This allotropic form of phosphorus is relatively nontoxic and, unlike white phosphorus, is not spontaneously flammable. Red phosphorus is an inorganic flame retardant with high performance. Compared to other flame retardants, adding lesser amounts of red phosphorous can reach the same levels of flame retardancy; therefore, it has little impact on the physical and mechanical properties of materials. However, ordinary red phosphorus has one disadvantage; it easily absorbs moisture oxidation and releases toxic phosphine gas in the oxidation process. Common red phosphorus has poor compatibility with plastics and is difficult to disperse in plastic. Besides, red phosphorus is deep red, which limits its use in polymers. In order to improve some of these shortcomings, the surface treatment of phosphorus is the main focus of research, and the most effective method is microencapsulation. The mechanism of flame retardant of red phosphorus is that when it is heated to decomposition to form strong dehydration of meta-phosphoric acid, the polymer surface is burned and carbonized. Carbonized layers can reduce the release of flammable gases, as well as the heat effect. In addition, PO• free radicals which are formed by red phosphorus and oxygen go into the gas phase and capture large H•, HO• free radicals.

Red phosphorus is the most concentrated source of phosphorus for flame retardancy. In fact, it is very effective in some polymers, such as thermoplastic polyesters or polyamides, where self-extinguishing UL-94 V-0 performance can be achieved at loadings of less than 10 wt%. Despite the apparent chemical simplicity of this additive, its mechanism of action is not completely understood. Most researchers agree that in oxygen- or nitrogen-containing polymers, red phosphorus reacts with the polymer and induces char formation. Although it is believed that red phosphorus is oxidized and hydrolyzed by water before it reacts with the polymer, (Ballistreri et al., 1983). There is strong evidence that red phosphorus can react directly with polyesters or polyamides in an inert atmosphere (Kuper et al., 1994) and in the absence of moisture (Levchik et al., 2000). There is also some evidence in favor of a free radical mechanism of interaction between red phosphorus and polyamide-6 (Levchik et al., 1998).

Red phosphorus shows relatively weak flame retardant effects in hydrocarbon polymers (e.g., polyolefins or polystyrene). It is believed that in these polymers red phosphorus depolymerizes to white phosphorus (P_4), which volatilizes and provides gas-phase action.

The red allotropic form of phosphorus is relatively nontoxic and, unlike white phosphorus, is not spontaneously flammable. Red phosphorus is, however, easily ignited. It is a polymeric form of phosphorus, thermally stable up to ~ 450°C. Elemental red phosphorus is a highly efficient flame retardant, especially for oxygen-containing polymers such as polycarbonates and poly (ethylene terephthalate). Red phosphorus is used on polyamide 6,6 working at a high processing temperature of about 280°C–the most phosphorus compounds fail in such cases (Levchik et al., 1996). In addition, coated red phosphorus is used to flame-retard polyamide electrical parts, mainly in Europe and Asia (Weil, 1999).

Red phosphorus may react with atmospheric moisture to form toxic phosphine gas and ignites readily in air. As a result, the commercial product is often encapsulated in an appropriate polymer matrix. Suitable stabilization and encapsulation have led to commercial concentrates containing 50% red phosphorus (Joseph and Ebdon, 2010).

Red phosphorus is one of the most useful agents for creating low-toxicity, halogen-free flame retardants. It is a very economical and effective flame-retardant agent, which works well not only in the condensed phase, but also in the gaseous phase. The flame-retarding mechanism for red phosphorus and oxygen-containing polymers is believed to be through the formation of phosphorus–oxygen bonds. Red phosphorus is widely used as a synergistic agent in many kinds of polymer matrices such as cotton, polyesters, polyurethanes, polyamides, epoxy resins. and plastics (Chen and Wang, 2010a).

However, red phosphorus has some serious shortcomings as a starting material, including the ready absorption of moisture to form highly toxic phosphine and fast oxidation (and loss of flame retardant performance). It is easily ignited by heat, friction, static electrical sparking, oxidizing agents or physical impact, and reignites (at about 260°C), even after the fire is extinguished. Application of red phosphorus to garments is minimal due to irritability to the skin, the evolution of corrosive phosphoric acids, and its red color (Levchik and Weil, 2005).

Due to their inherent drawbacks, including the potential safety hazard of red phosphorus during compounding with polymer matrices, reduced hydrolytic stability and plasticizing effect, particularly caused by the small-molecular and oligomeric phosphorus-containing organics, these compounds are undesirable in many application fields.

6.5   ORGANOPHOSPHOROUS FRs

There are a vast number of compounds in which phosphorus is bound to carbon, since it can replace both nitrogen atoms and CH groups. The resulting compounds are collectively termed organophosphorus compounds. These include several classes such as aliphatic and aromatic phosphines, phosphine oxides, phosphites, phosphates, phosphinites, phosphinates, phosphonate esters, and phosphonium salts (Greenwood and Earnshaw, 1984). Several of the aforementioned classes of compounds have been very successfully used, both as additives and as reactive flame retardants, for a wide variety of polymer-based systems.

Organic FRs usually exhibit excellent comprehensive properties, such as high FR efficiency, good compatibility with the epoxy matrix, and outstanding flexibility for their molecular designs, compared with their inorganic counterparts.

The various phosphorous compounds, their chemical structures, and oxidation states are shown in Table 6.1 (Nair, 2001).

The oxidation state, sometimes referred to as oxidation number, describes the degree of oxidation (loss of electrons) of an atom in a chemical compound. Organophosphorous durable flame retardants usually have functional groups, namely, phosphates, phosphonates, amido-phosphates, phosphazenes (phosphorous + nitrogen) or phosphonium salts. They are most effective when they are in the highest oxidation state (P^{+5}).

The phosphazenes have received considerable attention from both the academic and commercial communities (Chen-Yang et al., 1991); they contain both phosphorus and nitrogen atoms. Such compounds display enhanced flame retardancy when compared to similar compounds-containing phosphorus alone.

TABLE 6.1

Phosphorous compounds and their oxidation states

Sr. No.	Phosphorous Compounds	Chemical Structure	Oxidation State
1	Phosphate	$(RO)_3PO$	+5
2	Phosphite	$(RO)_3P$	+3
3	Phosphonate	$(RO)_2R'PO$	+3
4.	Phosphinate	$(RO)R'_2PO$	+1
5	Phosphine oxide	R_3PO	-1
6	Phosphine	R_3P	-3
7	Phosphonium salt	R_4PX	-3

Three different oxidation states of phosphorus (phosphite, phosphate, and phosphine oxide) additives, with different thermal stabilities at a constant phosphorus content (1.5 wt.%) were utilized by Mariappana et al. (2013). Thermal and flame retardant properties were studied by TGA and cone calorimetry, respectively. The thermal stability of both polymers decreases upon the incorporation of PFRs irrespective of oxidation state and a greater amount of residue was observed in the case of phosphite. Phosphate was found to be a better flame retardant in polyurea, whereas phosphite is suitable for epoxy resin.

Erythritol is a sugar alcohol (or polyol) that is used as a food additive and sugar substitute. Its formula is $C_4H_{10}O_4$. Pentaerythritol phosphate (Structure 6.1) has an excellent char-forming ability owing to the presence of the pentaerythritol structure. The bis-melamine salt of the bis acid phosphate of pentaerythritol is also available commercially. This is a high melting solid that acts as an intumescent FR additive for polyolefins. Synergistic combinations with ammonium polyphosphates have also been developed primarily for urethane elastomers (Joseph and Ebdon, 2010).

6.5.1 ORGANO-PHOSPHATES OR PHOSPHATE ESTER

Organophosphates (also known as phosphate esters, or OPEs/PEs, Structure 6.2) are a class of organophosphorus compounds with the general structure $O=P(OR)_3$. Lacking a P–C bond, these compounds are in the technical sense not organophosphorus compounds, but esters of phosphoric acid. Like most functional groups, organophosphates occur in a diverse range of forms, with

STRUCTURE 6.1 Pentaerythritol phosphate

STRUCTURE 6.2 General chemical structure of organophosphate important examples including key biomolecules such as DNA, RNA, and ATP, as well as many insecticides, herbicides, nerve agents and flame retardants.

Esterification of phosphoric acid reactions are shown in Equations 6.8, 6.9, and 6.10:

$$OP(OH)_3 + ROH \rightarrow OP(OH)_2(OR) + H_2O \qquad (6.8)$$

$$OP(OH)_2(OR) + ROH \rightarrow OP(OH)(OR)(OR) + H_2O \qquad (6.9)$$

$$OP(OH)(OR)(OR) + ROH \rightarrow OP(OR)(OR)(OR) + H_2O \qquad (6.10)$$

Organophosphites can also be readily oxidized to produce organophosphates (Equation 6.11):

$$P(OR)_3 + [O] \rightarrow OP(OR)_3 \qquad (6.11)$$

The phosphate esters bearing OH groups are acidic, and are partially deprotonated in aqueous solutions. For example, DNA and RNA are polymers of the type $[PO_2(OR)(OR')-]_n$. Polyphosphates also form esters; an important example of an ester of a polyphosphate is ATP, which is the monoester of triphosphoric acid ($H_5P_3O_{10}$).

Phosphate ester flame retardants are human-made chemicals that are typically liquids at room temperature, although some are solids. They are made up of groups of chemicals with similar properties but different structures. Phosphate esters are added to consumer and industrial products in order to reduce flammability. They are used in plasticizers, hydraulic fluids, solvents, extraction agents, antifoam agents, and coatings for electronics.

Phosphate esters (alkyl or aryl, or mixed) of phosphoric acid constitute an important family of organo-PFRs (Weil, 1993). There are several classes of amine phosphates commercially available to make a wide variety of polymeric substrates, both natural and synthetic, flame retardant. A classic example is provided by the three variations of melamine phosphate (Structure 6.3): melamine orthophosphate, dimelamine orthophosphate, and melamine pyrophosphate. Of these, the pyrophosphate is the least soluble and the most thermally stable.

STRUCTURE 6.3 Melamine phosphate

The synthesis of melamine phosphate, pyrophosphate and polyphosphate involves the following reactions (Equations 6.12 and 6.13) shown in a simplified form (Cichy et al., 2003):

Melamine phosphate (MP)

$$C_3N_6H_6 + H_3PO_4 \rightarrow C_3N_6H_6 \cdot H_3PO_4 \tag{6.12}$$

Melamine pyrophosphate (MDP)

$$2C_3N_6H_6 \cdot H_3PO_4 \rightarrow (250 - 300°C) \rightarrow (C_3N_6H_6HPO_3)_2 + H_2O \tag{6.13}$$

Melamine orthophosphate is converted to the pyrophosphate upon heating, with the loss of water. All the aforementioned variations are available as finally divided solids, are used commercially in coatings, and have utility in a wide variety of thermoplastics and thermosets.

Subsequent products of the condensation of phosphates differ mainly in terms of thermal resistance. MPP is designed mainly for polymers processed at higher temperatures, e.g., polyamide. In terms of environmentally-friendly and human-friendly flame retardants, melamine phosphates are an appealing option on the market. Their practical significance is on the rise. Most promising results in terms of the reduction of the flammability of polymer materials can be seen for polypropylene and polyamide.

Different from the traditional silicone materials, which are not easily ignited, silicone thermoplastic elastomer (Si-TPE) has poor flame retardant properties due to the existence of the hard segments in its molecular chains. Melamine phosphate (MP), a kind of halogen-free flame retardant, was adopted to improve the flame retardancy of Si-TPE. The results showed that MP played the role of flame retardant in both gas phases and condensed phases due to its nitrogen–phosphorus-containing structure. Inert gases, including nitrogen, steam, and ammonia which were released by the degradation of melamine during burning, could take away the heat and dilute the oxygen in the gas phase, and work further with the phosphoric acid, which was generated in the condensed phase, to form a denser and firmer char layer (Xu et al., 2018).

Triethyl phosphate, a colorless liquid boiling between 209°C and 218°C, and containing 17 wt% phosphorus, has been used commercially as an additive for polyester resins/laminates and in cellulosic materials. Blends of triaryl phosphates and pentabromodiphenyl oxide are extensively used as flame-retardant additives for flexible urethane foams.

Mixed esters, such as isopropylphenyl diphenyl phosphate and tert-butylphenyl diphenyl phosphate, are also widely used as both plasticizers/flame retardants for engineering thermoplastics and hydraulic fluids (Green, 1996). These esters generally show slightly less flame retardant efficacy when compared to triaryl counterparts; however, they have the added advantage of lower smoke production when burned.

Resorcinol bis(diphenyl phosphate) (RDP) (Structure 6.4) is a nonhalogen aromatic oligomeric phosphate flame retardant and flow modifier. Its high thermal stability and low volatility, compared with the triaryl phosphates, make it ideal for use in applications in which high processing temperatures are required (Bright et al., 2004).

Aryl phosphates were introduced into commercial use early in the 20th century for flammable plastics such as cellulose nitrate, and later for cellulose acetate (Green, 2000). In vinyls (plasticized), aryl phosphates are frequently used with phthalate plasticizers. Their principal applications are in wire and cable insulation, connectors, automotive interiors, vinyl moisture barriers, plastic greenhouses, furniture upholstery, and vinyl forms. Triarylphosphates are also used, on a large scale, as flame retardant hydraulic fluids, lubricants, and lubricant additives. Smaller amounts are used as nonflammable dispersing media for peroxide catalysts. Blends of triarylphosphates and pentabromodiphenyl oxide are extensively used as flame retardant additives for flexible urethane foams. They have been also used as a flame retardant additive for engineering thermoplastics such as polyphenylene oxide, high impact polystyrene, and ABS-polycarbonate blends.

Some novel oligomeric phosphate flame retardants (based on tetraphenyl resorcinol diphosphate) are also employed to retard flame in polyphenylene oxide blends, thermoplastic polyesters, polyamides, vinyls, and polycarbonates.

STRUCTURE 6.4　Resorcinol bis(diphenyl phosphate) (RDP)

Dimethyl methylphosphonate (Structure 6.5), a water soluble liquid (boiling point: 185°C), made by the rearrangement of trimethyl phosphite, has a phosphorus content of 25 wt %, the highest possible for an alkyl phosphonate ester (Price et al., 2002). Applications of dimethyl methylphosphonate include use as a viscosity depressant and as a flame retardant in alumina trihydrate-filled polyester resins, and as a flame suppressant for halogenated polyesters. It is also used in flame retardant rigid polyurethane foams, and as a component in blends with triarylphosphates to make flame retardant plasticizers for synthetic rubber and cellulosics. Diethyl ethylphosphonate, a higher member of this family of alkyl phosphonates, has a higher boiling point and is believed to be less susceptible to undesirable interactions with halogenated aliphatic components such as blowing agents, or with amine catalysts.

Even phosphate esters may lose FR activity due to the formation of calcium salts during laundering or rinsing with tap water.

6.5.2 Melamine Polyphosphate

The production of Melamine polyphosphate (MPP) is shown in Equation 6.14.

$$C_3N_6H_6 + H_3PO_4 \longrightarrow C_3N_6H_6 \cdot H_3PO_4 \xrightarrow[-H_2O]{250\text{-}300\,^\circ C} (C_3N_6H_6HPO_3)_2 \xrightarrow{300\text{-}330\,^\circ C} (C_3N_6H_6HPO_3)n \qquad (6.14)$$

melamine phosphate   melamine pyrophosphate   MPP

MPP can be applied as reactive and additive flame retardant for thermally resistant polyurethane foams. MPP was hydroxyalkylated with ethylene and propylene carbonates to get oligoetherols with 1,3,5-triazine ring and phosphorus. The structure and physical properties of the products were studied. The foams, (PUFs) obtained from this oligoetherols, were self-extinguishing. The polyurethane addition of powdered MPP into foaming mixture resulted in further decrease of flammability modified PUFs. The MPP-modified PUFs were characterized by physical methods adequate for thermal resistance and flammability of the PUFs. The best MPP-modified PUF showed an oxygen index 24.6. All the modified PUFs were remarkably thermally resistant; they could endure long-lasting thermal exposure, even at 200°C (Lubczak and Lubczak, 2016).

$$\begin{array}{c} O \\ \| \\ H_3CO \diagup \overset{\displaystyle P}{\underset{\displaystyle CH_3}{|}} \diagdown OCH_3 \end{array}$$

STRUCTURE 6.5 Dimethyl methylphosphonate

Organophosphorus-based flame retardants are quite versatile in their flame-retardant action, and often exhibit both condensed-phase and gas-phase activity. Depending on the substrate and phosphorus chemistry, there could be chemical interactions in the condensed phase at elevated temperatures, which lead to changes in the decomposition pathway of the polymer and possible formation of carbonaceous char residues on the surface of decomposing polymer, hence preventing its further oxidation. The char thus formed acts as a protective thermal barrier. In other instances, the phosphorus compounds and some of their decomposed products preferably volatilize from the polymer substrate when heated. These phosphorus species further decompose to release reactive phosphorus species, which then interact with the combustion intermediates in the gas phase as flame inhibitors (Liang et al., 2015, 2017). In most cases, such interactions lead to recombination of the H and OH radicals and prevent their oxidation P-C bond containing phosphorus compounds like any organophosphorus flame retardants can exhibit, either condensed-phase or gas-phase flame inhibition, and are very specific to a polymer/flame retardant combination. For polymers (cellulose, wool), with relatively high concentration of hydroxyl or amino groups, the phosphorus compounds primarily work in the condensed phase. In the case of synthetic polymers containing oxygen and nitrogen atoms, catalytic hydrolysis of the ester or amide groups by phosphorus acids promotes enhanced melt-dripping and fast shrinkage from the flame. Regarding olefin-based polymers, the phosphorus compounds mainly act in the gas phase by recombining the key fuel species such as H and OH radicals and by preventing their oxidation. Some minor physical effects due to volatilization of phosphorus compounds and dilution of the fuel can also occur (Salmeia et al., 2016).

The effects of phosphate esters on environment are:

- They can change chemical composition in the environment.
- Some phosphate esters deposit on wet and dry surfaces and others are broken down by water.
- They are poorly soluble in water and adsorb strongly to soil.
- They are considered emerging pollutants.
- Many *organophosphates* are potent nerve agents, functioning by inhibiting the action of acetylcholinesterase (AChE) in nerve cells. They are one of the most common causes of poisoning worldwide, and are frequently intentionally used in suicides in agricultural areas.
- Even at relatively low levels, organophosphates may be hazardous to human health.
- The International Agency for Research on Cancer (IARC), found that some organophosphates may increase cancer risk.
- They biodegrade in aquatic and terrestrial environments.

The *C-O-P* ester bond of ester phosphate undergoes thermal decomposition when heated to temperatures of 200°C or higher. Under these conditions, phosphoric acid remains on the surface of the substrate, although the ester structure is no longer preserved.

Norzali et al. (2014) have synthesized phosphate ester FRs from soybean oil. They have found that the phosphate esters improved the flammability of the PU. Palm-based polyurethane containing phosphate ester (PE) as fire retardant has been developed. This study was conducted to investigate the effect of PE inclusion into palm-based PU systems onto the mechanical and burning property of the PU. The PE was synthesized via ring-opening hydrolysis between phosphoric acid (H_3PO_4) and epoxy. Burning properties showed that the addition of PE has decreased the burning rate of the PU foam. PU with PE synthesized from 5.0 wt.% H_3PO_4 showed excellent burning properties with the lowest burning rate (0.047 cm/s), compared to control PU which was at 0.119 cm/s (Norzali and Badri, 2016).

6.5.3 PHOSPHONATES

Phosphonates and phosphonic acids are organophosphorus compounds containing $C-PO(OH)_2$ or $C-PO(OR)_2$ groups (where R = alkyl, aryl). Phosphonic acids, (Structure 6.6, pm = picometer = 10^{-12}m) typically handled as salts, are generally nonvolatile solids that are poorly soluble in organic solvents, but soluble in water and common alcohols. Many commercially important compounds are phosphonates.

Phosphonates feature tetrahedral phosphorus centers. They are structurally closely related to (and often prepared from) phosphorous acid. Phosphonate salts are the result of deprotonation of phosphonic acids, which are diprotic acids donating two proton or hydrogen atoms per molecule in aqueous solution (Equation 6.15).

$RPO(OH)_2 + NaOH \rightarrow H_2O + RPO(OH)(ONa)$ (monosodiumphosphonate)
$RPO(OH)(ONa) + NaOH \rightarrow H_2O + RPO(ONa)_2$ (disodiumphosphonate)

$$(6.15)$$

Phosphonate esters are the result of condensation of phosphonic acids with alcohols. Phosphinates or hypophosphites are a class of phosphorus compounds that are conceptually based on the structure of hypophosphorous

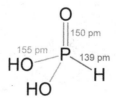

STRUCTURE 6.6 Phosphonic-acid

acid. IUPAC prefers the term *phosphinate* in all cases; however in practice, hypophosphite is usually used to describe inorganic species (e.g., sodium hypo-phosphite), while phosphinate typically refers to organophosphorus species.

6.5.4   CYCLIC OLIGOMERIC PHOSPHONATES

Cyclic oligomeric phosphonates (Structure 6.7) with the varying degrees of structural complexity are available in the market (Weil, 1993). They are widely used as flame retardant finishes for polyester fabrics.

After the phosphonate is applied from an aqueous solution, the fabric is heated to swell and soften the fibers, thus allowing the phosphonate to be absorbed and strongly held. It is also a useful retardant in polyester resins, polyurethanes, polycarbonates, polyamide-6, and in textile back coatings. A bicyclic pentaerythritol phosphate has been more recently introduced into the market for use in thermosets, as well as for polyolefins, preferably, in combination with melamine or ammonium polyphosphate (Structure 6.8a). Cyclic neopentyl thiophosphoric anhydride (Structure 6.8b), a solid additive, is used to flame-retard viscose rayon, especially in Europe. In spite of the anhydride structure, it is remarkably stable, surviving addition to the highly alkaline viscose, the acidic coagulating bath, and also resisting the multiple launderings of the rayon fabric.

STRUCTURE 6.7   Cyclic oligomeric phosphonates

STRUCTURE 6.8 A   A bicyclic pentaerythritol phosphate

STRUCTURE 6.8 B   Cyclic neopentyl thiophosphoric anhydride

6.5.5 ARYL POLYPHOSPHONATES

Aryl polyphosphonates (ArPPNs) have been demonstrated to function in wide applications as flame retardants for different polymer materials, including thermosets, polycarbonate, polyesters, and polyamides, particularly due to their satisfactory thermal stability compared to aliphatic flame retardants, and to their desirable flow behavior observed during the processing of polymeric materials (Chen and Wang, 2010b).

Compared with aliphatic polyphosphonates (AlPPNs) and aromatic polyphosphates (ArPPA), ArPPNs usually exhibit better thermal stability than AlPPNs during processing and molding, and higher hydrolytic stability than ArPPAs due to the partly hydrolyzable P-O-C bond substituted by the hydrophobic P-C bond. Therefore, applications of ArPPNs in flame retardation are more prevalent than the others, particularly for aromatic polycondensates with high processing temperature (Maiti et al., 1993).

For two decades, polymer-type phosphonates (polyphosphonates, PPNs) have received increasing interest from both academia and industry. PPNs have excellent flame retardancy and transparency and suitable melting points or flow temperatures, which can match the processing temperatures of different polymers that are flame-retarded. The melted flame retardants are very helpful to the compounding of PFR systems. In addition, due to their polymeric nature, they do not migrate out of the matrices when used as additives. Moreover, in some polymer systems, the addition of PPNs can improve the polymers' properties, such as heat distortion temperature (Chen and Wang, 2010b). The development history of PPNs goes back in patent literature to 1948, but they were not commercialized because they had low strength, low glass transition temperature, and poor hydrolytic stability. In the 1980s, Bayer AG in Germany discovered a way to make PPNs with better properties; however, it was based on expensive ingredients that made it uneconomical (Maiti et al., 1993; Schut, 2009).

6.5.6 HALOGENATED PHOSPHATES AND PHOSPHONATES

These belong to an important class of additives; the halogen contributes to some extent to the flame retardancy, although this contribution is offset by the lower phosphorus content. The halogens generally reduce vapor pressure and water solubility, thus aiding retention of these additives. Thus, more efficient and effective blending/manufacturing processes involved usually lead to a favorable economics of polymer flame retardation (Green, 2000).

Tris(2-chloroethyl)phosphate, tri(1-chloroethyl)phosphate, 1,3-dichloro-2-propanol phosphate, bis(2-chloroethyl) 2-chloromethyl phosphonate are the prime members of this class of FRs and are quite compatible with polymers containing polar groups. These compounds have different viscosities, solubilities, hydrolytic stabilities, and boiling points. Their flame retardant efficacies generally depend on structural features, as well as on the phosphorus–halogen ratios (Joseph and Ebdon, 2010).

Tris(2-chloroethyl) phosphate (TCEP) (Structure 6.9) is a chemical compound used as a flame retardant, plasticizer, and viscosity regulator in various types of polymers including polyurethanes, polyester resins, and polyacrylates. It is widely used in rigid polyurethane and polyisocyanurate foams, most classes of thermosets, cast acrylics, and in wood-resin composites. Tri(1-chloroethyl) phosphate, owing to the presence of a branched alkyl group, has lower reactivity to water and bases than the 2-chloroethyl homologue.

Because of its suspected reproductive toxicity, it is listed as a substance of very high concern under the EU REACH regulations.

The reaction of phosphoryl chloride and ammonia (Equation 6.16) results in phosphorylamide that polymerizes and reacts with cellulose.

$$POCl_3 + 3NH_3 \longrightarrow NH_2-\overset{\overset{\displaystyle O}{\big\uparrow}}{\underset{\underset{\displaystyle NH_2}{\big\downarrow}}{P}}-NH_2 + 3HCl \qquad (6.16)$$

However, a commercial product of this type (Flame Retardant PA) has been reported to have unsatisfactory durability.

6.5.6.1 THPC FRs

Tetrakis(hydroxymethyl)phosphonium chloride (THPC) is an organophosphorus compound. The cation $P(CH_2OH)_4+$ is four-coordinate, as is typical for phosphonium salts. THPC has applications as a precursor to fire-retardant materials, as well as a microbicide in commercial and industrial water systems. THPC has industrial importance in the production of crease-resistant and flame retardant finishes on cotton textiles and other cellulosic fabrics.

Tetrakis(hydroxymethyl) phosphonium chloride, $(HOCH_2)_4PCl$ popularly abbreviated as THPC (Structure 6.10) was the first flame retardant finish for cotton fabric durable towards multiple home and industrial launderings. The cation $P(CH_2OH)_4^+$ is four-coordinate, as is typical for phosphonium salts. THPC has applications as a precursor to fire-retardant materials.

STRUCTURE 6.9 Tris(2-chloroethyl)phosphate

STRUCTURE 6.10   Tetrakis(hydroxymethyl) phosphonium chloride (THPC)

THPC is of reducing nature. It reacts with many active hydrogen-containing chemicals, e.g. N methylol compounds, phenols, polybasic acids and amines to form insoluble polymers on cellulose. THPC is made by reacting phosphine, formaldehyde and hydrochloric acid (Equation 6.17).

$$PH_3 + 4HCHO + HCl \rightarrow \left[HOCH_2\!-\!\overset{\displaystyle CH_2OH}{\underset{\displaystyle CH_2OH}{\overset{\displaystyle |}{\underset{\displaystyle |}{P}}}}\!-\!CH_2OH \right]^+ Cl^- \qquad (6.17)$$

It is applied with urea, dried and cured. Control of pH and the oxidation state of the phosphorus is important in determining the flame retardant properties and the durability of the finish. The release of HCl may cause the fabric to tender during curing unless pH is controlled. The final step in finishing requires oxidation of P^{3+} to P^{5+} with hydrogen peroxide. This step too must be controlled to prevent excessive tendering of the fabric. An alternative to the THPC is tetrakis-(hydrxymethyl)phosphonium sulfate or THPS. Sulphuric acid is used instead of HCl and the corresponding phosphine sulphate is formed in place of the phosphine chloride.

A few popular compositions of THPC are as follows:

- THPC, methylol melamine and urea
- THPC and caustic soda. THPC reacts with caustic soda at pH 7.5–7.8 to produce a mixture of tetrakis-(hydroxymethyl)phosphoniumhydroxide (THPOH) and tetrakis-(hydroxymethyl) phosphonium (THP). A durable finish based on THPOH, trimethylolmelamine (TMM) and urea can be applied by pad-dry-cure (150 °C, 3 minutes)
- THPC, amide and caustic soda

THPC finishes are applied by the pad-dry-cure technique and add-ons of 25%–30% are required for its durability for multiple launderings. Lightweight fabrics need higher add-ons as compared to heavier ones.

Methylol groups of THPC readily react with amines and amides, but react slowly with the hydroxyl group of cellulose. Reactions with polyfunctional amines and amides produce highly insoluble polymers when the ionic phosphonium structure of THPC changes to phosphonium oxide structure, the latter being stable towards hydrolysis. An excellent flameproof finish is obtained with THPC and a phosphonamide by the pad-dry-cure process. Both formulations use nitrogenous resins to enhance the flame resistance and durability to laundering. Its methylol groups are capable of reacting with melamine resin to form a durable bond with cellulose. To minimize degradation of cellulose by hydrochloric acid (by-product), both urea and triethanolamine are added in the impregnation bath.

A typical formulation for heavy to medium-weight fabrics as follows:

THPC	16%
Trimethylol melamine	9–10%
Urea	10%
Triethanolamine	3%
Water	62%

The use of sodium sulfite (15%), sodium metabisulfite (2%), and ammonium sulfate are also suggested. After padding, the fabric is dried at 85°C and cured at 140°C for 4–5 minutes and washed.

The THPC treated fabric shows durable flame retardancy and retention of tensile strength. However, the treated fabric is slightly stiffer.

6.5.6.2 THPC-Urea Precondensate

A flame retardant finish can be prepared from THPC by the PROBAN Process (Rhodia, 2016) in which THPC is treated with urea. The urea condenses with the hydroxymethyl groups on THPC. The reaction converts phosphonium compound into phosphine oxide.

PROBAN is a registered trademark of Rhodia Operations. The PROBAN process uses Solvay's patented and licensed technology to form a special cross-linked polymer inside the cotton fibers. The process of PROBAN polymer formation is irreversible. The polymer is completely insoluble and is embedded in the body of the fiber. PROBAN fabrics are halogen-free and contain no antimony. The PROBAN polymer is approved by Oeko-Tex. PROBAN fabric suppliers can have fabric tested for OEKO-TEX standard 100 "Confidence in Textiles", which certifies the skin-friendliness of the end articles. OEKO-TEX testing for harmful substances is based on the actual use of textile: "If the product has more intensive skin contact, it should meet stricter human ecological requirements" (Solvay, 2017).

In the PROBAN process of Albright and Wilson, heat curing is replaced by an ammonia gas curing at ambient temperature. This minimizes fabric tendering associated with heat and acids. A precondensate of THPC with urea (mole ratio 1: 1) is applied and dried, and the fabric is passed through an ammonia gas reactor. An exothermic reaction creates a polymeric structure within the voids of the cotton fiber. The ammonia cure gives a P–N ratio of 12. Weight percentages of the respective elements should be P, N > 2% (Tomasino, 1992).

The phosphonium structure is converted to phosphine oxide as a result of the following reaction (Equation 6.18).

$$[P(CH_2OH)_4]Cl + NH_2CONH_2 \rightarrow (HOCH_2)_2POCH_2NHCONH_2$$
$$+ HCl + HCHO + H_2 + H_2O \ldots \ldots$$

$$(6.18)$$

This reaction proceeds rapidly, forming insoluble high molecular weight polymers. The resulting product is applied to the fabrics in a pad-dry process. This treated material is then treated with ammonia and ammonia hydroxide to produce fibers that are flame retardant.

The characteristics of Modified THPC-urea finish (Proban type) are (Schindler and Hauser, 2004):

- Demands moisture-controlled ammonia treatment and oxidation,
- Less shrinkage from washing,
- Better stability to hydrolysis,
- Better tearing strength,
- Less odor,
- Suitable for low-cellulose-content blends, and
- Preferred for larger runs to minimize process costs and for better machinery utilization.

Another successful durable finish can be obtained by using N-methylol phosphonopropionamide (20–30% add-on) along with trimethylol melamine and phosphoric acid as a catalyst in the pad-dry-cure process.

Tris(1-aziridinyl)phosphine oxide (APO), in conjunction with THPC or thiourea, provides flameproofing with excellent durability, but the use of APO has been discontinued due to its suspected carcinogenicity.

6.5.6.3 Toxicity of THPC

Some form of THPC finish is released from the fabric in an aqueous medium. It is possible that common solutions such as sweat, urine, and saliva might extract some free THPC, and hence, there is a possibility that some percentage of children could be at risk from long-term, low-level exposure to THPC from treated sleepwear.

Moderate to severe skin effects characterized by tissue changes were observed in male white rats and rabbits treated topically for 8 days with 0.75 mL of 15%, 20%, or 30% aqueous THPC (Aoyama, 1975).

Afanas'eva and Evseenko (1971) reported that many (number not reported) of the mice treated with aqueous extracts from fabrics treated with a THPC-based FR developed leukopenia (decrease in the number of white blood cells). When mice are treated with aqueous extracts of THPC-based,FR-treated fabrics, they became sluggish (neurological effects), and had reduced static working capacity and 20%–40% lower cholinesterase activity levels.

Loewengart and Van Duuren (1977) found that THPC had tumor-promoting activity in a mouse skin carcinoma assay. After treatment of female ICR/Ha Swiss mice with 2 mg THPC dissolved in 0.2 mL acetone/water, 3 times/wk for 496 days, no evidence of carcinogenicity was found by Van Duuren et al. (1978).

Thirteen-week studies with THPS and THPC were conducted (NIH, 1987) to identify affected organs, characterize toxic effects, and to determine doses to be used for the 2-year studies. Doses for the THPS studies ranged from 5 to 60 mg/kg for both rats and mice. Doses for the THPC studies ranged from 3.75 to 60 mg/kg for rats and 1.5 to 135 mg/kg for mice. Clinical signs, which included rough hair coats, labored breathing, swollen abdomens, tremors, arched backs, hind limb paralysis, and diarrhea, occurred in rats and mice dosed with THPS or THPC. In the 14-day and 13-week studies of THPS and THPC, the liver was the primary site affected in both rats and mice.

Doses for the 2-year studies were selected on the bases of mortality, decreased body weight, and hepatocellular lesions that occurred in the three highest dose groups in the 13-week studies. The observed toxicity varied with species and sex, resulting in a fourfold range in dose selection for the 2-year studies of THPS and THPC. Organ toxicity was mainly restricted to the liver, and the predominant nonneoplastic lesions were similar to those observed in the 13-week studies.

Concerns about the possible chronic toxicity of THPC were due, in part, to the potential decomposition of this compound to formaldehyde and hydrochloric acid, which might react to form bis(chloromethyl)ether (BCME), a known carcinogen. BCME is carcinogenic in mice by the dermal route (Van Duuren et al., 1972), causing squamous cell carcinomas of the skin, and in both rats and mice, by the inhalation route causing squamous cell carcinomas of the lung (Van Duuren et al., 1972).

After two years of studies, the carcinogenicity of THPS was not observed in either sex of F344/N rats or B6C3F1 mice given 5 or 10 mg/kg. There was no evidence of carcinogenicity of THPC in either sex of F344/N rats given 3.75 or 7.5 mg/kg, in male B6C3F1 mice given 7.5 or 15 mg/kg, or in female B6C3F1 mice given 15 or 30 mg/kg (NIH, 1987).

THPC and THPS were not carcinogenic in rats and mice in two-year bioassays. Dermal studies have shown that THP salts are promoters of skin cancer but not initiators. THPC and THPS have mutagenic potential in vitro, but THPS is not mutagenic in vivo. Limited mutagenicity data for THPC-urea suggest that it is not mutagenic in vivo. THPO is nongenotoxic. There is no convincing evidence to suggest that fabrics treated with THP salts are mutagenic. Available information indicates that there is no genotoxic hazard to humans. In short-term (up to 28 days) studies in rats and mice, the main toxic effect for both THPC and THPS is decreased body weight. The No Observed Adverse Effect Level (NOAEL) for both chemicals in both species is approximately 8 mg/kg body weight per day. In longer-term studies (13 weeks), the main target organ for toxicity is the liver. The NOAEL for this effect ranged from 3 to 7 mg/kg body weight per day for both salts in both species (Solenis, 2017).

Key information is needed on the types, amounts (including ratios), and toxicity of THPC derivatives present in THPC-treated cloth. It is important to note that THPC polymerizes within the fiber and fabric structure and may also react with other FR-formulation components present, so it might undergo other chemical changes that would alter its chemical properties and toxicity. It is also highly likely that oxidized forms of THPC are present in or on the aged THPC-treated fabric (NRC, 2000).

Hazard indices for the inhalation of particles and oral exposure to THPC are less than one and therefore, these route of exposure are not anticipated to be a concern. Dermal exposure to THPC through contact with treated material is not expected to occur, since THPC is chemically bound to the fabric (Bruchajzer et al., 2015).

6.5.6.4   Tetrakis(hydroxymethyl)phosphonium Hydroxide (THPOH)

THPC is usually partly neutralized with amines, amides, and/or alkali. Complete neutralization of THPC with sodium hydroxide yields a compound referred to as THPOH. The distinction between THPC used in a partially neutralized condition and THPOH is difficult to define. If the curing agent is basic as is ammonia, the distinction becomes meaningless.

THPOH-ammonia has received a great deal of commercial attention. The major advantage over THPC is reduced fabric tendering and reduced stiffness. Fabrics padded with THPOH give off formaldehyde during drying (Schindler and Hauser, 2004).

6.5.7   ORGANOPHOSPHORUS FRs CONTAINING P-C BOND

Organophosphorus compounds with P-C bonds are increasingly investigated as flame retardants due their high thermal and hydrolytic stability (P-C bond), ease of synthesis, and suitability of processing at high temperature. One should pay attention to other properties such as miscibility in the polymer matrix, reactivity with polymer or monomer, moisture absorption, tendency for blooming, and toxicological properties.

P-C bonds are thermally quite stable and exhibit typical bond energies of ~272 kJ/mol (Liang et al., 2012). Although the bond energies are lower than P-O (360 kJ/mol) and P-N (290 kJ/mol), the carbon substituents are not good leaving groups and are thus resistant to possible nucleophilic attacks, which becomes relevant in high-temperature polymer processing (Butnaru et al., 2015). However, hydrolysis of P-C bonds can help in their biodegradation and reduce their possible harmful impact to the environment. Recent studies have provided insights into possible enzymatic hydrolysis of suitable P-C bonds (Agarwal et al., 2011; Kamat et al., 2011).

P-C bond creation and its use in synthetic chemistry offer a wide range of possible design and development of exciting organophosphorus flame retardants. Conventional synthetic methodologies, namely Michael addition, Michaelis–Arbuzov, Friedels–Crafts, and Grignard reactions are the most commonly used by academics for development of P-C bonded flame retardants. Use of, for example, green reagents, catalytic reactions, microwave technology, and photo-initiated reactions. are increasingly used by researchers to synthesize P-C containing organophosphorus compounds. Recently, aromatic DOPO-based phosphinate flame retardants have been developed with relatively higher thermal stabilities (>250°C). Such compounds have potential as flame retardants for high-temperature-processable polymers, such as polyesters and polyamides. A vast variety of P-C bond containing efficient flame retardants are being developed; however, further work in terms of their economical synthetic methods, detailed impact on mechanical properties and processability, long term durability, and their toxicity and environmental impact is much needed for their potential commercial exploitations. (Wendels et al., 2017).

The phosphite esters, P(OR)$_3$ have a lone electron pair on the phosphorus center, and can thus react with various electrophiles (Stawinski and Kraszewski, 2002). When these phosphite esters are protonated or bear electron withdrawing groups, they react with nucleophiles. Based on the chemical reactivity of the organophosphorus compounds, diverse synthesis approaches to P-C bond formation have been explored and can be categorized (Troev, 2006).

The P-C bond formation has been reviewed in the past by various researchers (Michal et al., 2015), and some of the conventional synthetic strategies such as P-H transformation into P-C bond, rearrangement/transformation reactions, and C-H to C-P transformation are listed next:

P-H transformation into P-C bond

1. Pudovik reaction (Equation 6.19) (Kamath et al., 2017),
2. Kabachnik–Fields reaction
3. Michaelis–Becker reaction
4. Nucleophilic ring opening of epoxides
5. C-O/P-H cross coupling reaction.

$$R_1O-\overset{\overset{O}{\|}}{\underset{\underset{OR_1}{|}}{P}}-H \ + \ \overset{\overset{X}{\|}}{\underset{R_2 \quad R_3}{}} \ \xrightarrow{\text{base}} \ R_1O-\overset{\overset{O}{\|}}{\underset{\underset{OR_1 R_2}{|}}{P}}\overset{XH}{\underset{}{-}}R_3, \qquad X = NR_4, O \qquad (6.19)$$

Rearrangement/transformation reactions

1. Michaelis–Arbuzov reaction (Equation 6.20) (Shaikh et al., 2016),
2. Sandmeyer-type reaction
3. Abramov reaction.

$$R_1O-P\overset{OR_1}{\underset{OR_1}{}} \ + \ R_2-X \ \longrightarrow \ R_1O-\overset{\overset{O}{\|}}{\underset{\underset{OR_1}{|}}{P}}-R_2 \qquad (6.20)$$

C-H to C-P transformation

1. Friedel–Crafts reactions (Equation 6.21) (Montchamp, 2015).
2. Via Aryne formation
3. Nucleophilic substitution.

$$(6.21)$$

In the last decade, there has been an increased focus on development of new synthetic methodologies for P-C bond formation. As a result, new novel and green methodologies in organic synthesis gave impetus to replace toxic and harmful phosphorus starting materials with more benign alternatives (Montchamp, 2014). However, the commercial availability of the starting chemicals and their cost determines their potential commercial exploitation.

Due to electronic nature of the phosphorous atom, the organophosphopous compounds can be classified into different chemical classes based on the coordination number and the oxidation state of the phosphorus atom (Troev, 2006). Accordingly, the chemical reactivity of the organophosphorus compounds can be influenced by their structural geometry and valency of the phosphorus atom. For example, H-phosphonates with P(O)H bonds are commonly used as starting materials for P-C bond formation, and thus, their chemical reactions have been investigated in detail (Michal et al., 2015). It is reported that presence of the P=O bond is the driving force for their reactions (Stawinski and Kraszewski, 2002). In general, the phosphorus center of the H-phosphonates is electrophilic and can be modified to be nucleophilic (Guin et al., 2015).

Various substitution reactions are utilized in research to create P-C$_{aliphatic}$ bonds containing useful flame retardants (Horrocks and Zhang, 2002). The reaction of but-3-1-yl 4-methylbenzenesulfonate (3-Butynyl p-toluenesulfonate, $C_{11}H_{12}O_3S$) and butyl 4-methylbenzenesulfonate ($C_{11}H_{16}O_3S$), with triphenylphosphine producing two phosphonium salts, are shown in Structures 6.11A and 6.11B (Hou et al., 2016). It was found that the polycarbonate (PC) composite containing flame retardant with the alkene group (6.11A) exhibited a significant increase of LOI value (32.8%) compared to composite with compound 6.11B (30.8%) and a lower peak heat release rate (PHRR). The LOI value of the neat PC is 25%. The char yield at 700°C was found to be higher for both compounds (17 wt % and 15.6 wt % for compounds 6.x1 (A) and 6.x1 (B), respectively) than for the virgin PC (13.9 wt %), which means that the two compounds improved the flame retardancy of the PC and act in the condensed phase. It was also noted for both compounds that 5% in the composite was enough to obtain UL 94 V-0 rating at a concentration of 5%, together with an antidripping agent.

Addition reactions are also commonly used to create P-C bond–containing flame retardants.

Cyclic tert-butyl phosphonate (Structure 6.12) was synthesized by reaction of *tert*-butyl phosphonic dichloride and neopenthyl glycol (Lee et al., 2008). For a resin formulation with a composition of 85 wt % aromatic vinyl resin, 15 wt % polyphenylene ether resin, and 20 wt % of the aforementioned compound showed good impact strength and a high melt viscosity index. In addition, this formulation also exhibited UL 94 V-0 rating.

6.5.8 PHOSPHORUS-NITROGEN FRS (PNFRS)

As halogen-free flame retardant substances, the researchers have reported that PNFRs must cooperate with other synergistic agents to prepare excellent flame-retarding thermoplastic esters (Fei et al., 2010; Gao et al., 2006).

Phosphoramidate derivatives are promising flame retardant additives, due to the synergistic effect provided by phosphorus and nitrogen. A phosphoro-diamidate is a phosphate that has two of its OH groups substituted by NR_2 groups, e.g., $(HO)_2PONH_2$.

Gaan and co-workers et al. (2009) synthesized different phosphoramidate derivatives as flame retardant additives for cotton fabrics, following the Atherton-Todd reaction in the presence of dialkylphosphites as starting materials (Equations 22, 23 and 24, R = methyl or ethyl).

STRUCTURE 6.11 Two phosphonium salts produced by substitution reactions (see text)

STRUCTURE 6.12 Cyclic tert-butyl phosphonate

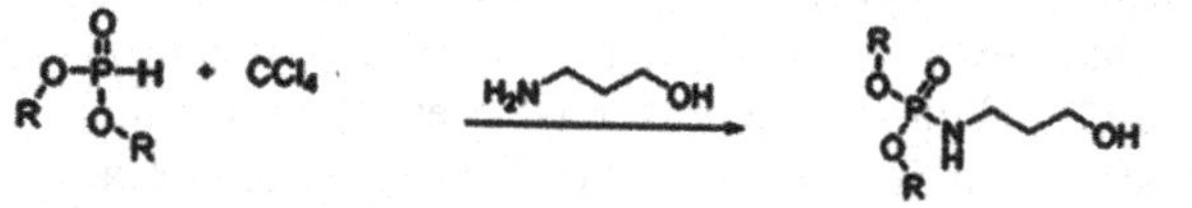

[6.22]

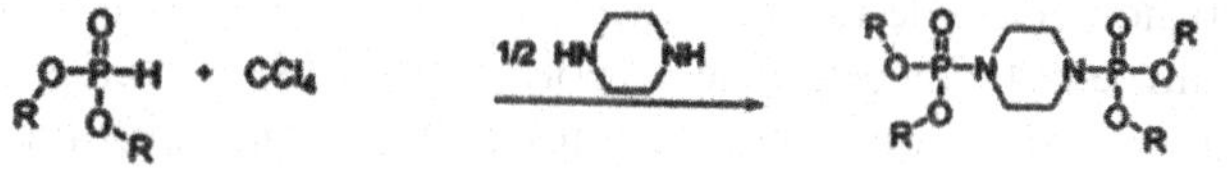

[6.23]

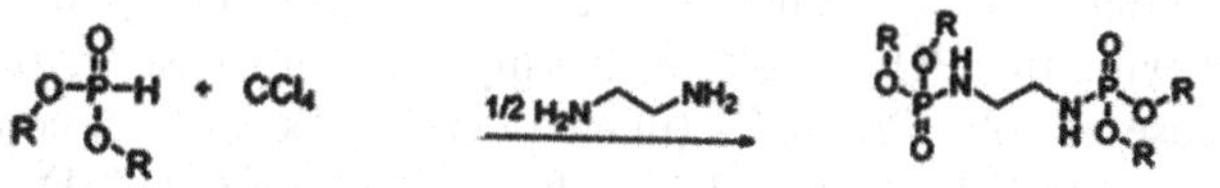

[6.24]

The Atherton-Todd reaction (2018) is exemplified by Equation 6.25 using the reactant dimethyl phosphate.

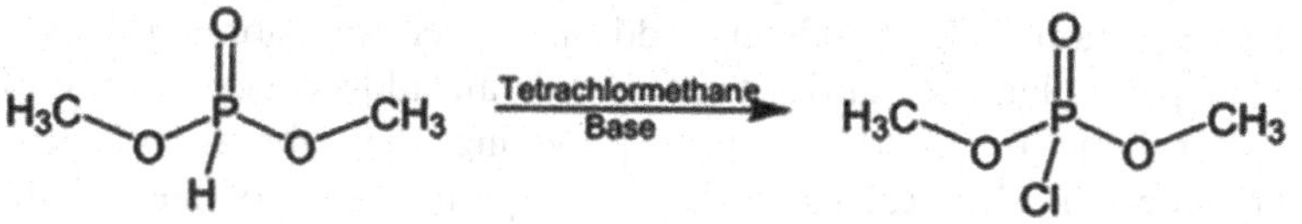

(6.25)

The reaction takes place after the addition of tetrachloromethane and a base. This base is usually a primary, secondary, or tertiary amine. Instead of methyl groups, other alkyl or benzyl groups may be present (Atherton et al., 1945).

LOI values were found to increase by increasing the phosphorus content of the treated cotton.

The Atherton-Todd reaction is named after British chemists F. R. Atherton, H. T. Openshaw and A. R. Todd. These described the reaction for the first time in 1945 as a method of converting dialkyl phosphites into dialkyl chlorophosphates (Atherton et al., 1945). The diacyl chlorophosphates formed are very much reactive, and it is difficult to isolate them, however. For this reason, the synthesis of phosphates or phosphoramidates can follow the Atherton-Todd reaction in the presence of alcohols or amines.

A phosphorodiamidate (or diamidophosphate) is a phosphate that has two of its OH groups substituted by NR_2 groups to result in a species with the general formula $HOPO(NH)_2$. The substitution of all three OH groups results in phosphoric triamides ($P(=O)(NR_2)_3$), which are commonly referred to as phosphoramides. In recent years, nitrogen–phosphorus (P–N)–based compounds, such as phosphoramidates (PAMDT), have been attracting more and more attention in the scientific community. Compared to the halogenated and some phosphorus-based compounds, their main advantages are their relatively low volatility, ease of synthesis, and low evolution of toxic gases and smoke in the event of fire. The versatility of flame retardation of various PAMDTs for application in different substrates (e.g., cotton, poly(butylene terephthalate), epoxy resins, cellulose, and cellulose acetate) has been demonstrated by researchers. An efficient single additive-type PAMDT is suitable for a wide range of polymeric materials, from highly charrable (polycarbonate), moderately charrable (PBT) to noncharrable matrices, e.g. ethylene-vinyl acetate polymer (EVA) and acrylonitrile-butadiene-styrene terpolymers (ABS). A series of model monosubstituted secondary phosphoramidates (PRs), monosubstituted tertiary PRs, and trisubstituted secondary PRs have been synthesized and incorporated into polyurethane (PU) foams. Overall, the PAMDT compounds had good compatibility with the PU matrix, with only some increase in the density of foam in the case of solid PAMDTs (phenyl derivatives) at high concentration (5% and 10%). Fire test results indicate that methyl ester PAMDTs, because of their higher phosphorus content, exhibit better flame retardancy compared to analogous phenyl ester derivatives at equal weight percentage in PU foams (Neisius et al., 2013).

Nguyen et al. (2014) further investigated the flame retardant performance of the product in Equation 6.22 at different add-ons on cotton fabrics (Nguyen et al., 2013): in particular, the fabrics treated with an add-on beyond 5 wt% showed no after-flame or afterglow time, providing the fabrics with self-extinction. The effect of chemical structure on the performance of these flame retardants was studied using micro combustion calorimeter (MCC). It was found that Total Heat Release (THR) values decreased with increases in the add-on of product of Equation [6.22, R = methyl]. Conversely, THR values were found to increase by increasing the add-on of Equation [6.22, R = ethyl]. The product of Equation [6.23, R = ethyl] promoted char formation and decreased the THR of the treated samples with increases in the add-on on the treated fabrics. However, vertical flammability tests on cotton fabrics treated with the product of Equation [6.23, R = ethyl] showed no afterglow phenomena.

Gao et al. (2015) synthesized a novel, halogen-free, formaldehyde-free, organic phosphorus-based, flame retardant, ammonium salt of hexamethylenediamine-N, N,N/,N/-tetra(methylphosphonic acid) (AHDTMPA), using urea and hexamethylenediamine-N,N,N/,N/-tetra(methylphosphonic acid) (HDTMPA) (Structure 6.13). AHDTMPA reacts with the O_6 hydroxyls of cotton fabric forming P–O–C covalent bonds. This ability was confirmed by Fourier-transform infrared (FTIR)

STRUCTURE 6.13   Chemical structure of HDTMPA

spectroscopy, attenuated total reflection FTIR spectroscopy, and X-ray diffraction. Cotton fabric treated with 70 g/L AHDTMPA had a limiting oxygen index (LOI) value of 36.0 %, which remained relatively stable after 50 laundering cycles with a 28.0 % LOI value. TGA showed that AHDTMPA can provide flame retardancy at temperatures below the pyrolysis temperature of cotton fabrics. The whiteness of the treated cotton fabric was maintained; however, the tensile strength of the cotton fabric grafted with AHDTMPA decreased slightly at 200° C, but was strong enough to be applied to clothing and indoor decorations.

Zheng et al. (2016) synthesized ammonium salt of ethylenediamine tetramethylenephosphonic acid (Structure 6.14) (AEDTMPA) by reacting ethylenediamine, phosphoric acid, and formaldehyde. It is claimed that it does not contain halogens or formaldehyde. The structure of AEDTMPA was confirmed using nuclear magnetic resonance (NMR) spectroscopy. The phosphorus content of AEDTMPA was higher than that of other typical phosphorus-containing flame retardant agents, exhibiting more efficient flame-retardant performance. The AEDTMPA molecule contains several–$PO(O^-NH_4^+)_2$ groups that can react with the–OH groups of cellulose and produce finished cotton fabrics with durable flame retardant performance. The finished cotton fabrics had excellent flame retardancy and durability, and no after-flame and afterglow were noted in the vertical flammability test. The char length of finished cotton fabrics decreased with increasing AEDTMPA concentration and the weave structure was well maintained after burning.

An efficient and environmentally friendly FR, consisting of an ammonium salt of pentaerythritol (Structure 6.15) tetraphosphoric acid (Structure 6.16) (APTTP) having reactive $P=O(O^--NH_4^+)_2$ groups was successfully synthesized by Jia et al. (2017). The word *pentaerythritol* is derived from words *penta* and

STRUCTURE 6.14 Chemical structure of AEDTMPA

STRUCTURE 6.15 Pentaerythritol

$$HO-\overset{\overset{\displaystyle O}{\|}}{\underset{\underset{\displaystyle OH}{|}}{P}}-O-\overset{\overset{\displaystyle O}{\|}}{\underset{\underset{\displaystyle OH}{|}}{P}}-O-\overset{\overset{\displaystyle O}{\|}}{\underset{\underset{\displaystyle OH}{|}}{P}}-O-\overset{\overset{\displaystyle O}{\|}}{\underset{\underset{\displaystyle OH}{|}}{P}}-OH$$

STRUCTURE 6.16   Tetraphosphoric acid

erythritol–penta in reference to the number of carbon atoms and erythritol, which also possess four alcohol groups. The cotton fabric treated with 110 g/L of APTTP could be used as a semidurable FR fabric, while the cotton fabric treated with 140 g/L of APTTP could be employed as a durable FR fabric. The flame on the treated cotton fabric extinguished once the fire source was removed, whereas the control cotton was consumed completely.

TG results demonstrated that the decomposition temperature of the treated cotton fibers decreased by approximately $100°C$, and the remaining residues significantly increased, compared to those of the control fibers. SEM images revealed that the APTTP-treated cotton fibers had nearly the same morphology as the original cotton fibers, while the XRD results indicated that the APTTP finishing may have slightly affected the crystalline particles. The remarkable decrease in peak heat release rate (PHRR) and THR of the treated cotton fabric demonstrated that APTTP is efficient in hindering fire spreading by favoring the formation of char. In addition, TG-IR results show the flame retardant APTTP decreased absorption intensity of flammable species and its flame retardancy mechanism occurred by condensation phase. The cotton fabric treated with APTTT exhibited high flame retardancy and durability characteristics (Jia et al., 2017).

6.5.8.1 N-Methyloldimethyl Phosphonopropioamide (PYROVATEX CP)

PYROVATEX CP is a reactive FR. It contains a P-C bond and provides a method of attaching phosphorus to cellulose that makes use of N-methylol reactivity with cellulose. PYROVATEX CP is applied with a methylolated melamine resin using a phosphoric acid catalyst by a pad-drycure process. The high nitrogen content of melamine provides a synergistic activity to the phosphorus of the flame retardant.

PYROVATEX CP NEW/CP-LF (Huntsman) is a dialkylphosphonocarboxylic acid amide with the general formula (Structure 6.17).

$$\begin{matrix} R-O \\ \\ R-O \end{matrix}\overset{\overset{\displaystyle O}{\|}}{P}-C_nH_{2n}-\overset{\overset{}{}}{\underset{\underset{\displaystyle O}{\|}}{C}}-NHCH_2-OH$$

STRUCTURE 6.17   Dialkylphosphonocarboxylic acid amide

PYROVATEX CP NEW/CP-LF, is preferably used together with methylola-mine precondensates, reacts with the hydroxyl groups of the cellulose. Alkaline washing is an integral step in the finishing process. It removes unfixed PYRO-VATEX CP NEW/CP-LF or acid from the catalyst.

6.5.8.2 Phosphine Oxides

These compounds have hydrolytically stable P–C bonds, whereas the P–O–C bond of phosphate esters hydrolyzes readily. They also generally have higher phosphorus contents than the corresponding aromatic phosphate esters and, therefore, are more effective flame retardants. Triphenylphosphine oxide is dis-closed in many patents as an FR, and may find some limited use as such a vapor-phase flame inhibitor. A phosphine oxide (Structure 6.18), incorporat-ing hydroxyl groups, has been proposed as a flame retardant for use in poly-propylene (PP). In spite of its high hydroxyl content, it is a relatively stable, high-melting solid additive.

Another possible strategy, differing from the conventional FRs, exploits the design of phosphorus-based polymeric flame retardants, rather than low molecular weight counterparts. Indeed, the polymeric FRs created more inter-est because of their durability and eco-friendliness.

6.5.9 POLYMERIC FRs

Another possible strategy, differing from the conventional FRs, exploits the design of phosphorus-based polymeric flame retardants, rather than low molecular weight counterparts. Indeed, the polymeric FRs created more inter-est because of their durability and eco-friendliness. Different polymeric flame retardants are employed as additives for either textiles or engineering plastics. In addition, polymeric flame retardants bearing hydroxyl or carboxyl groups were found to be very effective for cellulosic materials.

Different polymeric flame retardants are employed as additives for either textiles or engineering plastics. In addition, polymeric flame retardants bearing hydroxyl or carboxyl groups were found to be very effective for cellulosic materials.

Liu et al, (2012) reacted spirocyclic pentaerythritol bisphosphorate dispho-sphoryl chloride (SPBDC) with tartaric acid (TA) to form poly(1,2-dicarboxyl ethylene spirocyclic pentaerythritol bisphosphonate) (PDESPB) as shown in Equation 6.26.

STRUCTURE 6.18   Phosphine oxide flame retardant for use in polypropylene (PP)

$$(6.26)$$

SPBDC   TA   PDESPB

After caustic soda wash, cotton fabric was soaked in 30% PDESPB and catalyst. After drying and curing at 160°C, the treated fabric was washed with soda
ash and dried. LOI values showed an increase up to 33.8% by increasing the
add-on up to 21.2 wt %. Furthermore, vertical burning tests performed on the
treated samples showed a reduction of the afterglow time and of the char length,
without after-flame phenomena, compared to the untreated cotton fabric.

Dong et al. (2012) reacted phenyl dichlorophosphate with ethylenediamine,
in order to obtain poly(phoshorodiamidate) as shown in Equation 6.27.

$$(6.27)$$

Phenyl dichlorophosphate + ethylenediamine   Poly(phoshorodiamidate)

The product was applied onto cotton fabrics at different concentrations, in
the presence of tetraethyl orthosilicate employed as a cross-linker to permanently link the polymer to the fabric substrate. The samples were then dried
and cured at 160°C. All samples were then washed before testing. The flame-
retardant performance of the treated cotton was investigated through vertical
burning tests against untreated fabrics; shorter afterglow time and shorter
char lengths were found by increasing the add-on of the treated samples.

A polymeric additive shown in Structure 6.19 was also synthesized using
poly (4-iodo-n-butoxy) methylsiloxane and guanidine sulphamate (Bounor-
Legar'e et al., 2002). LOI values of the treated samples did not show
a significant drop after several washing cycles, hence proving the durability of
the flame retardant. The treated fabrics also achieved water repellence, due to
the reaction of the flame retardant with cotton fibers, leading to methyl group
orientation on the fiber surface.

6.5.10   SILICON FRs

Silicone is eco-friendly, widely available in nature. It is easy to prepare flame
retardants from silicone. The silica ash layer can also prevent oxygen from
reaching the matrix. A novel, simple, and low-cost flame retardant containing phosphorus, nitrogen, and silicon was synthesized by Zhou et al. (2015).
LOI was improved from 18.0 to 27.1, wrinkle recovery angle increased from
80° to 98°.

STRUCTURE 6.19 A polymeric additive

Recently, silicon-containing monomers have been successfully used as reactive-type HFFRs, particularly polyhedral oligomeric silsesquioxanes (POSS), due to their notable reinforcing effect (Bourbigot et al., 2005).

6.5.11 HYBRID FRs (HYFRs)

Compared with halogen-containing flame retardants, PFRs are more environmentally friendly and have good flame retardancy. However, if a flame retardant contains only phosphate, it is necessary to remarkably increase the phosphorus content to achieve good fire retardancy (Li and Liu, 2012). Therefore, the other fire retardant elements are always integrated into the phosphate-containing flame retardants to improve performance. Usually, researchers prefer to design nitrogen and phosphate into one flame retardant system for further study because of the phosphorus-nitrogen (P-N) synergistic effect. Silicone flame retardants, due to their particular superior qualities, such as being friendly to the environment, having a high content in nature, and easy preparation, have become the research focus at present.

When silicone flame retardants are exposed to elevated temperatures under oxygen, -Si-C- bonds can be formed from the -Si- in the silicone molecules. As a result, an inorganic combustion silica residue is left behind that serves as a mass transport barrier, delaying the volatilization of decomposition product and insulating the underlying polymer surface from incoming external heat flux. The silica ash layer can also prevent oxygen from reaching the matrix and a physical network formed by such additives of a high surface area (e.g., aerosol) in the polymer melt can reduce dripping (Carosio et al., 2011). Based on these fire properties, silicones offer significant advantages for flame retardant applications. However, compared with other antiflame agents, silicone flame retardants express a poor flame-resistant effect when used in isolation. If the flame-resistant agent contains both the elements, phosphorus and silicon, its performance is greatly enhanced. This phenomenon is called the phosphorus silicon synergistic effect. Phosphorus can catalyze the formation of char at high temperatures, and silicone can improve the thermostability of the char layer. Furthermore, if siloxane is used instead of silicone, the synergistic effect can be further strengthened, mainly because the silica layer produced by degrading siloxane can prevent char from oxidating, thus improving the thermostability of the char layer (Lin et al., 2011). Zhou et al., (2015) used phosphorus oxychloride, methanol, propylamine, and γ-amine triethoxysilane (silane coupling agent) as raw materials, and synthesized a fire-retarding agent containing phosphorus/nitrogen/siloxane, FR-PNSI (Structure 6.20). The flame retardant was synthesized and applied to cotton fabrics via the pad-dry-cure method. The FR-PNSI can form covalent bonds with cellulose macromolecular chains to obtain a firm combination, which leads to an improved anticrease property, due to the cross-linking in the amorphous form of cotton fabrics. The optimal parameters of the anti-flame finishing process conditions were:

- Padding with 200 g/L FR-PNSI and 25 g/L catalyzer at pH 5,
- Drying followed by curing at 170 °C for 3 min.

The treated fabric showed excellent flame retardancy: LOI improved from 18.0 to 27.1 and its anticrease property was obviously improved, The TGA curves also suggested that the flame retardant had a good ability of char formation due to the synergistic effect of phosphorus, nitrogen, and silicon. The LOI value of the treated sample remained at 23.3 after the washing fast test, compared to 18.0 for the untreated sample, which meant a durable flame-retarding effect.

In addition to phosphorus derivatives, silicon derivatives are also used as flame retardant additives (Khandual, 2016). In the study by Gawłowski et al. (2017), the flame retardant modification of PET POY fibers was successfully achieved by using aqueous sodium silicate solution–the so-called water-glass–a modifier that is inexpensive and, above all, 100% eco-friendly. The same was applied using a high-temperature bath method similar to dyeing fibers with disperse dyes. The applied conditions were as follows:

STRUCTURE 6.20   Chemical structure of FR-PNSI

- Temperature–115°C
- Heating rate–2.5°C/min
- Treatment time–1 hour

The PET POY fibers were treated in the water-glass solution of 0.3, 0.5, 1.0, 5.0 and 10.0 wt %. 1.5 g/dm^3 Dyspergator NNO (nonionic dispersing agent) was added into the modification bath. After the modification treatment, the investigated samples were washed in a solution of detergent Pretepon G (Organika-Rokita, Poland) in the amount of 5 g/dm^3. The washing time was 30 min, and the temperature was 60°C. Application of the bath method and water-glass (as the flame retardant), significantly reduces the flammability of the examined fibers. The obtained flame retardant effect is permanent. Flammability reduction reached by surface modification is slightly lower than for chemical modification (during the polycondensation process) or physical modification (mix of flame retardant and the polymer in the melt) (Baseri et al., 2012). However, both chemical and physical methods create techno-logical challenges, and are much more expensive that the surface modification method presented by Gawłowski et al. (2017). The flame retardant concentra-tion in the bath (0.5 wt %) was determined on a level that assured the max-imum flame-resistant effect (LOI = 28.2 %).

Recent interest is to find the potential for combining phosphorus, nitrogen, and silicon with cellulose substrates in order to create carbonaceous and sili-caceous char formation. (Horrocks, 2011). Zhao et al. (2017) synthesized a novel halogen-free, phosphorus–nitrogen–silicon FR monomer with reactive siloxy groups N-(diphenylphosphino)-1,1-diphenyl-N-(3-(triethoxysilyl)propyl) phosphinamine (DPTA) that was applied on cotton fabrics via the impregna-tion method.

In recent years, coating nanoparticles such as silica and carbon nanotubes with compounds containing phosphorus on the substrate has decreased flam-mability (Zhang et al., 2013a). Grafting and polymerizing flame retardant groups that form covalent bonds in cotton is the most efficient and durable technique to decrease the flammability of fabrics (Karim-nejad et al., 2015).

Silicone flame retardants, due to their particular superior qualities, such as being friendly to the environment, having a high content, and being easy to pre-pare, have become the research focus at present. When silicone flame retardants are exposed to elevated temperatures under oxygen, -Si-C- bonds can be formed from the -Si- in the silicone molecules. As a result, an inorganic combustion silica residue is left behind that serves as a mass transport barrier, delaying the volatilization of the decomposition product and insulating the underlying poly-mer surface from incoming external heat flux. The silica ash layer can also pre-vent oxygen from reaching the matrix.

A novel flame retardant containing phosphorus, nitrogen, and silicon was synthesized by Zhou et al. (2015). The preparation technique was simple and of low cost. The novel flame retardant can react with cellulose directly without using other cross-linking agents, and there is no formaldehyde and no halogen in the process of preparation and application; it is therefore an environmentally friendly agent. Cotton fabrics can be treated by the pad-dry-cure (170°C, 3 min, at pH 5) method. The treated sample showed a durability flame-retarding effect. LOI was improved from 18.0 to 27.1, while wrinkle recovery angle increased from 80° to 98°.

A novel phosphorous-silicone-nitrogen ternary flame retardant, [(1,1,3,3-tetramethyl-1,3-disilazanediyl)di-2,1-ethanediyl] bis(diphenyl phosphine oxide) (Pin) (Structure 6.21) was synthesized with high yield via a one-step procedure by the reaction of diphenyl phosphine oxide and vinyl-terminated simazine with triethyl borane as the catalyst system. The developed PSiN FR was applied in the flame retardancy of o-cresol Novolac epoxy (CNE)/phenolic novolac (PN) hardener system. Flow ability of the CNE/PN systems was improved by the addition of PSiN, but the Tg values of the epoxy thermosets decreased slightly with the increasing of PSiN loadings. Furthermore, the incorporation of PSiN favored the thermal stability of epoxy thermosets, which might be attributed to the char-yielding effects of P-containing and Si-containing components in PSiN. LOI values of the epoxy system increased with the PSiN contents, while heat release value was reduced. An UL 94 V-0 rating was achieved when the weight contents of PSiN in the epoxy composites reached 20 wt %. Characterization of the residue char implied that silica tended to migrate to the surface of the polymer for its low surface energy. Nitrogen components turned into gas in the combustion process. Phosphorus components played a role of FR in both the condensed phase and the gas phase. All the results indicated that current P-Si-N ternary FR might be a good candidate for epoxy flame retardancy (Li et al., 2014a).

All results indicated that current P-Si-N ternary FR might be a good candidate for epoxy flame retardancy (Li et al., 2014b).

Due to the banning of halogenated products, there is a strong interest in the development and utilization of more environmentally friendly polymeric materials with good flame retardancy and high performance. Rice husk (RH) is a naturally occurring organic/inorganic material predominantly consisting of lignocellulose and silica. Due to the presence of high silica content in RH, the addition of RH into synthetic polymers may cause synergistic behavior in the flammability of the resulting composites (Zhang et al., 2013b).

STRUCTURE 6.21   Chemical structure of PSiN

6.6   REACTIVE FLAME RETARDANTS

Reactive flame retardants, that is, those that are covalently attached to polymer chains, offer several advantages over those that are merely additives. They are inherently immobile within the polymer matrix and are therefore not susceptible to loss during service through migration to the polymer surface (blooming) or solvent leaching. Reactive flame retardants incorporated at the time of the synthesis of the polymer also can be homogeneously dispersed throughout the polymer at the molecular level, and may consequently be required to give a desired level of flame retardance at lower concentrations than comparable additives. The lower levels of incorporation of flame retardant groups may bring with it an added advantage that the overall properties of the polymer (chemical, physical, and mechanical) are less likely to be adversely affected when compared with those of the non-flame-retarded counterpart. Moreover, reactive flame retardants are prevented from forming a separate phase within the polymer matrix; this is particularly important in uses of polymers as fibers in which a heterogeneous phase structure is likely to bring with it problems during fiber-spinning and during reductions in the modulus of the resultant fibers.

The reactive incorporation of flame-retarding groups can bring such problems as relative difficulty and higher expense over well-established commercial methods for manufacturing of the existing unmodified variants. The extensive reactive modification of a partly crystalline polymer is likely to lead to a significant loss of crystallinity, whereas if an additive is introduced to a partly crystalline polymer, it would most probably end up in the amorphous phase and have little impact upon crystallinity, but may plasticize the amorphous regions. In general, the reactive modification of chain reaction polymers (e.g., acrylics, polystyrene, polyolefins) is less readily accomplished than is the case for step reaction polymers (e.g., polyesters, polycarbonates, polyamide), unless the reactive modification is applied after the manufacture of the primary chain, for example, through a post-polymerization grafting reaction. Grafting reactions may also be useful for covalently attaching flame retardant groups to the surfaces of polymer-based plastics moldings, films, and fibers, where they will be particularly effective at the point of first impingement of any flame. The flame retardants applicable on synthetic fibers are discussed in Chapter 9.

6.7   PHOSPHOROUS FRs FOR VARIOUS TEXTILE FIBERS AND MATERIALS

Phosphorus FRs can be effectively applied on various textile fibers, namely, cotton, rayon, wool, and synthetic fibers, as well as on plastics, resins etc. Phosphorous FRs applicable on natural and regenerated fibers are discussed next:

6.7.1   PHOSPHORUS FRs FOR COTTON

C-P-type flame retardants are used for flame retardant finishing of cotton fabrics to obtain flame retardant fabrics with low formaldehyde emission. The use of polycarboxylic acid instead of formaldehyde containing conventional cross-linker reduces the formaldehyde content of the whole flame retardant system, frees formaldehyde release on the flame retardant fabric on the one hand, reduces the amount of flame retardant and crosslinker on the other hand, and improves the application performance of PYROVATEX CP flame retardant (Hopechem, 2019).

The resistance to washing of PYROVATEX CP NEW/CP-LF is higher than those of other flame retardant finishing systems, mainly due to enlargement of the molecules of the finishing agent. Moreover, PYROVATEX CP NEW/CP-LF also differ from flame retardant finishing products for synthetic fibers in its chemistry and in the way it bonds to the fiber.

If flames or heat come into contact with flame retardant finished cellulose, a carbon scaffold is formed by dehydration of the cellulose, counteracting heat penetration and the spread of fire. Almost no flammable gases are formed by the pyrolysis of FR-treated fabrics, and this stops combustion (Huntsman, 2012). Fabric stiffening occurs when sufficient chemical is applied to give 2%–3% phosphorus on the weight of the fabric. The acid may cause loss of high strength if left in the fabric after curing; therefore, it is desirable to wash the fabric following the curing step.

The finish tends to produce smoke in the curing oven. The smoke is composed of volatile fragments of the finish that condense in the cooler reaches of the oven. The condensate may drip back onto the fabric causing unsightly spots (Tomasino, 1992).

Flame retardant finishing of cotton fabric is the most common method to produce flame retardant textile fabrics. For flame retardancy in cotton, the applied compounds should prevent the decomposition of cotton into flammable volatiles and promote the formation of carbon (Cheng and Yang, 2009). Thus, many have attempted to impart flame retardancy in cotton by grafting flame retardant groups, coating with flame retardant layers, or directly adding fire retardants to polymer materials.

Pyrovatex CP (N-Methyloldimethyl Phosphonopropioamide) is the most applicable flame retardant for cotton fabrics owing to its high efficiency and durability. Phosphorous compounds are capable of forming cross-linked structures on fibers or react with the cellulose fibers (Horrocks, 2011). These compounds affect the thermolysis, prevent the formation of levoglucosan and flammable volatiles, and promote the formation of char; however, they may release formaldehyde (Yang et al., 2012). In recent reports, many compounds containing phosphorus, such as phosphorylated chitin (Pan et al., 2015), biomolecules (Bosco et al., 2015), phytic acid (Laufer et al., 2012), and vinyl phosphonic acid (Nooralian et al., 2016), were used as cotton flame retardants. The results suggested that phosphorus in cellulosic fibers can act as a flame retardant.

A major route of development of FRs for cotton has been the introduction of FR materials or moieties within cellulose molecules, not necessarily by chemical reaction with it. Insoluble deposits such as deposition of stannic oxide by stannate-phosphate may be a quite effective step. Formation of semi-interpenetrating polymer network (SIPN) within cellulose structure through in situ polymerization, as in the case of THPC (tetrakis (hydroxymethyl) phosphonium chloride) finish is based on a similar principle.

A few durable (about 50 washes) phosphorous-based commercial FRs are listed in Table 6.2 (Nair, 2001). However, most of them cause high strength losses, stiff handle, and release of toxic formaldehyde.

It is difficult to finish 50/50 or greater, polyamide–cotton fabric with a flame retardant and have acceptable flame resistance properties coupled with durability after multiple washings.

A process for imparting flame resistance to polyamide/cotton blend fabrics containing at least 40% by weight polyamide, comprises the successive steps of:

1) Applying to the fabric a flame-retarding amount of a prepolymer condensate of urea and a tetrakis (hydroxymethyl) phosphonium salt flame retardant that fixes to the cotton fibers, drying the fabric to at most about 20% by weight of moisture, exposing the prepolymer

TABLE 6.2
Phosphorous-based Commercial FRs and Their Chemical Compositions

Sr. No.	Commercial name	Brand	Chemical Composition
1	Pyrovatex CP	formerly Ciba, now Huntsman	N-methylol dimethyl phosphonopropionamide
	Pyrovatex CP new	formerly Ciba, now Huntsman	Dialkyl-phophonocarboxylic acid amide,$C_6H_{14}NO_5P$.
2	THPC finishes (Pyroset TKC)	American Cyanamid, Hooker, Albright & Wilson and many Chinese manufacturers	Tetrakis(hydroxymethyl) phosphonium chloride,$(HOCH_2)_4PCl$
3	Pyroset TKP	American Cyanamid	Tris(hydroxymethyl) phosphine or THP salts ($C_3H_9O_3P$) with mixed phosphate and acetate anions
4	Proban	Albright & Wilson, Rhodia Specialities, UK	THP salt-urea pre-condensate
5	Proban CC	Albright & Wilson, UK	Tetrakis (hydroxymethyl)phosphonium chloride, $C_4H_{12}ClO_4P$-urea
6	THPOH-NH_3		THP salts at pH 7
7	MCC 100/200/300	Sigma-Aldrich, Pubchem, Molbase,	Trimethyl phosphoric triamide, $C_{12}H_{24}N_3OP$
8	Fyrol 76	Stauffer,Molbase	Condensate of bis(beta-chloroethyl) vinyl phosphonate,$C_9 H_{18}Cl_2O_6P_2{}^-$
9	DAP-urea-Ti		Diammonium hydrogen phosphate-urea-Titanyl sulphate

condensate containing fabric to a source of ammonia prepolymer to form a flame retardant polymer network within the cotton fiber structure, then oxidizing the fabric to further improve the flame resistance and enhance durability to multiple launderings and,

2) Applying an additional flame-retarding amount of (i) a cyclic phosphonate ester as shown in Structure 6.22 in which x is 0 or 1, that fixes onto the polyamide fibers in combination with (ii) a tetrakis (hydroxymethyl) phosphonium salt and urea to the fabric, heating the fabric to form an insoluble phosphorus-containing polymer in and on the cotton fibers, and then oxidizing the fabric to further improve the flame resistance and enhance durability to multiple launderings. (Hansen, 1989).

6.7.2   PHOSPHORUS FRS FOR RAYON

Though FRs can be applied on rayon materials by finishing methods, these compounds are usually added into the dope prior to spinning. The additives should be compatible with viscose solution, withstand spinning conditions, should not volatize during processing, and be smaller in particle size, but can be retained by the fiber during processing.

Besides a large number of phosphates, phosphonates and phosphazenes are extensively used for rayon manufacture. However, these liquid additives may migrate on the fiber surface during spinning. One successful FR additive, based on an alkyl dioxaphosphorinane disulfide (Structure 6.23), is used at ~ 20 % for effective flame retardancy.

An alternate FR additive for rayon based on thiophosphate is 2, 2/-oxybis (5,5-dimethyl-1, 2, 3 dioxaphosphorinane-2, 2/-disuphide) (Structure 6.24)

STRUCTURE 6.22   A cyclic phosphonate ester for nylon

STRUCTURE 6.23   FR additive for rayon based on an alkyl dioxaphosphorinane disulfide

STRUCTURE 6.24   FR additive for rayon based on thiophosphate

(Sandoflam 5060, Clariant); about 20% of this material is required for full FR effect (Levin, 1984).

Bychkova and Panova (2016) used the following combinations (mixed PFRs) to improve the fire resistance of viscose materials:

- FRS-1: a mixture of tricresyl phosphate, urea, and boric acid (BA) with the addition of soda (to neutralize the medium) and glycerol (as a plasticizer)
- FRS-2: A mixture of phosphoric acid, salt of polyethylene polyamine with formalin
- FRS-3: A mixture of diamide methyl phosphonate with ammonium chloride
- FRS-4: Methazine, and phosphoric acid (PA)
- T-2: A mixture of pyrofax (PF), diammonium chloride salt of diamide methyl phosphonic acid
- FRS-2 and FRS-3 are most similar in most of the tested parameters and exhibit better flame resistance.

Modification with FRS-3 is a highly effective method for reducing the flammability of viscose rayon materials characterized by technological simplicity. The modification of fibers with FRS-3 requires 8%–10 wt.% content thereof in baths, which provides high levels of P and N on the fiber and LOI values that do not change with a further increase in the concentration of FR in the bath.

According to pyrolysis gas chromatography (PGS) data, dehydration of viscose rayon fibers in the presence of a FR is shifted by 50–100°C to lower temperatures and occurs with a large output of water, compared to unmodified samples. The greatest amount of water is released during thermolysis of FRS-2 and FRS-3, making them less flammable. In addition, the effect of phosphorus and nitrogen compounds in the fiber manifests in a more intensive formation of noncombustible carbon dioxide. At the same time, the oxygen content in the gas flow decreases, and hence, the probability of formation of flammable oxygen-containing compounds in thermolysis and combustion decreases. There is a correlation between LOI values, changes in the yield of water (H_2O), and dehydration temperature (T_{max}). The developed fire-resistant fiber materials can be used not only for textile manufacturing purposes, but also as precursors for carbon fiber production, and as fillers for polymer composites.

6.7.3 PHOSPHORUS FRS FOR WOOL

Wool is naturally flame-resistant because of the way that wool fiber is structured. Wool requires more oxygen than is available in the air to become flammable. Wool is accordingly an excellent fiber when it comes to fire safety. Furthermore, it does not melt, drip, or stick to the skin when it burns. While most textile fibers are polymers containing mainly carbon and hydrogen that can burn easily, wool also contains high levels of nitrogen and sulfur. In fact, many fire retardant additives used for other materials are high in nitrogen. Wool therefore requires higher levels of oxygen in the surrounding atmosphere to accelerate combustion. The limiting concentration of oxygen required to support combustion of wool in standard tests is higher than the ambient oxygen concentration in air (21%). Therefore, it is difficult to ignite wool, but once ignited, the flame spreads slowly and it is easy to extinguish.

The evolution of increasingly severe mandatory flammability regulations in many aspects of domestic, social, and business life has meant that even inherently low flammability fibers, such as wool, require a flame retardant treatment for some applications. Typical applications subject to mandatory flammability requirements include children's nightwear, domestic and commercial furnishings, public transportation, and protective clothing.

Zirpro treatments are based on the exhaustion of negatively charged zirconium or titanium salts, under acid conditions, onto positively charged wool. This results in the deposition of only about 3% of flame retardant inside the fiber with negligible effect on such properties as handle. These treatments stabilize and further crosslink the protein structure. The best treatments are colorless; do not alter wool's natural properties, such as handle and moisture adsorption, and tend to be deposited near the surface of the fiber. These treatments tend to increase and strengthen the insulating foam produced as wool is decomposed by heat. Zirpro-treated wool also has good durability for washing and dry-cleaning.

The need to boil wool for exhaustion of titanium and zirconium chelates leads to felting, and lower dye fastness, and the process is energy intensive. To overcome these deficiencies, during the 1970s and 1980s, the International Wool Secretariat (IWS) developed a process based on titanium and zirconium hexafluoride known as the Zirpro Process (Benisek, 1971). Hexafluorotitanates and hexafluorozirconates are extremely stable in acid solutions and exhaust on wool well below the boiling point. The titanates turn wool yellower; hence, the zirconates are preferred commercially. K_2ZrF_6 at a pH < 3 gives 77% exhaustion at $50°C$ for 30 minutes and good levelling is achieved at $70°C$. A bath containing 3% K_2ZrF_6, 10% HCl (37%) for 30 minutes at $75°C$, gives rise to a washfast, lightfast, and improved heat- and flame-resistant fabric. Pad-batch and pad-dry application processes may also be used. It is accepted that the hexafluorozirconate ion is bound to cationic wool in the same manner as acid dyes.

The Low Smoke Zirpro treatment considerably improved the flame-resistance of wool woven carpets, while decreasing smoke emission and the concentration of some toxic gases. The solid phase mechanism of this flame retardant treatment, producing increased char volume, which restricts combustion, could be responsible for this advantageous burning behavior. (Benisek and Phillips, 1983).

Since the introduction of Zirpro, several different classes of flame retardants have been developed for wool, one of which is based on the application of highly effective halogen donors. The halogens in the treatments tend to interfere with free radical processes that maintain the flame. The use of halogen donors is now restricted by changes to environmental legislation stressing the need to develop alternatives. In spite of the environmental shortcomings, halogens and their derivatives still form a diverse and important role in the flame-resistant treatment of many products, not just textiles. Alternative treatments are based on phosphorous compounds, which tend to lower the thermal decomposition temperature of the textile, allowing the volatile fuel to escape before the ignition temperature is reached.

A water-soluble flame retardant monomer dimethyl-2-(methacryloyloxyethyl) phosphate (DMMEP) (Structure 6.25), initiated by potassium persulfate (KPS), was successfully applied onto wool fabric Adding ~ 50%–100% DMMEP on the weight of wool fabric yielded grafted wool fabric, which exhibited good flame retardancy with an LOI above 35% and could pass the vertical flammability test. FR wool changed the thermal decomposition mode, while being heated, and was prone to produce more nonflammable solid char under combustion. FR wool had a slight drop of moisture regain and permeability, but had no evident effect on wearing comfort. However, grafted wool tensile strength dropped and needs to be improved.

Recent research into flame retardants has focused on the development of intumescent agents. These agents combine the attributes of flame retardancy with the formation of a high thermal resistance insulating char layer. Although originally developed for cellulosics, wool specific intumescents have now been formulated to enhance the natural flame resist and char formation properties of wool (CSIRO, 2008).

6.7.4 PHOSPHORUS FRs FOR EPOXY RESINS

The epoxy (EP) resins possess excellent physical, chemical, dielectric and adhesion properties; they shrink less upon curing. They are widely used in many fields, such as composite materials, coatings, adhesives, laminates, and encapsulation for electronic packaging. However, inflammability is one of the drawbacks of EP resins that limit their applications when high thermal and fire

STRUCTURE 6.25 Chemical structure of DMMEP

resistance are required. The improvement of flame retardancy of epoxy resin is a subject of research for a large number of researchers.

Epoxy resin systems can be rendered fire-retardant, either by the incorporation of fire- retardant additives (e.g., triphenyl phosphate, decabromo diphenylether, aluminium ethyl phosphinate) or by copolymerization with reactive fire retardants (e.g., diglycidylether of tetrabromobisphenol A). However, they cannot exhibit UL94 V-0 rating for highly combustible polymers such as styrenic polymers because of their low P content. Current challenges are halogen-free, self-extinguishing composites that are characterized by the UL 94 classification V_0 for electronical engineering and electronics applications such as printed circuit boards. Furthermore, halogen-free flame-retarded materials are in demand that can comply with new advanced standards for use in new applications in transportation, such as railway vehicles and aviation. In the search for effective halogen-free solutions, a main focus is on organophosphorus compounds.

Fire-retardant epoxy resins can be made by incorporating fire-retardant additives or by copolymerizing with reactive fire retardants. Traditional HFRs are limited and some were even forbidden to be used, mainly because of the halogenated hydrogen gas problem and the potential to produce dangerous compounds (Ding et al., 2009). Thus, organophosphorus-based FRs, which are free of the halogen environmental problems and show good compatibility with EP resins, have been widely studied as a flame retardant for EP resins (Shan et al., 2017). Among various organic FRs, the species containing phosphorus elements have been extensively used in epoxy flame retardancy applications (Qian et al., 2014).

PFRs have the advantages of low toxicity, good compatibility, and negligible negative impact on polymer physical properties. Until now, polymeric flame retardants are mostly represented by polyphosphates (PPEs); however, high loadings were always required to meet the desired flame retardant grade because of their poor flame retardancy. As polymeric flame retardants, polyphosphonamides (PPDAs) have been scarcely investigated to date. In addition to nitrogen, PPDAs have shown a synergistic effect of P and N to give high thermal stabilities and high flame retardancy. Zhao et al. (2017) prepared a series of PPDAs with pendent P–C structure by solution polycondensation of phenylphosphonic dichloride and aromatic diamines. The PPDAs have been characterized by H NMR and P NMR spectroscopy and applied into flame retardant epoxy resins. To prove the structure–property relationships and widely-use of the phosphonamide structure with pendent P–C structure, aromatic diamines with methylene group, sulfone group, and ether group were employed as staring materials.

Three PPDAs, namely poly(4,4′-diamino diphenyl sulfone phenyl phosphonamide) (POS), poly(4,4′-diamino diphenyl methane phenyl phosphonamide) (POM), and poly(4,4′-diamino diphenyl ether phenyl phosphonamide) (POA) were synthesized by reacting phenylphosphonic dichloride with 4,4′-Diaminodiphenyl sulfone (DDS), 4,4′-diaminodiphenyl methane (DDM), 4,4′-diaminodiphenyl ether (DDE). Three reactions are shown in Equation 6.28.

$$(6.28)$$

Where, R= (O=S=O), for POS and DDS,

R = (CH$_2$), for POM and DDM,

R = (O), for POA and DDA.

The study showed that the PPDAs exhibit high thermal stability, high glass transition temperature, and low flammability, associated with their backbone structures. The following order of thermal stability and char-forming efficiencies of the incorporated groups were obtained: -SO$_2$ → -O- → -CH$_2$-. From TGA-FTIR results, the pyrolysis gases of the PPDAs have been found to be mainly aromatic compounds and phosphorus-containing gases. The high char yield, low flammability, and generation of phosphorus-containing gases indicated the potential high flame retardancy of the PPDAs.

Cone calorimeter, LOI, and UL-94 vertical burning tests all demonstrated that the addition of PPDAs into EP have greatly enhanced flame retardancy of the composites. Due to the incorporation of heteroatom, POS and POA show better flame retardant performances in LOI and UL-94 tests than POM. EP/POS shows a similar T$_g$ with pure EP, which is attributed to the hydrogen–bond interaction. The heat release rate and thermal properties of flame retardant EP were highly dependent on chemically surrounding phosphorus atoms in PPDAs. The flame-retardant mechanism of PPDAs in EP can be devised by the charring effect, barrier, and protective effect in the condensed phase, and in the flame-inhibition effect in the gas phase. In addition, the incorporation of heteroatom into PPDAs could enhance the flame retardancy of EP composites, such as the sulfone group, in promoting char-forming processes, and the ether group in flame-inhibition effects in the gas phase (Zhao et al., 2017).

As part of a program to develop more survivable aircraft, flame retardant epoxy resins were investigated for their potential as fire-resistant exterior composite structures for future subsonic commercial and general aviation aircraft. Four poly(phosphonamides) were synthesized by separate polycondensation of phenyl phosphonic dichloride (PPDC) with 3,3$'$-diaminobenzophenone, 4,4$'$–thiodianiline, 4,4$'$-diaminodiphenyl sulfone and 4,4$'$-oxydianiline. The poly-(phosphonamides) were used as toughening agents (with 4,4/-diaminodiphenyl sulfone) to partially cure a commercially available unmodified liquid epoxy resin. The formulations show good flame retardation with phosphorus content as low as 1.6% by weight, but high moisture uptake, compared to the baseline epoxy. The fracture toughness of the cured formulations showed no detrimental effect due to phosphorus content (≈1.5% P) (Gordon et al., 2010).

Nevertheless, it is desirable to develop thermally latent hardeners that comply with specialized demands. It is possible that a highly additive flame retardant efficiency between phosphorus and boron elements may induce the formation of a glasslike char acting as insulation to prevent heat and mass transfer (Braun et al.,

2007). A borate ester containing the phosphaphenanthrene group with N→B coordination structure (PBN), which was described in a patent (Liu et al., 2011), is employed as a latent hardener/flame retardant for diglycidyl ether of bisphenol An epoxy resin (E51). Application of PBN as a latent hardener/flame retardant would permit the development of environmentally friendly, one-pot, flame-retardant EPs. Borate ester containing the phosphaphenanthrene group with an intramolecular N→B bond (PBN) was synthesized by transesterification of tribu-tyl borate (B(OC$_4$H$_9$)$_3$) with 2-(6-Oxido-6H-dibenz⟨c,e⟩⟨1,2⟩oxaphos-phorin-6-yl) methanol (ODOPM) and N,N-dimethylethanolamine (DMEA). PBN can weaken the scission of cured EPs as revealed by the TGA-FTIR results. A higher char yield and a more consolidated char layer were responsible for the superior flame retardance of E51-PBN20, according to TGA analysis and the SEM morphologies of the residue (Liu et al., 2011).

Halogen-free FR epoxy resins based on hexaglycidyl tris(3-(bis(oxiran-2-ylmethyl)amino) phenyl)phosphine oxide (HGE) (Structure 6.26) were synthe-sized by Kheyrabadi et al. (2018). These high-performance polymers prepared from the reaction of HGE monomer with tris(3-aminophenyl) phosphine oxide (TAPO), 4,4′-diaminodiphenylsulfone (DDS), and 1,5-diaminonaphthalene (DAN) as hardeners achieved a cured polymer with high thermal stability and high char yield. The incorporation of phosphorus and nitrogen atoms in the backbone chain imparts nonflammability to these epoxy resins, due to the unique combination of synergistic effects of phosphorus and nitrogen heteroa-toms on flame retardancy. SEM micrographs showed a microporous morph-ology for the char surface after the flame test. The introduced HGE polymers met the UL-94 V0 classification.

However, there are still many challenges or risks for flame-retardation of epoxy resins using phosphorus-based FRs such as those associated with phosphorus resource efficiency, potential eutrophication from phosphorous,

STRUCTURE 6.26   Chemical structure of HGE

and flame-retarding performance in terms of UL 94 V-0 rating (Conley et al., 2009).

Compared with halogen-containing flame retardants, PFRs show more environmentally friendly performance and have good flame retardancy. However, if a flame retardant contains phosphate only, it is necessary to remarkably increase the phosphorus content to achieve good fire retardancy. 9,10-dihydro-9-oxa-10-phosphaphenanthrene 10-oxide (DOPO)-based flame retardants have embodied a significant improvement in the flame retardancy, glass transition temperature (Tg), mechanical properties, and adhesion of epoxy resins (Gu et al., 2014).

9,10-dihydro-9-oxa-10-phosphaphenanthrene-10-oxide (DOPO) (Structure 6.27a), a phosphorus-containing heterocyclic compound, has attracted intense attention due to its high flame-retarding efficiency and active reactivity. DOPO and its derivates were first synthesized by Saito (1972), and are now widely applied in electric/electronic cast resins and polyester fibers (Rieckert et al., 1997). Some researchers suggested that the flame retardant mechanism of DOPO and its derivatives can be explained by the release of low-molecular-weight, phosphorous-containing species that are able to scavenge the H· and OH· radicals in the flame (Schäfer et al., 2007).

Various epoxy resins, curing agents, and additives based on DOPO have been prepared and reported intensively. Diphenylphosphine oxide (DPPO) (Structure 6.27b) is another important phosphorus-containing compound, which has a similar chemical structure to DOPO. However, it has been revealed that the P-C bonds in DPPO usually exhibit a superior hydrolysis resistance to that of P-O-C bonds in DOPO (Shieh and Wang, 2002).

Based on the Pudovik reaction, and by introducing DOPO chemically into various monomers, phosphorus-containing HFFRs are developed for the epoxy resin. Since the late 1990s, a series of flame retardants and hardeners covalently bonded with the DOPO group have been prepared and tested in epoxy resins by Wang and Shieh (2000), among which, one reactive monomer-containing DOPO group displayed the highest phosphorus resource efficiency. An epoxy thermoset using this HFFR reached UL 94 V-0 rating, with the phosphorus content as low as 0.81 wt%. Such high flame retardant systems with low phosphorus content thereby limit processing difficulties.

However, the introduction of phosphorus-containing HFFRs affects the curing behavior and thermal stability of the cured resin (Ratna, 2007). Lin et al. (2010) have studied the curing sites of epoxy, which had been adjusted

(a)   (b)

STRUCTURE 6.27   Chemical structures of (a) DOPO; (b) DPPO

by the sites of – OH–NH$_2$ or – NH–, leading to a different activation of curing procedures as well as Tg. To sum up, optimizing flame-retarding performance and resource efficiency by selecting functional groups and adjusting their sites of the DOPO-based reactive monomer is feasible.

9,10- dihydro-9-oxa-10-phosphaphenanthrene-10-oxide (DOPO) is not employed as an addition.type FR, and is normally incorporated into the backbones of epoxy resins or the amine hardeners via appropriate chemical reactions.

However, some problems associated with the DOPO-modified EP resins still exist, such as:

1. Required dosage of DOPO is quite high.
2. Glass Transition Temperature (T$_g$) decreases.
3. Humidity resistance property is poor.
4. Decomposition temperature decreases.

In order to overcome the aforementioned problems, many studies have been carried out on designing special DOPO derivatives. The incorporation of DOPO derivatives into the polymer network would greatly improve the flame retardancy of EP resins, and the thermal stability and Tg of the EP resins would also benefit. For example, the derivatives prepared by the reaction of DOPO with a Schiff base (named after Hugo Schiff) can be used as intumescent reactive flame retardants because both phosphorus and nitrogen exist in one structure (Wang et al., 2016); a Schiff base is a compound with the general structure R$_2$C=NR' (R' ≠ H) e.g. R$_1$R$_2$C=NR$_3$. These belong to a subclass of imines, being either secondary ketimines or secondary aldimines on the basis of their structure. The derivatives from the reaction of DOPO and Schiff base rendered varied properties to the EP resins, and some of the properties are related to the structure of the flame retardants.

Quite recently, Vasiljević et al. (Vasiljević et al., 2015) synthesized a DOPO-VTS (Equation 6.29) by addition reaction of DOPO to vinyltrimethoxysilane (VTS) in the presence of Azo-bis-isobutyronitrile (AIBN) as an initiator.

[6.29]

Different concentrations of the obtained flame retardant were prepared in ethanol and were hydrolyzed using HCl. The cotton fabrics were then treated by the pad-dry-cure method at 20°C, subsequently dried at 120°C, and then cured at 150°C. TG analyses showed a decrease in the thermal stability and an increase in the char formation with increases in the amount of dry solid on the treated cotton fabrics. The flame retardancy of treated cotton fabrics was tested by vertical flame spread tests. With the increase in the solid dry add-on, an increased flaming combustion after the removal of the flame was found; conversely, LOI values were improved.

The Si-DOPO (DOPO-VTS) coating increased the thermo-oxidative stability of the cotton material and inhibited degradation of cellulose fibers. The Si-DOPO coating did not decrease the time of flaming combustion, but did completely stop the vigorous combustion of the fibers. The results also suggest that the flame retardation by the Si-DOPO coating is due to three phenomena listed below:

- Quenching of active radicals from the decomposing cellulose,
- The phosphorylation of cellulose by the DOPO component, and
- The siliconoxide formation by the silsesquioxane component on the fiber surface.

These findings indicate that the flame retardant efficiency of the Si-DOPO coating is due to the combined activity of phosphorus acting in both gas and condensed phases, and silicon acting in the condensed phase.

The combination of the sol–gel precursor (Equation 6.29) and tetraethoxysilane (TEOS) was utilized as a sol–gel finishing for polyamide 6 (PA6) fabrics (Šehić et al., 2016). The obtained hybrid system decreased the Total Heat Release (THR) values of the treated fabrics, while increasing the char yield with respect to the use of precursor (Equation 6.29) alone, hence proving the synergistic behavior of the combined hybrid system.

There is a trend to develop novel flame retardant systems with higher flame-retardant efficiency by constructing multiple flame retardant compositions (Qian et al., 2015). A few known chemical structures or functional groups including maleimide, s-triazine, triazine-trione, phosphaphenanthrene, cyclotriphosphazene, and silsesquioxane have been used to prepare novel flame retardant systems by physical mixing or chemical synthesis. As reported in some literatures (Yang et al., 2015) the maleimide/phosphaphenanthrene, s-triazine/phosphaphenanthrene, triazine-trione/phosphaphenanthrene, cyclotriphosphazene/phosphaphenanthrene, silsesquioxane/phosphaphenanthrene, maleimide/triazinetrione/phosphaphenanthrene, maleimide/s-triazine/phosphaphenanthrene, and maleimide/cyclotriphosphazene systems have synergistic flame retardant effect on polymer materials. The combination of different flame retardant elements or functional groups is very instructive for developing high-efficiency flame retardants. A phosphorus–nitrogen-containing flame retardant, DOPO-T ˋ(Structure 6.28) was synthesized by nucleophilic substitution of cyanuric chloride by 9,10-dihydro-9-oxa-10-phosphaphenanthrene-10-oxide (DOPO). DOPO-T was then blended with diglycidyl ether of bisphenol-A, DGEBA (Structure 6.29) to prepare FR epoxy resins (Yang et al., 2018). DOPO-T showed high flame retardant efficiency on epoxy resin due to the phosphorous–nitrogen synergistic effect. The combustion parameters indicated that the flame retardancy of the EP thermosets was significantly enhanced with the incorporation of DOPO-T. The LOI value and UL94 rating were dramatically increased under low loading of DOPO-T. In addition, the heat-releasing strength was obviously

STRUCTURE 6.28   Chemical structures of DOPO-T

STRUCTURE 6.29   Chemical structure of DGEBA

decreased. During combustion, DOPO-T decomposed to release PO_2 and HPO_2 free radicals with a quenching effect. DOPO-T also decomposed to generate phosphorus-containing compounds, which promoted the carbonization of EP matrix to form a protective char layer in the condensed phase.

There are several studies concerning DOPO derivatives, typically reporting on producing a PO• radical by a gaseous-phase mechanism (Buczko et al., 2014). Angell et al. (2012) reported the synthesis of an ethylene-bridged DOPO derivative, DOPO$_2$-ethylene. The results from the UL-94 test indicated a good flame retardant effect for EPs containing DOPO$_2$-ethylene. Qiu et al. (2016) synthesized a novel DOPO derivative containing two different chemical bridge bonds between phosphaphenanthrene and triazine-trione groups. The newly synthesized DOPO derivative was only 4 wt%, with a load-enabled EP thermoset that reached the UL94 V-0 rating with a limiting oxygen index (LOI) value of 35.9%.

Gu et al. (2017) synthesized a novel DOPO-based HFFR (DP-DDM) (Structure 6.30) via a one-pot procedure. DP-DDM is characteristic of the building block based on the performance of a superior curing agent, 4,4/-methylenedianiline (MDA) and resembles the excellent property of understanding and control of epoxy-amine reactions. DP-DDM with phenolic hydroxyl groups can accelerate the epoxy-amine curing rate, as the thermal curing reaction takes place at a relatively low temperature (about 80°C). Thus,

STRUCTURE 6.30   Chemical structure of DP-DDM

DP-DDM may be considered as a co-occurring agent that may covalently join rigid DOPO into the epoxy thermosets. An optimized curing procedure, in accordance with the property of a co-occurring agent such as DP-DDM, was suggested based on the analyses of thermal curing behaviors and curing kinetics. DP-DDM was synthesized in good yield (90.3%) through a one-pot procedure, without an external catalyst. DP-DDM modified epoxy resin thermosets with a phosphorus content of 0.75 wt% cured with MDA as hardener achieved a high thermal resistance because of their high T_g over 157°C, and a higher Tg for the thermosets under the optimized curing process. Importantly, DP-DDM-modified epoxy thermosets meet the UL 94 V-0 flammability rating and a high LOI value with a phosphorus content of only 0.75 wt%.

Two kinds of 9,10-dihydro-9-oxa-10-phosphaphenanthrene-10-oxide(DOPO)-based phosphonamidate flame retardants named 4,4′-diamino-diphenyl methane, (DDM-DOPO) (Structure 6.31) and MPL-DOPO (Structure 6.32) (M stands for morpholine) is successfully synthesized by the classic Atherton–Todd reaction.

The three common tetrahedral bond replacements have both similarities and differences in their chemical and biological behavior. The phosphonamidates are the most analogous to peptides because they remain hydrogen bond donors, but they are less stable toward hydrolysis than are the analogues. Phosphonates share some similarities with phosphonamidates in the way they are synthesized, the final products are more stable, and they may show similar

STRUCTURE 6.31   Chemical structure of DDM-DOPO

STRUCTURE 6.32   Chemical structure of MPL-DOPO

or quite different biological activity compared to the phosphonamidates. Phosphonamidates are usually prepared by coupling an activated phosphonate monoester with the amine of interest (Boger Dale, 2014).

Two phosphonamidates (DDM-DOPO and MPL-DOPO) and modified epoxy composites still retain relatively high transparency compared with pure EP. With the introduction of two phosphonamidates into EPs, although the thermal stability of the composites is lowered, fire performance of the composites is greatly improved. A high LOI value of 30% and an UL-94 V-0 rating are obtained for DDM-DOPO-modified epoxy composites containing only 0.25 wt% phosphorus content. Meanwhile, two kinds of DOPO-based phosphonamidates play an effective role in suppressing heat release and reducing smoke release. The two kinds of DOPO-based phosphonamidates show a gaseous-phase effect on reducing the release of combustible volatiles as confirmed by the results of TGA, cone calorimeter test, and TG-IR (Wang et al., 2018b).

A novel flame retardant (coded as BNP) was successfully synthesized by Yang and others (2017) through the addition reaction between triglycidyl isocyanurate, 9,10-dihydro-9-oxa-10-phosphaphenanthrene-10-oxide (DOPO), and phenylboronic acid. BNP was blended with diglycidyl ether of bisphenol-A to prepare flame retardant epoxy resin (EP). Thermal properties, flame retardancy, and combustion behavior of the cured EP were studied by thermogravimetric analysis, limited oxygen index (LOI) measurement, UL94 vertical burning test, and cone calorimeter test. The results indicated that the flame retardancy and smoke suppressing properties of EP/BNP thermosets were significantly enhanced. The LOI value of EP/BNP-3 thermoset was increased to 32.5% and the sample achieved UL94 V-0 rating. Compared with the neat EP sample, the peak of heat release rate, an average of heat release rate, total heat release, and total smoke production of EP/BNP thermosets were decreased by 58.2%–66.9%, 27.1%–37.9%, 25.8%–41.8%, and 21.3%–41.7%, respectively. The char yields of EP/BNP thermosets were increased by 46.8%–88.4%. The BNP decomposed to produce free radicals with quenching effect and enhanced the charring ability of EP matrix. The multifunctional groups of BNP with flame retardant effects in both gaseous and condensed phases were responsible for the excellent flame retardancy of the EP/BNP thermosets.

Long et al. (2016) performed studies on a phenethyl-bridged DOPO derivative (bisDOPO) incorporated to poly(lactic acid). An LOI value of 28.2% and UL-94 V-0 rating was achieved with 15 wt% $DOPO_2$-phenethyl load.

A series of flame retardant epoxy resins (EPs) containing either phenethyl-bridged 9 or 10-dihydro-9-oxa-10-phosphaphenanthrene-10-oxide derivative (bisDOPO) was prepared by a method patented by Yao et al. (2013). The flame retardant properties of bisDOPO on EP composites were characterized by the limiting oxygen index (LOI), the UL-94 vertical burning, and the cone calorimeter test (CCT). The LOI increased from 21.8% to 38 .0%, on forming a composite of EP with bisDOPO composites, and by using 10 wt% bisDOPO, the V-0 rating in the UL94 vertical burning test was obtained. The char residue following the CCT showed intumescent structures with continuous and compact surfaces that can effectively suppress the spread of the flame and extinguish the fire. This was confirmed by both visual observation and scanning electron microscopy (SEM) measurements (Yan et al., 2017).

The cyclic organic P compound, 9,10-dihydro-9-oxa-10-phosphaphenanthrene -10-oxide (DOPO), was patented in 1972 (Wang and Shieh, 1998). Many DOPO derivatives have been considered as suitable alternatives to halogenated flame retardants because of their versatile flame extinguishing behavior in the gas phase and condensed phase (Wang and Shieh, 1999). Compared with additive flame retardants, the reactive type of epoxy resins has been gaining attention for several reasons: better compatibility, nonmigratory performance, less negative impact upon physical and mechanical properties, and more stable flame retardant effect. Among the DOPO-containing reactive flame retardants, many compounds with phenolic hydroxyl groups were reported and used as hardeners or co-occurring agents to form advanced epoxy resin by Wang and Shieh (1999) and others. However, deterioration of glass transition temperature (Tg) or low curing activity may ensue. Even worse, some advanced epoxy resin exhibited low solubility in acetone; a disadvantage for the printed circuit laminates industry. Wang et al. (2018a) synthesized another kind of DOPO- containing epoxy ether, 9,10-dihydro-9-oxa-10-[1,1-bis(4-glycidyloxyphenyl) ethyl]-10-phosphaphenanthrene 10-oxide (DPBAEP) (Structure 6.33), using a much more economical method. Furthermore, DPBAEP could be cured to thermosets directly, or be added to common epoxy resins, to obtain flame retardancy. DPBAEP has a unique structure that is similar to that of diglycidyl ether of bisphenol A (DGEBA) (Structure 6.29), so the excellent compatibility with common epoxy resins is deserved. Simultaneously, it can maintain the cross-link density and rigidity of the epoxy network. More importantly, DPBAEP has a high content of aromatic groups, which contributes to a higher thermal stability and better mechanical properties.

Xu et al. (2015) prepared a phosphorus/nitrogen-containing co-occurring agent called D-bp, which modified the EP thermoset, which in turn, exhibited excellent flame retardancy at low phosphorus content. However, the thermal stability of the EP resins decreased greatly with the increase in phosphorus content.

STRUCTURE 6.33   Chemical structure of DPBAEP

Recently, He et al. (2017) focused on the synthesis and application of diphenylphosphine oxide (DPO) (Structure 6.34) and its derivatives. The formula structure of DPO is similar to DOPO and the phosphorus content of DPO (15.32%) is little higher than that of DOPO (14.33%). Thus, novel flame retardants and curing agents for EP resins can be designed by replacing the DOPO group with DPO group in the DOPO derivatives (Tian et al., 2014).

Rice husk (RH) is a naturally occurring organic/inorganic material predominantly consisting of lignocellulose and silica. Due to the presence of high silica content in RH, their addition into synthetic polymers may cause synergistic behavior in the flammability of the resulting composites.

6.7.5   PHOSPHORUS FRs FOR POLYSTYRENE

In most practical applications of plastic materials, restricted flammability is required. The action of graphite as an FR agent in polymers consists mainly in the formation of the carbonaceous barrier that prevents access by the flame and oxygen to the polymeric material. This effect is achieved by the use of so-called expandable graphite (EG), which swells up on heating to several hundred percent of its volume with the formation of a foamed insulation layer protecting the burning polymer. The EG is employed to reduce flammability of various plastic materials, usually in combination with other FR additives (Xiang-Gheng et al., 2008), including halogen-containing compounds, e.g., decabromodiphenyl ethane (Ling et al., 2009), and it plays a supporting role in fire retardancy as an additional filler improving the insulation properties of the plastic. However, such solutions are not environmentally friendly because of the presence of undesirable halogen compounds.

STRUCTURE 6.34   Chemical structure of diphenylphosphine oxide (DPO)

Polystyrene (PS) is one of the most easily flammable polymers. Currently, the most commonly practiced method to decrease PS flammability is the use of additives based on organic halogen derivatives (Grause et al., 2010). Various graphite compounds have been also applied as FR additives for PS. These were introduced to polymers during the reaction of cationic or anionic (Xiao et al., 2002) polymerization or compounded with previously prepared polymers. (Uhl and Wilkie, 2004).

A method of preparation of halogen-free, fire-retardant additives for polystyrene was developed using in situ synthesis of melamine salts (cyanurate or phosphate) in the presence of expandable graphite. The obtained products, with various compositions, were introduced as FR additives during the suspension polymerization of styrene. The polymerization was optimized by changing the process parameters such as:

- The ratio of aqueous to the organic phase,
- Type and amount of suspension stabilizer (poly(vinyl alcohol) or bentonite/gelatine system), and
- Type and amount of FR additive and time of its addition to the reaction mixture.

The maximum FH-1 and V-0 classifications were obtained using 15 parts by weight of the additives per 100 parts of polystyrene. LOI values increased by about 44% for EG/melamine cyanurate (MG) compound and 50% for EG/melamine phosphate (MP) compound, compared to nonmodified polystyrene. When 20 parts by weight of the additive per 100 parts of polystyrene were used, the LOI increased by about 50% for EG/MG compound and 56% for EG/MP compound (Jankowski and Kędzierski, 2013).

EG swells during combustion to form a char layer, which prevents the access of oxygen to the combustion zone. It is apparent from the literature that, in most cases, graphite is not used alone. From a variety of solutions, to increase its efficiency, the modification of EG by the melamine salt can be carried out. The method of obtaining this type of HFFRs has been developed on the laboratory scale by Jankowski and Kijowska (2016). The production of HFFRs consists of in situ synthesis of melamine cyanurate in the presence of expandable graphite in a dissolver, drying the product in a long gap mill (LGM) and packing into plastic bags. Modified graphite was used as a flame-retardant additive for polystyrene, epoxy resin, and polyester resin. Syntheses of HFRs were done on a semi-industrial scale. HFRs have been used to reduce the flammability of selected polymers: polyester, epoxide, and polystyrene. The flame-retardant effect is greater when the grain size of HFR is larger and the tendency of the additives to expand is greater. Excessive fragmentation of HFRs in the course of their production causes a reduction in the expansion properties of expandable graphite contained therein. This phenomenon is particularly evident during the second stage of production when the product is dried in the LGM. It can be concluded that the manner of production of HFRs has a significant influence on the properties of expandable graphite contained therein, and thus, on the flame retardant effectiveness of HFRs in their final compositions with polyester, epoxide, or polystyrene.

6.8 FUTURE TRENDS

Elemental phosphorus and its various compounds have been used to flame-retard a wide variety of polymer-based materials for several decades. Environmental considerations, particularly in terms of the use of halogen-based systems, have paved the way in recent years, for the increased use of PFRs as alternatives to halogen-containing compounds. Furthermore, this has generated active research in identifying novel flame retardants based on phosphorus, as well as synergistic combinations with compounds of other flame retardant elements (such as nitrogen and the halogens) and with several inorganic nano-fillers (e.g., phyllosilicates and carbon nanotubes). This research on halogen alternatives has resulted, especially in recent years, in a wealth of literature (including patents), and in the market acceptance of several phosphorus-based flame retardants. Although phosphorus compounds can be highly effective flame retardants, they are not effective in some major classes of polymers such as styrenic resins and polyolefins. Furthermore, the basic mode of intervention of phosphorus compounds, regardless of the phase in which they are active, is to suppress the efficiency of combustion reactions that are mostly radical in nature and occur in the gas phase.

Different kinds of additive and reactive flame retardants containing phosphorus are increasingly successful as halogen-free alternatives for various polymeric materials and applications. Phosphorus can act in the condensed phase by enhancing charring, yielding intumescence, or through inorganic glass formation; and in the gas phase through flame inhibition. Occurrence and efficiency depend, not only on the flame retardant itself, but also on its interaction with pyrolyzing polymeric materials and additives. Flame retardancy is sensitive to modification of the flame retardant, the use of synergists/adjuvants, and changes to the polymeric material (Schartel, 2010).

Horrocks (2011) in a recent review, clearly depicted that the world of flame retardants for textiles is still experiencing some changes that are focused on such topics as improving effectiveness and replacing toxic chemical products with counterparts that have low environmental impact, and hence, are more sustainable. In this context, phosphorus-based compounds play a key role and may lead, possibly in combination with silicon- or nitrogen-containing structures, to the design of new, efficient flame retardants for fibers and fabrics.

REFERENCES

Afanas'eva L.V. and Evseenko N.S. (1971). Hygienic assessment of fireproof fabrics treated with an organophosphorus impregnating compound based on tetrahydroxymethylphosphonium chloride, *Gigiena i Sanitarija*, 36(3), 102–103.

Agarwal V., Borisova S.A., Metcalf W.W., van der Donk W.A., and Nair S.K. (2011). Structural and mechanistic insights into c-p bond hydrolysis by phosphonoacetate hydrolase, *Chemical Biology*, 18, 1230–1240.

Al-Mosawi A.I. (2016). *Flammability of Composites in Lightweight Composite Structures in Transport*, J. Njuguna, Ed. Elsevier, Woodhead Publishing, Cambridge, UK.

Angell Y.L., White K.M., Angell S.E., and Mack A.J. (2012). DOPO derivative flame retardants. Patent US20120053265A1, USA.

Aoyama M. (1975). Effect of anti-flame treating agents on the skin, *Nagoya Med. J.*, 20 (1), 11–19.

Aseeva R.M. and Zaikov G.E. (1986). *Combustion of Polymer Materials*. Carl Hanser Verlag, Munich, Germany, p. 149.

Atherton F.R., Openshaw H.T., and Todd A.R. (1945). 174. Studies on phosphorylation. Part II. The reaction of dialkyl phosphites with polyhalogen compounds in presence of bases. A new method for the phosphorylation of amines (in German), *Journal of the Chemical Society (Resumed)*, 660–663. DOI: 10.1039/jr9450000660.

Atherton-Todd reaction. (2018). https://commons.wikimedia.org/wiki/File:Atherton_Todd_reaction_(Dimethyphosphite).svg#/media/File:Atherton_Todd_reaction_(Dimethyphosphite).svg, accessed on 15.4.18.

Babushok V. and Tsang W. (2000). Inhibitor rankings for alkane combustion, *Combust. Flame*, 124, 488–506.

Back E.L. (1967). Thermal auto-crosslinking in cellulose material, *Pulp and Paper Mag. Canada, Tech. Sec.*, 68, T165–T171.

Ballistreri A., Montaudo G., Puglisi C., Scamporino E., Vitalini D., and Calgari S. (1983). Mechanism of flame retardant action of red phosphorus in polyacrylonitrile, *J. Polym. Sci. Polym. Chem.*, 21, 679–689.

Basch A. and Lewin M. (1973). Low add-on levels of chemicals on cotton and flame retardancy, *Textile Res. J.*, 43, 693–694.

Baseri S., Karimi M., and Morshed M. (2012). Study of structural changes and mesomorphic transitions of oriented poly (ethylene therephthalate) fibres in supercritical CO2[J], *European Polymer Journal*, 48(4), 811–820.

Benisek L. (1971). Use of titanium complexes to improve natural flame retardancy of wool, *J. Soc. Dyers Color.*, 87, 277–278.

Benisek L. and Phillips W.A. (1983). The effect of the low smoke zirpro treatment and wool carpet construction on flame-resistance and emission of smoke and toxic gas, *Journal of Fire Sciences*, 1(6). DOI: 10.1177/073490418300100603.

Boger Dale L. (2014). *Houben-Weyl Methods of Organic Chemistry Vol. E 22c, 4th Edition Supplement: Synthesis of Peptides and Peptidomimetics (Edited)*. Georg Thieme Verlag, 14-May-2014, first published in 2003.

Bosco F., Casale A., Mollea C., Terlizzi M.E., Gribaudo G., Alongi J., and Malucelli G. (2015). DNA coatings on cotton fabrics: Effect of molecular size and pH on flame retardancy, *Surf Coat Technol*, 272, 86–95. DOI: 10.1016/j.surfcoat.2015.04.019.

Bounor-Legar´e V., Ferreira I., Verbois A., Cassagnau P., and Michel A. (2002). A new route for organic-inorganic hybrid material synthesis through reactive processing without solvent, *Polymer*, 43, 6085–6092.

Bourbigot S., Le Bras M., Flambard X., Rochery M., Devaux E., and Lichtenhan J. (2005). Polyhedral oligomeric silsesquioxanes: Application to flame retardant textiles, in: M. Le Bras, S. Bourbigot, S. Duquesne, C. Jama, and C.A. Wilkie, Eds., *Fire Retardancy of Polymers: New Applications of Mineral Fillers*. Royal Society of Chemistry (Pub), Cambridge, UK, pp. 189–201.

Braun U., Schartel B., Fichera M.A., and Jager C. (2007). Flame retardancy mechanisms of aluminium phosphinate in combination with melamine polyphosphate and zinc borate in glass-fibre reinforced polyamide 6,6, *Polym Degrad Stab*, 92, 1528–1545.

Bright D.A., Dashevsky S., Moy P.Y., and Williams B. (2004). Resorcinol bis(diphenyl phosphate), a non-halogen flame-retardant additive. DOI: 10.1002/vnl.10184, First published: 16 April.

Bruchajzer E., Frydrych B., and Szymańska J.A. (2015). Organophosphorus flame retardants – Toxicity and influence on human health, *Med Pr*, 66(2), 235–264. DOI: 10.13075/mp.5893.00120.

Buczko A., Stelzig T., Bommer L., Rentsch D., Heneczkowski M., and Gaan S. (2014). Bridged DOPO derivatives as flame retardants for PA6, *Polym Degrad Stab*, 107, 158–165.

Butnaru I., Fernandez-Ronco M.P., Czech-Polak J., Heneczkowski M., Bruma M., and Gaan S. (2015). Effect of meltable triazine-dopo additive on rheological, mechanical, and flammability properties of PA6, *Polymer*, 7, 1541–1563.

Bychkova E.V. and Panova L.G. (2016). Fire-resistant viscose rayon fiber materials, *Fiber Chemistry*, 48(3) (Russian Original No. 3, May-June, 2016). DOI: 10.1007/s10692-016-9771-9.

Carosio F., Laufer G., Alongi J., Carosioa F., Lauferb G., Alongia J., Caminoa G., and Grunlanb J.C. (2011). Layer-by-layer assembly of silica-based flame retardant thin film on PET fabric, *Polym Degrad Stab*, 96, 745–755.

Chen L. and Wang Y.Z. (2010a). Review on flame retardant technology in China, *Polymers for Advanced Technologies*, 21(1), 1–26.

Chen L. and Wang Y.Z. (2010b). Aryl polyphosphonates: Useful halogen-free flame retardants for polymers, *Materials*, 3, 4746–4760. DOI: 10.3390/ma3104746.

Cheng X.Y. and Yang C.Q. (2009). Flame retardant finishing of cotton fleece fabric: Part V. Phosphorus-containing maleic acid oligomers, *Fire Mater*, 33(8), 365–375. DOI: 10.1002/fam.1008.

Chen-Yang Y.M., Cheng S.J., and Tsai B.D. (1991). Preparation of the partially substituted (phenoxy)chlorocyclotriphosphazenes by phase-transfer catalysis, *Ind Eng Chem Res*, 30, 1314.

Cichy B., Łuczkowska D., Nowak M., and Władyka-Przybylak M. (2003). Polyphosphate flame retardant with increased heat resistance, *Ind. Eng. Chem. Res.*, 42(13), 2897.

Conley D.J., Paerl H.W., Howarth R.W., Boesch D.F., Seitzinger S.P., Havens K.E., Lancelot C., and Likens G.E. (2009). Controlling eutrophication: Nitrogen and phosphorus, *Science*, 323, 1014–1015.

CSIRO (2008). Flame resistance of wool, www.csiro.au, Published: Jan 2008.

Ding J.P., Tao Z.Q., Zuo X.B., Fan L., and Yang S. (2009). Preparation and properties of halogen-free flame retardant epoxy resins with phosphorus containing siloxanes, *Polym Bull*, 62, 829–841.

Dong Z., Yu D., and Wang W. (2012). Preparation of a new flame retardant based on phosphorus-nitrogen (p-n) synergism and its application on cotton fabrics, *Adv. Mater. Res.*, 381–384. DOI: 10.4028/www.scientific.net/AMR.479-481.381.

Ebdon J.R., Hunt B.J., Joseph P., and Konkel C.S. (2001). Flame-retarding thermoplastics: Additive versus reactive approach, in: S. Al-Malaika, A. Golvoy, and C. A. Wilkie (Eds.), *Speciality Polymer Additives: Principles and Applications*. Blackwell Science, Oxford, UK, pp. 231–257.

Ebdon J.R., Hunt B.J., Joseph P., Konkel C.S., Price D., Pyrah K., Hull T.R., Milnes G. J., Hill S.B., Lindsay C.I., McCluskey J., and Robinson I. (2000). Thermal degradation and flame retardance in copolymers of methyl methacrylate with diethyl(methacryloyloxymethyl)phosphonate, *Polym. Degrad. Stab.*, 70, 425–436.

Economy J. (1978). Phenolic fibers, in: M. Lewin, S.M. Atlas, and E.M. Pearce, Eds., *Flame Retardant Polymeric Materials*, Vol 2. Plenum Press, New York, pp. 203–208.

Fei G., Wang Q., and Liu Y. (2010). Synthesis of novolac-based char former: Silicon-containing phenolic resin and its synergistic action with magnesium hydroxide in polyamide-6 (2010), *Fire and Materials*, 34, 407–419.

Gaan S., Viktoriya S., Ottinger S., Heuberger M., and Ritter A. (2009). Phosphoramidate flame retardants. PCT Patent WO 2009,1530,34 A1, 23 December 2009.

Gao F., Tong L., and Fang Z. (2006). Effect of novel phosphorus-nitrogen containing intumescent flame retardant on the fire retardancy and thermal behaviour of poly (butylene terephthalate), *Polym Degrad Stab.*, 91, 1295–1299.

Gao -W.-W., Zhang G.-X., and Zhang F.-X. (2015). Enhancement of flame retardancy of cotton fabrics by grafting a novel organic phosphorous-based flame retardant, *Cellulose*, 22, 2787–2796. DOI: 10.1007/s10570-015-0641-z.

Gawłowski A., Fabial J., Ślusarczyk C., Graczyk T., and Pielesz A. (2017). Flame retardant modification of partially oriented poly(ethylene terephthalate) fibres – Structural conditions of application, *Polimery*, 62(11–12), 848–854. DOI: 10.14314/polimery.2017.848.

Gordon K.L., Thompson C.M., and Lyon R.E. (2010). Flame retardant epoxy resins containing aromatic poly(phosphonamides), *High Performance Polymers*, 22, 945–958.

Grause G., Ishibashi J., Kameda T., Bhaskar T., and Yoshioka T. (2010). Kinetic studies of the decomposition of flame retardant containing high-impact polystyrene, *Polymer Degradation and Stability*, 95, 1129–1137.

Green J. (1996). A review of phosphorus containing flame retardants, *J. Fire Sci.*, 13, 353–366.

Green J. (2000). Phosphorus-containing flame retardants, in: A.F. Grand and C.A. Wilkie, Eds., *Fire Retardancy of Polymeric Materials*. Marcel Dekker, New York, pp. 147–170.

Greenwood N.N. and Earnshaw A. (1984). *Chemistry of Elements*. Pergamon Press, Oxford, UK, pp. 546–636.

Gu L., Chen G., and Yao Y. (2014). Two novel phosphorus-nitrogen containing halogen-free flame retardants of high performance for epoxy resin, *Polym Degrad Stabil*, 108, 68–75.

Gu L., Qiu J., and Sakai E. (2017). A novel DOPO-containing flame retardant for epoxy resin: Synthesis, nonflammability, and an optimized curing procedure for high performance, *High Performance Polymers*, 29(8), 899–912. DOI: 10.1177/0954008316664123.

Guin J., Wang Q., van Gemmeren M., and List B. (2015). The catalytic asymmetric abramov reaction, *Angew. Chem. Int. Ed.*, 54, 355–358.

Hansen J.H. (1989). US Patent WO1989000217A1 – Flame-resistant polyamide/cotton fabric, Google Patentsuche, accessed on 11.12.2017.

Hasegawa H.K. (1990). *Characterisation and Toxicity of Smoke*. Publication STP 1082, American Association for Testing and Materials, Philadelphia, PA.

He W., Hou X., Li X.J., Song L., Yu Q., and Wang Z. (2017). Synthesis of P(O)-S organophosphorus compounds by dehydrogenative coupling reaction of P(O)H compounds with arylthiols in the presence of base and air, *Tetrahedron*, 73, 3133–3138.

Hendrix J.E., Drake G.L., and Barker R.H. (1972). Pyrolysis and combustion of cellulose.III. Mechanistic bases for the synergism involving organic phosphates and nitrogenous bases, *J. Appl. Polymer Sci.*, 16, 257.

Hopechem (2019). www.hopechemical.cn/product/Pyrovatex-CP.html, accessed on 23.12.19.

Horrocks A.R. and Zhang S. (2002). Enhancing polymer flame retardancy by reaction with phosphorylated polyols. Part 2. Cellulose treated with a phosphonium salt urea condensate (proban cc) flame retardant, *Fire Mater.*, 26, 173–182. DOI: 10.1002/fam.794.

Horrocks A.R. (2011). Flame retardant challenges for textiles and fibers: New chemistry versus innovatory solutions, *Polym Degrad Stab*, 96, 377–392. DOI: : 10.1016/j.polymdegradstab.2010.03.036.

Hou S., Zhang Y.J., and Jiang P. (2016). Phosphonium sulfonates as flame retardants for polycarbonate, *Polym. Degrad. Stab*, 130, 165–172. DOI: 10.1016/j. polymdegradstab.2016.06.004.

Huntsman (2012). Textile Effects PYROVATEX® CP NEW PYROVATEX® CP-LF. Handbook for technicians – flame retardants. www.huntsman.com/textile_effects, February.

Jankowski P. and Kędzierski M. (2013). Polystyrene with reduced flammability containing halogen-free flame retardants, *Polimery*, 58(5), 342–349.

Jankowski P. and Kijowska D. (2016). The influence of parameters of manufacturing hybrid flame retardant additives containing graphite on their effectiveness, *Polimery*, 61(5), 327–333. DOI: 10.14314/polimery.2016.327.

Jia Y., Hu Y., Zheng D., Zhang G., Zhang F., and Liang Y. (2017). Synthesis and evaluation of an efficient, durable, and environmentally friendly flame retardant for cotton, *Cellulose*, 24, 1159–1170. DOI: 10.1007/s10570-016-1163-z.

Joseph P. and Ebdon J.R. (2010). Phosphorus-based flame retardants, in: A. Wilkie Charles and B. Morgan Alexander, Eds., *Fire Retardancy of Polymeric Materials*, 2nd edition. CRC Press, USA, pp. 107–128.

Kamat S.S., Williams H.J., and Raushel F.M. (2011). Intermediates in the transformation of phosphonates to phosphate by bacteria, *Nature*, 480, 570–573.

Kamath P., Rajan R., Deshpande S.H., Montgomery M., and Lal M. (2017). An efficient synthesis of 3-phosphorylated benzoxaboroles via the pudovik reaction, *Synthesis*. DOI: 10.1055/s-0036-1588739.

Kandola B.K., Horrocks A.R., Price D., and Coleman G.V. (1996). Flame retardant treatments of cellulose and their influence on cellulose pyrolysis, *J. Macromol. Sci. Rev. Macromol. Chem. Phys.*, C36, 721–794.

Karim-nejad M.M., Nazari A., and Davodi-roknabadi A. (2015). Efficient flame retardant of mercerized cotton through crosslinking with citric acid and ZnO nanoparticles optimized by RSM models, *The Journal of The Textile Institute*, 106(10), 1115–1126. DOI: 10.1080/00405000.2014.976958.

Khandual A. (2016). Green fashion Vol. 2, in: S.S. Muthu and M.A. Gardetti, Eds., *Environmental Footprints and Ecodesign of Products and Processes*. Springer, Singapore, pp. 171–227.

Kheyrabadi R., Rahmani H., and Najafi S.H.M. (2018). Flame-retardant halogen-free polymers using phosphorylated exaglycidyl epoxy resin, *High Performance Polymers*, 30(2), 202–210. DOI: 10.1177/0954008316688759.

Kuper G., Hormes J., and Sommer K. (1994). In situ x-ray absorption spectroscopy at theK-edge of red phosphorus in polyamide 6,6 during a thermo-oxidative degradation, *Makromol. Chem. Phys.*, 195, 1741–1753.

Laufer G., Kirkland C., Morgan A.B., and Grunlan J.C. (2012). Intumescent multilayer nanocoating, made with renewable polyelectrolytes, for flame-retardant cotton, *Biomacromolecules*, 13, 2843–2848. DOI: 10.1021/bm300873b.

Lee M.S., Ju B.J., Lee B.G., and Hong S.H. (2008). Non-halogen flameproof resin composition. 20100152345. U.S. Patent.

Levchik G.F., Levchik S.V., Camino G., and Weil E.D. (1998). Fire retardant action of red phosphorus in nylon 6, in: M. Le Bras, G. Camino, S. Bourbigot, and R. Delobel, Eds., *Fire Retardancy of Polymers: The Use of Intumescence*. Royal Society of Chemistry, London, pp. 304–315.

Levchik G.F., Vorobyova S.A., Gorbarenko V.V., Levchik S.V., and Weil E.D. (2000). Some mechanistic aspects of the fire retardant action of red phosphorus in aliphatic nylons, *J. Fire Sci.*, 18, 172–182.

Levchik S.V. (2007). Chapter 1. Introduction to flame retardancy and polymer flammability, in: A.B. Morgan and C.A. Wilkie, Eds., *Flame Retardant Polymer Nanocomposites*. John Wiley, pp. 1–29.

Levchik S.V., Levchik G.F., Balanovich A.I., Camino G., and Costa L. (1996). Mechanistic study of combustion performance and thermal decomposition behaviour of Nylon 6 with added halogen-free fire retardants, *Polym. Degrad. Stab.*, 54, 217–222.

Levchik S.V. and Weil E.D. (2005). Flame retardancy of thermoplastic polyesters – A review of the recent, literature, *Polym Int*, 54(1), 11–35.

Levchik S.V. and Weil E.D. (2006). A review of recent progress in phosphorus-based flame retardants, *Journal of Fire Sciences*, 24, 345. DOI: 10.1177/0734904106068426.

Levin M. (1984). Flame retardance of fabrics, in: M. Levin and S.B. Sello, Eds., *Handbook of Fibre Science and Technology, Vol. II, Chemical Processing of Fibres and Fabrics. Functional Finishes, Part B.* Marcel Dekker, New York, pp. 1–141.

Lewin M. and Weil E.D. (2001). Mechanisms and modes of action in flame retardancy of polymers, in: A.R. Horrocks and D. Price, Eds., *Fire Retardant Materials.* Woodhead Publishing, Cambridge, England, pp. 31–68.

Li G., Wang W., Cao S., Cao Y., and Wang J. (2014b). Reactive, intumescent, halogen-free flame retardant for polypropylene, *Journal of Applied Polymer Science.* DOI: 10.1002/APP.40054 (1-9).

Li J. and Liu G. (2012). Flame retardancy properties of ammonium polyphosphate with crystalline form II by non-P_2O_5process, *Polym Degrad Stab*, 97, 2562–2566.

Li Z.-S., Liu J.-G., Song T., Shen D.-X., and Yang S.-Y. (2014a). Synthesis and characterization of novel phosphorous-silicone-nitrogen flame retardant and evaluation of its flame retardancy for epoxy thermosets, *Journal of Applied Polymer Science*, 131 (24). DOI: 10.1002/app.40412.

Liang S., Hemberger P., Levalois-Grützmacher J., Grützmacher H., and Gaan S. (2017). Probing phosphorus nitride (p-n) and other elusive species formed upon pyrolysis of dimethyl phosphoramidate, *Chemistry*, 23, 5595–5601.

Liang S., Hemberger P., Neisius N.M., Bodi A., Grützmacher H., Levalois-Grützmacher J., and Gaan S. (2015). Elucidating the thermal decomposition of dimethyl methylphosphonate by vacuum ultraviolet (vuv) photoionization: Pathways to the po radical, a key species in flame-retardant mechanisms, *Chemistry*, 21, 1073–1080.

Liang S., Neisius M., Mispreuve H., Naescher R., and Gaan S. (2012). Flame retardancy and thermal decomposition of flexible polyurethane foams: Structural influence of organophosphorus compounds, *Polym. Degrad. Stab.*, 97, 2428–2440.

Lin C.H., Lin H.T., Sie J.W., Hwang K.W., and Tu A.P. (2010). Facile, one-pot synthesis of aromatic diamine-based phosphinated benzoxazines and their flame-retardant thermosets, *J Polym Sci A: Polym Chem*, 48(20), 4555–4566.

Lin H., Yan H., Liu B., Wei L., and Xu B. (2011). The influence of KH-550 on properties of ammonium polyphosphate and polypropylene flame retardant composites, *Polym Degrad Stab*, 96, 1382–1388.

Ling Y., Jian-Hua T., Zhong-Ming L., Xian-Yan M., and Xing-Mao L. (2009). Flameretardant and mechanical properties of high-density rigid polyurethane foams filled with decabrominated dipheny ethane and expandable graphite, *Journal of Applied Polymer Science* 111(5), 2372–2380. DOI: 10.1002/app.29242.

Liu S.M., Zhao J.Q., and Chen J.B. (2011). Flame-retardant latent epoxy curing agent and halogen-free flame-retardant one-component epoxy resin composition. China Patent CN102174173A. DOI: 10.1002/pi.4890

Liu W., Chen L., and Wang Y.-Z. (2012). A novel phosphorus-containing flame retardant for the formaldehyde-free treatment of cotton fabrics, *Polym. Degrad. Stab.*, 97, 2487–2491. DOI: 10.1016/j.polymdegradstab.2012.07.016.

Loewengart G. and Van Duuren B.L. (1977). Evaluation of chemical flame retardants for carcinogenic potential, *J. Toxicol. Environ. Health*, 2(3), 539–546.

Long L.J., Yin J.B., He W.T., Qin S., and Yu J. (2016). Influence of a phenethylbridged dopo derivative on the flame retardancy, thermal properties, and mechanical properties of poly(lactic acid), *Ind Eng Chem Res*, 55, 10803–10812.

Lubczak J. and Lubczak R. (2016). Melamine polyphosphate – The reactive and additive flame retardant for polyurethane foams, *Acta Chim. Slov*, 77(63), 77–87. DOI: 10.17344/acsi.2015.1924.

Lyons J.W. (1970). *The Chemistry and Uses of Flame Retardants*. Wiley-Interscience, Hoboken, NJ.

Maiti S., Banerjee S., and Palit S.K. (1993). Phosphorus-containing polymers, *Prog. Polym. Sci.*, 18, 227–261.

Mariappana T., Zhou Y., Hao J., and Wilkie C.A. (2013). Influence of oxidation state of phosphorus on the thermal and flammability of polyurea and epoxy resin, *European Polymer Journal*, 49(10), 3171–3180. DOI: 10.1016/j.eurpolymj.2013.06.009.

Mcwebmaster (2017). Reactive PIN FR for rigid polyurethane foam, 1 Feb, 2017, www.pinfa.eu/, accessed on 14.3.2018.

Michal S., Kraszewski K., and Stawinski J. (2015). Recent advances in h-phosphonate chemistry. Part 2. Synthesis of c-phosphonate derivatives, in: J.-L. Montchamp, Ed., *Topics in Current Chemistry*, Vol. 361. Springer, Berlin, Germany, pp. 179–216.

Moniruzzaman M. and Winey K.I. (2006). Polymer nanocomposites containing carbon nanotubes, *Macromolecules*, 39, 5194–5205. DOI: 10.1021/ma060733p.

Montchamp J.-L. (2014). Phosphinate chemistry in the 21st century: A viable alternative to the use of phosphorus trichloride in organophosphorus synthesis, *Acc. Chem. Res.*, 47, 77–87.

Montchamp J.-L. (2015). Carbon-hydrogen to carbon-phosphorus transformations, in: J.-L. Montchamp, Ed., *Topics in Current Chemistry*, Vol. 361. Springer, Berlin, Germany, pp. 217–252.

Morgan A.B. (2009). Polymer flame retardant chemistry, ABM, F.R. Chemistry, 30.9.2009. www.nist.gov/system/files/documents/el/fire_research/2-Morgan.pdf, accessed on 12.12.2019.

Muir D.C.G. (1984). Anthropogenic compounds, in: O. Hutzinger, Ed., *Handbook of Environmental Chemistry*, Vol. 3. Springer-Verlag, Berlin, Germany, pp. 41–66.

Murashko E.A., Levchik G.F., Levchik S.V., and Bright D.A. (1999). Dashevsky, S. Fire retardant action of resorcinol bis(diphenyl phosphate) in PC–ABS blend, II: Reaction in the condensed phase, *J. Appl. Polym. Sci.*, 71, 1863–1872.

Murashko E.A., Levchik G.F., Levchik S.V., Bright D.A., and Dashevsky S. (1998). Fire retardant action of resorcinol bis(diphenyl phosphate) in a PPO/HIPS blend, *J. Fire Sci.*, 16, 233–249.

Nair G.P. (2001). Flammability in textiles and routes to flame retardant textiles – XII, *Colourage*, XLVIII, August, 43–47.

Neisius M., Liang S., Mispreuve H., and Gaan S. (2013). Phosphoramidate-containing flame-retardant flexible polyurethane foams, *Ind. Eng. Chem. Res.*, 52, 9752–9762. DOI: 10.1021/ie400914u.

Nguyen T.-M., Chang S., Condon B., Slopek R., Graves E., and Yoshioka-Tarver M. (2013). Structural effect of phosphoramidate derivatives on the thermal and flame retardant behaviors of treated cotton cellulose. *Industrial & Engineering Chemistry Research*, 52, 4715–4724. DOI: 10.1021/ie400180f.

Nguyen T.-M., Chang S., Condon B., and Smith J. (2014). Fire self-extinguishing cotton fabric: Development of piperazine derivatives containing phosphorous-sulfur-nitrogen and their flame retardant and thermal behaviors, *Mater. Sci. Appl.*, 5, 789–802. DOI: 10.4236/msa.2014.511079.

NIH (National Institutes of Health) (1987). NTP technical report on the toxicology and carcinogenesis studies of tetrakis(hydroxymethyl)phosphonium sulfate (THPS) and tetrakis(hydroxymethyl)phosphonium chloride (THPC) in F344/N rats and B6C3F1 mice (gavage studies), NIH Publication No. 87-2552, February.

Nooralian Z., Gashti M.P., and Ebrahimi I. (2016). Fabrication of a multifunctional graphene/polyvinylphosphonic acid/cotton nanocomposite via facile spray layer-by-layer assembly, *RSC Adv*, 6, 23288–23299. DOI: 101039/c6ra00296j.

Norzali N.R.A. and Badri K.H. (2016). The role of phosphate ester as a fire retardant in the palm-based rigid polyurethane foam, *Polymers & Polymer Composites*, 24(9), 711–717.

Norzali N.R.A., Badri K.H., Ong S.P., and Ahmad I. (2014). Palm-based polyurethane with soybean phosphate ester as a fire retardant, *The Malaysian Journal of Analytical Sciences*, 18, 456–465.

NRC (2000). *National Research Council (US) Subcommittee on Flame-Retardant Chemicals*. National Academies Press (US), Washington, DC.

Pan H.F., Wang W., Pan Y., Song L., Hu Y., and Liew K.M. (2015). Formation of self-extinguishing flame retardant biobased coating on cotton fabrics via layer-by-layer assembly of chitin derivatives, *Carbohydr Polym*, 115, 516–524. DOI: 10.1016/j.carbpol.2014.08.084.

Price D., Bullett K.J., Cunliffe L.K., Hull T.R., Milnes G.J., Ebdon J.R., Hunt B.J., and Joseph P. (2005). Cone calorimetry studies of polymer systems flame retarded by chemically bonded phosphorus, *Polym. Degrad. Stab.*, 88, 74–79.

Price D., Pyrah K., Hull T.R., Milnes G.J., Ebdon J.R., Hunt B.J., and Joseph P. (2002). Flame retardance of poly(methyl methacrylate) modifi ed with phosphorus-containing compounds, *Polym. Degrad. Stab.*, 77, 227–233.

Qian L., Qiu Y., Liu J., Xin F., and Chen Y. (2014). The flame retardant group-synergistic effect of a phosphaphenanthrene and triazine double-group compound in epoxy resin, *J Appl Polym Sci.*, 131(3), 39709–39716.

Qian L., Qiu Y., Wang J., and Xi W. (2015). High-performance flame retardancy by char-cage hindering and free radical quenching effects in epoxy thermosets, *Polymer (Guildf)*, 68, 262–269.

Qiu Y., Qian L.J., and Xia W. (2016). Flame-retardant effect of a novel phosphaphenanthrene/triazine-trione bi-group compound on an epoxy thermoset and its pyrolysis behavior, *RSC Adv*, 6, 56018–56027.

Ratna D. (2007). *Epoxy Composites: Impact Resistance and Flame Retardancy*, Vol 3. iSmithers Rapra Publishing, Shawbury.

Rhodia (2016)."Frequently asked questions: What is the PROBAN® process?www. rhodia-proban.com/uk/faq.asp, accessed on 25.2.2016.

Rieckert H., Dietrich J., and Keller H. (1997). Adduct preparation used as flame retardant e.g., in polyester fibre(s). DE 19711523.

Saito T. (1972). Cyclic organophosphorus compounds and process for making them. US Patent 3,702,878.

Salmeia K., Gaan S., and Malucelli G. (2016). Recent advances for flame retardancy of textiles based on phosphorus chemistry, *Polymers*, 8, 319. DOI: 10.3390/polym8090319.

Sander M. and Steininger E. (1968). Chapter 3: Phosphorus-containing resins, *Journal of Macromolecular Science, Part C: Polymer Reviews*, 2(1), 1–30.

Schäfer A., Seibold S., Lohstroh W., Walter O., and Döring M. (2007). Synthesis and properties of flame-retardant epoxy resins based on DOPO and one of its analog DPPO, *J. Appl. Polym. Sci.*, 105, 685–696.

Schartel B. (2010). Phosphorus-based flame retardancy mechanisms – Old hat or a starting point for future development?, *Materials (Basel)*, 3(10), 4710–4745. Published online 2010 Sep 30. DOI: 10.3390/ma3104710.

Schindler W.D. and Hauser P.J. (2004). *Chemical Finishing of Textiles*. Woodhead, Cambridge, England.

Schut J.H. (2009). *Polyphosphonate: New Flame-retardant Cousin of Polycarbonate*, www.ptonline.com/articles/200907cu1.html, accessed on July, 2009.

Šehić A., Tomšič B., Jerman I., Vasiljević J., Medved J., and Simončič B. (2016). Synergisticinhibitory action of P- and Si-containing precursors in sol–gel coatings on the thermal degradation of polyamide 6, *Polym. Degrad. Stab.*, 128, 245–252. DOI: 10.1016/j.polymdegradstab.2016.03.026.

Shaikh R.S., Düsel S.J.S., and König B. (2016). Visible-light photo-arbuzov reaction of aryl bromides and trialkyl phosphites yielding aryl phosphonates, *ACS Catal*, 6, 8410–8414. DOI: 10.1021/acscatal.6b0259.

Shieh J.Y. and Wang C.S. (2002). Effect of the organophosphate structure on the physical and flame-retardant properties of an epoxy resin, *J. Polym. Sci. Part A: Polym. Chem.*, 40, 369.

Shan H., Xiao H., Jiaojiao L., Xiujuan T., Qing Y., and Zhongwei W. (2017). A novel curing agent based on diphenylphosphine oxide for flame-retardant epoxy resin, *High Performance Polymers*, 30(10), 1229–1239.

Sirdesai S.J. and Wilkie C.A. (1987). RhCl(PPh3)3: A flame retardant for poly(methyl methacrylate), *Polym. Preprints, Am. Chem. Soc. Polym. Chem. Ed.*, 28, 149.

Solenis (2017). Section 11 toxicological information, Biosperse 535C, 4. 7.2017.

Solvay (2017). PROBAN®: Flame retardant fabrics and garments, www.solvay-proban.com/, accessed on 20.11.17.

Stawinski J. and Kraszewski A. (2002). How to get the most out of two phosphorus chemistries. Studies on h-phosphonates, *Acc. Chem. Res.*, 35, 952–960.

Tian X.J., Wang Z.W., Yu Q., Wu Q., and Gao J. (2014). Synthesis and property of flame retardant epoxy resins modified with 2-(diphenylphosphinyl)-1,4-benzenediol, *Chem Res Chinese U*, 30, 868–873.

Tomasino C. (1992). *Chemistry & Technology of Fabric Preparation & Finishing*. College of Textiles, North Carolina State University, USA.

Troev K.D. (2006). Structure and spectral characteristics of h-phosphonates, in: *Chemistry and Application of H-Phosphonates*. Elsevier Science Ltd., Amsterdam, The Netherlands, pp. 11–22.

Uhl F.M. and Wilkie C.A. (2004). Preparation of nano-composites from styrene and modified graphite oxides, *Polym. Degrad. Stab.*, 84, 215.

Van Duuren B., Katz C., Goldschmidt B., Frenkel K., and Sivak A. (1972). Carcinogenicity of haloethers. II. Structure-activity relationships of analogs of bis(chloromethyl) ether, *J. Natl. Cancer Inst.*, 48, 1431–1439.

Van Duuren B.L., Loewengart G., Seidman I., Smith A.C., and Melchionne S. (1978). Mouseskin carcinogenicity tests of the flame retardants tris(2, 3-dibromopropyl)-phosphate, tetrakis(hydroxymethyl)phosphonium chloride, and polyvinyl bromide, *Cancer Res.*, 38(10), 3236–3240.

Vasiljević J., Jerman I., Jakša G., Alongi J., Malucelli G., Zorko M., Tomšič B., and Simončič B. (2015). Functionalization of cellulose fibres with DOPO polysilsesquioxane flame retardant nanocoating, *Cellulose*, 22, 1893–1910. DOI: 10.1007/s10570-015-0599-x.

Wang C.S. and Shieh J.Y. (1998). Synthesis and properties of epoxy resins containing 2-(6-oxid-6H-dibenz<c, e><1, 2oxaphosphorin-6-yl) 1, 4-benzenediol, *Polymer*, 39, 5819–5826.

Wang C.S. and Shieh J.Y. (1999). Phosphorus-containing epoxy resin for an electronic application, *J Appl Polym Sci*, 73, 353–361.

Wang C.S. and Shieh J.Y. (2000). Synthesis and properties of epoxy resins containing bis (3-hydroxyphenyl) phenyl phosphate, *Eur Polym J*, 36(3), 443–452.

Wang M., Fang S., and Zhang H. (2018a). Study on flame retardancy of TGDDM epoxy resin blended with inherent flame-retardant epoxy ether, *High Performance Polymers*, 30(3), 318–327. DOI: 10.1177/0954008317694077.

Wang P., Fu X., Kan Y., Wang X., and Hu Y. (2018b). Two high-efficient DOPO-based phosphonamidate flame retardants for transparent epoxy resin, *High Performance Polymers*, 1–12. DOI: 10.1177/0954008318762037.

Weil E.D. (1993). Flame retardants: Phosphorus compounds, in: M. Howe-Grant, Ed., *Kirk-Othmer Encyclopedia of Chemical Technology*, 4th edn., Vol 10. Wiley-Interscience, New York, pp. 976–979.

Weil E.D. (1999). Synergists, adjuvants and antagonists in flame retardant systems, in: A. F. Grand and C.A. Wilkie, Eds., *Fire Retardancy of Polymeric Materials*. Marcel Dekker, New York, pp. 115–145.

Weil E.D. (2001). An attempt at a balanced view of the halogen controversy – Update 2001, Proc. Conf. Recent Adv. Flame Retardancy Polym. Mater., Business Communications Corp., Stamford, CT, 12: 158–175.

Weil E.D. (2005). Patent activity in the flame retardant field, Proc. Conf. Recent Adv. Flame Retardancy Polym. Mater., Business Communications Corp., Stamford, CT.

Wendels S., Chavez T., Bonnet M., Salmeia K.A., and Gaan S. (2017). Retardants containing P-C Bond and their applications, *Materials*, 10, 784. DOI: 10.3390/ma10070784.

Xiang-Gheng B., Zhong-Ming L., and Jian-Hua T. (2008). Flame retardancy of whisker silicon oxide/rigid polyurethane foam composites with expandable graphite, *Journal of Applied Polymer Science*, 110(6), 3871.

Xiao M., Sun L., Liu J., Li Y., and Gong K. (2002). Synthesis and properties of polystyrene/graphite nanocomposites, *Polymer* 43(8), 2245–2248. DOI: 10.1016/S0032-3861(02)00022-8.

Xu D., Lu H., Huang Q., Deng B., and Li L. (2018). Flame-retardant effect and mechanism of melamine phosphate on silicone thermoplastic elastomer, *RSC Adv.*, 8, 5034. DOI: 10.1039/c7ra12865g.

Xu W.H., Wirasaputra A., Liu S.M., Yuan Y., and Zhao J. (2015). Highly effective flame retarded epoxy resin cured by DOPO-based co-curing agent, *Polym Degrad Stabil*, 122, 44–51.

Yan W., Yu J., Zhang M., Long L., Wang T., Qin S., and Huang W. (2017). Novel flame retardancy effect of phenethyl-bridged DOPO derivative on epoxy resin, *High Performance Polymers*, 1–10. DOI: 10.1177/0954008317716525.

Yang S., Hu Y., and Zhang Q. (2018). Synthesis of a phosphorus–nitrogen containing flame retardant and its application in epoxy resin, *High Performance Polymers*, 1–11. DOI: 10.1177/0954008318756496.

Yang S., Wang J., Huo S., Cheng L., and Wang M. (2015). Preparation and flame retardancy of an intumescent flame-retardant epoxy resin system constructed by multiple flame-retardant compositions containing phosphorus and nitrogen heterocycle, *Polym Degrad Stab*, 119, 251–259.

Yang S., Zhang Q., and Hu Y. (2017). Preparation and investigation of flame-retardant epoxy resinmodified with a novel halogen-free flame retardant containing phosphaphenanthrene, triazine-trione, and organoboron units, *Journal of Applied Polymer Science*, 134(37), 45291.

Yang Z., Fei B., Wang X., and Xin J.H. (2012). A novel halogen-free and formaldehyde-free flame retardant for cotton fabrics, *Fire Mater.*, 36, 31–39. DOI: 10.1002/fam.1082.

Yao Q., Wang J., and Mack A.G. (2013). Process for the Preparation of DOPO-derived Compounds and Compositions Thereof. Patent20130018128Al, USA.

Zhang T., Yan H.Q., Peng M., Wang L.L., Ding H.L., and Fang Z.P. (2013a). Construction of flame retardant nanocoating on ramie fabric via layer-by-layer assembly of carbon nanotube and ammonium polyphosphate, *Nanoscale* 5, 3013–3021. DOI: 10.1039/c3nr34020a.

Zhang X., He Q., Gu H., Wei S., and Guo Z. (2013b). Polyaniline stabilized barium titanate nanoparticles reinforced epoxy nano-composites with high dielectric permittivity and reduced flammability, *Journal of Materials Chemistry C*, 1, 2886.

Zhang Z., Horrocks A.R., and Hall M.E. (1994). Flammability of polyacrylonitrile and its copolymers IV. The flame retardant mechanism of ammonium polyphosphate, *Fire and Materials*, 18, 307–312.

Zhao P., Xiong K., Wang W., and Liu Y. (2017). Preparation of a halogen-free P/N/Si flame retardant monomer withreactive siloxy groups and its application in cotton fabrics, *Chinese J Chem Eng*, 25, 1322–1328.

Zheng D., Zhou J., Zhong L., Zhang F., and Zhang G. (2016). A novel durable and high-phosphorous-containing flame retardant for cotton fabrics, *Cellulose*, 23, 2211–2220. DOI: 10.1007/s10570-016-0949-3.

Zhou T., He X., Guo C., Yu J., Lu D., and Yang Q. (2015). Synthesis of a novel flame retardant phosphorus/nitrogen/siloxane and its application on cotton fabrics, *Textile Research Journal*, 85(7), 701–708. DOI: 10.1177/0040517514555801.

7 Intumescent FRs (IFRs)

7.1 INTRODUCTION

There has been increasing concern about the potential release of harmful substances during fabric care or burning associated with halogenated or heavy metal–containing treatments (Horrocks, 2011). Most victims of fires die from inhaled combustion fumes, smoke, or toxic gases, rather than from burns (Subbulakshmi et al., 2000). However, addition of, or reaction with, FRs may not offer protection from aftereffects of combustion. IFRs are a promising alternative. IFRs were initially used in paints and coatings. With such intumescent coatings, one can visualize the burning polymer as a block consisting of several separate layers. The top char layer is followed by the intumescent front where the foaming reactions take place. Following is an unburned polymer coating layer that still contains flame retardant. The bottom layer represents the polymer substrate that is being protected by the intumescent coating. The charfoam provides a physical barrier to heat transfer and mass transfer, and therefore interferes with the combustion process (Gilman et al., 1997).

Intumescent coatings may be applied to textile fabrics, which, once activated by heat, generate a foamed insulating layer on the fabric surface. This intumescent layer prevents any further ignition until the fire consumes the intumescent layer itself. Thus, the primary aim is to prevent or retard the rate at which flammable substrates ignite, or nonflammable, thermoplastic substrates soften. The stable and thermally insulating char barrier created by the mechanism of intumescence is resistant to both radiant heat and flame. The mechanism operates within the solid phase and is designed to delay the combustion or softening of the textile material (Horrocks, 2003). Back-coating describes a family of application methods where the flame retardant formulation is applied in a binding resin to the reverse surface of an otherwise flammable fabric. In this way, the aesthetic quality of the face of the fabric is maintained while the flame retardant property is present on the back or reverse face. Careful use of viscosity modifiers and general back-coating application variables ensures that *grin-through* is minimized. Application methods include doctor blade or knife-coating methods and the formulation is as a paste or foam. These processes and finishes are used on fabrics where aesthetics of the front face are of paramount importance, such as furnishing fabrics and drapes. Coating methods, generally, unlike those requiring impregnation lead to little or no waste of application chemical formulation, and hence, minimal effluent problems.

A number of volatile phosphorus-containing flame retardant species has been identified as possible replacements for bromine-containing formulations used in textile back-coatings because of the need for vapor-phase activity. The selected retardants include tributyl phosphate (TBP), a monomeric cyclic phosphonate Antiblaze CU (Rhodia Specialites), and the oligomeric phosphate-phosphonate Fyrol 51 (Akzo). When combined with an intumescent char-forming pentaerythritol derivative (NH1197, Chemtura) and applied as a back-coating on cotton and polypropylene substrates, significant improvements in overall flame retardancy are observed. One sample applied to cotton and comprising both TBP and intumescent passed the simulated match-ignition test, BS5852:1979: Part1 after a water soak at 408°C for 30 min. Determination of residual phosphorus within chars shows that there is significant volatile activity present in these formulations. Addition of volatile nitrogen as melamine also demonstrated improved flame retardancy in similar formulations (Horrocks et al., 2007).

Vandersall (1971) made a classical review in 1971 of intumescent coatings, presenting the early history and detailed course of development of commercial intumescent coatings. The word *intumescence* is derived from Latin word *intumescere* meaning "to swell up." IFRs behave in a similar way. It is essentially a special case of a condensed phase mechanism. Intumescent systems interrupt the self-sustained combustion of the polymer at its earliest stage. When heated beyond a critical temperature, the material begins to swell and then to expand. The process creates a foamed cellular charred layer on the surface that protects the underlying material from the action of the heat flux or flame. The formation of char can occur either by free-radical- or acid-catalyzed polymerization reactions from the compounds produced in the pyrolysis. Intumescence can be described as a fire-retardant technology that causes an otherwise flammable material to foam, forming an insulating barrier when exposed to heat. Intumescence is the generation of expanded, foamed char created by heat and special additives, such as char-formers (for example starch), and catalysts that yield inorganic acids at about 150°C, generators that provide nonflammable gases for the foam and binders for fixation to the fabric. This system provides a foamed insulation layer on the fabric surface, similar to the formation of char by cellulose and wool fibers. This porous char layer seems to fulfill many flame retardancy requirements, such as preventing or retarding further ignition by thermal and chemical insulation, creating a flame barrier, including reduction of material exchange (volatiles, oxygen). Additionally, smoke and toxic gas development are decreased. This double barrier function (for heat and material exchange) is very effective and avoids some ecological disadvantages of common flame retardants. The expanded, foamed char is formed by heat and additives, such as char-formers (e.g., starch), acid-liberating catalysts, nonflammable gas generators, and binders. Among alternative possibilities, intumescent materials have gained considerable attention because they provide fire protection with minimal overall health hazard. Although intumescent coatings are capable of exhibiting good fire protection for the substrate, they have several disadvantages such as water solubility, brushing problems, and relatively high cost.

The main advantages of intumescent coatings are their low toxicity in case of fire, the absence of dioxin and halogen acids, and their low evolution of smoke. Their efficiency lies between that of halogen compounds and that of aluminum trihydrate and magnesium hydroxide. The metallic hydroxides split off water and are environmentally friendly, but their low activity requires high concentrations which change the mechanical properties of the matrix to which they are applied. In contrast to many halogen compounds, flame retardants based on nitrogen do not interfere with the types of stabilizers added to all plastic materials (Camino and Lomakin, 2001).

IFRs consist of three components (Bourbigot and Duquesne, 2010), namely:

1. Acid source,
2. Carbonization agents, and
3. Blowing agents.

7.1.1 ACID SOURCES

- Inorganic acid sources: phosphoric, sulphuric, or boric acid;
- Organic acid sources: ammonium salts, phosphates, polyphosphates, borates, polyborates, sulfates, halides;
- Phosphates of amine or amide: products of the reaction of urea or guanidyl, urea with phosphoric acids, melamine phosphate, a product of the reaction of ammonia with phosphorous pentoxide; and
- Organophosphorus compounds: tricresyl phosphate, alkyl phosphates, haloalkyl phosphates.

7.1.2 CARBONIZATION AGENTS

Starch, dextrins, sorbitol, mannitol, pentaerythritol (PER–monomer, dimer, trimer), phenol-formaldehyde resins, methylol melamine, char former polymers (e.g., PA-6, PA-6/clay nanocomposite PU, PC).

7.1.3 BLOWING AGENTS CALLED SPUMIFIC

Urea, urea-formaldehyde resins, dicyandiamide, melamine.

Spumific (plural spumifics) is a chemical that decomposes to produce large quantities of gas.

The ratios in which the different compounds are present are also of utmost importance. The optimum ratio must be determined experimentally. One or more of these substances could be replaced with others of the same class or group. Further studies have shown that more effective intumescent systems are obtained when two or more of the elements required for intumescence are incorporated in the same molecular complex (Camino et al., 1989).

The carbon agent forms multicellular charred layers, the char may be soft or hard (Xiao et al., 2014).

- Soft chars: IFR composed of a carbon source, pentaerythritol (PER), acid source (ammonium polyphosphate) and a gas-blowing additive (melamine)

- Harder chars: IFR composed of sodium silicates and graphite. These are suitable for use in plastic pipe fire-stops, as well as in exterior steel fireproofing.

PER is quite costly. A possible substitute is green carbon agent is chitosan (CS), obtained by the alkaline deacetylation of abundantly naturally occurring chitin. A good synergistic effect has been observed when chitosan/urea compound based phosphonic acid melamine salt (HUMCS), was added to an IFR system for polypropylene (PP) (Xiao et al., 2014).

Typically, phosphorus-based FRs are designed to develop its activity just before the start of the decomposition of the specific polymers for which they are used. They offer a partial gas-phase contribution to the flame extinguishing effect, which is comparable to bromine- or chlorine-containing FRs. However, the main feature is char-forming activity such as intumescence. The phosphorus-based FRs can be used in numerous polymers and resins such as polyamides, polyesters, polyolefins, epoxy, and styrene derivatives. The main area of application for the compounded materials is for injection-molded electrical and electronic (E&E) parts. Besides the E&E market, a very important market is FR fabrics for public buildings and public transportation seating. Compounds other than those of phosphorus mentioned previously can be used in intumescent systems, but they are not widely employed, except for boron-based compounds for nondurable textile applications and in intumescent coating for fire protection, and also expandable graphite for applications including intumescent seals, foams, and strips.

Today their main applications are: melamine for polyurethane flexible foams; melamine cyanurate in nylons; melamine phosphates, ammonium polyphosphate-pentaerythryol, or ethylene-urea formaldehyde polymers in polyolefins; melamine and melamine phosphates, or dicyandiamide in intumescent paints; guanidine phosphates for textiles; and guanidine sulfamate for wallpapers (Camino and Lomakin, 2001).

The most important inorganic nitrogen–phosphorus compound used as a flame retardant is ammonium polyphosphate, which is applied in intumescent coatings and in rigid polyurethane foams. The world demand for ammonium polyphosphate is 10^7 kg per annum. The most important organic nitrogen compounds used as flame retardants are melamine and its derivatives, which are added to intumescent varnishes or paints. Melamine is incorporated into flexible polyurethane cellular plastics and melamine cyanurate is applied to unreinforced nylons. Guanidine sulfamate is used as a flame retardant for PVC wall coverings in Japan. Guanidine phosphate is added as a flame retardant to textile fibers, and mixtures based on melamine phosphate are used as flame retardants to polyolefins or glass-reinforced nylons.

The following sequences of events take place in the development of the intumescent phenomena (Bourbigot and Duquesne, 2010):

1. Inorganic acid liberates typically between 150°C and 250°C depending on its source and other components.

2. The acid esterifies the carbon-rich components at temperatures slightly above the acid release temperature.
3. The mixture of materials melts prior or during esterification.
4. The ester decomposes via dehydration resulting in the formation of a carbon-inorganic residue.
5. Released gases from the aforementioned reactions and degradation products (in particular those resulting from the decomposition of the blowing agent) cause the carbonizing material to foam.
6. As the reaction nears completion, gelation and finally solidification occurs. This solid is in the form of multicellular foam.

7.2 GRAPHENE AND ITS OXIDE

Graphene is a material made of carbon atoms that are bonded together in a repeating pattern of hexagons. Graphene is so thin that it is considered two-dimensional. Graphene's flat honeycomb pattern gives it many extraordinary characteristics, such as being the strongest material in the world, as well as one of the lightest, most conductive, and transparent. Graphene has endless potential applications, in almost every industry (e.g., electronics, medicine, aviation, and much more).

The single layers of carbon atoms provide the basis for many other materials. Graphite, like the substance found in pencil lead, is formed by stacked graphene. Carbon nanotubes are made of rolled graphene and are used in many emerging applications from sports gear to biomedicine.

7.2.1 GRAPHENE OXIDE

As graphene is expensive and relatively hard to produce, great efforts are made to find effective yet inexpensive ways to make and use graphene derivatives or related materials. Graphene oxide (GO) is one of those materials – it is a single-atomic layered material, made by the powerful oxidation of graphite, which is cheap and abundant. GO is an oxidized form of graphene, laced with oxygen-containing groups. It is considered easy to process because it is dispersible in water (and other solvents), and it can even be used to make graphene. GO is not a good conductor, but processes exist to augment its properties. It is commonly sold in powder form, as a coating on substrates, or dispersed.

Graphite oxide is a compound made up of carbon, hydrogen, and oxygen molecules. The main difference between graphite oxide and graphene oxide is the interplanar spacing between the individual atomic layers of the compounds, caused by water intercalation. GO is effectively a by-product of the oxidization process, such as when the oxidizing agents react with graphite, and the interplanar spacing between the layers of graphite is increased. The completely oxidized compound can then be dispersed in a base solution such as water, and GO is then produced.

Very recently, many groups reported that the GO sheets could act as an intumescent flame retardant additive in polymer composites. GO is a promising flame retardant nano-additive; however, it displays thermal instability and does not disperse well into polymers, because GO sheets are monolayers of sp2-hybridized carbon atoms derivative with various oxygen-containing functionalities (e.g., hydroxyl, carboxyl, and epoxy) on the basal plane and along the edges. These hydrophilic groups provide graft points on the basal plane and along the edges of the GO sheets, with a large amount of functional substances grafted on, and so the hydrophilic groups may be replaced by some hydrophobic functional groups. Many agents, such as alkylchlorosilanes or isocyanates (Salavagione et al., 2009; Huang et al., 2010), have been used for the chemical functionalization of GO sheets with the objective of improving the solubility and compatibility with polymers by changing its surface properties. The alcohols or the modified products are considered to be good carbon sources (Peng et al., 2008). The hydroxyl is sufficiently reactive to combine with other oxygenic functionalities (e.g., hydroxyl, carboxyl, and epoxy) to reduce the hydrophilicity of GO. In addition, the introduction of organic molecular segments on GO sheets is of significance in elevating the thermal stability of the modified GO owing to the buildup of strong intermolecular and intramolecular interactions among graphene and the polymers. This results in an increased quantity and enhanced quality of the residual char from combustion of polymers into which the modified GO is incorporated. The physical barrier effect of modified GO enhances the quality of the residual char and slows the diffusion of gases and degradation products.

Silk fabrics were coated by graphene oxide hydrosol in order to improve its flame retardancy and ultraviolet resistance. In addition, montmorillonoid was doped into the graphene oxide hydrosol to further improve the flame retardancy of silk fabrics. The synergistic effect of graphene oxide and montmorillonoid on the thermal stabilization property of the treated silk fabrics was also investigated. The results show that the treated silk fabrics simultaneously have excellent flame retardancy, thermal stability, smoke suppression, and ultraviolet resistance (Ji et al., 2017).

Montmorillonite is a very soft phyllosilicate group of minerals that form when they precipitate from water solutions as microscopic crystals, known as clay. It is named after Montmorillon in France. Montmorillonite, a member of the smectite group, is a 2:1 clay, meaning that it has two tetrahedral sheets of silica sandwiching a central octahedral sheet of alumina.

7.3 CHEMICAL MECHANISM OF INTUMESCENCE

The main mode of action of an intumescent coating is to limit heat transfer from the flame to the underlying material.

The formation of char can occur either by free-radical- or acid-catalyzed polymerization reactions from the compounds produced in the pyrolysis. Blowing agent or spumific compounds create huge gasses which form carbon foam.

A suggested mechanism for char formation is a simple acid-catalyzed, dehydration reactions. This is shown in the four reactions (Equations 7.1–7.4) below:

$$-CH - CH_2 - OH + H^+ \rightarrow -CH - CH_2 - OH_2{}^+ \qquad (7.1)$$

$$-CH - CH_2 - OH_2{}^+ \rightarrow -C = CH_2 + H_2O + H^+ \qquad (7.2)$$

$$-CH - CH_2 - OH + H_3P(OH)_2O_4 \rightarrow -CH - CH_2 - O - P(OH)_2 = O + 2H_2O \qquad (7.3)$$

$$-CH - CH_2 - O - P(OH)_2 = O \rightarrow H_3P(OH)_2O_4 + -C = CH_2 \qquad (7.4)$$

The first two reactions show the depolymerization catalyzed by an acid. The second two reactions show the dehydration of the polymer when phosphoric acid is present. Both reactions essentially lead to the same result: producing -C=CH$_2$ fragments at the polymer chain ends. These fragments condense to form carbon-rich char residues.

Briefly stated, the way the phosphorous compounds work is that they phosphorylate carbonifics such as pentaerythritol to make polyol phosphates. These polyol phosphates can then break down to form char (Weil, 1992).

Inorganic acids such as Ammonium polyphosphate (APP) must be available for the dehydration action with the char former (PER) at a temperature below the temperature at which the degradation of the polymeric materials begins. Then, the formation of the effective char must occur via a semiliquid phase (high viscous material) that coincides with gas formation and expansion of the surface (bubble-gum effect). This action must occur before the charring liquid solidifies. Gases released from the degradation of the intumescent material and/or of the polymeric material should be trapped and have to be diffused slowly in the highly viscous molten material in order to create a layer with the morphological properties of interest. Thus, it is essential to examine carefully the dynamic viscoelastic properties of the intumescent shield, because control of the melt rheology is necessary to obtain a multicellular and highly expanded structure. Moreover, the mechanical integrity of the char is a crucial parameter because it has to resist external stress. In short, an intumescent formulation has to be optimized in terms of physical (e.g., char strength, expansion, viscosity) and chemical (thermal stability, reactivity) properties in order to form an effective protective char that can protect its host polymer (reaction to fire) or a substrate such as steel or wood (resistance to fire) (Bourbigot and Duquesne, 2010).

7.3.1 Organic-Based Systems

The use of polyols such as pentaerythritol, mannitol, or sorbitol as *classical* char formers in intumescent formulations for thermoplastics is associated with migration and water solubility problems. Moreover, these additives are often not compatible with the polymeric matrix and the mechanical properties of the formulations are consequently very poor. Those problems can be solved (at least partially) by the synthesis of additives that concentrate the three intumescent FR elements in one material, as suggested by the pioneering work of Halpern (1984). The b-MAP (melamine salt of 3,9-dihydroxy-2,4,8,10-tetraoxa-3,9-diphosphaspiro[5,5]-undecane-3,9-dioxide) and Melabis (melamine salt of bis (1-oxo-2,6,7-trioxa-1-phosphabicyclo[2.2.2]octan-4-ylmethanol) phosphate) were synthesized from pentaerythritol, melamine, and phosphoryl trichloride. They were found to be more effective to fire-retard PP than standard halogen-antimony FR.

Fontaine et al. (2008) suggested the synthesis of eco-friendly PEPA (1-oxo-4-hydroxymethyl-2,6,7-trioxa-1-phosphabicyclo[2.2.2]octane) and bis(PEPA) phosphate-melamine salt derivatives prepared by a novel and safe protocol (one step–one pot process). These new salts were incorporated in PP and exhibit high fire performance via an intumescence mechanism (LOI of 32 vol % and V-0 rating in the UL-94 on bars of 3.2 mm thick). Recently, Liu and Wang proposed to use the catalytic action of phosphotungstic acid (Liu and Wang, 2006) in the synthesis of melamine salts of pentaerythritol phosphate (called MPP or the same molecule as b-MAP) in order to solve the problems of conventional preparation methods (the use of $POCl_3$; the high temperature of the reaction, and thus, the high energy consumption). It was shown that the acid catalysis can enhance the conversion degree of the reaction and decrease the reaction temperature while keeping a satisfactory conversion, thus greatly controlling energy consumption in the preparation process of melamine salts of pentaerythritol phosphate (MPP or b-MAP) (Structure 7.1).

STRUCTURE 7.1   Melamine salts of pentaerythritol phosphate (MPP or b-MAP)

Other authors also suggested the synthesis of organic compound "3 in 1" containing the three main ingredients of intumescence. Wang et al. (2006) reported the synthesis of a novel intumescent FR, poly(2,2-dimethylpropylene spirocyclic pentaerythritol bisphosphonate) to yield PET with both excellent flame retardancy and anti-dripping properties. A novel phosphorus–nitrogen containing intumescent FR was prepared via the reaction of a caged bicyclic phosphorus (PEPA) compound and 4,4-diamino diphenyl methane (DDM) in two steps (Gao et al., 2006). Incorporated in poly(1,4-butylene terephtalate) (PBT) and in combination with thermoplastic polyurethane (TPU) as an additional char former, the intumescent PBT exhibits V-0 rating in the UL-94 test (3.2 mm) and an enhanced thermal stability. It is suggested that P–N bonds detected in the charred structure might play a role in its efficiency.

To substitute the polyol-based char formers, Li and Xu (2006) synthesized a novel char former for intumescent system based on triazines and their derivatives. It is a macromolecular triazine derivative containing hydroxyethylamino, triazine rings and ethylenediamino groups (Structure 7.2). They showed that the new char former in an intumescent formulation containing Ammonium Polyphosphate (APP) and a zeolite as synergist can achieve low flammability at only 18 wt % loading in PP (LOI of 30 vol % and V-0 rating (3.2 mm) in the UL-94 test). No tentative mechanism is suggested.

$$NH_2CH_2CH_2OH$$

STRUCTURE 7.2   A char former for intumescent system based on triazine

7.3.2   Inorganic-based Systems

Organic intumescent systems represent a large part of the studies focusing on intumescence, but the processing of intumescent mineral systems is even older without much reference in the literature (Arthur, 1912). Mineral intumescent systems are based on alkali silicates. The swelling of the material upon heating or on contact with a flame is due to an endothermic process and is associated with the emission of water vapor that is ionically hydrated in the silicate system (Langille et al., 2003). The solid foam formed is rigid and consists of hydrated silica. The structure remains solid until it reaches its glass-softening point. Since only water vapor is released, toxic fumes that may be released from organic-based systems are eliminated. Intumescent alkali have serious limitations, however; in particular, they are sensitive to carbon dioxide and water, which are present in the atmosphere, causing the silicate coating to gradually lose its intumescence, to become brittle, and to lose its adhesion.

7.3.3   Intumescent Inorganic Polymer

Inorganic polymers are inherently FR and are used *as is*. In the particular case of inorganic–organic polymers, they are either already flame retardant, or they are used in applications that do not require high levels of flame retardancy. The only published work devoted to intumescent inorganic polymer concerns polyphosphazene (Bourbigot and Duquesne, 2010).

While polyphosphazene exhibits some level of flame retardancy, its performance is not enough to fulfill the requirements of the FAA for aircraft interiors. In order to make ultra-fire-resistant elastomers, the use of expandable graphite was investigated by Lyon et al. (2003).

Expandable graphite (EG) represents another class of inorganic intumescent systems. EG swells during combustion to form a char layer, which prevents the access of oxygen to the combustion zone. Expandable graphite is one of the intumescent flame retardant additives that is produced by intercalation of sulfuric acid into graphite in the presence of a strong oxidizing agent. At elevated temperatures, expandable graphite decomposes with an emission of volatile products. This causes the formation of a foamed char layer, which is a physical barrier that reduces heat and mass transfer between burning materials and the environment. EG has been used advantageously in PU coatings to develop fire protective coating for polymeric substrates.

EG is an intumescent additive constituted by chemicals trapped between the graphite layers. The expansion can be more than 100 times its original thickness, resulting in a nonburnable, insulating layer. The addition of the expandable graphite to the polyphosphazene rubber reduces its HRR to a level approximating that of the fire-resistant engineering plastics or that of thermoset resins currently used in aircraft interiors (<100 kW/m^2). In contrast, the polyphosphazene leaves a char residue equal to 35% of the original weight. The PU and polyphosphazene formulated with the expandable graphite leave a light friable char on the order of 20–50 times the original sample volume. The expanded graphite char insulates the underlying polymer from burning. The high thermal efficiency of the expanded char results in a peak HRR that is five and seven times lower than that for the virgin polyphosphazene and PU polymers, respectively.

7.3.4 SYNERGISTS

Performance of intumescent systems in terms of LOI, UL-94, or cone calorimetry was enhanced dramatically by adding a small amount of an additional compound leading to a synergistic effect. The researchers showed that adding small amounts of minerals such as zeolites (Bourbigot, et al., 1996), natural clays (Le Bras and Bourbigot, 1996a), and zinc borates (Samyn et al., 2007) in intumescent systems, the FR performance can be drastically enhanced. Concurrently, Levchik et al. (1995) proposed the use of small amounts of talc and manganese dioxide combined with APP in PA-6 to promote charring and to enhance insulation properties of the intumescent coating leading to a significant improvement of the flammability performance.

Another recent approach using borosiloxane elastomer in PP allows also one to obtain a very large synergistic effect in intumescent systems (Anna et al., 2002). Large synergistic effects are observed when incorporating nanofillers in intumescent formulation, and most of the recent works on synergy are devoted to this. The presence of a nanofiller can modify the chemical (reactivity of the nanofiller vs. the ingredients of the intumescent system) and physical (expansion, char strength, and thermophysical properties) behavior of the intumescent char when exposed to a flame or heat flux, leading to enhanced performance. Lewin (2005) describes this phenomenon as a catalytic effect. It is noteworthy that the catalyst (the nanofiller) is a crucial ingredient (reactant) in the development of intumescence forming additional species stabilizing the structure and modifying the rheological behavior. The nanofiller is incorporated at an amount as low as 1 wt % (sometimes less as in the case of the incorporation of nanoparticles of copper at an amount as low as 0.1 wt % in epoxy resin containing APP), and it permits the formation of active species selecting chemical reactions in the condensed phase and yielding char with the dynamic properties of interest.

Different methods are usually involved to provide flame retardancy by intumescence for natural and synthetic fibers.

7.4   IFRs FOR NATURAL FIBERS

The intumescents, which are based on ammonium and melamine phosphate-containing intumescents applied in a resin binder, can raise the fire barrier properties of flame-retarded viscose and cotton fabrics to levels associated with high performance fibers such as aramids (Horrocks and Kandola, 1998). Furthermore, charred residues are considerably stronger than those from flame-retarded fabrics alone (Kandola and Horrocks, 1999). Applications of these same intumescents to wool-containing fabrics have shown that enhanced barrier properties are possible and this occurs for both flame-retarded (Zirpro) and non-flame-retarded wool fibers present (Davies, Horrocks and Miraftab, 2000; Horrocks and Davies, 2000). Current research is now focused on enabling these intumescent formulations to be applied to conventional fabric structures using normal application technologies. In addition, some success has been achieved at creating durable intumescent finishes based on the reaction of phosphorylated polyols (for example, pentaerythritol diphosphoryl chloride) with cellulose, flame-retarded cellulose, wool, and nylon (Horrocks and Zhang, 1999, 2001). Horrocks et al. (2005) undertook considerable work on intumescents applied by substantive fibre treatments for cellulose. Based on the work of Halpern et al. (1984), on cyclic organophosphorus molecules, they developed a phosphorylation process for cotton fibres achieving intumescent cotton fabric with considerable durability. Char enhancement is as high as 60 wt% compared to pure cotton, which is associated with very low flammability.

Coating (or back-coating) on fabric is another way to provide flame retardancy to cotton. Davies et al. (2005) used intumescent back-coatings based on APP as the main FR, combined with metal ions as a synergist. The thermal degradation of APP was promoted by metal ions at lower temperatures enabling FR activity at lower temperatures in the polymer matrix thereby enhancing FR efficiency. The advantage of this concept is to reduce the water solubility of the phosphate and to produce textile back-coatings with good durability. The flame-retarding behavior of these coated cotton fabrics was evaluated with the cone calorimeter and it was shown that they exhibit a significant FR effect. The development of intumescent char at the surface of the fabric was observed, confirming the expected mechanism.

Among natural fibers, wool has the highest inherent nonflammability. It exhibits a relatively high LOI of about 25 and low flame temperature of about 680°C. The IFR activity is due to char-forming reactions that may be enhanced by applying a number of flame retardants. Horrocks and Davies (2000) studied intumescent formulations based on melamine phosphate (MP) to improve flame retardancy of wool. They proposed a comprehensive model on the mechanism of protection via an intumescent process, involving the formation of cross-linked char by P–N and P–O bonds resistant to oxidation. Later, they used spirocyclic pentaerythritol phosphoryl chloride (SPDPC, Structure 7.3) phosphorylated wool to achieve intumescent wool, which exhibits large char expansion and good flame retardancy (Horrocks and Zhang, 2004).

STRUCTURE 7.3   Chemical structure of SPDPC

SPDPC has the same chemical structure as that of spirocyclic pentaerythritol bisphosphorate disphosphoryl chloride (SPBDC) shown in Equation 7.5.

(7.5)

SPDPC was applied to the Proban-treated cotton fabric, which created a further higher level of char formation owing to the presence of Proban, and enhanced flame retardancy (Horrocks and Zhang, 2002). The intumescent finishes must be substantively attached to the textile fabric to achieve wash durability without losing its important textile properties.

Ghosh and Joshi (2016) developed an FR for cotton apparel fabric using spirocyclicpentaerythritol di (phosphoryl chloride) (SPDPC), which was further modified into bisdiglycolspirocyclicpentaerythritolbisphosphorate (BSPB) by reacting with ethylene glycol. The flame retardant agent was then attached to the fabric using a sol–gel process. The treated fabric showed high flame retardancy,compared to the untreated cotton per vertical flame testing.

7.5 IFRs FOR SYNTHETIC FIBERS

Several FRs have been designed for polyester extrusion (bisphenol-S-oligomer derivatives from Toyobo, cyclic phosphonates (Antiblaze CU and 1010) from Albemarle, or phosphinate salts such as OP950, from Clariant).

Chen et al. (2005) proposed the use of a novel antidripping flame retardant, poly(2-hydroxy propylene spirocyclic pentaerythritol bisphosphonate) (PPPBP, Structure 7.4) to impart flame retardancy and dripping resistance to PET fabrics. PPPBP is synthesized by reacting SPDPC with glycerol.

STRUCTURE 7.4 Chemical structure of PPPBP

The treatment of PET fabrics with PPPBP showed significantly enhanced FR property in vertical flame burning of the flame retardancy producing a nonignitable fabric and either a significant reduction of melt dripping at low levels or an absence of dripping at higher levels of PPPBP. PPPBP produces phosphoric or polyphosphoric acid during thermodecomposition leading to the formation of phosphorus-containing complexes at higher temperatures. On burning, it forms char following a condensed-phase mechanism via char promotion. It was suggested that the high yields of char are protected from thermo-oxidation by the presence of phosphoric acid contained in the charred residue and because of the high thermal stability of C=C groups in the char. A new halogen-free FR, only 5% of which was incorporated in master batches for polyester, enabled PET to obtain classification according to several standards, such as NF P 92 501 or NF P 92 503 (M classification), or FMVSS 302 or BS 5852 (Crib 5). In this case, an intumescent behavior is observed, but its mechanism of formation should be investigated.

Sulfur is a flame retardant element which may act as the acid source in the intumescent flame retardant system (Bourbigot and Duquesne, 2007). A sulfur-containing main chain aromatic polyphosphonates, named poly(sulfonyldiphenylene phenylphosphonate) (PSPPP, Structure 7.5) was synthesized successfully through melt (Wang et al., 1999) or solution polycondensation (Masai et al., 1973; Wang, 1997) between phenylphosphonic dichloride and bisphenol S (4,4'-sulfonyldiphenol).

STRUCTURE 7.5   Chemical structure of PSPPP

PSPPP has been used to prepare the flame rerdant Terylene by adding it to PET before spinning for many years (Masai et al., 1973; Wang, 1997). Granzow (1978) pointed out that the choice of phosphorus-containing flame retardants in PET should be restricted to those which are thermally stable during the processing temperature of PET (around 300°C); to be effective, however, the additive has to decompose rapidly around 400°C, the surface temperature of burning polyester. Aside from these thermal stability constraints, good compatibility with the polyester melt and the absence of any detrimental effects on the spinnability are mandatory. PSPPP have been found to be particularly suitable for the spinning of PET and to meet those various requirements (Masai et al., 1973; Wang, 1997).

Balabanovich and Engelmann (2003) investigated the flame retardancy and charring effect of PSPPP in PBT alone or in combination with polyphenylene oxide (PPO) or 2-methyl-1,2-oxaphospholan-5-one-2-oxide (OP). The UL-94 test V-0 rating could be achieved by addition of 10 wt % PSPPP, 10 wt % PPO, and 10 wt % OP. The fire-retardant effect was attributed to promoting char yield by involving the polymer in charring. Pyrolysis gas-chromatography /mass spectrometry results suggested that PSPPP was shown to induce the formation of thermally stable polyarylates and phenolic functionalities in PBT (Chen and Wang, 2010).

Like polyester, polyamides are synthetic fibers made from semicrystalline polymers that find use in a variety of applications in textiles similar to polyester fibers. Recent work of Davies et al. (2005) has investigated the effect of adding selected intumescent FRs based on APP, MP, pentaerythritol phosphate, cyclic phosphonate, and similar formulations into polyamide 6 and 6.6 in the presence and absence of nanoclay. The authors found that in polyamide 6.6, all of the systems containing the nanoclay demonstrated significant synergistic behavior, except for MP, because of the agglomeration of the clay.

7.6   IFRs FOR POLYOLEFINS

Polypropylene (PP) is presently one of the fastest-growing fibers for technical end uses when high tensile strength coupled with low cost are essential features. Halogen-containing flame retardants for polyolefin flame retardation have been more and more restricted in recent years because of environmental concerns, and halogen-free flame retardants are highly desired. IFRs have provided an effective method because of their low toxicity and low smoke production during burning. IFRs, which combine an acid source, a blowing agent, and a char-forming agent within one molecule, have received more and more attention. A halogen-free IFR, pentaerythritol spirobisphosphoryl dicyandiamide (SPDC), which has a ratio of phosphorus, nitrogen, and carbon of 1.00:1.81:2.13, was reported to have a char yield of 39.2 wt % at $600^{\circ}C$ in nitrogen. Polypropylene (PP) with 30 wt % SPDC addition passed the UL-94 V-0 rating and showed a limiting oxygen index (LOI) value of 32.5 (Li et al., 2014).

Presently, the use of phosphorus-based, halogen-free FRs in PP fibers is prevented by the need to have at least 15–20 wt % additive. Because they are char-promoting while all halogen-based systems are essentially non-char-forming in PP. the flammability of PP is reduced by the addition of small amounts of clay in conjunction with a conventional phosphorus-containing FR and a hindered amine. By adding small amounts of minerals such as zeolites, natural clays, and zinc borates in intumescent systems, the FR performance can be drastically enhanced.

A reactive IFR, 2-({9-[(4,6-diamino-1,3,5-triazin-2-yl)amino]-3,9-dioxido-2,4,8,10-tetraoxa-3,9-diphosphaspiro [5.5]undecan-3-yl}oxy) ethyl methacrylate (EADP) (Structure 7.6), was synthesized from phosphorus oxychloride, pentaerythritol, hydroxyethyl methacrylate, and melamine. EADP exhibited excellent thermal stability and char-forming ability without affecting tensile strength of polypropylene (PP) (Li et al., 2014).

STRUCTURE 7.6 Chemical structure of EADP

Catalysis

Some divalent and multivalent metal compounds influence the flame retardancy of intumescent FRs–LOI increased by 7–9 units and grade in vertical burning tests (UL-94) improved from V-2 to V-0. Various organic and inorganic have been chosen as catalysts with conventional FRs or alone. Metal compounds convert the flammable polymer to graphite under fire conditions.

Piperazine spirocyclic phosphoramidate (PSP) (Structure 7.7), a novel halogen-free intumescent flame retardant, was synthesized 3,9-dichloro-2,4,8,10-tetraoxa-3,9-diphosphaspiro[5.5]undecane-3,9-dioxide (SPDPC) (Structure 7.3) and piperazine and used to flame-retard polypropylene independently or when combined with ammonium polyphosphate (APP) and a triazine polymer charring-foaming agent (CFA). The optimum flame retardant formulation (PSP-IFR) was PSP: APP: CFA:: 53: 6: 2 (weight ratio). The PSP-IFR-PP was tested for limiting oxygen index (LOI), vertical burning tests (UL-94), thermogravimetric analysis, and cone calorimetry. The results indicated that the PSP-IFR-PP had both excellent flame retardancy and antidripping ability. The optimum flame retardant formulation gave an LOI value of 39.8 and a UL-94 V-0 rating to PP. Moreover, both the heat release rate and the total heat release of the IFR-PP with the optimum formulation decreased significantly relative to those of pure PP, according to the cone calorimeter analyses (Li et al., 2014).

STRUCTURE 7.7   Chemical structure of PSP

In an intumescent system on polypropylene (PP) containing ammonium polyphosphate (APP, $(NH_4PO_3)_n$, n = 700) and pentaerythritol (PER) in 3:1 ratio (wt/wt), the reaction of the acidic species (APP and its degradation products, orthophosphates, and phosphoric acid) with the char forming agent (PER) takes place in the first stage at temperature less than 280°C with formation of esters mixtures. The carbonization process then takes place at about 280°C (mainly via a free radical process) (Le Bras et al., 1996b).

In a second step, the blowing agent decomposes to yield gaseous products (i.e., evolved ammonia from the decomposition of APP) which cause the char to swell (280°C ≤ T ≤ 350°C). The intumescent material then decomposes at higher temperatures and loses its foamed character at about 430°C (Bourbigot and Duquesne, 2010). The heat conductivity of the char decreases between 280°C and 430°C and the insulation of the underlying material increases.

Thermoplastic polyolefin (TPO) refers to a family of blends of isotactic polypropylene (iPP) with various polyolefin such as ethylene copolymers with propylene (EPR), butene (EBR), and others. These polyolefin form a dispersed phase that enhances the physical and mechanical properties of iPP in the solid state. Because of their light weight, low cost, good mechanical properties, and relative ease of molding into complex shapes, they have become an important class of materials for the automotive industry. Thermoplastic polyolefins as one of thermoplastic elastomers can be completely recyclable. However, TPO would burn easily once exposed to fire due to its chemical structure.

Utility of IFR in polyolefin is the most promising approach with the objectives of achieving halogen free flame retardancy, lower density, and better processability. An FR system including ammonium dihydrogen phosphate (ADP, acid source and blowing agent) and starch (char-forming agent) for TPO was studied by Wang et al. (2012a). The LOI and UL-94 test indicated that the best fire-retardant behavior (V-0 rating and LOI reach to 28.1%) was obtained at the formulation of TPO/ADP/starch (100/60/20). With the further loading of starch, flame retardant properties became relatively worse. TGA demonstrated that the presence of ADP/starch promoted the esterification and carbonization process in lower temperature ranges, while enhancing the thermal stability of IFR-TPO in high temperature ranges (Wang et al., 2011).

7.7 IFRs FOR EPOXY

IFR is one of the widely used halogen-free flame retardants in epoxy composites (Lu et al., 2002). Epoxy composites can release a large amount of smoke and toxic gases in a fire. The fire smoke contains not only carbon monoxide, but also a complex mixture of gases. The inhalation of fire smoke is the main cause of heavy casualties on fire accidents (Hu et al., 2007). In addition, the high-temperature smoke containing a high amount of heat can accelerate the spread of fire and do serious thermal damage to people. Moreover, the visibility impairing and narcotic irritating effect because of the evolution of smoke, toxic gases, and irritant compounds can prevent many people from escaping (Purser, 2000). Consequently, flame retardancy and smoke suppression properties are equally important for epoxy composites. It has been reported that iron-containing compounds, such as iron montmorillonite, iron oxide, iron orthophosphate and ferrocene, have been widely used in the flame retardant system and as smoke suppression agents (Wang et al., 2012b).

A new intumescent flame retardant epoxy (IFREP) resin-containing organic-modified montmorillonite (Fe-OMT) was successfully synthesized and studied by Chen et al. (2015a). The flame retardance of cured EP materials increases with increasing content of IFR and Fe-OMT. The results of the cone calorimeter test indicated that with the IFR and Fe-OMT content increasing, the heat release rate (HRR), total heat release (THR), and smoke production rate (SPR) decreased greatly, whereas the char residue amount increased with the increasing content of IFR and Fe-OMT. The TGA results showed that IFR and Fe-OMT decreased the initial decomposition temperatures of cured EP samples and enhanced the residue amount at high temperature.

A series of intumescent flame retardant epoxy resins (IFREPs) were prepared based on EP as a matrix resin, ammonium polyphosphate (APP) and pentaerythritol (PER) as IFRs, and ferrite yellow (goethite), α-FeO(OH) as a smoke suppressant. Goethite, α-FeO(OH), is a commonly available iron oxide mineral. It is also the most common ingredient of iron rust. It was named in 1806 for J.W. von Goethe, a German poet and philosopher with a keen interest in minerals.

Iron(III) oxide-hydroxide occurs naturally as four minerals, the polymorphs denoted by the Greek letters α, β, γ, and δ. Goethite, α-FeO(OH), has been used as a pigment since prehistoric times. The air oxidation of $Fe(OH)_2$ particles in highly alkaline media leads to the epitaxial growth of α-FeOOH, a process that should be facilitated by the similarity between the anion arrangements in both phases. By changing the supersaturation, one can modify the morphology and the size distribution of the α-FeOOH particles obtained at the end of the oxidation reaction.

The smoke suppression properties of α-FeO(OH) on IFR epoxy composites were evaluated using the cone calorimeter test (CCT) and scanning electron microscopy (SEM). The results showed that goethite can significantly reduce HRR, THR, and SPR, as well as total smoke release (TSR). There are obvious synergistic flame retardant and smoke suppression effects between goethite and IFRs in epoxy composites (Chen et al., 2015b).

Conversely, ammonium polyphosphates are relatively water-insoluble, non-melting solids with very high phosphorus contents (up to about 30%). They are also widely used as components of intumescent paints and mastics where they function as the acid catalyst (i.e., by producing phosphoric acid upon decomposition). They are used in paints with pentaerythritol (or with a derivative of pentaerythritol) as the carbonific component, and with melamine as the spumific (A chemical that decomposes to produce large quantities of gas) compound (Caze et al., 1998). In addition, the intumescent formulations typically contain resinous binders, pigments, and other fillers. These systems are highly efficient in flame-retarding hydroxylated polymers.

Ammonium sulfate and hexamethylene tetramine are used as buffers. Additions of dicyandiamide, urea, and borates have also been suggested. DAP is very effective in suppressing glowing. Alkaline and alkali earth salts of phosphorous do not have flame-retarding activities.

Flame retardancy properties and structure of the charred layer (the charry layer and the multicellular intumescent layer) of intumescent flame retardant thermoplastic polyolefin (IFR-TPO) composites were investigated by Wang et al. (2012a). The LOI and UL-94 test indicated that the best fire-retardant behavior (V-0 rating and LOI reach to 28.1%) was obtained at the formulation of TPO/ADP/starch (100/60/20). With further loading of starch, flame retardant properties became relatively worse. TGA demonstrated that the presence of ADP/starch promoted the esterification and carbonization process in lower temperature ranges, while enhancing the thermal stability of IFR-TPO in higher temperature ranges. In addition, as can be seen in SEM and OM, with combustion time prolonged, the intumescent layers obtained greater numbers of cells, and the charry layer became more compact, while the size of the carbon granules became smaller on the surface. Introduction of starch has an obvious effect on the structure of the multicellular intumescent and charry layers. With increased char-forming agent starch loading, the size of the carbon granules on the surface of the charry layer increased. At a longer combustion time of 60 s, cells of TPO/ADP/starch occurred to cross each other on the section. The charry layer of IFR-TPO-20 became harder and more compact than that of IFR-TPO-0 and IFR-TPO-50. The results were attributed to the weight ratio of ADP to starch in IFR, which was fixed as 3:1, indicating good cooperation between ADP and starch to promote a compact charry layer and to obtain the better flame retardancy properties. Furthermore, the better the charred layers produced, the better flame retardant properties they obtained.

7.8   IFRs FOR POLYURETHANE FOAM

Rigid polyurethane foam (RPUF) is highly flammable and releases toxic gases while burning. Traditional IFRs are mixtures, and are not compatible with the RPUF matrix (Kim and Nguyen, 2008). A phosphorous-nitrogen-containing

IFR, toluidine spirocyclic pentaerythritol bisphosphonate (TSPB) (Structure 7.8), exhibited better compatibility with RPUF and a less negative influence on the mechanical properties of TSPB-RPUF, forming compact and smooth char (Wu et al., 2014).

STRUCTURE 7.8   Chemical structure of TSPB

Giraud et al. (2005) developed the concept of microcapsules of ammonium phosphate embedded in PU and polyurea shells to make an intrinsic intumescent system compatible in normal PU coating for textiles.

Waterborne Polyurethane Dispersions (PUDs) are coatings and adhesives that use water as the primary solvent. With increasing federal regulation on the amount of volatile organic compounds (VOCs) and hazardous air pollutants (HAPs) that can be emitted into the atmosphere, PUDs are being used in many industrial and commercial applications.

Today, PUDs are typically stable, easy-to-use and exhibit properties similar to solvent-based systems. These systems may be formulated as ambient-cured (air-dried) or baked coatings for flexible and rigid substrates, such as flooring, fabric, leather, metal, plastics and paper. Additionally, PUDs are graphic art inks, adhesives for shoes and textiles, and can also be the coatings on many products.

Waterborne polyurethane (WPU) has good applicability, including fabric coating as against conventional solvent borne polyurethanes. Octahydro-2,7-di (N,N-dimethylamino)-1,6,3,8,2,7-dioxadiazadiphosphecine (ODDP) (Structure 7.9), with biphosphonyl in a cyclic compound, was synthesized. The ODDP reacted waterborne PU (DPWPU) has excellent flame retardancy due to the presence of ODDP. The LOI value of DPWPU is 30.6%, with UL-94, V-0 classification obtained at 15 wt % ODDP (Gu et al., 2015).

STRUCTURE 7.9   Chemical structure of ODDP

A novel diphosphaspiral structure-containing IFR, ethanolamine spirocyclic pentaerythritol bisphosphonate (EMSPB) (Structure 7.10), was synthesized and characterized. The EMSPB has favorable compatibility with the RPUF matrix and does not deteriorate the mechanical properties of the RPUF. The EMSPB can improve the flame retardancy of RPUF effectively. The RPUF system containing 25 wt % EMSPB can obtain LOI value of 27.5 and reach V-0 rating of UL-94. TGA curves indicate that the char yield of the EMSPB (25%)–RPUF is higher than that of the pure RPUF. The SEM observations of the residues of EMSPB (25%)–RPUF confirmed the formation of the compact charred layers, which could inhibit the transmission of heat and heat diffusion during contacting fire (Wu et al., 2013).

STRUCTURE 7.10   Chemical structure of EMSPB

7.9   FUTURE TRENDS

The intumescent behavior resulting from a combination of charring and foaming of the surface of burning polymers is being widely developed for fire retardance because it is characterized by a low environmental impact. However, the fire retardant effectiveness of intumescent systems is difficult to predict because the relationship between the occurrence of the intumescence process and the fire protecting properties of the resulting foamed char is not yet understood. The characterization of the char is quite complex and requires special techniques for solid state characterization (Camino and Lomakin, 2001).

Research work in intumescence is very active. New commercial molecules, as well as new concepts, have appeared. The quick overview of the mechanisms of action reveals that the formation of an expanded charred insulation layer acting as thermal shield is involved. The mechanism of action is not completely revealed, especially the role of the synergist.

There is increasing interest in the development and use of IFRs across the whole spectrum of flame retardant polymeric materials as a result of increased need to decrease the concentrations and usage of common Sb–Br formulations, coupled with the superior fire barrier and decreased toxic combustion gas properties that IFRs generally confer. In the next few years, there will be increased use of these materials in the textile and related sectors (Horrocks, 2003).

REFERENCES

Anna P., Marosi G.Y., Csantos I., Bourbigot S., Le Bras M., and Delobel R. (2002). Intumescent flame retardant systems of modified rheology, *Polymer Degradation and Stability*, 77, 243–251.

Arthur W. (1912). Making intumescent alkali silicate, U.S. 1041565.

Balabanovich A.I. and Engelmann J. (2003). Fire retardant and charring effect of poly (sulfonyldiphenylene phenylphosphonate) in poly(butylene terephthalate), *Polymer Degradation and Stability*, 79, 85–92.

Bourbigot S. and Duquesne S. (2007). Fire retardant polymers: Recent developments and opportunities, *Journal of Materials Chemistry*, 17, 2283–2300.

Bourbigot S. and Duquesne S. (2010). Chapter 6. Intumescence-based fire retardants, in: C.A. Wilkie and A.B. Morgan, Eds., *Fire Retardancy of Polymeric Materials*, 2nd Edition. CRC Press, Boca Raton, FL, pp. 129–162.

Bourbigot S., Le Bras M., Trémillon J.-M., Bréant P., and Delobel R. (1996). Zeolites: New synergistic agents for intumescent thermoplastic formulations—Criteria for the choice of the zeolite, *Fire and Materials*, 20, 145–158.

Camino G., Costa L., and Martinasso G. (1989). Intumescent fire-retardant systems, *Polymer Degradation and Stability*, 23(4), 359–376.

Camino G. and Lomakin S. (2001). Chapter 10. Intumescent materials, in: A. R. Horrocks and D. Price, Eds., *Fire Retardant Materials*. Woodhead Publishing, Cembridge, UK, pp. 318–336.

Caze C., Devaux E., Testard G., and Reix T. (1998). New intumescent systems: An answer to the fl ame retardant challenges in the textile industry, in: M. Le Bras, G. Camino, S. Bourbigot, and R. Delobel, Eds., *Fire Retardancy of Polymers: The Use of Intumescence*. Royal Society of Chemistry, London, pp. 363–375.

Chen D., Wang Y., Hu X., Wang D., Qu M., and Yang B. (2005). Flame-retardant and antidripping effects of a novel char-forming flame retardant for the treatment of poly(ethylene terephthalate) fabrics, *Polymer Degradation and Stability*, 88, 349–356.

Chen L. and Wang Y.Z. (2010). Aryl Polyphosphonates: Useful halogen-free flame retardants for polymers, *Materials*, 3, 4746–4760. DOI: 10.3390/ma3104746.

Chen X., Liu L., Jiao C., Qian Y., and Li S. (2015a). Influence of ferrite yellow on combustion and smoke suppression properties in intumescent flame-retardant epoxy composites, *High Performance Polymers*, 27(4), 412–425. DOI: 10.1177/0954008314553644.

Chen X., Liu L., Jiao C., Qian Y., and Li S. (2015b). Influence of ferrite yellow on combustion and smoke suppression properties in intumescent flame-retardant epoxy composites, *High Performance Polymers*, 27(4), 412–425. DOI: 10.1177/0954008314553644.

Davies P.J., Horrocks A.R., and Alderson A. (2005). The sensitisation of thermal decomposition of ammonium polyphosphate by selected metal ions and their potential for improved cotton fabric flame retardancy, *Polymer Degradation and Stability*, 88, 114–122.

Davies P.J., Horrocks A.R., and Miraftab M. (2000). Scanning electronmicroscopic study of wool/intumescent char formation, *Polymer International*, 49(10), 1125–1132.

Fontaine G., Bourbigot S., and Duquesne S. (2008). Neutralized fl ame retardant phosphorus agent: Facile synthesis, reaction to fi re in PP and synergy with zinc borate, *Polymer Degradation and Stability*, 93, 68–76.

Gao F., Tong L., and Fang Z. (2006). Effect of a novel phosphorous-nitrogen containing intumescent flame retardant on the fire retardancy and the thermal behaviour of poly(butylene terephthalate), *Polymer Degradation and Stability*, 91, 1295–1299.

Ghosh S. and Joshi V. (2016). Development of eco-friendly flame Retardant fabric using phosphorous based intumescences chemistry, *International Journal of Advances in Chemistry (IJAC)*, 2(1), February 2016 DOI: 10.5121/ijac.2016.210645.

Gilman J.W., Ritchie S.J., Kashiwagi T., and Lomakin S.M. (1997). Fire-retardant additives for polymeric materials—I. Char formation from silica gel–potassium carbonate, *Fire and Materials*, 21(1), 23–32.

Giraud S., Bourbigot S., Rochery M., Vroman I., Tighzert L., Delobel R., and Poutch F. (2005). Flame retarded polyurea with encapsulated ammonium phosphate for textile coating, *Polymer Degradation and Stability*, 88, 106–113.

Granzow A. (1978). Flame retardation by phosphorus-compounds, *Accounts of Chemical Research*, 11, 177–183.

Gu L., Ge Z., Huang M., and Luo Y. (2015). Halogen-free flame-retardant waterborne polyurethane with a novel cyclic structure of phosphorus–nitrogen synergistic flame retardant, *Journal of Applied Polymer, Science*, 132(41), 288.

Halpern Y., Mather M., and Niswander R.H. (1984). Fire retardancy of thermoplastic materials by intumescence, *Industrial & Engineering Chemistry Product Research and Development*, 23, 233–238.

Horrocks A.R. (2003). Chapter 6, Flame-retardant finishes and finishing, in: D. Heywood, Ed., *Textile Finishing*. Society of Dyers and Colourists, Bradford, UK, pp. 214–250.

Horrocks A.R. (2011). Flame retardant challenges for textiles and fibres: New chemistry versus innovatory solutions, *Polymer Degradation and Stability*, 96, 377–392.

Horrocks A.R., Davies P.J., Kandola B.K., and Alderson A. (2007). The potential for volatile phosphorus-containing flame retardants in textile back-coatings, *Journal of Fire Sciences*, 25, 523–540. DOI: 10.1177/0734904107083553.

Horrocks A.R. and Kandola B.K. (1998). Flame retardancy cellulosic textiles, in: M. Le Bras, G. Camino, S. Bourbigot, and R. Delobel, Eds., *Fire Retardancy of Polymers: The Use of Intumescence*. Royal Society of Chemistry, London, p. 343.

Horrocks A.R., Kandola B.K., Davies P.J., Zhang S., and Padbury S.A. (2005). Developments in flame retardant textiles: A review. Analyse des Porphyrs von Kreuznach im Nahethale, *Polymer Degradation and Stability*, 88, 3–12.

Horrocks A.R. and Zhang S. (1999). Flame retardant treatment for polymeric materials, UK Patent Application 0017592.9 (28 July).

Horrocks A.R. and Zhang S. (2001). Enhancing char formation by reaction with phosphorylated polyols I cellulose, *Polymer*, 42, 8025–8033.

Horrocks A.R. and Zhang S. (2002). Enhancing polymer flame retardancy by reaction withphosphorylated polyols.part 2. cellulose treated with a phosphonium salt urea condensate (proban cc) flame retardant, *Fire and Materials*, 26, 173–182.

Horrocks A.R. and Zhang S. (2004). Char formation in polyamides (polyamides 6 and 6.6) and wool keratin phosphorylated by polyol phosphoryl chlorides, *Textile Research Journal*, 74, 433–441.

Horrocks A.R. and Davies P.J. (2000). Char formation in flame-retarded wool fibres. Part 1. Effect of intumescent on thermogravimetric behaviour, *Fire and Materials*, 24, 151–157.

Hu L.H., Fong N.K., and Yang L.Z. (2007). Modeling fire-induced smoke spread and carbon monoxide transportation in a long channel: Fire dynamics simulator comparisons with measured data, *Journal of Hazardous Materials*, 140(1), 293–298.

Huang Y.J., Qin Y.W., Zhou Y., Niu H., Yu Z.Z., and Dong J.Y. (2010). Polypropylene/ graphene oxide nanocomposite via in-situ Ziegler-Natta polymerization, *Chemistry of Materials*, 22, 4096–4104.

Ji Y.-M., Cao -Y.-Y., Chen G.-Q., and Xing T.-L. (2017). Flame retardancy and ultraviolet resistance of silk fabric coated by graphene oxide, *Thermal Science*, 21(4), 1733–1738.

Kandola B.K. and Horrocks A.R. (1999). Complex char formation in flame retarded in fibre-intumscent combinations – III Physical and chemical nature of char, *Textile Research Journal*, 69, 374–381.

Kim J. and Nguyen C. (2008). Thermal stabilities and flame retardancies of nitrogen–phosphorus flame retardants based on bisphosphoramidates, *Polymer Degradation and Stability*, 93, 1037–1043.

Langille K., Nguyen D., and Veinot D.E. (2003). Inorganic intumescent coatings for improved fire protection of GRP, *Fire Technology*, 35, 99–110.

Le Bras M. and Bourbigot S. (1996a). Mineral fillers in intumescent fire retardant formulations—Criteria for the choice of a natural clay, filler for the ammonium polyphosphate/pentaerythritol, *Fire and Materials*, 20, 39–49.

Le Bras M., Bourbigot S., Delporte C., Siat C., and Le Tallec Y. (1996b). New intumescent formulations of fire retardant polypropylene: Discussion about the free radicals mechanism of the formation of the carbonaceous protective material during the thermo-oxidative treatment of the additives, *Fire and Materials*, 20, 191–203.

Levchik S.V., Camino G., Costa L., and Levchik G.F. (1995). Mechanism of action of phosphorus-based fl ame retardants in nylon 6. I. Ammonium polyphosphate, *Fire and Materials*, 19, 1–10.

Lewin M. (2005). Unsolved problems and unanswered questions in fl ame retardance of polymers, *Polymer Degradation and Stability*, 88, 13–19.

Li B. and Xu M. (2006). Effect of a novel charring-foaming agent on flame retardancy and thermal degradation of intumescent fl ame retardant polypropylene, *Polymer Degradation and Stability*, 91, 1380–1386.

Li G., Wang W., Cao S., Cao Y., and Wang J. (2014). Reactive, intumescent, halogen-free flame retardant for polypropylene, *Journal of Applied Polymer Science*, 1–9. DOI: 10.1002/APP.40054.

Liu Y. and Wang Q. (2006). Catalytic action of phospho-tungstic acid in the synthesis of melamine salts of pentaerythritol phosphate and their synergistic effects in flame retarded polypropylene, *Polymer Degradation and Stability*, 91, 2513–2519.

Lu S.-Y. and Hamerton I. (2002). Recent developments in the chemistry of halogen-free flame retardant polymers, *Progress in Polymer Science*, 27(8), 1661–1712. DOI: 10.1016/S0079-6700(02)00018-7.

Lyon R.E., Speitel L., Walters N., and Crowley S. (2003). Fire-resistant elastomers, *Fire and Materials*, 27, 195–208.

Masai Y., Kato Y., and Fukui N. (1973). Fireproof, thermoplastic polyester-polyaryl phosphonate composition. US Patent 3,719,727.

Peng H.Q., Zhou Q., Wang D.Y., Chen L., and Wang Y.Z. (2008). A novel charring agent containing caged bicyclic phosphate and its application in intumescent flame retardant polypropylene systems, *Journal of Industrial and Engineering Chemistry*, 14(5), 589–595.

Purser D.A. (2000). Toxic product yields and hazard assessment for fully enclosed design fires, *Polymer International*, 49(10), 1232–1255.

Salavagione H.J., Gomez M.A., and Martinez G. (2009). Polymeric modification of graphene through esterification of graphite oxide and poly(vinyl alcohol), *Macromolecules*, 42, 6331–6334.

Samyn F., Bourbigot S., Duquesne S., and Delobel R. (2007). Effect of zinc borate on the thermal degradation of ammonium polyphosphate, *Thermochim Acta*, 456, 134–144.

Subbulakshmi M.S., Kasturiya N., Hansraj N., Bajaj P., and Agarwal A.K. (2000). Production of Flame-Retardant Nylon 6 and 6.6, *Journal of Macromolecular Science, Part C*, 40(1), 85–104.

Vandersall H.L. (1971). Intumescent coating systems, their development and chemistry, *Journal of Fire and Flammability*, 2, 97–140.

Wang D.Y., Ge X.G., Wang Y.Z., Wang C., Qu M.H., and Zhou Q. (2006). A novel phosphorus-containing poly(ethylene terephthalate) nanocomposite with both flame retardancy and anti-dripping effects, *Macromolecular Materials and Engineering*, 291, 638–645.

Wang L., Yang W., and Wang B. (2012b). The impact of metal oxides on the combustion behavior of ethylene–vinyl acetate copolymers containing an intumenscent flame retardant, *Industrial & Engineering Chemistry Research*, 51, 7884–7890.

Wang X., Feng N., Chang S., Zhang G., Li H., and Lv H. (2011). Intumescent flame retardant TPO Composites: Flame retardant properties and morphology of the charred layer, *Journal of Applied Polymer Science*, 124, 2071–2079. (2012).

Wang X., Feng N., Chang S., Zhang G., Li H., and Lv H. (2012a). Intumescent flame retardant tpo composites: flame retardant properties and morphology of the charred layer, *Journal of Applied Polymer Science*, 124, 2071–2079.

Wang Y.-Z. (1997). *Flame-Retardation Design of PET Fibers.* Sichuan Science & Technology Press, Chengdu, Sichuang, China.

Wang Y.-Z., Zheng C.-Y., and Yang -K.-K. (1999). Synthesis and characterization of polysulfonyl diphenylene phenyl phosphonate, *Polymeric Materials Science and Engineering*, 15, 53–56.

Weil B.D. (1992). Phosphorus-based flame retardants, in: R. Engel, Ed., *Handbook of Organophosphorus Chemistry.* Marcel Dekker, New York, pp. 683–738.

Wu D., Zhao P., Zhang M., and Liu Y. (2013). Preparation and properties of flame retardant rigid polyurethane foam with phosphorus–nitrogen intumescent flame retardant, *High Performance Polymers*, 25(7), 868–875. DOI: 10.1177/0\954008313489997.

Wu D.-H., Zhao P.-H., Liu Y.-Q., Liu X.-Y., and Wang X.-F. (2014). Halogen free flame retardant rigid polyurethane foam with a novel phosphorus-nitrogen intumescent flame retardant, *Journal of Applied Polymer Science.* DOI: 10.1002/APP.3958101.

Xiao Y., Zheng Y., Wang X., Chen Z., and Xu Z. (2014). Preparation of a Chitosan-based flame-retardant synergist and its application in flame-retardant polypropylene, *Journal of Applied Polymer Science*, 131, 1–8.

8 Nanomaterial-based FRs

8.1 NANOMATERIALS

A composite material (also called a composition material or shortened to composite, which is the common name) is a material made from two or more constituent materials with significantly different physical or chemical properties that, when combined, produce a material with characteristics different from the individual components. The individual components remain separate and distinct within the finished structure, differentiating composites from mixtures and solid solutions.

The new material may be preferred for many reasons. Common examples include materials which are stronger, lighter, or less expensive when compared to traditional materials. Typical engineered composite materials include:

- Reinforced concrete and masonry,
- Composite wood such as plywood,
- Reinforced plastics, such as fiber-reinforced polymer or fiberglass,
- Ceramic matrix composites (composite ceramic and metal matrices),
- Metal matrix composites, and
- Other advanced composite materials.

Advanced composite materials (ACMs) are also known as advanced polymer matrix composites. These are generally characterized or determined by unusually high strength fibers with unusually high stiffness, or modulus of elasticity characteristics, compared to other materials, while bound together by weaker matrices. These are termed advanced composite materials (ACM) in comparison to the composite materials commonly in use such as reinforced concrete, or even concrete itself. The high strength fibers are also low density while occupying a large fraction of the volume.

Textile-reinforced composites are increasingly used in various industries such as aerospace, construction, automotive, medicine, and sports, due to their distinctive advantages over traditional materials such as metals and ceramics. Fiber-reinforced composite materials are lightweight, stiff, and strong. They have good fatigue and impact resistance. Their directional and overall properties can be tailored to fulfill specific needs of different end uses by changing constituent material types and fabrication parameters such as fiber volume fraction and fiber architecture. A variety of fiber architectures can be obtained by using two-dimensional (2D) and three-dimensional (3D) fabric production techniques such as weaving, knitting, braiding, stitching, and nonwoven methods. Each fiber architecture/textile form results in a specific configuration of mechanical and performance properties of the resulting composites and determines the end-use possibilities and product range.

Nanocomposites are composites in which at least one of the phases shows dimensions in the nanometer range. These are high performance materials that exhibit unusual property combinations and unique design possibilities and are thought of as the materials of the 21st century. With an estimated annual growth rate of about 25% and huge demand for engineering polymers, their potential is so promising that they are useful in several applications ranging from packaging to biomedical. A literature survey reveals that about 18,000 publications, including papers and patents, have been published on nanocomposites in the last two decades. It has been reported that at the nanoscale (below about 100 nm), a material's property can change dramatically. With only a reduction in size and no change in the substance itself, materials can exhibit new properties such as electrical conductivity, insulating behavior, elasticity, greater strength, different color, and greater reactivity characteristics that the very same substances do not exhibit at the microscale or macroscale (Anandhan and Bandyopadhyay, 2011). For example:

1. By the time gold crystals are just 4 nm across, the melting point drops to 700 K from its *encyclopedia value* of 1337 K (Mulvaney, 2001).
2. White crystals such as those of ZnO and TiO_2 are used as paint pigments or whitening agents, but they become increasingly colorless as the crystals shrink in size, and ZnO and TiO_2 colloids become invisible to the human eye below about 15 nm (Mulvaney, 2001).
3. Aluminum can spontaneously combust at the nanoscale and has been used as rocket fuel.

Nanotechnology has been a known field of research for many decades. Since *nanotechnology* was presented by Nobel laureate Richard P. Feynman during his famous 1959 lecture "There's plenty of room at the bottom" (Feynman, 1960), there have been various revolutionary developments in the field of nanotechnology. Nanotechnology produced materials of various types at nanoscale level. Nanoparticles (NPs) are wide class of materials that include particulate substances, which have at least one dimension less than 100 nm (Laurent et al., 2010). Depending on the overall shape these materials can be 0D, 1D, 2D or 3D (Tiwari et al., 2012). The importance of these materials realized when researchers found that size can influence the physiochemical properties of a substance, e.g., the optical properties. A 20-nm gold (Au), platinum (Pt), silver (Ag), and palladium (Pd) NPs have characteristic wine red color, yellowish gray, black and dark black colors, respectively.

In ISO/TS 80004 (European Commission, March 29–March 30, 2011), nanomaterial is defined as the "material with any external dimension in the nanoscale or having internal structure or surface structure in the nanoscale", with nanoscale defined as the "length range approximately from 1 nm to 100 nm". This includes both nanoobjects, which are discrete pieces of material, and nanostructured materials, which have internal or surface structure on the nanoscale; a nanomaterial may be a member of both these categories.

On October 18, 2011, the European Commission adopted the following definition of a nanomaterial:

> A natural, incidental or manufactured material containing particles, in an unbound state or as an aggregate or as an agglomerate and for 50% or more of the particles in the number size distribution, one or more external dimensions is in the size range 1 nm–100 nm. In specific cases and where warranted by concerns for the environment, health, safety or competitiveness the number size distribution threshold of 50% may be replaced by a threshold between 1% to 50%.

Nanomaterials may be obtained from engineered, incidental, or natural sources, which are described next.

8.1.1 ENGINEERED

Engineered nanomaterials have been deliberately engineered and manufactured by humans to have certain required properties (DHHS (NIOSH), 2013). They include carbon black and titanium dioxide nanoparticles.

8.1.2 INCIDENTAL

Nanomaterials may be incidentally produced as a by-product of mechanical or industrial processes. Sources of incidental nanoparticles include vehicle engine exhausts, welding fumes, and combustion processes from domestic solid fuel heating and cooking.

8.1.3 NATURAL

Biological systems often feature natural, functional nanomaterials. The structure of foraminifera (mainly chalk) and viruses (protein, capsid, or the protein shell of a virus), the wax crystals covering a lotus or nasturtium leaf, spider and spider-mite silk, the blue hue of tarantulas, the "spatulae" on the bottom of gecko feet, some butterfly wing scales, natural colloids (milk, blood), horny materials (skin, claws, beaks, feathers, horns, hair), paper, cotton, nacre, corals, and even our own bone matrix are all natural organic nanomaterials.

Meanwhile, different, novel strategies have been designed for the use of nanomaterials; in particular, three approaches have shown the most interesting results:

1. The use of nanocomposite synthetic fibers,
2. The introduction of nanoparticles in traditional back-coatings, and
3. The deposition of (nano) coatings on the fabric substrates. Up to now, the (nano) coating approach has mainly focused on the use of ceramic protective layers or flame retardant species, either alone or coupled together.

Thus, it has embraced different methods, such as nanoparticle adsorption, layer-by-layer assembly, sol–gel and dual cure processes, and plasma deposition (Alongi et al., 2014).

8.2 CARBON-BASED NPS

Fullerenes and carbon nanotubes (CNTs) represent two major classes of carbon-based NPs. Fullerenes are a form of carbon having a large spheroidal molecule consisting of a hollow cage of sixty or more atoms, of which Buckminster fullerene was the first known example. Fullerenes are produced chiefly by the action of an arc discharge between carbon electrodes in an inert atmosphere. Fullerenes contain nanomaterials that are made of globular hollow cage such as allotropic forms of carbon. They have created noteworthy commercial interest due to their electrical conductivity, high strength, structure, electron affinity, and versatility (Astefanei et al., 2015). These materials possess arranged pentagonal and hexagonal carbon units, while each carbon is sp2 hybridized.

Carbon nanotubes (CNTs) are elongated, tubular structure, 1–2 nm in diameter. These can be predicted as metallic or semiconducting, depending on their diameter telicity. These structurally resemble graphite sheet rolling upon it. The rolled sheets can be single, double or many walls and therefore they named as single-walled (SWNTs), double-walled (DWNTs) or Multiwalled carbon nanotubes (MWNTs), respectively. They are widely synthesized by deposition of carbon precursors, particularly the atomic carbons, vaporized from graphite by laser or by electric arc onto metal particles. Lately, they have been synthesized via chemical vapor deposition (CVD) technique. Due to their unique physical, chemical, and mechanical characteristics, these materials are not only used in pristine form, but also in nanocomposites for many commercial applications such as fillers, efficient gas adsorbents for environmental remediation, and as support medium for different inorganic and organic catalysts (Khan et al., 2019).

One advantage of nanoparticles (NPs), as polymer additives when compared with the traditional additives, is that their loading requirements are quite low. The dispersibility of NPs largely depends on the force between them (Mallakpour and Zadehnazari, 2013). Organic modifiers are used as dispersants to increase the repulsive force between NPs in suspension. These additives are adsorbed onto the surface of NPs, improving the interparticle electrostatic repulsion and modifying them into hydrophobic particles (Hanaor et al., 2012).

Nanoparticles (NPs) act in the condensed phase where they influence the dripping behavior of the polymer in low loadings (<10%). NPs reduce the burning temperature, prevent the supply of oxygen from the atmosphere, and promote char formation. NPs do not have an influence on the limiting oxygen index (LOI) and are only added as synergists to other flame retardants. The disadvantages of NPs create difficulties in achieving uniform dispersion. As the viscosity of the polymer decreases during heating, nanoclays migrate to the surface and create a protective layer (Šehić et al., 2016b).

Nanomaterials can be used as:

- Nanopolymer composites,
- Nanoparticles in traditional back-coatings, and
- Deposition of (nano) coatings on the fabric substrates.

Carbon nanotubes (CNTs) are allotropes of carbon with a cylindrical nanostructure. These cylindrical carbon molecules have unusual properties, which are valuable for nanotechnology, electronics, optics, and other fields of materials science and technology. Owing to the material's exceptional strength and stiffness, nanotubes have been constructed with length-to-diameter ratio of up to 132,000,000:1 (Wang et al., 2009) – significantly larger than for any other material.

Multiwalled nanotubes (MWNTs) consist of multiple rolled layers (concentric tubes) of graphene. There are two models that can be used to describe the structures of multiwalled nanotubes. In the Russian Doll model, sheets of graphite are arranged in concentric cylinders, e.g., a (0,8) single-walled nanotube (SWNT) within a larger (0,17) single-walled nanotube. In the Parchment model, a single sheet of graphite is rolled in around itself, resembling a scroll of parchment or a rolled newspaper. The interlayer distance in multiwalled nanotubes is close to the distance between graphene layers in graphite, approximately 3.4 Å.

Multiwall carbon nanotubes (MWNTs) were first tested as nanoparticle additives in flame retardant polymers in 2002 by Kashiwagi et al. (2002). Well-dispersed MWNT enhanced the thermal stability of polypropylene (PP) in nitrogen and air and greatly reduced the heat release rate and mass loss rate during combustion (Yin et al., 2015). It is proposed that the nanotube network formed a protective layer on the sample surface during burning. The protective layer works as a shield reducing the exposure of the polymer to external radiant heat flux and slowing its pyrolysis (Kashiwagi et al., 2004). MWNT increases the melt viscosity and prevent dripping and flowing of the burning material (Schartel et al., 2005).

The EU DEROCA Project (2016) studies the synergic effect of carbon nanotube (CNT) with phosphorus-based flame retardants and other new additives in intumescent or carbon crust formation systems (DEROCA Project, 2016). Multiwalled carbon nanotubes (MWCNT) act as synergistic coadditive with some FR applications. Hence, by using MWCNT the loading of FRs in products can be reduced and FR performance can be enhanced.

The main FR mechanism of CNTs during thermal decomposition is that they promote the formation of a char layer covering the polymer surface, acting as an insulating barrier. This barrier limits heat transfer and diffusion of oxygen into the material and reduces the escape of volatile, combustible degradation products to the flame (Kashiwagi, 2007). In addition CNTs can increase the strength of the material, make it more fatigue-resistant and improve its electrical properties. CNTs act synergistically to other FRs and have, for example, been suggested for application in textiles or polymer composites (US-EPA, 2013).

In a study by Xue et al. (2017), phosphorylated chitosan (PCS) was successfully loaded on the surface of multiwalled carbon nanotubes (MWCNTs) by a chemical deposition cross-inking method, affording novel flame retardant PCS-MWCNTs that showed better dispersion and efficient flame retardancy. PCS-MWCNTs led to the enhancement in the onset temperature of PET and increase in the char residue formation. The char residue amount of PCS-MWCNTs/PET with a MWCNTs (loading of 1 wt% increased from 12.62% (pure PET) to 15.46%. PCS-MWCNTs not only retained the effect of the alternating couplet C and physical barrier by MWCNTs, but also formed phosphorous–carbon compounds, improving the flame retardancy of PET. The total combustion time was shortened by 98 sec., from 388 sec. to 290 sec., indicating that PCS-MWCNTs could extinguish the fire.

Buckypaper is a macroscopic aggregate of carbon nanotubes (CNT), or *buckytubes*. It owes its name to the buckminsterfullerene, the 60 carbon fullerene (an allotrope of carbon with similar bonding that is sometimes referred to as a *Buckyball* in honor of R. Buckminster Fuller) (Yahoo!, 2008).

Hybrid buckypapers (HBP) were developed and showed potential as efficient fire- retardant materials by implementing multiple fire retardance mechanisms. The fabrication of HBP was performed using multiwalled carbon nanotubes (MWCNTs) and magnesium hydroxide (Mg(OH)) nanoparticles. The Mg(OH) nanoparticles were well dispersed throughout the CNTs network, as revealed by scanning electron microscopy and Energy Dispersive X-ray spectroscopy. Thermogravimetric analysis and differential scanning calorimetry both confirmed the decomposition of magnesium hydroxide in the HBPs and heat absorption under elevated temperatures. The initial results indicated that when used as a skin layer, the HBP has the potential to significantly improve the fire-retardant properties of epoxy carbon fiber composites (Knight et al., 2012). With any sort of nanotechnology, the potential for harm is associated with the size and shape of the particles (Betts Kellyn, 2008).

8.3 BLACK PHOSPHORUS OR PHOSPHORENE

Black phosphorus is the thermodynamically stable form of phosphorus at room temperature and pressure. It is obtained by heating white phosphorus under high pressures (12,000 atmospheres). The appearance, properties, and structure of black phosphorus are very similar to those of graphite, both being black and flaky, electrical conductor and having puckered sheets of linked atoms. Black phosphorus has an orthorhombic structure and is the least reactive allotrope, a result of its lattice of interlinked six-member rings where each atom is bonded to three other atoms (Structure 8.1). As black phosphorus is very similar to graphite, black phosphorus may form scotch-tape delamination (exfoliation) resulting in phosphorene, a graphene-like 2D material with excellent charge and thermal transport properties. Phosphorene can be viewed as a single layer of black phosphorus, much in the same way that graphene is a single layer of graphite. Phosphorene is a strong competitor to graphene. Unlike graphene, phosphorene has a direct band gap from 0.3 to 2.0 eV (Carvalho et al., 2016). Phosphorene was first isolated in 2014 by mechanical exfoliation (Liu et al., 2014).

STRUCTURE 8.1 Orthorhombic structure of black phosphorous

Monolayer or multilayer black phosphorus (BP), also named black phosphorene or phosphorene. It is a type of two-dimensional (2D) nanomaterial with physical/chemical properties due to the dimensionality effect.

Red phosphorus and some phosphorus-containing compounds have been used as flame retardants in the early past, because they can be formed into phosphoric acid by high-temperature thermal decomposition, which promotes the polymers to form into a heat-resistant carbonaceous protective layer, consequently interferes with the transport of oxygen to the burning zone. Phosphorus may also react with H or OH radicals, to reduce the energy of the flame in the gas phase. However, most of the phosphorus-containing flame retardants are required to add a large proportion in order to improve the flame retardant performance, and they have a low compatibility between flame retardant and matrix materials. It is interesting to find that phosphorene is a type of 2D nanomaterial, which has the same feature as the other nanoflame retardants, such as graphene and carbon nanotubes, leading to a high flame-resistant efficiency with a low additive amount. They can also distribute uniformly into matrix materials and qualify a good compatibility with polymers, due to the small size. Thus, phosphorene has a potentiality to be an efficient FR.

Waterborne polyurethane (WPU) dispersion is a binary colloidal system in which the particles of PU are dispersed in a continuous water phase. Generally, WPU is prepared by incorporating hydrophilic segments or ionic groups, which act as internal emulsifiers, into the polymer backbone. Waterborne polyurethanes (WPU) were specifically developed to replace solvent based PUs, used for coatings and adhesives, to prevent environmental pollution from organic volatile compounds (VOCs). Due to the more severe regulations of volatile organic compound (VOC) release and the increasing prices of solvents, WPU, which possesses prominent environmental advantages over solvent-based PU, has been becoming one of the most rapidly developing and active branches of PU chemistry and technology over the past two decades. In addition to its eco-friendliness, WPU has many superior properties which are inherent in PU materials and could be readily tailored in a wide range by varying its molecular weight, composition, and proportion of hard and soft segments. All these factors have made WPU commercially important materials with wide applications such as adhesives and coatings for a large variety of substrates (e.g., textile, metal, plastic, and wood). However, some properties of WPU still need to be further improved to meet various demands, such as water and solvent resistance, thermal stability, and mechanical strength. In this context, in order to overcome some drawbacks of WPU, numerous technologies have been developed over the past decade, such as copolymerizing or grafting of other polymers, rendering external or internal cross-linkings, simple blending or forming interpenetrating networks (IPNs), and modifying with nanoparticles.

Waterborne polyurethanes (WPU) have been widely used in the films, adhesives, paints, varnishes, and coatings, but they are inflammable, leading to a non-negligible disadvantage that threaten people's safety (Engels et al., 2013).

Xinlin et al. (1998) demonstrated a novel application of black phosphorene as a flame retardant by adding a small amount (0.2 wt %) of it into WPU polymer. Phosphorene can effectively improve the thermostability, flame-resistance, and LOI of WPU; moreover, it can significantly decrease the HRR and restrict the degradation of the WPU. Thus, black phosphorene has proved to be an excellent flame retardant, which gets distributed uniformly into the WPU polymer and also provide outstanding flame-resistant performances. Compared to the pure WPU, the limiting oxygen index (LOI) of BP-WPU increased by 2.6%, the heat flow determined by thermal analysis significantly decreased by 34.7%; moreover, the peak heat release rate (PHRR) decreased by 10.3%.

8.4   SILICA NPs FROM RICE HUSK (RH-SNP)

Rice husk is considered an agricultural waste product. The abundance, low price, and high yield of silica in rice husk encourage scientists to produce cost-effective silica nanoparticles from rice husk (RH-SNP) for various applications (Chaudhary and Jollands, 2004). Silica nanoparticles that originated from agriculture waste rice husk were prepared through one-pot thermal method. The rice husk was heated in the oven at $700°C$ for 3 h. at heating rate $20°C/min$. The obtained silica was washed several times, followed by drying at $150°C$. Afterwards, the silica particles were grinded using mortar for 1 h. and finally sieved to obtain RH-SNP. Organic borate, diborate malonate (DBM), was produced by mixing 1 mol of diethyl malonate with 2 mol of boric acid and then, refluxing at $120°C$ for 2 h. The flame retardant back coating formulations were prepared by mechanical mixing of the binder with RH-SNP and DBM.

The historical importance of archaeological textiles in museums force the scientists to find a solution to various challenges, such as high combustibility, ease of ignition, light, relative humidity, and temperature effects (Ahmed and

Ziddan, 2011). Linen fabrics are one of the textiles used extensively in historic conservation, due to their unique properties, used to as a lining layer to support the historic textiles in the restoration process.

The back coating paste was spread on the back surface of linen fabrics and then dried for 10 min. at 110°C for curing. Flame retardancy of the back-coated linen samples improved. The coated textile fabric achieved a high degree of flame retardancy textile fabrics of zero rate of burning, compared to 80.3 mm/min for blank (Attia et al., 2017).

8.5 POLYMER NANOCOMPOSITE

Polymer-layered silicates are the commonest group of nanocomposites. Although first reported by Blumstein (1961), the real exploitation of this technology started in the 1990s. Because of their nanometer size dispersions, nanocomposites exhibit superior properties in comparison with pure polymer constituents or conventionally filled polymers. The main advantages of nanocomposites are lightweight, high modulus and strength, decreased gas permeability, increased solvent resistance, and increased thermal stability. Their mechanical properties are superior to unidirectional fiber-reinforced polymers because reinforcement from the inorganic layers occurs in two, rather than in one, dimension (Lee et al., 1997).

Because of the length scale involved that minimizes scattering, nanocomposites are usually transparent (Giannelis, 1996). They also exhibit significant increases in thermal stability, as well as a self-extinguishing character. Layered silicate clays, because of their chemically stable siloxane surfaces, high surface areas, high aspect ratios, and high strengths are the most widely used for the formation of organic-inorganic nanocomposites (Ren et al., 1998). Their high aspect ratios and high strengths make them very good reinforcing elements as well. Their two particular characteristics exploited for the formation of nanocomposites are:

1. The rich intercalation chemistry used to facilitate exfoliation of silicate nanolayers into individual layers. As a result, an aspect ratio between 100–1000 nm can be obtained (compared to 10 for poorly dispersed particles).
2. Layer exfoliation maximises interfacial contact between organic and inorganic phases (Kandola and Horrocks, 1998).
3. The ability to modify finely their surface chemistries through ion exchange reactions with organic and inorganic cations.

The silicates most commonly used in nanocomposites are layered silicates (clay minerals) or phyllosilicates (rock minerals). Clay minerals are built of two structural units. One is a sheet of silica tetrahedra arranged as a hexagonal network in which the tips of the tetrahedra all point in the same direction; this is the same unit as for phyllosilicates. The other structural unit consists of two layers of closely packed oxygen or hydroxyl groups in which aluminium, iron, or magnesium atoms are embedded so that each is equidistant from six oxygen atoms or hydroxyl groups.

Polymer nanocomposites have recently attracted extensive attention in materials science because they often exhibit properties quite different from those of their counterpart polymer microcomposites the matrices of which contain the same inorganic components. The surface areas of nanofillers are drastically increased so that polymer nanocomposites show macro/micro/nanointerfaces. Nanofiller-based flame retardants show high flame-retardant efficiencies. Adding only a small amount (i.e., <5 %) of nanofiller can reduce the peak heat release rates (PHRRs) of polymers and thus reduce the speed at which flames spread throughout them. Further, the small amount of nanofiller does not reduce polymer processability and can improve the mechanical properties of polymers. However, adding only nanofiller cannot produce self-extinguishing (V-0, −1, and −2) polymers, which are required for most fire-retardant products. The nanofillers should be combined with other conventional flame retardants to give a better balance of flammability/mechanical properties.

Not all polymer and inorganic additive combinations form nanocomposites: the compatibility and interfacial properties between polymer matrix and inorganic additives significantly influence the essential characteristics of materials (Kandola, 2001). Generally, inorganic additives have poor compatibility with the polymer matrix, except for water soluble polymers. Therefore, inorganic additives must be organically-modified, using organic surfactants, to improve compatibility. The organic surfactants in the organically-modified additives play the important role of lowering the surface energy of the inorganic host, improving the wetting characteristics and miscibility with the polymer matrix (Sinha Ray and Masami, 2003).

Nanocomposites are multiphase, solid materials derived from the combination of two or more components, including a matrix (continuous phase) and a discontinuous nanodimensional phase with at least one nanosized dimension (i.e., with less than 100 nm) (Arao, 2015a).

Nanocomposites (NCs) are the complex of nanophase materials and other materials that can optimize the performance of traditional substances. Zirconium oxide (ZrO_2) is one of the NPs having flame retardant property. Zirconium oxide (ZrO_2) nanoparticles (NPs) could be an ideal building block for NCs due to their advantages, such as excellent thermal stability, chemical inertness, and high hardness. The relatively high coefficient of thermal expansion and low thermal conduction of ZrO_2 make it a suitable material for thermal barrier coating on metal components. Zirconia crystals can be organized in three different patterns (Mallakpour and Khadem, 2014):

1. Monoclinic, which is thermodynamically stable up to $1100°C$.
2. Cubic phase is found above $2370°C$.
3. Tetragonal phase exists in the range of $1100C°–2370°C$.

Nanocomposites can be classified into three categories according to the number of dimensions of the nanofillers (<100 nm) dispersed in polymers:

1. Lamellar,
2. Nanotubular, and
3. Spherical polymer nanocomposites.

8.5.1 FR Mechanism of Nanocomposites

According to Gilman (1999), the nanocomposite flame retardant mechanism is a consequence of high-performance carbonaceous–silicate char build-up on the surface during burning, which insulates the underlying material and slows down the mass loss rate of decomposition products. On burning of polymers, the residue layer forms and the silicate layers reassemble. Since there is little improvement in residue yields, once the presence of silicate is accounted for, this indicates that reduced flammability of these materials is not via retention of a large fraction of carbonaceous char in the condensed phase.

Most clay minerals are sandwiches of two structural units, tetrahedral and octahedral. The simplest type of sandwich is made of a single layer of silica tetrahedra with an aluminium octahedral layer on top: these are called 1: 1 minerals and are of the kaolinite family. The other main type of sandwich is that of the 2: 1 structure (smectite minerals), consisting of an octahedral filling between two tetrahedral layers (Kandola et al., 2001).

8.5.2 Varieties of Polymer Nanocomposites

Almost all types of polymers, such as thermoplastics, thermosets, and elastomers have been used to make polymer nanocomposites. A range of nanoreinforcements with different shapes have been used in making polymer nanocomposites. An important parameter for characterizing the effectiveness of reinforcement is the ratio of surface area (A) of reinforcement to volume of reinforcement (V) (McCrum et al., 1996). When the surface area to volume ratio A: V of a cylindrical particle of given volume, plotted against aspect ratio a=l/d (defined as the ratio of length (l) to diameter, d), the predicted optimum shape for the cylindrical reinforcement to maximize A:V is when a » 1 it is a fiber) and when a «1 it is a platelet. Therefore, it can be understood that the two main classes of nanoreinforcement are fibers (e.g., carbon nanotubes) and platelets (e.g., layered silicate clays).

Polymer-layered inorganic platelet nanocomposites are bio-inspired materials. Mother of pearl (nacre) is a bio-nanocomposite made of 95% aragonite (calcium carbonate), a brittle ceramic, and 5% flexible biopolymer (conchiolin). It is several times stronger than nylon; its toughness is almost equal to silicone. It is built like a *brick-and polymer mortar structure*, where millions of ceramic plates stacked on top of each other with each layer of plates glued together by thin layers of the biopolymer. Mixtures of brittle platelets and the thin layers of elastic biopolymers make the material strong and resilient. The *brickwork* arrangement also inhibits transverse crack propagation.

8.5.3 CARBON-BASED NANOCOMPOSITES

Carbon-based nanomaterials showing such morphologies are thus named graphene, carbon nanotubes (CNTs), and carbon black (CB), respectively. Graphene is the completely exfoliated structure of graphite (single layer). The method of producing graphene was established recently, so graphene has attracted significant research interest (Stabkovich et al., 2006). CNTs are commonly used as fillers to improve the mechanical, electrical, and flame retardancy properties of nanocomposites. Kashiwagi et al. revealed the flame retardancy mechanism of CNTs (Wen et al., 2013).

Adding (CNTs) can improve not only the mechanical properties, but also the functionalities, such as electrical, thermal, and flammable properties of composites. CNTs are one of the most typical nanomaterials used to give unique properties to polymers. Technology for the large-scale production of CNTs has recently been developed, decreasing the price of CNTs. Consequently, some CNT-based nanocomposites have started appearing. For example, Evonik Industries is producing molding PA12 CNT-containing compounds for fuel lines (Evonik-Industries, 2020). The main advantage of this material is that it can avoid ignition induced by electrostatic charges. Fire risk can be substantially reduced by producing percolation networks of CNTs in polymers. Adding CNTs to polymers also modifies their flammability.

CNT-containing nanocomposites absorb more radiation than polymers during fires; therefore, nanocomposite temperatures increase faster than those of polymers. The ignition temperatures of materials decrease because the CNTs absorb large amounts of radiation. Polymers begin to burn when they are heated to temperatures at which thermal degradation begins. The degradation products are superheated and nucleated to form bubbles. The bubbles burst at heated surfaces, evolving their contents as fuel vapor into the gas phase. There are a couple of possible mechanisms through which CNTs accumulate at material surfaces; the force of numerous rising bubbles during combustion pushes the CNTs to the material surface or the force of the polymer receding from the material surface during pyrolysis, leaving behind the CNTs. The nanoreinforced layers cause (1) thermal shielding and prevent absorption of radiation and thermal emission from the surface (2) reduction of fuel (decomposed gas) and oxygen diffusion (Arao, 2015a).

Kashiwagi et al. showed that close to 50 % of the incident flux was lost through emission from the hot nanotube surface layer and that the reminder of the flux was transferred to the nanotube-network layer and the virgin sample (Kashiwagi et al., 2005). The nanotube-network layer emits radiation from the material surface and acts as a barrier against the decomposed gas supplied from the bulk polymer and against oxygen diffusing from the air into the material, which accelerates polymer decomposition. The nanotube network layer must be smooth, crack-free, and opening-free so that it may act as an effective gas barrier (Kashiwagi et al., 2008). Surface-layer cracks deteriorate nanocomposite flame retardancy during combustion.

Uniformly dispersed nanocomposites show rheological properties similar to those of true solids. Therefore, the dispersion of carbon-based nanofillers determines the quality of the surface layer formed during combustion and thus affects the nanocomposite flame retardancy. Choosing appropriate CNTs is important (Arao, 2015a).

Other carbon-based materials, such as CB and graphene, have also recently been investigated as flame retardants. Dittrich et al. (2013) showed that graphene were the most effective carbon-based fire retardants. Interestingly, Wen et al. (2012) found new fire retardancy mechanism for CB. They showed that peroxy radicals, the chief factor affecting the thermal decomposition of polypropylene (PP), could be efficiently trapped in CB at elevated temperatures to form a gelled-ball crosslinked network. The PHRR was reduced 75% and the LOI improved from 18% to 27.6 % by combining CB (to trap the peroxy radicals) and CNT (to create the networked layer) (Wen et al., 2013). Surprisingly, adding CB and CNTs decreased the THRR; other nanocomposite systems do not show this tendency. The new fire-retardancy mechanism for CB has the potential to further improve the flame retardancy of carbon-based nanocomposites.

8.5.4 CLAY-BASED NANOCOMPOSITES

Inorganic fillers have conventionally been added to polymer matrices to enhance their mechanical strength and other properties, as well as to reduce the cost of the overall composites. Layered aluminosilicates, also popularly described as clays, are one such type of filler, which are responsible for a revolutionary change in polymer composite synthesis, as well as for transforming polymer composites into polymer nanocomposites. Aluminosilicate particles consist of stacks of 1-nm–thick aluminosilicate layers (or platelets) in which a central octahedral aluminum sheet is fused between two tetrahedral silicon sheets (Bailey, 1984). Owing to isomorphic substitutions, there is a net negative charge on the surface of the platelets that is compensated for by the adsorption of alkali or alkaline earth metal cations. Because of the presence of alkali or alkaline earth metal cations on their surfaces, the platelets are electrostatically bound to each other, causing an interlayer to form between them. The majority of the cations are present in the interlayers bound to the surfaces of the platelets, but a small number of cations are bound to the edges of the platelets. Although the use of layered aluminosilicates has been documented in some older studies that indicated their potential for substantially improving polymer properties, reports from Toyota researchers in the early 1990sattracted serious attention (Yano et al., 1993). In these studies, polyamide nanocomposites were synthesized by in situ polymerization in the presence of clay with organic modifiers (Mittal, 2011).

Nanoclays are nanoparticles of layered mineral silicates. They are classified depending on their element composition, electrical charge grade, and dimension.

When depending on elements, nanoclays are divided into many categories with the following main groups:

- Montmorillonite,
- Smectite,
- Kaolinite,
- Chlorite, and
- Illite.

Electrical divisions consist of two categories: cationic and anionic. On the basis of these categories, the surface treatment process and surfactant material should be changed. Also, clay shapes are categorized as nanolayer, nanoparticle, nanotube, and whisker. Finally, clay can be divided based on their creation pathway, of which there are three types: natural particles, incidental particles, and synthetic nanoparticles (Kumar et al., 2009).

Many kinds of clays exist. However, when making a flame retardant nanocomposite, one typically uses layered silicates that are synthesized or originate from nature. Among the clays, the cationic-layered silicate type of clay is domestically used for manufacturing composites. This clay's structure is specified as a layered crystal structure and consists of a tetrahedral structure, a silicon atom surrounded by four oxygen atoms and an octahedral sheet with metals (aluminum, iron, magnesium, and lithium) by eight oxygen atoms. The clay layer thickness is approximately 1 nm, and the lateral dimension range is dependent on the class of clay and synthesis process, from the smaller 20 nm to the larger micron scale (Ray and Okamoto, 2003).

Montmorillonite (MMT) and saponite are the most commercially and wildly used materials. The repeat unit of MMT, $(Na, Ca)_{0.33}(Al, Mg)_2(Si_4 O_{10})(OH)_2 \cdot nH_2O.$, has been measured and has a high aspect ratio in a well-dispersed nanocomposite with a high surface area of approximately 750 m^2g^{-1} (Manias et al., 2001). For these reasons, MMT has shown good flame-retardancy when added to a polymer-composited material as a nanofiller. Halloysite, a hollow tube type of clay that is rolled up to form a bilayer and has a chemistry division considered to be kaolinite, has recently attracted attention (Du et al., 2010). Halloysite nanotubes (HNTs) originate from nature as an aluminosilicate. Some clay exists in a layered state, but predominantly many clays exist as a hollow shape structure. In general, clay has an electrical interaction between clay layers, but the HNTs have no interaction between clay layers because they are rolled up and the electrical interaction cancels itself intraclay; however, some interactions exist via hydrogen bonds and van der Waals forces. Therefore, HNTs' dispersion is easier than layered silicate dispersion in a polymer nanocomposite (Liu et al., 2014). Other kinds of clay can also be used as a nanometric material in a polymer composite.

Nanoclays are clay minerals optimized for use in clay nanocomposites–multifunctional material systems with several property enhancements targeted for a particular application. Polymer-clay nanocomposites are an especially well-researched class of such materials. Nanoclays are a broad class of naturally occurring inorganic minerals, of which platelike montmorillonite is the most commonly used in materials applications. Montmorillonite consists of ~ 1 nm thick aluminosilicate layers that are surface-substituted with metal cations and stacked in ~ 10 μm-sized multilayer stacks. The stacks can be dispersed in a polymer matrix to form a polymer-clay nanocomposite. Within the nanocomposite, individual nm-thick clay layers are fully separated to form platelike nanoparticles with very high (nm × μm) aspect ratio. Even at low nanoclay loading (a few weight %), the entire nanocomposite consists of interfacial polymer, with the majority of polymer chains residing in close contact with the clay surface. This can dramatically alter the properties of a nanocomposite compared to the pure polymer. Potential benefits include increased mechanical strength, decreased gas permeability, superior flame-resistance, and even enhanced transparency when dispersed nanoclay plates suppress polymer crystallization (Morgan, 2007).

In polymer-layered silicates, composite properties are achieved at a much lower volume fraction of reinforcement in comparison with conventional fiber- or mineral-reinforced polymers. They can be processed by such techniques as extrusion and cast common to polymers, which are superior to the costly and cumbersome techniques used for conventional fiber- and mineral-reinforced composites; furthermore, they are adaptable to films, fibers and monoliths.

Polymer-layered-silicate nanocomposites (PLSNs) with silicate enhance mechanical properties, increase heat distortion temperatures, improve thermal stability, decrease gas/vapor permeability, and reduce flammability (Sinha Ray and Masami, 2003). The use of P–N flame retardants with montmorillonite (MMT) as a flame retardant synergist for flame-resistant thermoplastics such as PP, PA6, and PA66 has been reported (Gilman, 1999).

Montmorillonite clay minerals of this group make a very popular choice for nanocomposites because of their small particle size (< 2mm) and hence, easy polymer diffusion into the particles. They also possess high aspect ratios (10–2000) and high swelling capacity, which are essential for an efficient intercalation of the polymer (Kornmann et al., 1998).

Phyllosilicates include muscovite ($KAl_2(AlSi_3O_{10})(OH)_2$), talc ($Mg_3(Si_4O_{10})(OH)_4$), and mica. Chemically, montmorillonite is hydrated sodium calcium aluminium magnesium silicate hydroxide $(Na,Ca)_{0.33}(Al,Mg)_2(Si_4O_{10})(OH)_2 \cdot nH_2O$. Montmorillonites swell or expand considerably more than other clays due to water penetrating the interlayer molecular spaces and concomitant adsorption. By exchanging sodium cations for organic cations, the surface energy of MMT decreases and the interlayer spacing expands. The resulting material is called organoclay (Singla et al., 2012).

Montmorillonite is the most commonly used clay because it is naturally ever-present, can be obtained at high purity and low cost, and exhibits very rich intercalation chemistry, meaning that it can easily be organically modified. The natural clay surface is hydrophilic, so the clay easily disperses in aqueous solutions, but not in polymers. Natural clays are often modified using organic cations such as alkylammonium and alkylphosphonium cations, forming hydrophobic organomodified clays that can be readily dispersed in polymers.

Clay-based nanocomposites are usually classified into three categories because clay properties are unique:

1. Immiscible (also known as microcomposites),
2. Intercalated, and
3. Exfoliated (also known as delaminated).

Exfoliated nanocomposites are usually desired because they show improved mechanical properties (Paul and Robeson, 2008). Clay-based nanocomposites loaded with <5 % clay are already used as commercial flame retardants because of their improved mechanical properties and flame retardancy (Morgan, 2006).

The fire-retardancy mechanisms for clay- and carbon-based nanocomposites are almost identical. One fire-retardancy mechanism is the reduction in the peak heat release rates (PHRRs) due to the formation of a protective surface barrier/insulation layer consisting of clay platelets that accumulated with a small amount of carbonaceous char (Qin et al., 2005). The clay platelets accumulated because the clay remaining on the surface from polymer decomposition and clay migration was pushed by numerous rising bubbles of degradation products.

The surface quality appears to determine flame retardant efficiency. An explanation suggested by Wilkie et al. (Zhu et al., 2001) is that the paramagnetic iron in the matrix traps radicals, and thus enhances thermal stability. In fact, adding only 0.1 wt% iron-containing clay reduced the polystyrene (PS) PHRR by 60 %. This effect was not observed for carbon-based nanocomposites because most of their iron is not on the surface, and because their contact with the polymer is minimal (Kashiwagi et al., 2004).

Modifying clay surfaces is the most important parameter for improving the fire retardancies of clay-based nanocomposites. Microcomposites are obtained instead of nanocomposites when unmodified clays are incorporated into polymers. The flammabilities of the microcomposites are usually almost identical to, or sometimes worse, than those of pure polymers. Organo-modifying clays produce intercalated or exfoliated nanocomposites.

The second most important factor in improving the fire retardancy of nanocomposite is clay loading. Unlike CNT loading, increasing clay loading improves nanocomposite fire retardancy, and there is no optimal clay loading in the range <15 wt% (Liu et al., 2013). It is difficult to form a crack-free clay network layer. Therefore, the main flame retardancy mechanism is through the formation of a barrier against the heat source instead of gases. Clay-based nanocomposite flame retardancies could be further improved if polymer-clay nanocomposites could be tuned to form more stable crack-free networks during burning.

Although the heat release rate (HRR) of the polymer-nanocomposite are greatly reduced, the total amount of heat released (THRR) remains unchanged so that the nanocomposites burn slowly once they ignite, but do not self-extinguish. It is expected that combining polymer nanocomposites with conventional fire retardants can fully exploit the fire-retardancy mechanism of nanofillers; i.e., the slow burning and the mechanical reinforcement of char layers.

8.5.5   SURFACE MODIFICATION OF CLAY

Organoclay is another class of nanofillers used for flame retardant polymer nanocomposites due to its barrier effect. Montmorillonite (MMT) clay constitutes the most commonly studied layered silicate for producing clay-based nanocomposites. MMT reduces the peak heat release rate (PHRR) of polymer nanocomposites by the formation of protective and thermal-insulating layers (Kashiwagi et al., 2004). The concept of synergism between nanofillers was proposed in recent years to achieve better flame retardant performance. Higher enhancement of thermal stability, as compared to binary systems, as well as increased residual char, and lower peak heat release rate (PHRR) in different polymer systems were obtained by using the mixture of nanoclay and MWNT (Hapuarachchi and Peijs, 2010). The mechanism behind this synergism was hypothesized to be the sealing effect of MWNTs between clay platelets, creating a compact protective surface layer (Kashiwagi et al., 2008).

The flame retardant effect of organoclays has been reported previously (Zanettia et al., 2001). Acidic amine and α-olefin, produced due to Hoffman elimination of these organic compounds, accelerate the degradation of the polymer matrix. However, this becomes advantageous to the flame retardant effect of clay because the charring process is accelerated by the acidic catalytic sites of the layered silicates that are derived from Hoffman reaction of the organic alkyl ammonium cations (Zanettia et al., 2001). The char layer made of carbon is an excellent insulator and mass transport barrier slowing down the escape of the volatiles generated by decomposition. Hence, the organoclay may be used as flame retardant filler. However, the vigorous bubbling of the decomposition products produced during the combustion process results in the development of many cracks on the surface of char residue. This may limit its applications.

Isitman and Kaynak (2010) reported the combination of clay with the organic flame retardant compounds. However, the high density and requirement of high-loading content limit the use of organic flame retardants in various aspects. Subsequently, the multiwalled carbon nanotubes (MWNTs) have been found suitable as a flame retardant for polymer matrix, due to its high decomposition temperature, i.e., >500°C (Rahatkera et al., 2010). It was found that the MWNTs are capable of forming continuous but thin network structure protective layers without the formation of cracks that compromise flame-retardant effectiveness (Isitman and Kaynak, 2010). This results in a significant reduction in heat release rate (HRR) with a MWNT mass concentration as low as 0.5% (Kashiwagi et al., 2004). However, at high-loading content, MWNTs possess a tendency to bundle in the polymer matrix, which results in the formation of a discontinuous layer of char residue after combustion. This discontinuous char layer consists of fragmented islands rather than continuous network protective layer (Kashiwagi et al., 2005). Also, the low concentration of tubes may yield the similar fragmented island structure.

Therefore, the formation of the network structure protective layer during burning with cracks or openings becomes crucial for the flame retardancy of nanocomposites. Based on in-depth literature studies (Ma et al., 2007), it was found that the combination of MWNTs and organoclay provides another class of materials, with exploration as flame retardant additives (Peeterbroeck et al., 2004).

Ternary nanocomposites based on organoclay, MWNT, and PP matrix MAPP (maleic anhydride grafted PP) was prepared via the melt-blending technique. The interaction between clay and MWNT was noted from rheological assessment, since a much confined networked structure was present in ternary nanocomposites. A phenomenal increase in the thermal stability in terms of delayed decomposition temperature was noted for ternary nanocomposites, as compared to neat PP and its binary nanocomposites. The high flame retardancy of ternary nanocomposites from cone calorimeter analysis was observed in terms of much lowered heat release rate (HRR) and mass loss rate (MLR). The SEM micrographs of char residue revealed uniformly formed dense char layers in case of ternary nanocomposites. However, the char layer of PP/MAPP/organoclay, cloisite 15A (C15A) was found to be thicker, accompanied by large cracks and pores, and continuous but thin char layers were evident in the case of PP/MAPP/MWNT (Pandey et al., 2014).

Hu et al. (2003) reported the synergistic effect of nanoparticles with conventional flame retardants using a PA6/clay composite. Experimental results showed that the addition of conventional flame retardants to clay-polymer nanocomposites lead to a more satisfactory performance in both cone calorimeter and UL 94 tests (Hu et al., 2003). More evident synergism between nanoclay and intumescent flame retardants (IFRs) was also reported more recently by Chen et al. (2009). Carbon nanotubes as nanofillers enhanced flame retardancy of IFRs, by lowering PHRR and improving char strength.

When nanoclay is used in a composite as a flame retardant nanometric additive, dispersion is a very important factor to accomplish the flame retardant property. Layered silicates inevitably have electrical interactions; their chemical structure creates an imbalance in charge, so a treated silicate layer appears as a stacked clay colony in a polymer matrix. This stacked clay cluster reduces the specific surface area and consequently impoverishes the flame retardant property of the polymer composite (Alexandre and Dubois, 2000).

The clay dispersion state can be classified in three states:

- Phase separated,
- Intercalated, and
- Exfoliated.

Exfoliated clay is the ideal state to use as a polymer nanocomposite numerical additive. Their specific surface area increases following the silicate layer intercalation and exfoliation. For the interaction between the silicate and polymer, the specific surface area is an important factor influencing the polymer composite's mechanical and flammability properties (Kim et al., 2020).

Pristine layered cationic clay interlayers contain charged Na+ or K+ ions (Brindley and Brown, 1984). In the clay organomodification process, an organic surfactant is commonly used. Hence, the interlayer ions' replacement surfactant incises the clay d-spacing, and the d-spacing shows differences, depending on the replacement level and surfactant tail length (Hackett et al., 1998). Such an organic surfactant modified layered clay has a positive effect on dispersity in a composite. However, organic surfactants have a single carbon chain, and a single conjugation linkage inevitability has a low thermal stability. Therefore, when a nanocomposite is exposed to surrounding heat and consumption, the surfactant consequentially has a bad influence on the composite's thermal stability and flammability (Kim et al., 2020).

The use of polymer-clay nanocomposites for flame retardant applications is becoming more common, especially as it is realized that the clay nanocomposite can replace part of the flame retardant package while maintaining fire safety ratings at a lower flame retardant loading. This results in a better balance of properties for the nanocomposite material compared to the non-nanocomposite flame retardant product, and in some cases, better cost for the flame retardant resin, especially if the organoclay is cheaper than the flame retardant that it is replacing. It should be noted that the organoclay can replace traditional flame retardants on more than a 1:1 by weight basis, meaning 1 g of organoclay can replace more than 1 g of traditional flame retardant, resulting in a lighter material.

In fact, it appears that clay nanocomposite systems serve as a nearly universal synergist for flame retardant additives, with some exceptions. The synergistic enhancements of clay nanocomposites for fire safety applications has led to two commercial products: a Wire & Cable jacket material: (organoclay + aluminum hydroxide) produced by Kabelwerk Eupen AG, and a series of polypropylene + organoclay + flame retardant systems (Maxxam FR) produced by PolyOne. It is likely that other commercial materials will be released soon as more manufacturers begin to see the value of these nanocomposite systems (Morgan, 2007).

Polyamide 6 (PA6) was melt-blended with an intumescent flame retardant (FR), Multiwall carbon nanotubes (MWNTs), and nanoclay particles to produce multicomponent FR-PA6 nanocomposites. Various nanocomposites were made from FR-PA6 nanofibers. Both the bulk-form nanocomposites and the electrospun nanofiber membranes exhibited significantly improved combustion properties, including both heat release rate and total heat release. The thermal stability was not changed significantly.

Thermoplastic poly(ester ether) elastomers (TPEEs) are easily ignited and rapidly burned. TPEE nanocomposites with PNFRs and organic montmorillonite (o-MMT) were prepared by melt-blending. A significant fire-retardant effect was induced in TPEE that renders a V-0 classification in the UL 94 test (Zhong et al., 2014).

Organically treated layered silicates (clays), carbon nanotubes/nanofibers, or other submicron particles at low loading (1–10 wt%) are used for polymer nanocomposites. The polymer nanocomposites greatly lower the base flammability of a material. Recent research has suggested that combining nanoparticles with traditional fire retardants (e.g., intumescents) or with surface treatment (e.g., plasma treatment) effectively decreases flammability (Bourbigot et al., 2006).

When polymer nanocomposite is used, very little additive is needed (no significant cost increase) and polymer dripping/flow while burning is reduced. The composite may have multifunctional performance (e.g., electrical conductivity from carbon nanotubes) and a balance between flammability and mechanical properties may be maintained. However, careful design and analysis are required to set up a polymer nanocomposite structure. Many aspects of nanocomposite technology, (e.g., long-term aging, environmental hazards, conservation of fire safety principles etc.) are not still known.

Sometimes layered silicate–nanocomposites are used in combination with other flame retardants as a means of improving the mechanical properties of polymers as recently reviewed by Gilman et al (2000). The General Electric Company has used this approach for polybutylene terephthalate (PBT, Valox 315) (Takekoshi et al., 1998). The treated MMT (2%, dimethyl di (hydrogenated tallow) ammonium montmorillonite) in combination with poly(tetrafluoroethylene) (PTFE) dispersed on a styrene–acrylonitrile copolymer (50% PTFE) is used to replace 40% of the brominated polycarbonate–Sb_2O_3 flame retardant in PBT.

8.5.6   Advantages of Polymer-layered Silicate Nanocomposites

Polymer-layered silicate nanocomposites are environmentally friendly alternatives to some traditional flame retardants. Some of the advantages are:

- Not only do these nanocomposites provide a promising means of producing flame-retarding polymers, but they do not have the usual drawbacks associated with other additives.
- Relatively low concentrations of silicates are necessary, compared with the amounts used for conventional additive flame retardants in order to achieve similar or even superior levels of flame retardancy.
- Polymer nanocomposites can be processed with normal techniques used for polymers, such as extrusion, injection molding, and casting.
- Furthermore, the physical properties of the polymer are not degraded but are greatly improved. An additional advantage is that during combustion, the silicates remain intact at very high temperatures and act as insulating layers against the heat. They slow down the release of volatile decomposition products from the polymer, and thus impart a self-extinguishing character to it.

Clearly, nanocomposites offer novel means of developing flame retardant polymeric materials and increased research in this area continues to signify their potential (Kandola et al., 2001).

8.5.7 POLYHEDRAL OLIGOMERIC SILSESQUIOXANE (POSS)

Silsesquioxanes are inorganic-organic hybrid materials that combine the mechanical, thermal, and chemical stability of ceramics with the solution processing and flexibility of traditional soft materials.

Polyhedral oligomeric silsesquioxane (POSS), with a general structure of silicon-oxygen cage surrounding by organic R groups, (Structure 8.2), has been recently developed as a promising flame retardant material forming thermally stable silica layers under degradation that reaches to the surface acting as a protective layer.

STRUCTURE 8.2   Chemical structure of silsesquioxane

Polyhedral oligomeric silsesquioxane (POSS) molecules are typically stable to over 400°C, which is higher than the thermal degradation temperatures of most organic molecules (Phillips et al., 2004). The use of amine-functionalized POSS or amine-POSS enabled an ionic exchange reaction and the surfactant's incorporation into the clay interlayer spacing (Kuo and Chang, 2001). The resulting nanocomposites exhibited an enhanced thermal stability up to 300°C because of the presence of the silicate clay. The intercalation of amine-POSS salts into clay galleries was evidenced by an increase in the interlayer spacing from 1.26 to 1.61 nm. The polystyrene/clay nanocomposites, which were produced by incorporating the amine-POSS–clay hybrid, exhibited exfoliation of the clay platelets and enhanced thermal stability compared with pristine polystyrene (Chiu et al., 2014).

8.5.8 NANOCOMPOSITE PREPARATION

The polymer nanocomposite preparation methods are classified as three dispersion techniques, namely:

1. The solvent method,
2. In situ polymerization, and
3. Melt compounding.

8.5.9   SOLVENT METHOD

The polymer and nanometric additives are added to water and an organic solvent, and they are mechanically mixed by stirring or ultrasonification. The final step of the preparation method is that the solvent is totally removed from composite by using an oven or vacuum situation oven to precipitate vaporization. The solvents that are mainly used are xylene, toluene, benzene, ethylene acetate, benzene, and other organic solvents, and can be used to dilute an olefin polymer. Sometimes, water can be used to dilute water soluble polymers, for example, polyethylene glycol (PEG), polyvinyl alcohol (PVA), and polyethylene oxide (PEO). The solvent method is easy to adapt for many varieties of polymer nanocomposites because of the materials' solubilities. Nevertheless, as a strong point, it is not too easy to adopt on an industrial scale, because this method needs much time to process, a large amount of solvent, and much energy to evaporate the solvent; therefore, it affects the environment and is economically prohibitive (Ray and Okamoto, 2003).

8.5.10   IN SITU POLYMERIZATION

In situ polymerization is a very effective method for nanomaterial dispersion. Well-known research involving a polymer–nanoclay composite on PA6 by using in situ polymerization was performed by the Toyota Research Group (Kojima et al., 1993). After more than two decades, much research has been attempted. Normally, the in situ polymerization follows two steps. In the first step, the nanometric particles are added to the monomer. These nanometric composites show good flammability, compared with novel polymer resin. In addition, they can have synergetic effect on flammability when they are used together in a polymer matrix as additives. Sepiolite nanoclay (Sep) has a unique morphology in the class of nanoclays, and CNT creates a network, forming a tight char. Clay and CNT both create char layers.

When they are used together, the clay and CNT hybrid composite make a much higher-density network form, when compared with other CNT/PP or clay/PP composites, Cone calorimeter in situ polymerization has strong points regarding nanodispersion with many kinds of polymer and nanometric additives; however, this process has some drawbacks on the industrial scale. First, the materials need a long curing time of over 24 h. Second, sometimes the nanometric additives reaggregate during the subsequent processing step; therefore, they are not always thermodynamically stable. Finally, this process is suitable for a resin manufacturer who adopts the process on the production line (Sepehr et al., 2005). Novel poly(dimethylsiloxane)-based polyurethane nanocomposites (TPU-NCs) were synthesized using in situ polymerization with the nanoclay, Cloisite 30B. Cloisite 30B is an additive for plastics to improve various plastic physical properties, such as reinforcement, HDT, CLTE, and barrier. It is a quaternary, ammonium salt–modified natural montmorillonite polymer additive. This information is provided by Southern Clay Products, Inc. (BYK Additives, Inc). Houston, Texas.

Differential scanning calorimetry, thermogravimetric analysis, and dynamic mechanical thermal analysis showed that TPU-NCs with an organoclay content of ≤5 wt % exhibited increased thermal stability, storage modulus, and hard-segment melt temperatures, but decreased degrees of crystallinity. TPU-NCs displayed increased surface hydrophilicity and enhanced surface free energy with increasing organoclay content. Small- and wide-angle X-ray scattering confirmed intercalated formations of organoclays in the nanocomposites. Individual clay particles on the surfaces of TPUs with lower organoclay loadings (1 or 3 wt %), or organoclay agglomerates in TPUs with higher amounts of organoclay (≥5 wt %), were detectable using scanning electron microscopy. The relatively smooth and homogeneous character of pure TPUs, and the distinctly heterogeneous and rough surfaces of TPU-NCs were detected via atomic force microscopy. Among the nanomaterials prepared, TPU-NCs with 1 wt % organoclay provided the best balance between the organoclay concentration and the functional properties desired in biomedical applications (Pergal et al., 2017).

8.5.11 MELT COMPOUNDING

Melt blending is a well-known and widely used nanometric additive and polymer compounding process. Melted polymer resin is mixed with nanometric particles in a twin extruder or internal mixer, with forced shear stress and thermodynamic kinetics. Compared with the solvent method and in situ polymerization, melt compounding has the advantages of needing little time and being more economical when adopted to an industrial production line, and the formulation can be easily changed. This process does not need organic solvents, because the process is more eco-friendly compared with the previous two methods. Melt compounding is largely divided into two groups:

- Dynamic compounding and
- Static compounding.

Typically, static compounding uses an internal mixer, and dynamic compounding uses a twin extruder. For convenience and efficiency, the dynamic compounding method is widely used in industry and laboratories. During the melt compounding process, nanometric materials and polymer resin are subjected to shear forces; thus, a nanometric bundle is resolved, and they mix together. However, just using melt compounding is insufficient for achieving high dispersity and good clay intercalation, because the interaction between the matrix and the novel nanometric particles is weaker than that between the particles. To overcome these drawbacks, a coupling agent can be used, usually a malic anhydride-grafted polymer (Liang et al., 2004). Otherwise, surface-modified nanometric particles can be used, alone or with a coupling agent, for good dispersity in a polymer matrix (Ray and Okamoto, 2003).

Graphene has a similar morphology to nanoclay and shows similar results regarding thermal stability, when used for nanometrics in a polymer matrix, and many papers show an increased thermal stability of the graphene polymer composite (Li and Zhong, 2011).

Exfoliation involves the removal of the oldest dead skin cells on the skin's outermost surface. Exfoliation is involved in all facials, during microdermabrasion or chemical peels. Exfoliation can be achieved by mechanical or chemical means. This process involves physically scrubbing the skin with an abrasive.

Exfoliated graphite (EG) refers to graphite that has a degree of separation of a substantial portion of the carbon layers in the graphite. Graphite nanoplatelet (GNP) is commonly prepared by mechanical agitation of EG. The EG exhibits clinginess, due to its cellular structure, but GNP does not. The clinginess allows the formation of EG compacts and flexible graphite sheet without a binder. The exfoliation typically involves intercalation, followed by heating. Upon heating, the intercalate vaporizes and/or decomposes into smaller molecules, thus causing expansion and cell formation. The sliding of the carbon layers relative to one another enables the cell wall to stretch. The exfoliation process is accompanied by intercalate desorption, so that only a small portion of the intercalate remains after exfoliation. The most widely used intercalate is sulfuric acid. The higher concentration of residue in unwashed EG causes the relative dielectric constant (50 Hz) of the EG to be 360 (higher than 120 for KOH-activated GNP), compared to the value of 38 for water-washed EG. An EG compact is obtained by the compression of EG at a pressure lower than that used for the fabrication of flexible graphite. Compared to flexible graphite, EG compacts are mechanically weak, but they exhibit viscous character, out-of-plane electrical/thermal conductivity and liquid permeability. The viscous character (flexural loss tangent up to 35 for the solid part of the compact) stems from the sliding of the carbon layers relative to one another, with the ease of the sliding enhanced by the exfoliation process.

Nevertheless, graphene should be exfoliated to show a sufficient thermal property. When exfoliated graphite (EG) is compared with natural graphite (NG) in a polylactide matrix as a nanometric additive, the thermal stability shows a difference in the TGA curve. In the case of the untreated exfoliated graphite graphene (natural graphene polymer composite), the TGA curve is undistinguishable. Furthermore, 0.5 wt% of NG composite TGA curve is lower than that of the neat PLA resin. Only the residue of the composite at temperatures over 400°C amount wt% was increased proportionally with the initial NG feed content in the composite. EG and NG composites were compared with respect to their thermal stabilities, thermal degradation temperature at 5%, and 50% weight-loss point (T5% and T50%), evaluated by the TGA method. T5% and T50% of polylactide homopolymer were determined to be 350°C and 386°C, respectively. For instance, the T5% of the PLA/EG nanocomposite with 3.0 wt % EG was ~364°C, which is ~259°C higher than that of PLA homopolymer. This improved thermal stability of PLA/EG nanocomposites is believed to originate from the fact that graphite nanoplatelets of EG.

Polyamide 6 (PA6) was melt-blended with an intumescent flame retardant (FR), multiwall carbon nanotubes (MWNTs), and nanoclay particles to produce multicomponent FR-PA6 nanocomposites. FR-PA6 nanofibers were processed from varied nanocomposite formulations via electrospinning. Electrospinnability, morphology, and combustion and thermal properties of the nanofibers were investigated. Both the bulk-form nanocomposites and the electrospun nanofiber membranes exhibited significantly improved combustion properties, including both heat release rate and total heat release. On the other hand, thermal stability appeared compromised. With proper FR additive concentrations, synergism between MWNTs and nanoclay was observed (Yin et al., 2015).

Ambuken et al. (2012) studied the high-temperature flammability and mechanical properties of thermoplastic polyurethane (TPU) nanocomposites. From the UL 94 studies, it was found that high loading of nanoparticles is needed for all the formulations to have a stable char structure. From the cone calorimeter studies on TPU nanocomposites, the lowest PHRR was observed for TPU/Cloisite 30B formulation, while the highest PHRR was shown by TPU/CNF nanocomposites. The char obtained after firing TPU/Cloisite 30B nanocomposites was swollen and it retained its structural integrity, while TPU/MWNT and TPU/CNF had cracks.

DMA testing was performed on the same materials to study complex modulus as a function of temperature up to 300°C where the char was anticipated to have formed. TPU/MWNT displayed better char modulus, compared to TPU/Cloisite 30B. TPU/CNF tests could not be completed above 220°C because the sample turned into a viscous melt. All samples began to lose modulus (soften) in the range of 120°C–160°C. For TPU/MWNT and TPU/Cloisite 30B nanocomposite recovery in modulus (reinforced char formation) was observed at about 230°C with loadings above 7.5 wt%. The temperature range at which modulus had decreased but char formation was not yet significant enough to exemplify modulus recovery was termed the *reinforcement gap*. A correlation can be drawn between cross-over of dissipative versus elastic behavior (dominance of loss modulus versus storage modulus) and dripping in UL 94 tests.

8.6 NANOCOATING

Nowadays, significant scientific efforts focus on surface modifications as after-treatments capable of changing or conferring different properties to the investigated textiles. Since such aftertreatments should not almost affect the textile substrate properties, they have to involve surface modifications and the formation of micro- to nanosized coatings. As a consequence, the possibility of creating novel coatings that would be suitable to be applied to any type of fiber has to be considered. These novel coatings, which may exhibit a complete inorganic or hybrid organic–inorganic composition, can be generated by using different approaches (Alongi et al., 2013):

1. Nanoparticle adsorption,
2. Layer-by-layer assembly,
3. Sol–gel and dual-cure processes,
4. Plasma treatments, and
5. Biomacromolecule deposition.

Solution casting, which is a method of producing nanocomposites with high filler contents, has recently been developed. For instance, solution-casting can be used to produce 100 % clay paper. Paper produced from CNTs is called *CNT Bucky paper*. Such high-nanofiller-loaded materials may be applied as coatings to reduce substrate flammability. Solution-cast nanocomposites are different from bulk nanocomposites because the fillers do not migrate within the composites and because strong nanofiller networks can always be formed in nanocoatings. Preparing nanodispersed slurry is the most important step in nanocoating because nanocoating highly depends on slurry dispersion. The nanocoating can then be dipped, sprayed, or deposited layer by layer onto the substrate. Solution-casting or vacuum filtration can then be used to fabricate the nanofiller barrier film. Platelet materials such as clay or graphene are preferable for improving gas barrier properties because they force permeating molecules to travel extended paths referred to as tortuous pathways. The gas permeability of the thin coating layer composed of aligned platelets is some order of magnitude lower than that of the virgin material, depending on the aspect ratio and filler content. In fact, numerous researchers have used clay (Gusev, 2001) and graphene (Dikin et al., 2007) to produce nanodispersed films showing extraordinary mechanical and gas-barrier properties.

Two important fire-retardancy mechanisms involve reducing fuel-gas diffusion to the fire source and reducing oxygen diffusion inside materials. Therefore, multifunctional nanocoatings are expected to contribute to flame retardancy.

The PHRR was reduced by 62 % when only a 5-μm-thick nanolayer was coated onto the PA-6.

A >100-μm-thick IFR coating must be coated onto the PP to reduce the PHRR to the same level. The nanocoated polymer obviously exhibited highly efficient flame retardancy. Although the fire-retardancy mechanisms for the nanocoating and IFR coating are completely different, both act at the condensed phase, reducing the HRR without changing the THR. IFR swells during combustion and acts as insulation, reducing the amount of heat transported into the substrate. The nanocoating, on the other hand, does not swell like IFR. Therefore, the main nanocoating fire-retardancy mechanism reduces the amounts of fuel gas and oxygen transported to the fire and into the material, respectively. The clay-platelet shape reduces the gas and oxygen diffusion rates by several orders of magnitude owing to the tortuous effect. The nanocomposite layer cannot retard the fire if it fractures during combustion, and it has been concluded that the most important factor for producing flame-retardant nanocoatings is to form a uniform nanodispersed layer that does not fail during combustion.

Choosing the proper clay and polymer–clay (PC) binder and producing a sufficiently thick coating are all important factors in obtaining high-strength char during combustion. Laufer et al. (2012) used LBL deposition to develop a completely green coating composed of clay and chitosan. They applied the coating to a polylactic acid film and polyurethane (PU) foam to improve the oxygen barrier and fire retardancy. Notably, a <100-nm-thick clay-chitosan nanocoating reduced the oxygen permeability of a 0.5-mm-thick PLA film by four orders of magnitude. In addition, the 30-nm-thick green nanocoating reduced the PHRR by 52% and eliminated the secondary peak in the HRR curve. They demonstrated that environmentally benign nanocoatings can prove beneficial for application to new types of food packaging or for replacing environmentally persistent antiflammable compounds. In addition to clay, carbon-based nanomaterials such as CNT or graphene are candidate nanocoatings to improve substrate flame retardancies. Wu et al. (2010) tried using single-walled carbon nanotubes (SWNTs) and multiwalled carbon nanotubes (MWNTs) as ~20 μm thick buckypapers to protect carbon-fiber-reinforced epoxy composites from fire. Although the MWNT buckypaper was an effective flame retardant shield, the SWNT one was not. The MWNT buckypaper reduced the substrate PHRR by ~50 %; the SWNT one, on the other hand, reduced it by <10 %. The SWNT Bucky paper's effective air diffusivities were superior to the MWNT buckypaper ones, owing to the dense SWNT network. However, unlike the MWNT buckypaper, the SWNT buckypaper could not improve flame retardancy because SWNTs are less thermally stable than MWNTs. In fact, Wu et al. (2010) showed that the SWNT buckypaper had burned away during combustion, leaving only a red iron-catalyst residue. The MWNT buckypaper showed high thermal stability and had survived combustion. Liu et al. (2008) coated cotton fibers with CNTs and improved the fiber mechanical properties and flame retardancy. They used simple dip coating to fabricate CNT-network armors on the fiber surface and found that the CNT-coated cotton textiles exhibited enhanced mechanical properties and extraordinary flame retardancy because the CNTs had reinforced and protected the fibers.

Nanocoating technology drastically improves substrate flame retardancy. A thin nanomaterial coating can improve not only substrate fire retardancy, but also permeability, of a 0.5-mm-thick PLA film by four orders of magnitude. In addition, the 30-nm-thick green nanocoating reduced the PHRR by 52% and eliminated the secondary peak in the HRR curve. Arao demonstrated that environmentally benign nanocoatings can prove beneficial for application to new types of food packaging, or for replacing environmentally persistent antiflammable compounds (Arao, 2015b).

Although much research has been done concerning nanocomposite preparation, the homogeneous dispersion of nanofillers in a polymeric matrix is still a difficult task. The main drawbacks resulting from filler incorporation in a polymeric matrix are particle agglomeration and void formation at the interface of the particles and polymer matrix.

The sol–gel process is a wet-chemical technique used for the fabrication of both glassy and ceramic materials. In this process, the sol (or solution) evolves gradually towards the formation of a gel-like network containing both a liquid phase and a solid phase. In this chemical procedure, a *sol* (a colloidal solution) is formed that then gradually evolves towards the formation of a gel-like diphasic system containing both a liquid phase and solid phase, the morphologies of which range from discrete particles to continuous polymer networks. In the case of the colloid, the volume fraction of particles (or particle density) may be so low that a significant amount of fluid may need to be removed initially for the gel-like properties to be recognized. This can be accomplished in any number of ways. The simplest method is to allow time for sedimentation to occur, and then pour off the remaining liquid. Centrifugation can also be used to accelerate the process of phase separation. Removal of the remaining liquid (solvent) phase requires a drying process, which is typically accompanied by a significant amount of shrinkage and densification. The rate at which the solvent can be removed is ultimately determined by the distribution of porosity in the gel. Clearly, the ultimate microstructure of the final component is strongly influenced by changes imposed upon the structural template during this phase of processing. The sol–gel method is based on a hydrolysis–condensation reaction of a metal alkoxide, which allows the synthesis of nanoparticles that are well dispersed in the polymer matrix. Metal alkoxides are the precursors most widely used (Turova et al., 2002), because they can react rapidly with water to form hydroxyl compounds which, in turn, allow fast reactions.

Using the sol–gel method to synthesize hybrid nanocomposites, two different types of materials can be obtained, one containing only weak bonds between the two phases, without motion restriction of the matrix molecules, and the other containing covalent bonds between the polymer and metal that were formed during the hydrolysis-condensation reaction of the metal alkoxide. In the latter, the dispersion of the inorganic nanoparticles increases due to network formation. Therefore the molecular motion of the polymer is restricted and the material properties are improved (Bounor-Legar′e et al., 2004). Preparation of nanocomposites of ethylene-vinyl acetate (EVA) with the addition of inorganic compounds such as clays or metal hydroxides is a common practice that is well described in the literature (Bounor-Legar′e et al., 2002).

For applications such as wire and cables, fillers are frequently added. The EVA containing aluminium nanoparticles was produced in the melt, in a batch mixer, equipped with two rotors running in a counterrotating way. The rotor speed was 50 rpm and the set temperature was 90°C (melt temperature around 100°C).

Ethylene-vinyl acetate (EVA) nanocomposites with enhanced flame retardance were prepared by the sol–gel process in the melt. Two EVAs with different vinyl acetate (VA) contents and aluminium isopropoxide were used as organic and inorganic phases. Aluminium isopropoxide presented low activation energy, which allows the synthesis of the nanoparticles without a poststep treatment. Nanocomposites with smaller and well-dispersed metal nanoparticles were produced with an EVA of higher VA content. EVA nanocomposites achieve the requirements for 94 V-0 classification (Oliveira and Machado, 2013).

Several nanocomposites based on inorganic additives, such as metal oxides and hydroxides, nanomaterials (nanoclays (LS) and nanotubes) and phosphorous-based intumescent flame retardants, as well as their combinations, were used as FR polylactic acid (PLA). These FRs acted predominantly in the condensed phase by forming intumescent inorganic layers or char that reduced heat and fuel transfer.

Natural fiber-reinforced polymer biocomposites have received increasing attention in the development of new materials with improved mechanical and fire-retarded properties (Kandola and Toqueer-Ul-Haq, 2012). Among the different natural fibers, kenaf fibers are used in commercial applications because of their advantages, which include availability, short harvesting time, and good mechanical properties (Saba et al., 2017).

Besides many industrial and medical applications, there are certain toxicities that are associated with NPs and other nanomaterials, and basic knowledge is required for these toxic effects to encounter them properly. NPs surreptitiously enter the environment through water, soil, and air during various human activities (Khan et al., 2019).

8.7 TOXICITY

In spite of the wide applications of NPs, they may present a variety of hazards for environmental and human health. Because of the higher surface area of NPs, there is a correlation between a decrease in their size and an increase in toxicity. Hazards associated with NPs exposure can have an impact on the structure and function of the ecosystem. They can provide a surface that may easily bind and transport toxic pollutants. NPs can penetrate into the body and travel freely in the blood throughout the body, reaching such organs as the brain, liver, and lungs. They can cause generous changes inside the cells, from reacting with proteins to attaching to DNA. Some NPs show catalytic properties to generate highly reactive forms of oxygen that can cause tissue injury (Khan, 2013).

8.8 FUTURE TRENDS

Occasionally, nanometric additives have priority for dispersing in polymer composites to achieve high flame resistance. In the case of clay, it should be modified on the surface to make clay intercalate in an exfoliated state. To solve these problems, the nanometric additive's surface is grafted by polymer consisting of an intumescent agent system element. The composites that contain grafted nanometric additives show advanced flame retardancy. Using different kinds of nanometric additives together in the composite result in synergetic flame retardancy originating from the morphological properties. Additionally, reactions between the intumescent agent and the nanosilicate surface result in the formation of carbonic char, and this reaction is the reason for the synergetic effect on flammability. Finally, flammability has many complicated causes, including chemical properties and physical properties originating from morphology, and when additives are used together, they can have synergetic outcomes.

The polymer-nanocomposite provides safe, eco-friendly flame retardancy in order to replace halogenated flame retardants. Nanocomposites should be combined with proper flame retardants to achieve the same flame retardancies that halogenated compounds show. The main nanofiller flame retardancy mechanism involves nanofillers acting as barriers against gas flow and oxygen diffusion at the condensed phase. Therefore, producing strong, dense, crack-free nanocoating surface layers during combustion is the key factor in producing effective polymer nanocomposite flame retardants. Nanofiller dispersion and distribution are important factors contributing to flame retardant synergistic effects, which are unobtainable when nanofillers are incorporated into other flame retardants, because poor nanofiller distribution scatters surface layers render them unable to act as barrier layers. The technology used to produce reliable nanocomposites requires great care and skill because nanofillers are relatively new, the technology is not yet completely understood, and polymer nanocomposite structures are unique (Arao, 2015a).

Among many highly hyped technological products, polymer nanocomposites have lived up to expectation. Polymer nanocomposites exhibit superior properties, such as mechanical, barrier, optical, compared to micro- or macrocomposites. Owing to this, polymer nanocomposites have shown ubiquitous presence in various fields of application. Polymer nanocomposites for various applications could be synthesized by proper selection of matrix, nanoreinforcement, synthesis method, and surface modification of either the reinforcement or polymer (if required). Many products based on polymer nanocomposites have been commercialized (Anandhan and Bandyopadhyay, 2011).

In the future, nanodimensional materials will be a part of commercial fire-retardant additives, whereas in the past, fire retardants were usually a single compound (perhaps in combination with a synergist as with Sb_2O_3 and bromine compounds). The enhanced mechanical properties that arise when MMT is used, for instance, may be of value to offset the reduction in mechanical properties due to some additives.

In order to aid in the development of these products, researchers must identify the processes by which nanodimensional materials can effectively reduce the PHRR so that one can combine mechanisms to achieve fire retardancy. In the case of layered double hydroxide (LDHs), one must determine how to obtain a well-dispersed LDH in a polymer to determine whether dispersion plays any role in the fire retardancy of these materials. The most commonly investigated nanodimensional material is MMT and work must continue with this material to refine its use and identify the optimal loading at which it should be used. Carbon nanotubes are currently very popular, and are being more and more investigated as the price of this material falls. This will continue and CNTs may well become more important; their single disadvantage is the color. New growth will occur with other nanodimensional materials that are not now under serious investigation for fire retardancy. It is important to investigate these novel materials, such as metal oxides, sulphides, and phosphates, to determine whether they can be well-dispersed in polymers, as well as how they affect the properties of polymers (Wang et al., 2010).

Nowadays, due to environmental awareness and economic considerations, the natural fiber reinforced composite materials have been the subject of several studies to provide a possible material to replace synthetic fiber–reinforced polymer composites (such as glass, aramid, and carbon fibers). Natural fiber-reinforced (NFR) composites are environmentally friendly, renewable, recyclable, less abrasive, light-weight and, economically sound. NFR composites have a wide range of applications because of their better performance, higher ultimate strain, and good impact-resistance. The main advantages of natural fibers are availability, biodegradability, and CO_2 neutrality when they are incinerated.

Among all the natural fibers, jute appears to be the most useful, inexpensive, and commercially available. Jute can be molded into a variety of flat and complex-shaped components by exploiting their attractive reinforcing potential. In addition, jute is one of the best fibers in terms of resistance, mechanical properties, and flexibility. It can be combined with different polymer resins at lower processing temperatures. Thermoplastic resin-based composites are popular due to their processing advantages and satisfactory mechanical properties. Among commodity thermoplastics, polypropylene (PP) and polyvinyl chloride (PVC) possesses some outstanding properties.

PP is a semicrystalline, thermoplastic, linear-structured polymer. PP is widely used as a matrix material because it has some excellent characteristics for composite fabrication. It possesses some vital and useful properties, which include good resistance to fatigue, melting point of 170°C, transparency, dimensional stability, flame resistance, high-heat distortion temperature and high impact strength, low density, good flex life, good surface hardness, scratch resistance, very good abrasion resistance, and excellent electrical properties. PP can be used extensively for manufacturing automotive components, home appliances, and other industrial products because of its balanced mechanical properties, good process ability, and low cost.

An attempt was made to fabricate jute fiber/PVC-PP composite and to investigate the effects of using aluminum trihydrate (ATH) and zinc borate with antimony trioxide (ZB-AT) as fire-retardant materials on the mechanical behavior of the optimum composite, with a recommendation for their use. It was observed that the inclusion of ATH and ZB-AT in the composite influenced mechanical properties as a result of their lower compatibility with fibers, or PP or PVC. Therefore, tensile strength and impact strength of the composite decreased with the increase in the wt % of ATH and ZB-AT that loaded in the blend separately, with a concomitant increase in the bending modulus and tensile modulus. Bending strength decreased by increasing the ATH content while enhanced with the incorporation of ZB-AT. The new compounds are expected to have improved fire-resistance performance over the virgin matrix, but the most important aspect is that the enhancement of fire retardancy is sometimes at the expense of the mechanical strength of the composite laminates.

REFERENCES

Ahmed H.E. and Ziddan Y.E. (2011). A new approach for conservation treatment of a silk textile in Islamic art museum, Cairo, *Journal Cultural Heritage*, 12, 412.

Alexandre M. and Dubois P. (2000). Polymer-layered silicate nanocomposites: Preparation, properties and uses of a new class of materials, *Materials Science and Engineering*, 28, 1–63.

Alongi J., Bosco F., Carosio F., Di Blasio A., and Malucelli G. (2014). A new era for flame retardant materials?, *Materials Chemistry*, 17(4), 152–153. DOI: 10.1016/j.mattod.2014.04.005.

Alongi J., Frache A., Malucelli G., and Camino G. (2013). Multicomponent flame resistant coating techniques for textiles, in: F.S. Kilinc-Balci, Ed., *Handbook of Fire Resistant Textiles*. Woodhead Publishing, Cambridge (UK), p. 63.

Ambuken P., Stretz H., Koo J.H., Lee J., and Trejo R. (2012). High-temperature flammability and mechanical properties of thermoplastic polyurethane nanocomposites, in: A. Morgan, et al. (Eds.), *Fire and Polymers VI: New Advances in Flame Retardant Chemistry and Science*. ACS Symposium SeriesAmerican Chemical Society, Washington, DC, pp. 343–360.

Anandhan S. and Bandyopadhyay S. (2011). Polymer nanocomposites: From synthesis to applications, www.intechopen.com/, DOI: 10.5772/17039.

Arao Y. (2015a). Chapter 2. Flame retardancy of polymer nanocomposite, in: P.-M. Visakh and Y. Arao, Eds., *Flame Retardants, Engineering Materials*. DOI: 10.1007/978-3-319-03467-6_2.

Arao Y. (2015b). Flame retardancy of polymer nanocomposite, in: P.M. Visakh and Y. Arao (Eds.), *Flame Retardants: Polymer Blends, Composites and Nanocomposites*. Springer, Berlin, pp. 15–44.

Astefanei A., Nu´nez O., and Galceran M.T. (2015). Characterisation and determination of fullerenes: A critical review, *Analytica Chimica Acta*, 882, 121. DOI: 10.1016/j.aca.2015.03.025.

Attia N., Ahmed H., Yehia D., Hassan M., and Zaddin Y. (2017). Novel synthesis of nanoparticles-based back coating flame-retardant materials for historic textile fabrics conservation, *Journal of Industrial Textiles*, 46(6), 1379–1392. DOI: 10.1177/1528083715619957.

Bailey S.W. (1984). *Reviews in Mineralogy*. Virginia Polytechnic Institute and State University, Blacksburg.

Betts Kellyn S. (2008). New thinking on flame retardants, *Environ Health Perspectives May*, 116(5), A210–A213. PMCID: PMC2367656.

Blumstein A. (1961). Etude des polymerisations en couche adsorbee I, *Bulletin de la Société Chimique de France*, 98, 899–906.

Bounor-Legar´e V., Angelloz C., Blanc P., Cassagnau P., and Michel A. (2004). A new route for organic–inorganic hybrid material synthesis through reactive processing without solvent, *Polymer*, 45, 1485–1493.

Bounor-Legar´e V., Ferreira I., Verbois A., Cassagnau P., and Michel A. (2002). A new route for organic-inorganic hybrid material synthesis through reactive processing without solvent, *Polymer*, 43, 6085–6092.

Bourbigot S., Duquesne S., and Jama C., (2006). Polymer nano-composites: How to reach low flammability? *Macromolecular Symposia*, 233(1), 180–190.

Brindley G.W. and Brown G. (1984). Crystal structures of clay minerals and their x-ray identification, *Mineralogical Society Monograph*, 5, 504.

Carvalho A., Wang M., Zhu X., Rodin A.S., Su H., and Castro Neto A.H. (2016). Phosphorene: From theory to applications, *Nature Reviews Materials*, 1(11), 16061. Bibcode:2016NatRM.116061C DOI: 10.1038/natrevmats.2016.61.

Chaudhary D.S. and Jollands M.C. (2004). Characterization of rice hull ash. *Journal of Applied Polymer Science*, 93, 1–8.

Chen Y., Fang Z., Yang C., Wang Y., Guo Z., and Zhang Y. (2009). Effect of Clay dispersion on the synergism between clay and intumescent flame retardants in polystyrene, *Journal of Applied Polymer Science*, 115(2), 777–783.

Chiu C.W., Huang T.K., Wang Y.C., Alamani B.G., and Lin J.J. (2014). Intercalation strategies in clay/polymer hybrids, *Progress in Polymer Science*, 39, 443–485.

DEROCA Project (2016). http://cordis.europa.eu/project/rcn/105644_en.html, accessed on 10.8.16.

DHHS (NIOSH) (2013). Current strategies for engineering controls in nanomaterial production and downstream handling processes. November, DHHS (NIOSH) Publication Number 2014-102, US Department of Health and Human Services, www.cdc.gov/niosh/docs/2014-102/default.html

Dikin D.A., Stankovich S., Zimney E.J., Piner R.D., Dommett G.H., Evmenenko G., Nguyen S.T., and Ruoff R.S. (2007). Preparation and characterization of graphene oxide paper, *Nature*, 448, 457–460.

Dittrich B., Wartig K.A., Hofmann D., Mülhaupt R., and Schartel B. (2013). Flame retardancy through carbon nanomaterials; carbon black, multiwall nanotubes, expanded graphite, multilayer grapheme and grapheme in polypropylene, *Polymer Degradation and Stability*, 98, 1495–1505.

Du M., Guo B., and Jia D. (2010). Newly emerging applications of halloysitenanotubes: A review, *Polymer International*, 59(5), 574–582.

Giannelis E.P. (1996). Polymer layered silicate nano-composites, *Advanced Materials*, 8 (1), 29.

Gilman J.W., Jackson C.L., Morgan A.B., Harris R., Manias E., Giannelis E.P., Wuthenow M., Hilton D., and Phillips S. (2000). Flammability properties of polymer layered-silicate (clay) nano-composites, in: *Flame Retardants*. Interscience, London, p. 49.

Engels H.W., Pirkl H.G., Albers R., Albach R.W., Krause J., Hoffmann A., Casselmann H., and Dormish J. (2013). Polyurethanes: Versatile materials and sustainable problem solvers for today's challenges, *Angewandte Chemie International Edition*, 52, 9422–9441.

Evonik-industries (2020). Safe fuel lines, http://nano.evonik.com/sites/nanotechnology/en/technology/applications/cnt/pages/default.aspx, accessed on 6.2.2020.

Feynman R.P. (1960). There's plenty of room at the bottom, *Engineering Physics*, 22, 22–36.

Gilman J.W. (1999). Flammability and thermal stability studies of polymer-layered-silicate (clay) nano-composites, *Applied Clay Science*, 15, 31–49.

Gusev A.A. (2001). Lusti, R.: Rational design of nanocomposite for barrier applications, *Advanced Materials*, 13, 1641–1643.

Hackett E., Manias E., and Giannelis E.P. (1998). Molecular dynamics simulations of organically modified layered silicates, *The Journal of Chemical Physics*, 108, 7410.

Hanaor D., Michelazzi M., Leonelli C., and Sorrell C.C. (2012). The effects of carboxylic acids on the aqueous dispersion and electrophoretic deposition of ZrO_2, *Journal of the European Ceramic Society*, 32(1), 235–244.

Hapuarachchi T.D. and Peijs T. (2010). Multiwalled carbon nanotubes and sepiolite nanoclay as flame retardants for polylactide and its natural fibre reinforced composites, *Composites*, 41(8), 954–963.

Hu Y., Wang S., Ling Z., Zhuang Y., Chen Z., and Fan W. (2003). Preparation and Combustion Properties of Flame retardant polyamide 6/Montmorillonite Nanocomposite, *Macromolecular Materials and Engineering*, 288(3), 272–276.

Isitman N.A. and Kaynak C. (2010). Flame retardancy via nanofiller dispersion state: Synergistic action between a conventional flame-retardant and nanoclay in high-impact polystyrene, *Polymer*, 95(9), 1759–1768. DOI: 10.1016/j.polymdegradstab.2010.05.012Get.

Kandola B.K. (2001). Chapter 6. Nanocomposite, in: *Flame Retaardant Materials*. A. R. Horrock and D. Price (Ed.). Woodhead Publishing, Cambridge, UK, pp. 204–219.

Kandola B.K. and Horrocks A.R. (1998). Flame retardant composites, a review: The potential for use of intumescents, in: M.L. Bras, G. Camino, S. Bourbigot, and R. Delobel, Eds., *Fire Retardancy of Polymers–The Use of Intumescence*. The Royal Society of Chemistry, Cambridge.

Kandola B.K., Horrocks A.R., and Horrocks S. (2001). Char formation in flame-retarded fibre–intumescent combinations. Part V. exploring different fibre/intumescent combination, *Fire and Materials*, 25(4), 153–160.

Kandola B.K. and Toqueer-Ul-Haq R. (2012). The effect of fibre content on the thermal and fire performance of polypropylene–glass composites, *Fire and Materials*, 36(8), 603–613. DOI: 10.1002/fam.1120.

Kashiwagi T. (2007). Flame retardant mechanism of the nanotubes-based nano-composites, Philadelphia, PA, http://fire.nist.gov/bfrlpubs/fire07/art034.html.

Kashiwagi T., Mub M., Winey K., Cipriano B., Raghavan S.R., Pack S., Rafailovich M., Yang Y., Grulke E., Shields J., Harris R., and Douglas J. (2008). Relation between the viscoelastic and flammability properties of polymer nanocomposites, *Polymer*, 49, 4358–4368.

Kashiwagi T., Du F., Winey K.I., Groth K.M., Shields J.R., Bellayer S.P., Kim H., and Douglas J.F. (2005). Flammability properties of polymer nanocomposites with single-walled carbon nanotubes: Effects of nanotube dispersion and concentration, *Polymer*, 46, 471–481.

Kashiwagi T., Grulke E., Hilding J., Groth K., Harris R., Butler K., Shields J., Kharchenko S., and Douglas J. (2004). Thermal and flammability properties of polypropylene/carbon nanotubes nanocomposite, *Polymer*, 45(12), 4227–4239.

Kashiwagi T., Grulke E., Hilding J., Harris R., Awad W., and Douglas J. (2002). Thermal degradation and flammability properties of poly(propylene)/Carbon nanotube composites, *Macromolecular Rapid Communications*, 23(13), 761–765.

Khan F.H. (2013). Chemical hazards of nanoparticles to human and environment (a review), *Oriental Journal of Chemistry*, 29, 1399–1408.

Khan I., Saeed K., and Khan I. (2019). Nanoparticles: Properties, applications and toxicities, *Arabian Journal of Chemistry*, 12, 908–931.

Kim H., Ji-W P., and Kim H.-J. (2020). Chapter 1. Flame retardant nano-composites containing nano-fillers in science and applications of tailored nanostructures, www.onecentralpress.com/science-and-applications-of-tailored-nanostructures/accessed on 9.2.2020.

Knight C.C., Ip F., Zeng C., Zhang C., and Wang B. (2012). A highly efficient fire-retardant nanomaterial based on carbon nanotubes and magnesium hydroxide, *Fire and Materials*, 37(2). DOI: 10.1002/fam.2115.

Kojima Y., Usuki A., Kawasumi M., Okada A., Kurauchi T., and Kamigaito O. (1993). Synthesis of nylon 6-Clay hybrid by montmorillonite intercalated with ε-caprolactam, *Polymer Chemistry*, 31, 983–986.

Kornmann X., Berglund L.A., Sterte J., and Giannelis E.P. (1998). Nano-composites based on montmorillonite and unsaturated polyester, *Polymer Engineering & Science*, 38(8), 1351.

Kumar A.P., Depan D., Tomer N.S., and Singh R.P. (2009). Nanoscale particles for polymer degradation and stabilization—Trendsand future perspectives, *Progress in Polymer Science*, 34, 479–515.

Kuo S.W. and Chang F.C. (2001). POSS related polymer nanocomposites, *Progress in Polymer Science*, 36, 1649–1696.

Laufer D., Kirkland C., Cain A.A., and Grunlan J.C. (2012). Clay-chitosan nanobrick walls: Completely renewable gas barrier and flame-retardant nanocoatings, *ACS Applied Materials & Interfaces*, 4, 1643–1649.

Laurent S., Forge D., Port M., Roch A., Robic C., Vander Elst L., and Muller R.N. (2010). Magnetic iron oxide nanoparticles: Synthesis, stabilization, vectorization, physicochemical characterizations, and biological applications, *Chemical Reviews*, 110, 2574–2575. DOI: 10.1021/cr900197g.

Lee J., Takekoshi T., and E P G. (1997). Fire retardant polyetherimide nano-composites, *Mat. Res. Soc. Symp. Proc.*, 457, 513.

Li B. and Zhong W.H. (2011). Review on polymer/graphite nanoplatelet nanocomposites, *Journal of Materials Science*, 46, 5595–5614.

Liang G., Xu J., Bao S., and Xu W. (2004). Polyethylene/maleic anhydride grafted polyethylene/organic-montmorillonite nanocomposites. I. Preparation, microstructure, and mechanical properties, *Applied Polymer*, 91, 3974–3980.

Liu J., Fu M., Jing M., and Li Q. (2013). Flame retardancy and charring behavior of polystyreneorganic montmorillonite nanocomposites, *Polymers for Advanced Technologies*, 24, 273–281.

Liu M., Jia A., Jia D., and Zhou C. (2014). Recent advance in research on halloysite nanotubes polymer nanocomposite, *Progress in Polymer Science*, 39, 1498–1525.

Liu Y., Wang X., Qi K., and Xin J.H. (2008). Functionalization of cotton with carbon nanotubes, *Journal of Materials Chemistry*, 18, 3454–3460.

Ma H., Tong L., Xu Z., et al. (2007). Synergistic effect of carbon nanotube and clay for improving the flame retardancy of ABS resin. *Nanotechnology*, 18, 375602–375610.

Mallakpour S. and Khadem E. (2014). A green route for the synthesis of novel optically active poly(amide–imide) nano-composites containing N-trimellitylimido-L-phenylalanine segments and modified alumina nanoparticles, *High Performance Polymers*, 26, 392–400.

Mallakpour S. and Zadehnazari A. (2013). Functionalization of multiwalled carbon nanotubes with S-valine amino acid and its reinforcement onamino acid-containing poly(amide-imide) bionano-composites, *High Performance Polymers*, 25(8), 966–979.

Manias E., Touny A., Wu L., Strawhecker K., Lu B., and Chung T.C. (2001). Polypropylene/montmorillonite nanocomposites. Review of the synthetic routes and materials properties, *Chemistry of Materials*, 13, 3516–3524.

McCrum N.G., Buckley C.P., and Bucknall C.B. (1996). *Principles of Polymer Engineering*. Oxford Science, New York.

Mittal V. (2011). Polymer nanocomposites: Layered silicates, surface modifications, and thermal stability, in: V. Mittal, Ed., *Thermally Stable and Flame Retardant Polymer Nanocomposites*. Cambridge University Press, Cambridge, UK and New York, pp. 3–28.

Morgan A.B. (2006). Flame retarded polymer layered silicate nanocomposites: A review of commercial and open literature systems, *Polymers for Advanced Technologies*, 17, 206–217.

Morgan A.B. (2007). Polymer-Clay Nanocomposites: Design and Application of Multi-Functional Materials, *Material Matters*, 2(1), 20–23.

Mulvaney P. (2001). Not all that's gold does glitter, *MRS Bulletin*, 26(12), 1009–1014. DOI: 10.1557/mrs2001.258.

Oliveira M. and Machado A.V. (2013). Preparation and characterization of ethylene-vinyl acetate nano-composites: enhanced flame retardant, *Polymer International*, 62, 1678–1683.

Pandey P., Mohanty S., and Nayak S.K. (2014). Improved flame retardancy and thermal stability of polymer/clay nano-composites, with the incorporation of multiwalled carbon nanotube as secondary filler: Evaluation of hybrid effect of nanofillers, *High Performance Polymers*, 26(7), 826–836, DOI: 10.1177/0954008314531802.

Paul D.R. and Robeson L.M. (2008). Polymer nanotechnology: Nanocomposites, *Polymer*, 49, 3187–3204.

Pergal M.V., Poręba R., Steinhart M., Jovančić P., Ostojić S., and Pírková M. (2017). Influence of the organoclay content on the structure, morphology, and surface related properties of novel poly(dimethylsiloxane)-Based polyurethane/organoclay nanocomposites, *Industrial & Engineering Chemistry Research*, 56, 4970–4983. DOI: 10.1021/acs.iecr.6b04913.

Peeterbroeck S., Alexandrea M., Nagy J.B., et al. (2004). Polymer-layered silicate–carbon nanotube nano-composites: unique nanofiller synergistic effect, *Composites Science and Technology*, 64, 2317–2323.

Phillips S.H., HaddadT. S., and Tomczak S.J. (2004). Developments in nanoscience: Polyhedral oligomeric silsesquioxane (POSS)-polymers, *Current Opinion in Solid State and Materials Science*, 8(1), 21–29. DOI: 10.1016/j.cossms.2004.03.002.

Qin H., Zhang S., Zhao C., Hu G., and Yang M. (2005). Flame retardant mechanism of polymer/clay nanocomposites based on polypropylene, *Polymer*, 46, 8386–8395.

Rahatkera S.S., Zammaranoa M., Matkoa S., et al. (2010). Effect of carbon nanotubes and montmorillonite on the flammability of epoxy nano-composites, *Polymer Degradation and Stability*, 65, 870–879.

Ray S.S. and Okamoto M. (2003). Polymer/layered silicate nanocomposites: A review from preparation to processing, *Progress in Polymer Science*, 28, 1539–1641.

Ren X., Mei Y., Lian P., Xie D., Yang Y., Wang Y., Zirui Wang Z., and T J P. (1998). Hybrid organic–inorganic nano-composites: Exfoliation of magadiite nanolayers in an elastomeric epoxy polymer, *Chemistry of Materials*, 10, 1820.

Saba N., Safwana A., Sanyang M.L., Mohammad F., Pervaiz M., Jawaid M., Alothmand O.Y., and Sain M. (2017). Thermal and dynamic mechanical properties of cellulose nanofibers reinforced epoxy composites, *International Journal of Biological Macromolecules*, 102, September), 822–828. DOI: 10.1016/j.ijbiomac.2017.04.074.

Schartel B., Pötschke P., Knoll U., and Abdel-Goad M. (2005). Fire behaviour of polyamide6/multiwall carbon nanotubes nano-composites, *European Polymer Journal*, 41(5), 1061–1070.

Šehić A., Tavčer P.F., and Simončič B. (2016b). Flame Retardants and Environmental Issues, *Tekstilec*, 59(3), 196–205. DOI: 10.14502/Tekstilec2016.59.196-205.

Sepehr M., Utracki L.A., Zheng X., and Wilkie C.A. (2005). Polystyrenes with macro-intercalated organoclay. Part I. Compounding and characterization, *Polymer*, 46, 11557–11568.

Singla P., Mehta R., and Upadhyay S.N. (2012). Clay modification by the use of organic cations, *Green and Sustainable Chemistry*, 2, 21–25.

Sinha Ray S. and Masami O. (2003). Polymer/layered silicate nano-composites: A review from preparation to processing, *Progress in Polymer Science*, 28, 1539–1641.

Stabkovich S., Dikin D., Dommett G., Kohlhaas K.M., Zimney E.J., Stach E.A., Piner R.D., Nguyen S.T., and Ruoff R.S. (2006). Graphene-based composite materials, *Nature*, 442, 282–286. DOI: 10.1038/nature04969.

Takekoshi T., Fouad F., Mercx F.P.M., and De Moor J.J.M. (1998). US Patent 5 773 502. Issued to General Electric Company.

Tiwari J.N., Tiwari R.N., and Kim K.S. (2012). Zero-dimensional, one dimensional, two-dimensional and three-dimensional nanostructured materials for advanced electrochemical energy devices, *Progress in Materials Science*, 57, 724–803. DOI: 10.1016/j.pmatsci.2011.08.003.

Turova N., Turevskaya E., Kessler E., and Yanovskaya M. (2002). The Chemistry of Metal Alkoxides. Kluwer Academic Publishers, Dordrecht.

US-EPA (2008). Tracking Progress on U.S. EPA's Polybrominated Diphenyl Ethers (PBDEs), Project Plan: Status Report on Key Activities. Washington, DC, www.epa.gov/sites/production/files/2015-09/documents/pbdestatus1208.pdf.

US-EPA (2013). Comprehensive environmental assessment applied to: Multiwalled carbon.

Wang L., He X., and Wilkie C.A. (2010). The Utility of Nanocomposites in Fire Retardancy, *Materials (Basel)*, Sep; 3(9), 4580–4606, Published online 2010 Sep 3 (), . DOI: 10.3390/ma3094580.

Wang X., Li Q., Xie J., Jin Z., Wang J., Li Y., Jiang K., and Fan S. (2009). Fabrication of ultralong and electrically uniform single-walled carbon nanotubes on clean substrates, *Nano Letters*, 9(9), 3137–3141. Bibcode:2009NanoL.9.3137W. CiteSeerX 10.1.1.454.2744. DOI: 10.1021/nl901260b. PMID 19650638.

Wen X., Tian N., Gong J., Chen Q., Qi Y., Liu Z., Liu J., Jiang Z., Chen X., and Tang T. (2013). Effect of nanosized carbon black on thermal stability and flame retardancy of polypropylene/carbon nanotubes nanocomposites, *Polymers for Advanced Technologies*. DOI: 1002/pat.3172.

Wen X., Wang Y., Gong J., Liu J., Tian N., Wang Y., Jiang Z., Qiu J., and Tang T. (2012). Thermal and flammability properties of polypropylene/carbon black nanocomposites, *Polymer Degradation and Stability*, 97, 793–801.

Wu Q., Zhu W., Liang Z., and Wang B. (2010). Study of fire retardant behavior of carbon nanotube membranes and carbon nanofiber paper in carbon fiber reinforced epoxy composites, *Carbon*, 48, 1799–1806.

Xinlin R., Yi M., Peichao L., Delong X., Yunyan Y., Yongzhao W., Wang Z., and Pinnavaia T.J. (1998). Hybrid organic–inorganic nano-composites: Exfoliation of magadiite nanolayers in an elastomeric epoxy polymer, *Chemistry of Materials*, 10, 1820.

Xue B., Peng Y., Song Y., Bai J., Niu M., Yang Y., and Liu X. (2017). Functionalized multiwalled carbon nanotubes by loading phosphorylated chitosan: Preparation, characterization, and flame-retardant applications of polyethylene terephthalate, *High Performance Polymers*, 1–12. DOI: 10.1177/0954008317736375.

Yahoo! (2008). Future planes, cars may be made of 'buckypaper'. Tech News. 2008-10–17.accessed on 18.10.2008.

Yano K., Usuki A., Okada A., Kurauchi T., and Kamigaito O. (1993). Synthesis and properties of polyimide–clay hybrid, *Journal of Polymer Science, Part A: Polymer Chemistry*, 3, 2493–2498.

Yin X., Krifa M., and Koo J.H. (2015). Flame-retardant polyamide 6/carbon nanotube nanofibres: Processing and characterization, *Journal of Engineered Fibres and Fabrics*, 10(3), 1–11.

Zanettia M., Caminoa G., Thomann R., and Mülhaupt R. (2001). Synthesis and thermal behaviour of layered silicate–EVA nanocomposites, *Polymer* 42(10), 4501–4507. DOI: 10.1016/S0032-3861(00)00775-8.

Zhu J., Uhl F.M., Morgan A.B., and Wilkie C.A. (2001). Studies on the mechanism by which the formation of nanocomposites enhances thermal stability, *Chemistry of Materials*, 13, 4649–4654.

9 Flame Retardancy of Synthetic Fibers

9.1 INTRODUCTION

Most phosphorus-containing retardants, on heating, first release polyphosphoric acid, which phosphorylates the C_6 hydroxyl group in the anhydroglucopyranose moiety, and simultaneously acts as an acidic catalyst for dehydration of these same repeat units. The first reaction prevents formation of levoglucosan, the decomposition product of cellulose, the precursor of flammable volatile formation, and this ensures that the competing char-forming reaction is now the favored pyrolysis route. Phosphorus-containing flame retardants increase char formation as expected, but evidence suggests that those with a greater dehydrating power, such as ammonium polyphosphate, have a greater tendency to form aromatic chars than those based on organophosphorus. Furthermore, most of the original phosphorus remains in the char, some of which is believed to combine with the carbon present via P–O–C bonds. This has the effect not only of increasing the oxidation resistance, but also of mechanically toughening the structure. Surprisingly, the bromine-containing retardants studied also appeared to have slight char-promoting effects.

Char formation is not a simple process and the preceding discussion serves to illustrate that rarely do flame retardants function by a single mode. Furthermore, the general route to char requires the presence of functional groups that enable both dehydration and cross-linking reactions to occur as precursors to the formation of an aliphatic carbonaceous structure, and finally, an aromatic char structure. The presence of elements such as nitrogen and sulfur are known to synergistically enhance the performance of phosphorus-containing retardants by further increasing char-forming tendencies. While the chemistry of these actions is not well understood, it is considered that the presence of the elements by formation of P–N and C–N bonds not only influences char-forming chemistries, but also modifies the char structures and thermal stabilities. Such reactions also occur in wool fibers as a consequence of their complex protein (keratin) structure and in the non-thermoplastic aromatic fibers, which have wholly aromatic chains and behave as char-precursor structures.

The major problem lies, however, with the commonly available synthetic polymers–polyester, polyamide, polypropylene and polyacrylic–which, because of their tendencies to pyrolysis by chain scission or unzipping reactions and their general lack of reactive side groups, do not tend to be char-forming. The polyacrylic fibers are the only real exception here. This lack of reactive side groups is aggravated by their thermoplasticity. An ideal char-promoting flame retardant would have to promote cross-linking reactions before thermoplastic effects physically destroyed the coherent character of the textile. Few, if any, commercially available flame retardants, whether as additives, as treatments, or as copolymeric modifications, react with the conventional synthetic fiber structures in char-enhancing modes, unless a degree of prior cross-linking has been introduced as, for example, by radiation.

The conventional synthetic fibers are hydrophobic with physical structures inaccessible to the saltlike materials used for nondurable flame retardants. Synthetic fibers may be rendered flame retardant by mixing FR additives in the monomers, polymer melt, or solution during polymer production, thereby creating a degree of inherent flame retardancy. The conventional method of finishing, i.e., exhaust or padding methods, are not very effective due to their hydrophobic nature. The surface adhesion may be improved by pretreatment with adhesion promoter (AP) followed by back-coating. Otherwise, these FR additives may be applied by thermosol method at high temperature. The suitable FR additives include (Heywood, 2003):

- Cyclic oligomeric phosphonate (pad–dry (110°–135°C)–cure (185°–200°C), trade names: e.g., Antiblaze CU/CT (Rhodia), Aflammit PE (Thor), Flacavon AZ (Schill and Seilacher), Pekoflam PES new liq. (Clariant)
- Organic nitrogen and sulfur compound (probably a thiourea derivative) and a reactive cross-linking compound; for polyamide: cure at 150°–170°C for 45–60 sec. trade names: Aflammit NY
- Organic phosphorus and nitrogen-containing compound (+ binder), trade names: Flacavon H12/10
- Bromine compound and antimony oxide (+ binder); Flacavon H14/587 (Schill and Seilacher)
- Organohalogen compounds: Flame Out PE-60 (Emco), Apex Flameproof 1510 (Apex)

The common synthetic fibers–namely, polyamide, polyester, polyacrylic and polypropylene–are candidates for semidurable and durable flame-retarding if suitable finishes are available, unless alternative inherently flame retardant analogues are commercially acceptable.

In the case of acrylics, because of the difficulty of finding an effective flame retardant finish, modacrylic fibers are preferred unless a back-coating is considered as an acceptable solution, while back-coatings may be similarly effective on other synthetic fiber-containing fabrics and may offer sufficient char-forming character and char coherence to offset fiber thermoplastic and fusion consequences.

The relatively high application levels necessary for all flame retardants can adversely influence fabric handle, drape, and appearance; these effects are minimized by back-coating, which describes a family of application methods in which the flame retardant formulation is applied in a binding resin to the reverse surface of an otherwise flammable fabric. In this way, the aesthetic quality of the face of the fabric is maintained while the flame retardant property is present on the back or reverse face. Careful use of viscosity modifiers and general back-coating application variables ensures that *grin-through* is minimized. In addition, softening agents may be included within the formulations during application; careful selection of these is essential if compatibility with the formulation is to be assured and they are to have minimal effects on the resulting flame retardant property.

This is less easily achieved for polypropylene fabrics. The low melting point, nonfunctionality, and high hydrocarbon fuel content of polypropylene are three factors that have created problems in finding an effective durable flame retardant finish, and also pose difficulties in the design of effective back-coatings. This leaves only polyamides and polyesters as possible candidates for durable, flame retardant treatments (Schindler and Hauser, 2004).

9.2 POLYESTER

Polyester fabrics owing to unavailability of reactive functional groups are difficult to make flame retardant by conventional finishing treatments. The most common method for making flame retardant polyester is by applying either phosphorus or halogen-based flame retardants by the thermosol method (Weil and Levchik, 2008). The fabrics are generally treated with aqueous solutions of flame retardants (2%–15%), dried at around 100°C and cured at 185°–195°C (knitted fabrics) or 190°–205°C (woven fabrics). for 120–160 s. Durable flame retardant treatments can be obtained by this method. The flame retardants are believed to diffuse into the soft polymer during the curing operation. Once the fabrics are cooled, the flame retardant gets trapped and excess flame retardant on the surface is washed off. A common phosphorus flame retardant composition for polyester fabric is a mixture of cyclic phosphonate compounds (Structure 9.1, chemical structure of phosphonic acid and P-methyl-, (5-ethyl-2-methyl-2-oxido-1,3,2-dioxaphosphorinan-5-yl) methyl ester complex) and is sold as Kappaflam P 31 by Kapp-Chemie GmbH, Germany. Diphosphonate products are also available from Thor (Afflamit) or Zschimmer & Schwarz (Flammex). They are slightly acidic in nature, therefore an application solution needs buffering of the pH to 6.5. Usually a final fixation of 1.5% of the compound is quite sufficient to achieve satisfactory flame retardancy.

STRUCTURE 9.1   Chemical structure of cyclic phosphonate compounds

Several FRs have been designed for polyester extrusion(bisphenol-S-oligomer derivatives from Toyobo, cyclic phosphonates (Antiblaze CU and 1010) from Albemarle, or phosphinate salts such as OP950, from Clariant). A novel antidripping intumescent flame retardant, poly(2-hydroxypropylene spirocyclic pentaerythritol bisphosphonate) (PPPBP) was synthesized (Structure 7.4, see Chapter 7) and applied by the thermosol method to impart flame retardancy and dripping resistance to PET fabrics (Chen et al., 2005).

Halogen-containing novel organophosphorus compounds, such as dichloro-tribromophenyl phosphate (DCTBPP) have been synthesized and applied to polyester fabrics by the thermosol method to obtain self-extinguishing fabrics with durability of up to 50 laundry cycles (Yoo-Hun et al., 2001).

The PET melts at 245°–290°C, so it is not safe during the combustion due to the dripping which is considered as the new source of ignition (Smith and Hashemi, 2006). To overcome this, it is important to improve the ignition and antidripping properties of PET. An adhesion promoter (AP) is used to improve the cohesion between components and material surface (textile and plastics) in the presence of solvents such as methyl ethyl ketone, toluene, and butanol (Whelan, 1994).

APs such as aminosilane and epoxysilane have a significant effect on the synthesis of flame retardant coatings (Nagelsdiek et al., 2014). AP can be used by two methods:

- Immersing the specimen in AP then coating with different techniques (spraying or dipping) and
- Using the specimen as the main component during the stages of coating synthesis.

Another kind of AP is chlorinated polyolefin, used in many industrial fields such as inks on polyolefin plastics and adhesion of paint to plastic surface in the automotive industry (Ebnesajjad and Ebnesajjad, 2013).

The AP was prepared by mixing modified chlorinated polyolefin with a blend of toluene and xylene. The composition is mixed with slow speed and high torque mixer until the components are dispersed. Different coating techniques such as dipping, spraying, and brushing were used to coat the surface of the specimen with AP.

After immersing the specimens with AP they are subjected to drying in an oven for 4 min at a temperature of 80°C to remove the solvent by evaporation and the specimens are ready for adhesion with the coating. Durable flame-retardant coating (DFRC) was synthesized by mixing two solutions, namely A with B. Firstly, solution A was synthesized by·dissolving 0.01mol of zinc acetate dehydrate in 50 ml of methanol with stirring for 90 min at 50°C; while dissolving 0.02 mol sodium hydroxide in 50 ml methanol in a round flask for 60 min at 50°C with autostirring leads to synthesize solution B. Finally, ZnO nanosol has been synthesized when solution B was added to solution A with constant stirring for 90 min at room temperature, followed by heating for 90 min (Younis, 2012). The specimens coated with DFRC were washed at 60°C for 30 min in the presence of 2% NaHCO$_3$ and 5% nonionic detergent, and then dried in an oven at 110°C for 5 min to study the durability of the coating. Depending on internal force formed between the surface of the polyester fabric and the outer layer (durable or nondurable), this technique is applied to protect PET fabric from the ignition, to improve the rate of burning, and increasing the LOI by 50% compared to PA, PE, and PF specimens. The results show that AP improved the flame retardancy and dripping of PET fabric compared to blank. The char yield increased from 8% to 18%, and LOI increased from 17.5% to 27.5% (Younis, 2017).

Polyester has a moisture regain of only 0.4% at 20°C and 65% relative humidity, which makes polyester extremely hydrophobic. This hydrophobic nature of polyester results in a relatively low chemical absorbance during the finishing treatment, and the fabric induces static electricity that creates problems when the fabric is worn, cut, or sewn. Since the migration of flame retardants to the surface of the material during drying is still an unsolved problem, high concentrations of flame retardant agents have to be used in padding systems in order to achieve better flame retardant properties. To eliminate these problems and minimize the amounts of the finishing agent, energy, and costs, which would reduce the amount of waste and lower the environmental and health risks, an alternative technique is proposed, in which the textile fabrics are pre-treated with a low-temperature plasma (Cireli et al., 2007).

According to the wicking hydrophilicity test, the most hydrophilic fabric was achieved with a plasma treatment with 10 min of exposure time. Experiments were performed on polyester fabric with Diener vacuum plasma. Commercially available oxygen gas was discharged into the chamber. The samples were placed in the discharge chamber and treated with oxygen gas. the optimum conditions were determined to occur when the LF (low frequency) generator operated at a frequency of 40 kHz, at a power of 100 W, at a pressure of 2 mbar with an exposure time of 10 min and at 42°C. LOI results showed that a concentration of 50 g/L flame retardant agent would-be required after oxygen plasma treatment, instead of 100 g/L flame retardant agent on the untreated fabric in the padding system. Because of the fact that the oxygen plasma treatment improved the hydrophilic properties of polyester fabrics, reduced concentration of flame retardant agent was sufficient to achieve the same flame retardant effect without changing LOI significantly. However, the LOI% of all the polyester fabrics was reduced after the washing process (Ömeroğulları and Dilek, 2012).

Commercially FR polyesters are made in three ways:

- Using additives in polymer melts,
- Using FR comonomers during polymerisation, or
- Applying topical finishes.

Flame resistance may be imparted to polyester by incorporating FR additive prior to fiber spinning. All the methods use phosphorous- or halogen-containing compounds. The dope additives should be stable at 250°–300°C for a considerable time. Aromatic bromine derivatives are used for greater stability. Some of the additives for polyester are:

- Octabromo bisphenyl (Figure 5.1(c), cyclic phosphonate compounds, see Chapter 5), toxicity, LDLo (rat)- 2gm/kg, human toxicity–no adverse dermal effects have been found.
- Tetrabromo-bisphenol A (Figure 5.1(b), TBBPA, see Chapter 5). European Food Safety Authority (EFSA) in December 2011 says, "current dietary exposure to TBBPA in the European Union does not raise a health concern". Some other studies suggest that TBBPA may be an endocrine disruptor and immune-toxicant.
- Triphenylphosphine oxide (TPPO, Structure 6.18, see Chapter 6). The potential human carcinogenicity hazard for TPPO cannot be determined due to an inadequate database (US EPA, 2005).
- Decabromobiphenyl ether (DBDPO, Structure 5.2, see Chapter 5). The toxic effects of PBDEs can lead to thyroid hormone disruption, neurobehavioral toxicity, cancer, and other adverse health effects.

There are many halogenated compounds that improve polyester's flame retardancy. The challenge is to apply the finish so that it does not affect the fabric hand and still be durable to repeated launderings.

One current commercial FR for polyester consists of a mixture of cyclic phosphate/phosphonates that can be applied by pad-dry-heat set (190–210 ^{0}C for 0.05–2 min) method. An add-on of 3%–4% is sufficient.

9.2.1 Polylactic Acid

Among the biodegradable polymers, polylactic acid (PLA) is one of the most attractive polymers for replacing petrochemical polymers due to its biodegradability, abundant renewable sources, and excellent mechanical properties. PLA has various applications in biomedical, engineering and other fields (Garlotta, 2001). Although PLA is found to be suitable for different applications, its flammability and serious dripping during combustion limit its further application in some areas, especially in electronic and electrical fields. Therefore, the improvement in flame retardant performance of PLA is still an important and urgent task (Bourbigot and Fontaine, 2010).

Polylactic acid is biodegradable thermoplastic aliphatic polyester. It is derived from renewable resources, such as corn starch, tapioca roots, chips or starch, or sugarcane. A range of drying and curing application conditions was evaluated for imparting a durable flame retardant effect (based on a mixture of cyclic phosphonate esters) to the poly(lactic acid) fabrics. A mixture of cyclic phosphonate esters (Afflamit PE, Thor Chemicals, UK), biodegradable and low-toxic FR was applied, along with an oil- water- repellent fluorocarbon chemical FC-251 (based on a blend of a fluorochemical urethane and a fluorochemical acrylate) and an anionic fiber softener cum lubricant, Patsoft 1220 was applied on PLA fabric by padding, followed by drying and thermofixation. The best application conditions for applying the flame retardant with a fluorochemical and softener/lubricant, without causing significant color change and deterioration in handle properties, were drying by 110°C, followed by thermo-fixation at 135°C for 90 s. The flame-retardant performance of the polylactic acid fabrics was found to be durable, even after 50 washing cycles and the application of three different finishing chemicals (flame retardant, fluorochemical, and softener/lubricant) in the same bath did not adversely affect either the flame retardant performance or color change of the polylactic acid fabrics. Combined bath applications were preferred over separate bath applications due to the lower color change; it was also found that the softener cum lubricant used in this study had a deleterious effect on oil repellency recovery performance after hot pressing (Avinc et al., 2012).

A commercial cyclic phosphonate ester flame retardant (DP-150 supplied by Hangzhou Dawne Textile Tech Co. Ltd., China) originally designed for PET fiber was successfully used to improve the flame retardancy of PLA nonwoven fabric by a pad-dry-cure technique. The treated PLA fabric showed good flame retardancy with a high LOI of 35%, compared to 26.3% for the untreated fabric. The treated fabric exhibited higher thermal stability than the untreated fabric. Although a very small amount of char residues during the thermal degradation process of the treated fabric was found, the resulting condensed-phase flame retardant mechanism should be restricted to a lesser extent due to too few char residues. An obvious decrease in phosphorus content for the flame retardant fabric after burning was proved by the SEM-EDS analysis, indicating that the role of a cyclic phosphonate ester is predominately restricted to the gas phase (Cheng et al., 2016).

One current commercial flame retardant for polyester is a mixture of cyclic phosphate/phosphonates used in a pad–dry–heat set process. Heat set conditions of 190°–210°C for 0.5–2 min are adequate. This product when applied at ~ 3%–4 % add-on can provide durable flame retardancy to a wide variety of polyester textiles (Anderson et al., 1974).

The basic routes for using reasonable amounts of flame retardants and nanofillers in fully green, flame-retarded kenaf fiber–reinforced polylactic acid (K-PLA) biocomposites for technical applications are discussed by Sypaseuth and others (2017).

In keeping with growing demand for green flame-retarded polymeric materials, the application of halogen-free flame retardants, such as $Mg(OH)_2$, ammonium polyphosphate (APP), or expandable graphite (EG) is preferred because they belong to the few commercial flame retardants that are recommended to be used with respect to environmental concerns. Yet, to maintain the mechanical properties of flame-retarded polylactic acid (PLA) biocomposites, the flame retardant content must be limited.

Sypaseuth and others (2017) present a systematic study of kenaf reinforced PLA using multicomponent flame retardant systems with a maximum amount of 20 wt% additives based on MH/nanofiller or expandable graphite (EG)/ammonium polyphosphate (APP) flame retardants. Synergistic multicomponent systems containing EG and APP, or MH with adjuvants offer a promising route to green, flame-retarded, natural fiber–reinforced PLA biocomposites.

MAGNIFIN is a high-purity grade of magnesium hydroxide for applications in flame retardancy. It is especially recommended when low smoke emissions and high thermal stability of the halogen-free, flame retardant additive are required. The proprietary process results in high-purity magnesium hydroxide grades with the regular crystal form required for flame retardancy applications in plastics and rubber. The electron microphotograph of MAGNIFIN illustrates the very regular particles (hexagonal platelets) with a median particle size of about 1 micron. Low surface energy minimizes the tendency to form agglomerates. Four standard grades (MAGNIFIN H10, H7, H5 and H3) with different particle size distributions and specific surfaces are offered.

Magnifin H5 (MH) is a well-known, halogen-free flame retardant (High-purity fine precipitated magnesium hydroxide, surface treated with a special vinyl-silane coating) that is normally used in high amounts of up to 60 wt%, with consequent negative impacts on the material's mechanical properties (www.magnifin.com/magnifin_flame.htm). At the same time, only 3–10 wt% nanomaterials can improve fire behavior under forced-flaming conditions, because they protect the residual layer formed upon combustion. Naturally, it would be of great interest to combine these additives in a synergistic system exploiting the advantages of each component, while keeping the overall additive concentrations as low as possible. The application of this intriguing concept is to combine metal hydrates with nanoparticles for flame-retarded composites (Dittrich et al., 2014).

Polyester fabrics when burned exhibit melt–drip behavior. The fabric melts away from the flame; some polyester fabric may pass vertical flame tests without any flame retardant treatment waiving of melt–drip specifications (Schindler and Hauser, 2004).

9.2.2   REACTIVE FRs

A phosphorous-based comonomer can be introduced into PET, preferably as a diacidic component (or its diester derivative) or as a diol (containing two hydroxyl groups). Thus, flame retardancy has been achieved in commercial PET fibers by incorporation of 2-carboxyethyl(methyl) phosphinic acid (Structure 9.2), 2-carboxyethyl(phenyl) phosphinic acid (Structure 9.3), or their cyclic anhydrides.

STRUCTURE 9.2   (2)-Carboxyethyl(methyl) phosphinic acid

STRUCTURE 9.3 (2)-Carboxyethyl(phenyl) phosphinic acid

PETs based on the former are marketed under the trade name Trevira CS. It has been suggested that fibers containing either Structure 9.2 or Structure 9.3 might be further improved by the additional incorporation of aromatic dicarboxylic acid monomers to act as charring agents (Asrar, 1988). Another commercially utilized P-containing comonomer is the 9,10-dihydro-9-oxa-10-phosphaphenanthrenyl-10-oxide [DOPO (Structure 6.27a, see Chapter 6)] adduct of itaconic acid (Endo et al., 1978) (Structure 9.4). Filament fibers and fabrics based on PET copolymers containing this flame retardant comonomer are commercially available from Toyobo under the trade name, HEIM, and have LOIs (limiting oxygen index) of up to 28.

STRUCTURE 9.4 DOPO adduct of itaconic acid

All the P-modified PETs appear to be subject to both the vapor- and condensed-phase mechanisms of flame retardance, with the former predominating (Chang and Chang, 1999).

Aliphatic polyamides are not chemically modified to improve flame retardance, probably because the chemical modification of aliphatic polyamides disrupts intermolecular hydrogen bonding and hence crystallinity, thus reducing melting points.

The solubility parameter is a reliable way to study the adsorption phenomenon quantitatively. In this study, solubility parameters of five selected phosphorous flame retardants were computed by molecular dynamics (MD) simulations, the phosphorous groups of which were not available in the group contribution method, with the aim of addressing the intermolecular interactions of flame retardants with poly(ethylene terephthalate) (PET) while transferring from an aqueous bath to PET textiles. To verify the reliability of the MD strategy, the solubility parameter of flame retardant 1,2,5,6,9,10-hexabromocyclododecane (HBCD) was computed by group contribution and MD as well. The obtained solubility parameters of the five phosphorous flame retardants were found to have a linear correlation with their thermodynamic parameters, which describe their adsorption on PET fibers. The flame retardants with lower solubility parameter are more likely to adsorb on PET fibers (Chang et al., 2017).

9.2.3 Unsaturated Polyesters

Unsaturated polyesters (UP) are widely used as a matrix of composite materials for the building industry, transportation sector, and electrical industry, and others, because of their low cost, easy processability, low density, good corrosion resistance, and high strength-to-weight ratio. However, typical polyester resins are highly inflammable and produce large quantities of smoke and toxic acid when burning, limiting their industrial use (Atkinson et al., 2000).

Recently, halogen-free intumescent flame retardants (IFR) consisting of an acid source, a char-forming agent, and a blowing agent. are attracting more and more attention from both academic and industrial communities for their multifold advantages, including low toxicity, high efficiency, low smoke, low corrosion, and lack of corrosive gas (see Chapter 7).

A novel macromolecular IFR (MIFR), which contains an acid source, a gas source, and a char source, had been simultaneously synthesized by pentaerythritol, phosphoric acid, melamine, urea, and formaldehyde (Gao and Yang, 2010), and were applied to epoxy resin to acquire good flame retardant efficiency (Gao et al., 2013). While 24% of MIFR were doped into UP to get 30.5% of the limiting oxygen index and UL 94 V-0, its tensile strength and the impact strength decreased by only 7.2% and 7.0%, respectively. The results for UP containing MIFR, compared with untreated UP, show that the weight loss, thermal stability, and decomposition activation energy decreased, and the char yield increased, showing that MIFR can catalyze decomposition and carbonization of UP to form an effective charring layer to protect the underlying substrate.

9.2.4 Partially Oriented Polyester

During the application of flame retardants in the finishing process, the key problem is the accessibility or penetration of the fiber material by flame-retardant particles. In case of standard PET fibers, this is accomplished by raising the temperature of modifying bath by 40°C or more above the glass transition temperature of these fibers (in practice, a temperature of 120°–135° C). At these temperatures, there is a sufficient amount of so-called free volume that provides for optimal sorption of flame retardant particles.

Partially oriented fibers are obtained by increasing the forming rate several times as compared to the classic technology of manufacturing of PET filament fibers. The increasing of take-up velocity of the fibers makes it possible to eliminate the separate stretching operation. The take-up velocity of the fibers in POY technology is selected so that the final product has the specific tear strength equivalent to the one achieved for the fibers produced using the classic method. Despite the corresponding mechanical properties, the supermolecular structure of the POY fibers is different. PET POY fibers are characterized by low crystallinity and relatively small overall orientation; thus, a supermolecular structure of the fibers is extremely sensitive to any kind of heat treatments (Kim and Kim, 2016). These characteristics have an important influence on the sorption of the flame retardant particles by the fiber.

9.3 POLYAMIDE (PA)

PA 6.6 is widely used in the textile industry for the production of yarns, garments, and carpets. Nevertheless, these materials are highly flammable, whereas legislation very often requires low flammability. Various ways can be considered to improve the fire performance of this polymer. An efficient method consists of incorporating flame retardants directly in the bulk during processing, or covering the surface of the polymer by either a thick or thin flame retardant coating. In fact, surface treatments present several advantages because they do not modify the intrinsic properties of the materials. They are usually easily processed (in a small scale) and they can be applied on several materials. Due to their effectiveness, intumescent coatings are commonly used as flame retardant coatings. Different groups developed the layer-by-layer (LbL) technique (Iler, 1966) as a surface treatment to make thin and tunable IFR coatings to lower the flammability of textiles such as cotton, polyester, and polyester–cotton blends (Carosio et al., 2013). LbL technique consists of successively dipping a solid substrate into polycation and polyanion solutions to form polyelectrolyte multilayer (PEM) films. Grunlan et al. (Li et al., 2011) were the first to make an IFR system using the layer by-layer technique. This intumescent coating, inspired by Li et al.'s work (2011), consists of alternating polyallylamine (PAH) (charged positively) and sodium polyphosphates (PSP) (charged negatively) deposits on the PA6.6 fabrics. In this study, only the surface of polyamide fabrics was modified, with $(PAH-PSP)_n$ coatings at 5, 10, 15, and 40 bilayers, without spoiling the intrinsic properties of polyamide. The thickness, morphology, topography, thermal stability, and fire performance of these assemblies have been assessed (Apaydin et al., 2014). According to the SEM and mapping analyses, it could be observed that the coating covers only the surface of the external fibers of the polyamide fabrics, contrary to the coatings deposited on cotton or polyester–cotton blends that penetrate inside the textile. The thermal degradation of the polyamide fabric was catalyzed with 5, 20, and 40 bilayers (BL) coated films. All coated samples influence the mechanism of decomposition of polyamide fabric. Moreover, the amount of residue increases for N = 20 BL and 40 BL samples. PCFC results showed on the one hand, a decrease of the decomposition temperature (catalytic effect), and on the other hand, a significant decrease of peak of heat release rate (pHRR) (up 36%) for all coated fabrics.

Usually, polyamide fabrics are fireproofed by incorporating flame retardant additives in the bulk. The challenge to develop flame-retarded PA6 fibers is to find a system in which the total loading of additives does not exceed 10 wt%. The FRs also have to be stable at the rather high processing temperature of PA6; eventually the spinning ability of the formulation is required. Three types of additives were chosen for study by Coquelle (2014); the first type consists of commercial phosphorus fire retardants; the second type consists of sulfur-based FRs that were found effective at very low loadings in PA; the third type corresponds to nanoparticles, based on two discrimination properties: the spinnability of the formulation and their flame retardant properties.

Pyrolysis–Combustion Flow Calorimetry (PCFC) was chosen as the main fire test to evaluate the different formulations. PCFC is a relatively new method to characterize the potential flame retardancy of polymeric materials. Its main advantage is the possibility to test materials with only milligrams of samples, allowing a quick determination of the materials properties, which is particularly suitable for a screening step in which small amounts of materials are available.

The LOI of PA66 is 24.2%, which limits its applications. Li et al (2017) produced flame retardant PA66 by condensation polymerization of nylon salt and DOPO-based flame retardant, followed by melt spinning. Although the presence of DOPO-based flame retardant decreased the molecular weight, melt temperature, crystallinity, and mechanical properties of PA66, the flame retardancy properties improved. The flame retardant PA66 can achieve a V-0 rating, according to the UL94 criterion, with a LOI value of 32.9%, which showed that the flame retardant PA66 had relatively satisfactory flame retardancy. The breaking strength, elongation break, sound velocity, orientation factor, and modulus of flame retardant PA66 fiber decreased, compared to untreated PA66 fiber. The tenacity at break of flame retardant PA66 fiber still attained 2.82 cN·dtex-1, which satisfied the requirements for fabrics.

Sulfur-based FR compounds such as ammonium sulfamate (AS) were chosen due to their ability to raise LOI values in PA6 and improve the UL-94 rating at very low content. Ammonium sulfamate alone in PA6 also leads to a V0 with 2 wt.-% loading (Lewin et al., 2002). The addition of a small amount of ammonium sulfamate and dipentaerythritol, 2 wt. and 0.7 wt.-%, respectively in PA6 yielded a V0 at UL-94 rating on bars of 1.6 mm thickness, and a LOI value of 35.7 vol.-% (Lewin et al., 2007).

AS (Structure 9.5) and guanidine sulfamate (Structure 9.6) were evaluated as FRs in PA6 at various loadings (5 and 10 wt.%). These additives gave the most interesting results in terms of pHRR decrease. The screening results obtained with PCFC do not show any improvement on the curves of heat release rate. This means that the pHRR were not significantly decreased;

however, the total heat release is generally decreased. Therefore, either less material is burned, or the combustion of the fuels is less efficient. Sulfur-based additives led to the best screening results. Indeed, formulations containing sulfamate salts show the lower values of peak of heat release rate. It is possible to reach a decrease of more than 20% of the PHRR with only 5% of AS. However, AS leads to a degradation of the matrix when higher loadings (>10 wt.-%) are incorporated. Another sulfamate salt having a guanidinium molecular counterion has allowed drawing fibers with 10 wt.-% of additives and overcoming the degradation issue. However, in this case the pHRR reduction was not as important as was the case with AS (while keeping the same amount of FR in the PA6). Guanidine sulfamate-and AS-containing formulations both start to degrade earlier than PA6, but, as for the phosphorus - based FR, TG curves also show a different char oxidation. Thus, it is assumed that the sulfamate salts might also enhance char formation and modify its nature. Finally, nanoparticles (tubelike multiwall carbon nanotubes and Halloysite nanotubes, sheetlike nanoparticle graphene oxide, and modified MMT) have been melt-mixed in PA6 with different sets of parameters. Severe and soft conditions both gave microdispersions, as shown by SEM and optical microscopy. The different formulations were analyzed using the PCFC and TGA. No significant reductions of pHRR or THR were recorded as the results are within the margin of error of the instruments.

STRUCTURE 9.5   Chemical structure of ammonium sulfamate (AS)

STRUCTURE 9.6   Chemical structure of guanidine sulfamate (GAS)

Horrocks et al. (2016) studied incorporation of a nanoclay along with two types of flame retardant formulations–aluminium diethyl phosphinate (alpi) at

10 wt %, known to work principally in the vapor phase, and an ammonium sulfamate (AS)/dipentaerythritol (DP) system present at 2.5 and 1 wt %, respectively–believed to be condense-phase active. The nanoclay chosen is organically modified montmorillonite clay, Cloisite 25A. The effect of each additive system is analyzed in terms of its ability to maximize both filament tensile properties relative to 100% PA6 and flame retardant behavior of knitted fabrics in a vertical orientation. None of the alpi-containing formulations achieved self-extinguishability, although the presence of nanoclay promoted lower burning and melt-dripping rates. The AS–DP-containing formulations with total flame retardant levels of 5.5 wt % or less showed far superior properties and with nanoclay, showed fabric extinction times of 39 S. and reduced melt-dripping.

Most polyamide (PA) fabrics pass flammability standards as the polymer burns at a very slow rate. The polyamide materials can also be made flame-resistant by adding additives in dope or by finishing in the fabric form with phosphates, phosphonates, phosphine oxides (sometimes along with halogenated compounds), halogenated aliphatic and aromatic compounds with or without antimony, heavy metal salts, and many other complex compounds.

A number of flame retardant systems containing halogen, metal hydroxide, phosphorous, and nitrogen are used to provide flame retardancy to PA6. The halogen compounds are famous for being highly efficient flame retardants at low dosage, but the usages are limited because of the toxicity of their combustion products and potential corrosion to spinning equipment (Isitman et al., 2009). Among the metal hydroxides that can be used as flame retardants in polymer materials, magnesium hydroxide, $Mg(OH)_2$ and aluminum hydroxide, $Al(OH)_3$ are the most important and are widely used. Although they have low toxicity, good anticorrosion properties, low cost, and low smoke emission during processing and burning, metal hydroxides have some serious disadvantages, such as relatively low flame retardancy, low thermal stability, and deterioration of the physical and mechanical properties of the matrix. (Balakrishnan et al., 2012). NFRs and PFRs are environmentally friendly and of low dosage, and have acceptable flame retardancy, and both of them have been studied extensively in providing flame retardancy to polymer materials. By comparison, NFRs add no new elements to polyamides such as PA6, and show good compatibility with them.

To improve the flame retardancy of PA6 fibers, its composites with melamine cyanurate (MCA-PA6) were synthesized via in situ polymerization of α-caprolactam in the presence of adipic acid-melamine salt and cyanuric acid-hexane diamine salt. The flame retardant MCA/PA6 composite fibers were prepared by melt-spinning. Experimental results indicated that the MCA/PA6 composites loaded with 8 wt% of additives can achieve UL94 V-0 rating with an LOI value of 29.3%. The tenacity at break of PA6 fiber decreased from 4.85 to 3.11 cN·dtex^{-1} for MCA/PA6-8 composite fiber. However, the MCA/PA6 composites prevent propagation of flame in the fabric. This means that the in situ polymerization approach paves the way for the preparation of MCA/PA6 composites that have good spinnability and flame retardancy.

Due to lower reactivity and higher flammability, especially in the presence of non-thermoplastic materials, it is difficult to make polyamide flame-retardant. Phosphorous-based FRs, and halogen compounds at lower concentration are less effective as FRs for polyamide. At higher concentrations of halogen compounds, the gaseous phase mechanism sets in and polyamide becomes effectively flame retardant. Common practice for making FR polyamide carpets is to apply a bromine donor (e.g., DBDPO) with antimony trioxide as back-coating.

Decabromodiphenyl ether (decaBDE) continues to be a popular fire-retardant additive for polyamide 6, its cost, high bromine content, and good thermal stability make it an attractive product. In Europe and the United States, most of the banned FRs are replaced by brominated polystyrene in order to reduce environmental pollution. Brominated polystyrene is less expensive than the other polymeric fire retardants and has very good thermal stability.

Poly(pentabromobenzyl acrylate), another polymeric fire retardant, is particularly suitable for use with polyamides irrespective of whether they contain fiber reinforcement. Its advantages over other fire-retardant additives result from a combination of its polymeric nature, high bromine content, and thermal stability.

High-molecular-weight brominated epoxy polymers are efficient fire retardants for polyamides, which offer the following advantages: high thermal stability and thermal aging, excellent processability, nonblooming, high UV stability, and low corrosivity. It is important to mention that only brominated epoxies with negligible epoxy content are suitable for polyamide applications, in order to avoid adverse reactions between epoxies moieties and amine groups in polyamide.

Unlike brominated polystyrene, the chemical structure of dodecachloropentacyclooctadeca-7,15-diene (e.g., Dechlorane Plus, Oxychem, Structure 9.7) fits the ignition temperature of polyamides. This results in a high limiting oxygen index (LOI) value of 38, achieved in glass-reinforced polyamide 6. It is produced by the Diels–Alder reaction of two equivalents of hexachlorocyclopentadiene with one equivalent of cyclooctadiene. The syn and anti-isomer are formed in the approximate ratio of 1:3. Dechlorane plus was added to the list of REACH Substances of Very High Concern on January 15, 2018 (ECHA, 2018).

STRUCTURE 9.7 Chemical structure of Dechlorane Plus

There has been no successful commercial production of flame-retarded polyamides incorporating P-containing reactives, although both Monsanto and Solutia have patented P-containing diacids designed to replace some of the adipic acid in the manufacture of flame-retardant polyamide-6,6 fibers (Asrar, 1988).

9.4 ACRYLIC

Acrylic fibers burn readily with melting and sputtering, while modacrylic fibers burn very slowly with the formation of hard black beads and do not drip. Modacrylic fibers made by using halogen-containing copolymers have excellent flame retardancy and acceptable fiber properties.

FR additives for fiber spinning consist mainly of insoluble aromatic chlorine and bromine, together with insoluble organo-phosphorous derivatives. Ammonium polysulphide and polybromocyclo-hexane are topical FR additives. Fluoboric acid has been suggested as a sol for acrylonitrile.

A variety of P-containing acrylates, methacrylates (Structure 9.8), and styrene derivatives (Structure 9.9) have been synthesized and copolymerized using free-radical initiators with methyl methacrylate (MMA) (Ebdon et al., 2000) and acrylonitrile (AN) (Crook, 2004). Copolymers of MMA with diethyl-(methacryloyloxy) methyl phosphonate (DEMMP, Structure 9.5 with R = C_2H_5 and R' = CH_2), for example, have higher LOIs, lower rates of heat release, and lower overall heats of combustion, and give a V0 rating in a UL94 protocol (Price et al., 2000).

STRUCTURE 9.8 Chemical structure of P-containing acrylates, methacrylates

STRUCTURE 9.9 Chemical structure of P-containing styrene derivatives

Studies of the thermal degradation and combustion of some of these reactively modified PMMAs indicate that flame retardance involves both vapor-phase and condensed-phase activity, with the latter leading to significant char production. Char production probably involves trans-esterification reactions and interchain cyclization catalyzed by phosphonic acid groups liberated during combustion. The reactive route to flame retardance has also been shown to offer significant advantages over routes using analogous P-containing additives, in that additives tend to plasticize PMMA significantly and to give rise only to vapor-phase modes of flame retardance (Price et al., 2000).

Polyacrylonitrile fabrics unlike modacrylics are rarely used for manufacturing fabrics that might be of use in areas that require flame protection. Nevertheless, one can find several research articles related to flameproofing of polyacrylonitrile fabrics with various phosphorus- and halogen-containing compounds. Several compounds, such as decabromodiphenyl oxide and ammonium bromide, urea and phytic acid, hydrazine hydrate, hydroxylamine sulfate, and metal salts have been used as flame retardants (Bajaj et al., 2000).

More recent investigations have shown that polyacrylonitrile fabrics can be made flame retardant by grafting acrylic-based organophosphorus compounds using environmentally friendly plasma technologies. In this research, argon plasma-induced graft polymerization of four acrylate monomers containing phosphorus, diethyl(acryloyloxyethyl)phosphate (DEAEP), diethyl-2-(methacryloyloxyethyl)phosphate (DEMEP), diethyl(acryloyloxy methyl) phosphonate (DEAMP), and dimethyl(acryloyloxymethyl)phosphonate (DMAMP) was performed. The treatments resulted in moderate flame retardant properties for the fabrics with limited durability to several laundry cycles (Tsafack and Levalois-Grützmacher, 2006).

9.5 POLYPROPYLENE

Polypropylene has few fire-retarded applications, because of the difficulty of reaching the high performance required for applications. The required high loading of fire-retardant tends to increase brittleness and decrease mechanical properties (Smith et al., 1996).

Aromatic bromine compounds such as decabromodiphenyl oxide (DBDPO), and alicyclic chlorine derivatives such as Dechlorane Plus (Structure 9.7), with or without synergist, Sb_2O_3 and phosphorous compounds are added to the melt before spinning. Bromine and aliphatic compounds are more effective than chlorine and aromatic compounds, respectively. When PP fibers are used in carpets, they are FR-treated with polyvinyl chloride (PVC) or chlorinated paraffin, together with Sb_2O_3.

The primary action of halogen fire-retardant action for polypropylene is in the gaseous phase; thus, the fire-retardant additives for polypropylene are often based on aliphatic bromine compounds in order to develop bromine at its low ignition temperature.

Tris(tribromoneopentyl) phosphate ((ICL's FR-370, Structure 9.10) combines bromine and phosphorus in the same molecule; it has been successfully incorporated into polypropylene. Studies have dealt with the question of synergism between bromine and phosphorus present in the same molecule (Weil and Levchik, 2008). Fire-retardant efficiency without the need for antimony oxide opens the door for this product in the field of PP fibers and textiles.

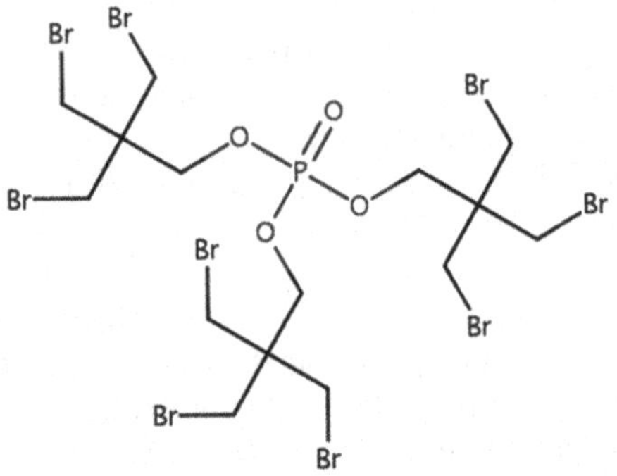

STRUCTURE 9.10   Chemical structure of Tris(tribromoneopentyl) phosphate

A suspected carcinogen, Decabromodiphenyl ether (decaBDE) (Figure 5.1 (d), m = n = 5, see Chapter 5) is also used for polyethylene wire and cable applications included at a level of 20–24 wt%. Chlorinated paraffin with a loading of about 25 wt % is also used.

The reactive modification of polyolefins via copolymerization with P-containing monomers is not compatible with conditions that apply during the polymerization of olefin monomers, especially using organometallic coordination catalysts. However, both polyethylene (PE) andpolypropylene (PP) can be chemically modified after polymerization to introduce P-containing side groups.

For example, PE has been oxidatively phosphonylated to give a polymer containing about 5 wt % P in $[-P(O)(OCH_3)_2]$ side groups. This polymer has an LOI significantly greater than that of unmodified PE, but unfortunately, the crystallinity is greatly reduced, leading to the inferior physical and mechanical properties (Banks et al., 1993).

9.6   PVC

Liquid chlorinated paraffins are the main halogen-containing fire-retardant additives used for polyvinyl chloride often in combination with a phosphate ester. In this case, the chlorinated paraffins have the secondary function of plasticizers. The thermal degradation mechanism of chlorinated paraffins is similar to that of PVC) so in this case polyvinylchloride stabilizers also have the secondary function of stabilizing chlorinated paraffins.

For more demanding applications, tetrabromophthalate ester (Structure 9.11), a thermally stable, liquid fire-retardant additive with a bromine content

of approximately 45 wt %, is used. A suspected carcinogen, decabromodiphenyl ether (decaBDE) (Figure 5.1(d), m = n = 5) is used for foamed soft PVC
for thermal insulation, even if diphenyl ether-free systems have been developed
because of environmental concerns.

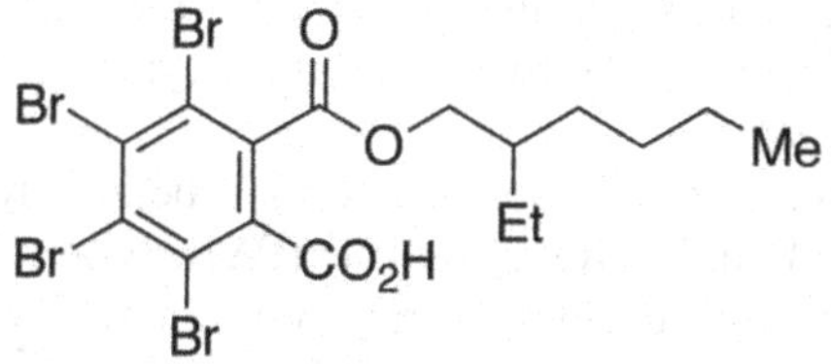

STRUCTURE 9.11 Chemical structure of tetrabromophthalate ester

9.7 FIBER BLENDS

Experience has shown that flame retardants that are effective on one fiber,
when in contact with a second differently flame-retarded fiber, may prove to
be antagonistic and render the blend flammable. Consequently, the current
rules for the simple flame-retarding of blends are either to apply flame-
retardant only to the majority fiber present or apply halogen-based back-
coatings, which are effective on all fibers because of their common flame
chemistries in the vapor phase.

The widespread use of cotton–polyester blends coupled with the apparent
flammability-enhancing interaction in which both fiber components participate
(the so-called scaffolding effect (Horrocks, 1986) means that cotton/polyester has
promoted greater attention than any other blend). However, because of the
observed interaction, only halogen-containing coatings and back-coatings find
commercial application to blends that span the whole blend composition range.

The materials prepared from the blending of cotton–viscose and polyester
are used widely. In spite of considerable efforts, the flammability problem of
these blends is yet to be solved satisfactorily. The physical and chemical properties of these two types of fibers are quite different. Cotton decomposes at
a much lower temperature with the formation of char, while polyester melts at
about $260°C$ with dripping. In case of blends, the molten polymer cannot drip
or flow away from the flame. It remains entrapped in the char that act as
candle wick and continues to burn. Though cotton decomposes first, it burns
after burning of the polyester melt that coats the cotton fiber. The burning of
cotton provides additional heat to the flame. When diammonium phosphate is
used, 2% phosphorous is needed to achieve LOI value of 28 for 100% cotton,
as opposed to 3–3.5 % for 50:50 blends and 4–5% for 35:65 blends (Yeh and
Birky, 1973). Cotton, polyester and their 50:50 blends have LOI values of
18–19, 20–21, and 18 respectively (Schindler and Hauser, 2004). An opposite
tendency (higher LOI of 35) was, however, observed when 40–60% cotton is

blended with modacrylic (LOI of 33). It is difficult to find a single FR which would penetrate both the fibers at the same time to render a durable effect. The formulations may consist of two-component FRs suitable for each type of fiber, but they must be compatible and should have same fixation conditions.

High amounts of FRs are required to make blends of natural and synthetic fibers flame retardant. The necessary amount of FRs may also be applied as fabric coating with the help of latex binder and softener. DBDPO and antimony trioxide can provide excellent flame retardancy; the add-on required, however, is as high as 37%.

The fabrics from mixtures of wool (W) and synthetic polyamide (PA, Capron), polyester (PE, Lavsan), and polyacrylonitrile (PAN, Nitron) fibers in ratios from 10:90 to 90:10 (mass%) were modified by fire-retardant technology using laser-radiation energy (Besshaposhnikova et al., 2006). The modification was carried out by an aqueous solution (10%) of the phosphorus- containing fire retardants (FR) dimethylmethylphosphonate (DMMP), methylphosphonamide, and phosdiol (PD). The CO_2-laser radiation (LR) power density for the mixed cloths was 4.5–5.0 W/cm^2 as the most optimal (Pulina and Besshaposhnikova, 2013).

The FR chemicals that are commercially used for CO/PET blends are not sufficient to comply with the European standards and they have application difficulties, such as extra equipment requirement for binding. FRs based on phosphorous compounds have shown that they have lower toxicity profiles, as compared to halogen-based counterparts (Hirsch et al., 2017). However, the phosphorus–nitrogen PNFR systems (N/P-CH$_2$OR(H)-type FRs or crosslinkers) such as Fyrol 76, Fyroltex HP, Pyrovatex CP New, and Proban have a formaldehyde-release problem, despite showing a good N–P synergism (Holme and Pater, 1980).

With regard to cotton, polyester, and CO/PET blends, research to develop efficient formaldehyde-free, char-former FR systems is ongoing (Atakan et al., 2018). Li et al. (2010) synthesized a novel polymeric FR with P–N pendent groups (PVP-P-DCA) by polyvinyl alcohol (PVA), phosphoric acid, and dicyandiamide (DCA). The steps are as follows:

1. PVA (25g) + hydroxyl-functional polyester (5g), is mixed in 120 ml water in 1 hour.
2. 55mL (85%) orthophosphoric acid is added in the mixture for 30 min.
3. 1.2g urea added and boiled for 3 hours.
4. The suspended solution is precipitated in 250 ml ethanol.
5. (Step 4) is repeated three times.
6. The white product (PVA(PR)-POH) is dried at 60°C for 8 hours.
7. 35 g PVA(PR)-POH is mixed with 45 ml water.
8. Temperature is raised to 70°C, and 13g DCDA is added.
9. The solution is boiled for 2 hours.
10. The solution is cooled and precipitated in 100mL ethanol, and is then filtered.
11. Process is repeated three times.
12. The product (PVP-P-DCA) is filtered and dried for 8 h at 60°C.

This FR was applied to CO/PET fabrics with 30:70 blend ratio by pad-dry thermosol finishing process. As a result, it was observed that PVP-P-DCA is a reactive FR agent on CO/PET blends with high efficiency. The LOI value was about 26.4 after washing 10 times; however, the washing process used in the study is extremely mild and cannot meet the current standards of washing conditions. Li et al. (2011) further developed a hydroxyl-functional, organo-phosphorus FR with the chemical structure of 2,2-dihydroxymethylpropane-1,3-diolylbis (hydrogen phenylphosphonate) (DHDBP) and implemented to 30%/70% CO/PET blends. The results indicated that DHDBP was reactive and a good char-forming agent for the CO/PET blended fabrics. However, it was not durable against 20 laundering cycles due to insufficient fixation of DHDBP on the fiber. Developing phosphorus-based, UV curable FR coating systems have been attempted in the literature since 2011. In 2015, Mayer-Gall et al. (2015) developed a UV curable FR agent (allyl-functionalized polypho-sphazenes) and applied this product onto cotton and different cotton–polyester blended fabrics, using an impregnation method followed by UV-grafting to achieve the permanent effect. Results showed that the treated fabrics exhibited good flame retardancy (LOI: 24–27 for CO/PET blends) with char formation. The UV-cured FR coatings showed also a good washing fastness. However, this product needs improvement in terms of solubility in aqueous systems to be a commercial textile-finishing product.

Apart from grafting flame retardant films on textile surfaces, an efficient approach for the covalent bonding of phosphorus-based (Yoshioka-Tarver et al., 2012) and phosphorus–nitrogen-based flame retardants (Nguyen et al., 2012) were developed for cotton. On subjecting to treatments, compound fab-rics with LOI values of more than 30% were obtained.

With regard to fabrics made of PET, a mixture of a bis-phosphonic acid derivative and ammonium sulfamate shows not only efficient flame retardancy, but also anti-dripping properties when applied to PET (Feng et al., 2012).

These compounds were applied onto PET fabrics by padding and curing pro-cess in different ratios. LOI values were more than 28%. Furthermore, the after-flame time was 0 s and the fabrics showed an intumescent char structure after burning. Instead of the current phosphorus-based synthetic products, eco-friendly and sustainable FRs have been sought since 2014. For example, Carosio et al. (2014) studied the FR effect of caseins on polyester and CO/PET (35/65%) fabrics. Fabrics were treated with an aqueous suspension of caseins through the pad-dry-cure method. The results showed that caseins had enhanced the resist-ance of cotton, polyester, and CO/PET blends against a methane flame and strongly reduced the burning rate in the case of polyester and CO/PET blends. The studies related to developing P–N-based FR systems also increased because of the synergetic effects of P and N. As an example, Jiang et al. (2015) developed a P–N-based intumescent FR system which was applied to cotton, nylon, and polyester fabrics using padding method. As a result of the vertical flammability test, treated nylon exhibited high flame retardancy, while treated cotton and poly-ester exhibited moderate flame retardancy.

Li et al. (2010) developed a formaldehyde-free P–N FR chemical with multiple reactive groups, dioxo(3-triethylphosphite-5-chlorine-1-triazine) neopentyl glycol (DTCTNG) for cotton fabrics.

In order to enhance the durability of FR properties, polyester resin (PR) was used in addition to PVA and dicyandiamide (DCDA) in the formulation. 100% polyester, 100% cotton, and 50/50% CO–PET blends were treated via pad-dry-thermosol process with this novel polymeric flame retardant PVP (PR)-P-DCDA. FR performance and thermal properties of treated fabrics were investigated.

FTIR results verified that FR was successfully transferred onto the fabrics. The thermal stability of the treated fabrics in nitrogen, as well as their resistance to a flame application has proven to be strongly affected by PVP (PR)-P-DCDA treatment. DSC showed that the treated polyester fabrics tend to melt in lower temperatures and crystalize quickly in higher temperatures than the untreated ones. TGA results indicated that decomposition temperature of the treated fabrics were reduced by treatment and PVP(PR)-P-DCDA are able to protect all three types of fabrics against their thermal degradation and favoring the formation of a stable char, which differs from the untreated ones. These results suggested that the addition of FRs onto the cotton, polyester, and CO/PET might have reduced the flammability via PVP (PR)-P-DCDA dehydration into char. With regard to the FR performance of fabrics, the treated fabrics were not ignited and did not show afterglow in the vertical flammability test. Indeed, in the case of polyester, a significant increase in LOI (from 22.5% to 33%) has been reached, even at 8–9% add-ons. In the case of cotton and CO/PET blends, PVP (PR)-P-DCDA turned out to slow down in char length and weight loss and FR properties (LOI values of greater than 26) were obtained in higher concentrations (Atakan et al., 2018).

At a larger scale, Eksoy Chemical Company commercialized PVP (PR)-P-DCDA under the name Fire-off EBR. In conclusion, Fire-off finish applied to lightweight fabrics is suitable for daily use and finds applications in many areas including curtains, upholstery, and garment fabrics. While commercial FR products such as Pyrovatex CP (N-methylol phosphonopropionamide) can be used for cotton or its polyester blends of up to 30% synthetic fiber content, Fire-off can be applied to cotton, polyester, and CO/PET fabrics in any blending ratios (Hicklin et al., 2008).

9.8 THERMOSETTING RESINS

In thermosetting systems, the reactive flame retardant can be incorporated either in one or more of the principal chain-forming components, or in the cross-linking agent. Both strategies have been employed with P-containing flame retardants in a variety of thermosets.

Since the P–H bond (like N–H, O–H, and S–H bonds) can add across an epoxy group with consequent ring-opening, hydrogen phosphonates or phosphinates can readily be incorporated covalently into epoxy resins. This reaction is currently widely used to incorporate DOPO into epoxies destined for electronic applications. Reacting DOPO first with reagents such as p-benzoquinone or naphthaquinone gives (after mild reduction) diphenolic derivatives (e.g., DOPO-HQ) that can be incorporated in an epoxy resin through a conventional chain-extension process with no net loss of functionality (Wang and Lee, 2000).

9.9 POLYURETHANES

Nowadays, the automotive, construction, and aviation industries focus on reducing the weight of vehicles and machines. Thus, many structures are largely made from lightweight polymeric materials as well as polymer-based composites. The rigid polyurethane (PUR) foams play an important role in construction (Borreguero Ana et al., 2013). Due to their low fire resistance, they have to be appropriately modified. One of the basic methods for the preparation of PUR with reduced flammability is adding flame retardants.to them So far, the most commonly applied flame retardants have contained chlorine and bromine. During combustion of the foam containing halogen compounds, toxic gases and fumes cause environmental pollution, which threatens the health and life of humans. Regarding these risks, the EU introduced restrictions on their application (Gu et al., 2015).

ECHA (European Chemical Agency) has recommended that a restriction proposal is prepared on the flame retardants Tris(2-carboxyethyl)phosphine hydrochloride (TCEP), Tris (chloroisopropyl) phosphate (TCPP)and Tris (1,3-dichloro-2-propyl)phosphate (TDCP) in flexible PUR foams in childcare articles and residential upholstered furniture.

TCPP and TDCP are used as flame retardants in flexible PUR foams in products such as baby mattresses, car safety seats, baby slings, and residential upholstered furniture. Although TCEP is not in use in the EU, ECHA says it may be present as an impurity in other commercial flame retardants or in imported articles. The three substances are being treated as a group because they have similar uses and are structurally and toxicologically similar (Chemical watch, 2018).

Due to environmental aspects, industry began to apply environmentally friendly flame retardants such as nanofillers, expanded graphite and phosphorus/nitrogen compounds to PUR. Environmentally friendly flame retardant additives help to reduce the amount of emitted toxic gases and fumes. Depending on the type of flame retardant, there are several mechanisms for reducing the flammability of PUR foams. Phosphorus–nitrogen compounds are inhibitors of the combustion in the gas phase and form a glassy coating, limiting heat transfer to the interior of foam. Nanofillers (for example, montmorillonite) in the composition form a barrier for emitted gaseous decomposition products of the polymer, whereas expanded graphite during the combustion forms a foamed carbon coating, hampering heat flow. To obtain the desired effect of fire resistance, even 40% of the flame retardant should be introduced to PUR foams, what adversely affects the mechanical properties and increases the manufacturing costs. Utilizing a synergic effect occurring between different types of flame retardants, the flammability of PUR polymer matrix can be successfully limited by introducing a smaller amount of flame-retardant additives (Cheng-Qun et al., 2013).

Czech-Polak et al. (2016) prepared new compositions of PUR foams with reduced flammability containing environmentally friendly flame retardants. Specimens of rigid PUR foams were manufactured in the hot pressing process using a press which is part of the tooling in long fiber injection (LFI) process. At the first step, the polyol component was mechanically mixed for 5 min with selected flame retardant additives. Then, the modified polyol component and the isocyanate were stirred together for 25 s. In the final step, the PUR mixture was poured into a mold heated to 65°C. The mold was closed under the pressure of 50 MPa, for 360 s. The same method was used to prepare the reference polyurethane sample without flame retardant. As flame retardants, the following compounds were used:

- Ammonium polyphosphate (APP)
- Melamine pyrophosphate (MPYP)
- Triethyl phosphate (TEP)
- Expanded or expandable graphite (also known as *intumescent flake graphite*),a synthesized intercalation compound of graphite that expands or exfoliates when heated, and is produced by treatment of flake graphite with various intercalation reagents that migrate between the graphene layers in a graphite crystal and remain as stable species
- Bentonite modified benzyl(hydrogenated tallow alkyl) dimethyl ammonium chloride
- Bentonite from Russian deposits modified with butyl(triphenyl) phosphonium chloride

All prepared PUR foams containing flame retardants are characterized by acceptable resistance to flame. Burning time values obtained during the UL 94 HB test lead to the conclusion that the foam containing flame retardants in liquid form (triethyl phosphate) has the best flame resistance, most likely due to the easier and better homogenization of flame retardant additives and polyol. Modification of polyurethane foams using suitably selected composition of flame retardants may also improve the transparency of the fumes released during the combustion of the material. In this test, the highest values were obtained for the foam containing melamine pyrophosphate and expanded graphite. The addition of flame retardant caused only slight worsening of the tensile strength of the investigated polyurethane foams. The utilizing of suitably selected flame retardants for PUR foams allows manufacturing lightweight foam materials characterized by high flame resistance with satisfying mechanical properties. Analysis of all tests leads to the conclusion that the best is a composition marked as 1.9% bentonite modified butyl(triphenyl) phosphonium chloride, 4.0% ammonium polyphosphate, and 3% expanded graphite. This PUR foam was characterized by increased fire resistance and decreased smoke density with only a small decrease in mechanical properties.

A common approach with segmented polyurethanes (PUs) has been to incorporate a P-containing group (phosphate, phosphonate, phosphine oxide, or phosphazene) into a diol that is then chain-extended with a diisocyanate (e.g., Structure 9.12. P-containing group with extended diisocyanate) (Lee et al., 1984). Phosphonate was found to be a more effective flame retardant than phosphate.

STRUCTURE 9.12   P-containing group with extended diisocyanate

The most important application of PUs is as foams for which flammability can be particularly high, owing to their high surface areas and, in the case of flexible foams, their open-cell structures. Commercially, such foams are still flame-retarded mainly with organophosphorus additives, albeit sometimes of high molecular weight. The use of high molecular weight additives is particularly important for foams used in automotive applications in which the use of more volatile, low molecular weight, additives can lead to the excessive fogging of glass surfaces. Reactive organo-PFRs such as P-containing diols (e.g., Structure 9.2) have clear potential for such applications.

9.10   CARBON FIBER

With respect to tensile strength, carbon fiber significantly exceeds steel of an equivalent weight. Importantly, such performance is accompanied by outstanding stiffness and resistance to high temperature, chemical inertness, and fatigue characteristics. Carbon and graphite fibers are high-carbon-content materials that are generally produced using a continuous process of controlled pyrolysis of PAN fiber at temperature 1000°–3000°C. Carbon fibers used in the composite may be 6–8 mm in diameter.

The thermal stability of carbon fiber has been known since at least 1879, when Thomas Edison used it as the filament in the light-producing bulb. The important properties of carbon fibers significant for the aviation industry include strength, stiffness, high strength-to-weight ratio, outstanding fatigue characteristics, flame retardancy and stability at high temperature, and chemical resistance (Anonymous, 2019).

Carbon–carbon fiber composites were produced by bonding carbonaceous fibers together using a binder (such as phenolic resin) and heating to higher carbonizing temperatures. The carbon–carbon composite obtained showed high strength-to-weight ratio. It may be stable to high temperatures of 800°C–3000°C without any significant reduction of strength (Taylor, 1981).

FR fiber composites will remain of significant research interest, particularly those meeting the requirements for durable performance. Carbon fiber, para-aramid fiber, and ceramic fiber have provided the inherent flame retardancy and heat resistance coupled with several other high-performance effects including high strength characteristics and resistance to environmental effects and fatigue. Therefore, these fibers occupy significant consumption in producing composite structures for spacecraft (Uddin, 2016).

9.11   FUTURE TRENDS

Cellulosic-based natural and man-made fibers are highly flammable. Based on this fact, FRs have been developed and used since ancient times. The most popular and inexpensive halogen compounds, primarily bromide, have been affordable and have satisfied our needs; they are based on vapor-phase. However, studies have shown that these compounds and substrates made flame-resistant by using them are harmful to human beings, and a few of them are even carcinogenic. Consequently, phosphorous-based FRs have started dominating as FRs. They dehydrate and form cross-links within substrates, and ultimately convert substrates into char. As discussed, synthetic fibers are not easy to char. Furthermore, hydrophobicity and melting are two disadvantages of making synthetic fibers flame-resistant. To address these problems, back-coating and IFRs are alternative ways to make synthetic fibers flame-resistant. In this book, flame retardancy and fire retardancy of various nontextile materials (e.g., plastics, resins) have been discussed with the objective of h helping researchers to find newer FRs for synthetic fibers.

REFERENCES

Anderson J.J., Camacho V.G., and Kinney R.E. (1974). (a) 'Cyclic phosphonate esters and their preparation', US Patent 3,789,091, 1974; (b) 'Fire retardant polymers containing thermally stable phosphonate esters', US Patent 3,849,368, 1974; both patents assigned to Albright & Wilson Inc.

Anonymous (2019). A brief history of carbon fiber (2019). Brief History of Carbon Fiber, https://dragonplate.com/a-brief-history-of-carbon-fiber, 25th June, accessed on 12.5.2020.

Apaydin K., Laachachi A., Ball V., Jimenez M., Bourbigot S., Toniazzo V., and Ruch D. (2014). Intumescent coating of (polyallylamine-polyphosphates) deposited on polyamide fabrics via layer-by-layer technique, *Polymer Degradation and Stability*, 106, 158–164.

Asrar J., Solutia Inc. (1988). Polymer-bound non-halogen fi re resistant compositions, U.S. Patent, 5 750 603.

Atakan R., Bical A.R., Celebi E., Ozcan G., Soydan N., and Sarac A.S. (2018). Development of a flame retardant chemical for finishing of cotton, polyester, and CO/PET blends, *Journal of Industial Textiles*, 1–21. DOI: 10.1177/1528083718772303.

Atkinson P.A., Haines P.J., Skinner G.A., and Lever T.J. (2000). Studies of fire-retardant polyester thermosets using thermal methods, *Journal of Thermal Analysis and Calorimetry*, 59, 395–408.

Avinc O., Day R., Carr C., and Wildin M. (2012). Effect of combined flame retardant, liquid repellent and softener finishes on poly(lactic acid) (PLA) fabric performance, *Textile Research Journal*, 82(10), 975–984.

Bajaj P., Agrawal A.K., Bajaj P., Agrawal A.K., Dhand A., Kasturia N., and Hansraj (2000). Flame retardation of acrylic fi bers: An overview, *Polymer Reviews*, 40, 309–337.

Balakrishnan H., Hassan A., Isitman N.A., Kaynak C. et al. (2012). On the use of magnesium hydroxide towards halogen-free flame retarded polyamide-6/polypropylene blends, *Polym Degrad Stabil*, 97, 1447–1457.

Banks M., Ebdon J.R., and Johnson M. (1993). Influence of covalently bound phosphorus-containing groups on the flammability of poly(vinyl alcohol), poly(ethylene-co-vinyl alcohol) and low-density polyethylene, *Polymer*, 34, 4547–4556.

Carosio F., Di Blasio A., Alongi J., and Malucelli G. (2013). Green DNA-based flame retardant coatings assembled through layer by layer, *Polymer*, 54(19), 5148–5153.

Besshaposhnikova V.I., Kulikova T.V., et al. (2006). Method for production of fire-retardant polyester fibrous material, RU Pat. 2,281,992, Aug. 10, 2006; Appl. No. 20050105921.

Borreguero Ana M., Sharma P., Spiteri C., Velencoso M.M., Carmona M.M., Moses J. E., and Rodríguez J.F. (2013). A novel click-chemistry approach to flame retardant polyurethanes, *Reactive and Functional Polymers*, 73(9), 1207–1212.

Bourbigot S. and Fontaine G. (2010). Flame retatdancy of polylactide: An overview, *Polymer Chemistry*, 1, 1413–1422.

Carosio F., et al. (2014). Flame retardancy of polyester and polyester–cotton blends treatedwith caseins, *Industrial & Engineering Chemistry Research*, 53, 3917–3923.

Chang S., Zhou X., and Xing Z. (2017). Computing solubility parametersof phosphorous flame retardants by molecular dynamics andcorrelating their interactionswith poly(ethylene terephthalate), *Textile Research Journal*, November 13 1–9. DOI: 10.1177/0040517517741161.

Chang S.J. and Chang F.C. (1999). Synthesis and characterization of copolyesters containing the phosphorus linking pendent groups, *Journal of Applied Polymer Science*, 72, 109–122.

Chemical watch (2018). Echa recommends restriction on flame retardants in polyurethane foams, 12 April 2018., https://chemicalwatch.com.

Chen D., Wang Y., Hu X., Wang D., Qu M., and Yang B. (2005). Flame-retardant and antidripping effects of a novel char-forming flame retardant for the treatment of poly(ethylene terephthalate) fabrics, *Polymer Degradation and Stability*, 88, 349–356.

Cheng X.-W., Guan J.P., Tang R.-C., and Kai-Qiang Liu K.-Q. (2016). Improvement of flame retardancy of poly(lactic acid) nonwoven fabric with a phosphorus-containing flame retardant, *Journal of Industrial Textiles*, 46(3), 914–928. DOI: 10.1177/1528083715606105.

Cheng-Qun W., Feng-Yan G., Jie S., and Zai-Sheng C. (2013). Effects of expandable graphite and dimethyl methylphosphonate on mechanical, thermal, and flame-retardant properties of flexible polyurethane foams, *Journal of Applied Polymer Science*, 130, 916. DOI: 10.1002/app.39252.

Cireli A., Kutlu B., and Mutlu M. (2007). Surface modification of polyester and polyamide fabrics by low frequency plasma polymerization of acrylic acid, *Journal of Applied Polymer Science*, 104(4), 2318–2322.

Coquelle M. (2014). Flame retardancy of polyamide 6 fibers: The use of sulfamate salts, Ph.D Thesis, Université Lille 1, France.

Crook V.L. (2004). Flame retardant acrylonitrile copolymers, Ph.D. Diss., University of Sheffield, Sheffield, UK.

Czech-Polak J., Przybyszewski B., Heneczkowski M., Czulak M., and Gude M. (2016). Effect of environmentally-friendly flame retardants on fire resistance and mechanical properties of rigid polyurethane foams, *Polimery Journal (Industrial Chemistry Research Institute, Warsaw, Poland)*, 6(2), 113–116. DOI: 10.14314/polimery.2016.113.

Dittrich B., Wartig K.A., Mülhaupt R., and Schartel B. (2014). Flame-retardancy properties of intumescent ammonium poly(phosphate) and mineral filler magnesium hydroxide in combination with graphene, *Polymers*, 6, 2875–2895.

Ebdon J.R., Hunt B.J., Joseph P., Konkel C.S., Price D., Pyrah K., Hull T.R., Milnes G. J., Hill S.B., Lindsay C.I., McCluskey J., and Robinson I. (2000).Thermal degradation and flame retardance in copolymers of methyl methacrylate with diethyl(methacryloyloxymethyl)phosphonate, *Polymer Degradation and Stability*, 70, 425–436.

Ebnesajjad S. and Ebnesajjad C. (2013). *Surface Treatment of Materials for Adhesive Bonding*, 2nd edition. Elsevier.Ec., Europa, EU. (2017).

ECHA (European Chemicals Agency) (2018). https://echa.europa.eu/candidate-list-table/-/dislist/details/0b0236e181f392bf. accessed on 16.1.2018.

Endo S., Kashihara T., Osako A., Shizuki T., and Ikegami T. (1978). Toyo Boseki Kabushiki Kaisha, Phosphorus-containing compounds, U.S. Patent: 4 127 590.

Feng Q., et al. (2012). An antidripping flame retardant finishing for polyethylene terephthalatefabric, *Industrial & Engineering Chemistry Research*, 51, 14708–14713.

Gao M., Wu W., and Xu Z. (2013). Thermal degradation behaviors and flame retardancy of epoxy resins withnovel siliconcontaining flame retardant, *Journal of Applied Polymer Science*, 27, 1842.

Gao M. and Yang S.S. (2010). A novel intumescent flameretardant epoxy resins system, *Journal of Applied Polymer Science*, 115(4), 2346–2351.

Garlotta D. (2001). A literature review of poly(lactic acid), *Journal of Polymers and the Environment*, 2001(9), 63–84.

Heywood D. (2003). *Textile Finishing*. SDC, Bradford, UK.

Hirsch C., Striegl B., Mathes S., Adlhart C., Edelmann M., Bono E., Gaan S., Salmeia K.A., Hoelting L., Krebs A., et al. (2017). Multiparameter toxicity assessment of novel DOPO derived organo-PFRs, *Archives of Toxicology*, 91(1), 407–425. DOI: 10.1007/s00204-016-1680-4.

Holme I. and Pater S. (1980). A study of nitrogen–phosphorus synergism in the flame–retardant finishing of resin–treated polyester–cotton blends, *Coloration Technology*, 96, 224–237.

Horrocks A.R. (1986). Flame-retardant finishing of textiles, *Review of Progress in Coloration*, 16, 62.

Horrocks A.R., Sitpalan A., Zhou C., and Kandola B.K. (2016). Flame retardant polyamide fibres: The challenge of minimising flame retardant additive contents with added nanoclays, *Polymers*, 8, 288. DOI: 10.3390/polym8080288.

Iler R.K. (1966). Multilayers of colloidal particles, *Journal of Colloid and Interface Science*, 21(6), 569–594.

Isitman N.A., Gunduz H.O., and Kaynak C. (2009). Halogen-free flame retardants that outperform halogenated counterparts in glass fibre reinforced polyamides, *Journal of Fire Sciences*, 28, 87–100.

Jiang W., Jin F.-L., and Park S.-J. (2015). Synthesis of a novel phosphorus-nitrogen-containingintumescent flame retardant and its application to fabrics. *Journal of Industrial and Engineering Chemistry*, 27, 40–43.

Kim H.A. and Kim S.J. (2016). Moisture and thermal permeability of the hollow textured PET imbedded woven fabrics for high emotional garments, *Fibers and Polymers*, 17, 427. DOI: 10.1007/s12221-016-5942-9.

Lee F.T., Green J., and Gibilisco R.D. (1984). Recent developments using phosphorus-containing diol as a reactive combustion modifi er for rigid polyurethane foams 3, *Journal of Fire Sciences*, 2, 439–453.

Lewin M., Brozek J., and Martens M.M. (2002). The system polyamide/sulfamate/dipentaerythritol: Flame retardancy and chemical reactions, *Polymers for Advanced Technologies*, 13(10–12), 1091–1102.

Lewin M., Zhang J., Pearce E., and Gilman J. (2007). Flammability of polyamide 6 using the sulfamate system and organo-layered silicate, *Polymers for Advanced Technologies*, 18(9), 737–745.

Li Q.-L., Wang X.-L., Wang D.-Y., Xiong W.-C., and Zhong G.-H. (2010). A novel organophosphorus flame retardant: Synthesis and durable finishing ofpoly (ethylene terephthalate)/cotton blends, *Journal of Applied Polymer Science*, 117, 3066–3074.

Li Y.-C., Liu K., and Xiao R. (2017). Preparation and characterizations of flame retardant polyamide 66 fiber IOP Conf, *Series: Materials Science and Engineering*, 213, 012040.

Li Y.-C., Mannen S., Morgan A.B., Chang S., Yang Y.-H., Condon B., and Grunlan J.C. (2011). Intumescent all-polymer multilayer nanocoating capable of extinguishing flame on fabric, *Adv Mater*, 23(34), 3926–3931. DOI: 10.1002/adma.201101871.

Mayer-Gall T., et al. (2015). Permanent flame retardant finishing of textiles by allyl-functionalizedpolyphosphazenes, *ACS Applied Materials & Interfaces*, 7, 9349–9363.

Nagelsdiek R., Gobelt B., Omeis J., Freytag A., Greefrath D., and Assignee C., BYK Chemie GmbH (2014). Adhesion promoter for coatings on different substrate surfaces. USA: Patent, Patent number is 8778458.

Nguyen T.M.D., Chang S., Condon B., et al. (2012) Development of an environmentally friendly halogen-free phosphorus-nitrogen bond flame retardant for cotton fabrics, *Polymers for Advanced Technologies*, 23, 1555–1563.

Ömeroğulları Z. and Dilek K. (2012). Application of low-frequency oxygen plasma treatment to polyester fabric to reduce the amount of flame retardant agent, *Textile Research Journal*, 82(6), 613–621. DOI: 10.1177/0040517511420758.

Price D., Pyrah K., Hull T.R., Milnes G.J., Wooley W.D., Ebdon J.R., Hunt B.J., and Konkel C.S. (2000). Ignition temperatures and pyrolysis of a fl ame-retardant methyl methacrylate copolymer containing diethyl(methacryloyloxymethyl)-phosphonate units, *Polymer International*, 49, 1164–1168.

Pulina K.I. and Besshaposhnikova V.I. (2013). Fire-retardant features of wool-containing multi-component cloths for special clothing, *Fibre Chemistry*, 45 (1), 25–30. May (RussianOriginal No. 1, January-February, 2013).

Schindler W.D. and Hauser P.J. (2004). *Chemical Finishing of Textiles*. Woodhead, Cambridge, England.

Smith R., Georlette P., Finberg I., and Reznick G. (1996), Development of environmental friendly multifunctional flame retardants for commodity and engineering plastics. *Polymer Degradation and Stability*, 54(2–3), 167–173.

Smith W.F. and Hashemi J. (2006). *Foundations of Materials Science and Engineering*. 4th.McGraw-HillMcGraw-Hill Book Company, New York, pp. 373–378.

Sypaseuth F.D., Gallo E., Çiftci S., and Schartel B. (2017). Polylactic acid biocomposites: Approaches to a completely green flame retarded polymer, *e-Polymers*, 17(6), 449–462. DOI: 10.1515/epoly-2017-0024.

Taylor G.J. (1981). Process for producing carbon-carbon fibre composites suitable for use as aircraft brake discs. Patent application no. 06/053535, USA.

Tsafack M.J. and Levalois-Grützmacher J. (2006). Plasma-induced graftpolymerization of flame retardant monomers onto PAN fabrics, *Surface and Coatings Technology*, 200, 3503–3510.

Uddin F. (2016). Flame-retardant fibrous materials in an aircraft, *Journal of Industrial Textiles*, 45(5), 1128–1169. DOI: 10.1177/1528083714540700.

U.S. EPA (2005). *Guidelines for Carcinogen Risk Assessment*. Risk Assessment Forum, Washington, DC; EPA/630/P-03/001F. *Federal Register* 70(66), 17765–17817, www.epa.gov/raf.

Wang C.S. and Lee M.C. (2000). Synthesis and properties of epoxy resins containing 2-(6-oxid-6H-dibenz(c,e) (1,2) oxaphosphorin-6-yl) 1,4-benzenediol (II), *Polymer*, 41, 3631–3638.

Weil E.D. and Levchik S.V. (2008). Flame retardants in commercial use or development for textiles, *Journal of Fire Sciences*, 26, 243–281.

Whelan T. (1994). *Polymer Technology Dictionary*. Chapman & Hall, London.

Yeh K.-N. and Birky M.M. (1973). Calorimetric study of flammable fabrics. II. Analysis of flameretardant-treated cotton. *Journal of Applied Polymer Science*, 17(1), 255.

Yoo-Hun K., Jinho J., Song K.-G., Lee E.-S., and Ko S.-W. (2001). Durable flame-retardant treatment of polyethylene terephthalate (PET) and PET/cotton blend using dichlorotribromophenyl phosphate as new flame retardant for polyester, *Journal of Applied Polymer Science*, 81, 793–799.

Yoshioka-Tarver M., et al. (2012). Enhanced flame retardant property of fiber reactive halogenfreeorganophosphonate, *Industrial & Engineering Chemistry Research*, 51, 11031–11037.

Younis A.A. (2012). Protection of aluminum alloy (AA7075) from corrosion by sol-gel technique. PhD Thesis, Chemnitz University of Technology.

Younis A.A. (2017). Protection of polyester fabric from ignition by a new chemical modification method, *Journal of Industrial Textiles*, 47(3), 363–376. DOI: 10.1177/1528083716648761.

10 Flame Retardants and the Environment

10.1 IMPACT OF FRs ON ENVIRONMENT

While FRs could ensure the production of fire safety products, their application is of considerable economic importance in different fields, e.g., transportation, building, electronics, electrical engineering, textiles, wood, and paper industries. There are more than 175 different types of FRs on the market, which contain bromine, chlorine, phosphorus, nitrogen, boron, and antimony compounds or their combinations. FRs are comprised of inorganic and organic compounds with gas-phase and/or condensed-phase mechanisms of FR action. According to their chemical structure, FRs that include reactive groups, can be chemically bonded to the polymer material, whereas nonreactive FRs can only be physically incorporated into the polymer structure, which may easily penetrate the environment (Šehić et al., 2016).

The period between the years 1950 and 1980, is known as the *golden age of flame retardant research* (Horrocks, 2011). During this period, numerous non-durable and durable halogen-, phosphorous- and phosphorus-halogen-containing commercial products were synthesized and commercialized without concerns about their potential toxicity and environmental unacceptability. The production expansion of FRs was also influenced by the continuously increasing use of synthetic polymers and plastics that replaced other materials in numerous applications, due to their excellent functional properties and aesthetic comfort. As the flammability of these materials represents an important limitation in their use, their protection against combustion was of great importance.

Since 1977, articles have been published with warnings about the carcinogenicity and mutagenicity of the brominated FRs, which were widely used in the production of children's sleepwear. The toxicological problems of hazardous FRs and their long-term effects on human health and the environment have been increasingly exposed and discussed (Jurgen, 1998). The market for FR additives is increasing due to raising safety standards worldwide and currently, frequently used additives include organophosphorus compounds. In addition, nanocomposite technology has been recently studied using various nanoparticles, such as organoclay and carbon nanotubes (Šehić et al., 2016).

Flame retardants may undergo pyrolysis and contribute their own new toxic ingredients, but many studies have not been done in this field. The toxicity of some flame retardant components and of their combustion gases is a particular concern for flame retardant finishes, especially if based on halogens and several heavy metals. Therefore, aircraft textile equipment has to fulfill special requirements, for example, smoke density and toxicity tests.

The major single cause for fire fatalities is the release of carbon monoxide, which reacts with blood haemoglobin, produces carboxy-hemoglobin and causes lower oxygen content in the blood and reduced rate of blood flow. Carbon monoxide causes:

- Nausea at 100 ppm,
- Severe poisoning at 1000 ppm and,
- Death within 1–3 minutes at 12,000 ppm, and
- Damaged tissues by diffusing into them.

HCN, nitrogen oxides form methoglobin and their dangerous concentration levels are 100 and 150 ppm, respectively. SO_2 is irritating at long exposure above 50–100 ppm. H_2S is rapidly fatal above 1000 ppm. Carbonyl sulfide (COS), formed during decomposition of wool, is highly toxic and its limiting concentration is 0.15 ppm. COS is the chemical compound with the linear formula OCS, normally written as COS as a chemical formula that does not imply its structure; it is a colorless flammable gas with an unpleasant odor. COS is a linear molecule consisting of a carbonyl group double bonded to a sulfur atom. Carbonyl sulfide can be considered to be intermediate between carbon dioxide and carbon disulfide, both of which are valence isoelectronic with it.

Toxicity problems include (Schindler and Hauser, 2004):

- Halogenated compounds, especially aromatics, are capable of generating polyhalogenated dioxins and furans, which are highly toxic and can cause reproductive problems and damage the immune system. The cause of dioxin formation is often connected with the use of brominated FRs in combination with antimony oxides as synergist additives.
- No determination has been made regarding whether hexa-bromium or penta-bromium compounds, e.g., hexabromocyclododecane (HBCD), or decabromodiphenyl oxide (DBDPO) is more dangerous.
- Waste water contains dust composed of antimony oxide, phosphorous, antimony, and zirconium compounds.
- Halogenated organic flame retardants, especially aromatic ones, get into waste water; often, they are only slowly biodegradable and cause high AOX (halogenated adsorbable organics) values.
- Formaldehyde is released during curing of the permanent flame retardant finishes of cellulose and free formaldehyde of finished fabrics (storage, transport).
- Different brominated substances have been bioaccumulated in the food chain, which represents a serious health risk for the general population.
- The most pressing health and environmental problems associated with the use of triaryl-phosphates arises from their ability to induce delayed polyneuropathy. This rare neurotoxic toxic effect in humans

and experimental animals has been determined for tricresyl-phosphate (TCP), whereas in the literature, there are contradictory arguments about the toxicity of triphenyl phosphate (TPhP) (Anonymous, 2016); namely, some papers report possible neurotoxicity of TPhP, while other papers deny these claims (Reemtsma et al., 2008). TPhP is reported to cause allergenic effects on humans and to possibly cause toxicity to water organisms. The toxicological studies of other triaryl-phosphates are very rare with insufficient data available in the literature.

When brominated FRs are physically incorporated into textiles and plastics, they can easily leach during products' manufacturing, use, and laundering. Therefore, significant concentrations of brominated compounds have been detected in the environment (Horrocks, 1997).

10.2 BANNED AND RESTRICTED FRs

During the last few decades, knowledge about the toxicity and environmental impact of chemicals has rapidly grown, along with public awareness of asso-ciated potential dangers. A large number of flame retardant compounds, mostly halogen-based, are on special lists of national or international envir-onmental committees because of their harmful properties. Many of these substances are either banned, or their use has been restricted. Scientific and political discussions have focused on these products because of their growing worldwide presence in sediments, building dust, microorganisms, fish and animals such as polar bears, seals, raptors and their eggs, and even in human blood, tissue, and breast milk (D'Silva et al., 2004; Hardy et al., 2009; Schecter et al., 2009).

The halogen-based fire retardants are very effective in reducing fire risk, i.e., the probability of occurrence of a fire, but they present high fire hazard, that is, the probability of producing toxic, corrosive, obscuring smokes while performing the fire-retardant action. The radical trapping in the gas phase performed by hydrogen halides (HX) is bound to increase production of carbon monoxide that would otherwise be oxidized by OH radicals. Further-more, restriction of oxidation increases the quantities of nonoxidized products that may condense into droplets or particles when they leave the flame, increasing the optical density of the smoke. Finally, HX and metal halides are highly corrosive. The ensuing threat to people, structures, and goods involved in fires may discourage the use of these fire retardants in spite of their high effectiveness and versatility, which is far ahead of any other system developed so far. Moreover, the environmental impact of halogenated fire retardants during their entire life cycle, including end-of-life disposal, has been of growing concern since planetary contamination by bioaccumulation of synthetic halo-genated compounds such as 1,3,7,8-tethrachlorodibenzodioxine (TCDD, dioxin) (Structure 10.1), benzofurans, (Structure 10.2) and PCBs (Structure 10.3) were discovered.

STRUCTURE 10.1 (1,3,7,8)-Tethrachlorodibenzodioxine (dioxin)

STRUCTURE 10.2 Benzofurans

STRUCTURE 10.3 Polychlorobyphenyls

Phosphorous, antimony, and zirconium compounds remains in the waste water in the FR- manufacturing and FR-finishing industries. Phosphorus is a common constituent of agricultural fertilizers, manure, and organic wastes in sewage and industrial effluent. It is an essential element for plant life, but when there is too much of it in water, it can speed up eutrophication (a reduction in dissolved oxygen in water bodies caused by an increase in mineral and organic nutrients) of rivers and lakes. Soil erosion is a major contributor of phosphorus to streams (USGS, 2018).

10.2.1 Dioxins

Dioxins are found throughout the world in the environment and they accumulate in the food chain, mainly in the fatty tissue of animals. They are environmental pollutants. Dioxins belong to a group of dangerous chemicals known as persistent organic pollutants (POPs), also named *dirty dozen*. Dioxins are of concern because of their highly toxic potential. Dioxins are highly toxic and can cause reproductive and developmental problems, damage the immune system, interfere with hormones, and cause cancer.

The toxicity of polychlorinated hydrocarbons, including polychlorinated naphthalenes were known from very early days due to a variety of industrial accidents (Drinker et al., 1937).

However, the first evidence of bioaccumulation and toxicity on animals was noted in 1966 when emaciated seabird corpses with very high PCB body burdens washed up on beaches (Jensen, 1966). The concern over the toxicity and persistence (chemical stability) of PCBs in the environment led the United States Congress to ban their domestic production in 1977. PCBs are persistent organic pollutants and despite the production ban in the 1970s, they still persist in the environment and remain a focus of attention (EHP, 1978). Their use as fire retardants was discontinued when their toxicity was discovered.

Nowadays, the main concern on brominated aromatic flame retardants such as PBDE is due to an industrial accident, known as the *Seveso disaster* that focused public attention for the first time on the very high toxicity of dioxins. The halogenated aromatic fire retardants, such as PCBs, PBBs, and PBDEs, on heating under various laboratory conditions produce supertoxic halogenated dibenzofurans and dibenzodioxins (Buser, 1987).

In February 2003, the Restriction of Hazardous Substances Directive (RoHS) was adopted by the EU (ec.europa.eu, 2017). This directive is on the restriction of the use of certain hazardous substances in electrical and electronic equipment. The RoHS directives were implemented on July 1, 2006, and are required to be enforced and included in the laws of each member state. This directive restricts the use of six hazardous materials, included brominated flame retardants, in the manufacture of various types of electronic and electrical equipment. RoHS is also connected with the Waste Electrical and Electronic Equipment (WEEE) Directive2002/96/EC. This Directive sets collection, recycling, and recovery targets for electrical goods and is part of a legislative initiative to solve the problem of huge amounts of toxic e-waste. The RoHS restricts the use of PBBs and PBDPEs. The maximum concentrations allowed are 0.1 wt % of the homogeneous material. The WEEE directive imposes a separated collection of polymer materials fire-retarded with brominated aromatics, unless the size is below 10 cm^2 for printed circuits. While PBBs have not been used for many years as a consequence of the results of the EU risk assessments, PBDPEs, and pentabromodiphenyl and octabromodiphenyl ethers have been banned, but decabromodiphenylether (DBDPE) was shown to necessitate no risk reduction measures. Thus, following the positive outcome of the risk assessment, in a Commission Decision published on October 15, 2005, the EU has exempted DBDPE from the RoHS Directive. The brominated flame retardants were one of the first categories investigated by EU's Registration, Evaluation and Authorisation of Chemicals (REACH), and at the end of 2008, the registration was practically complete. In contrast to RoHS, the REACH regulation has the tendency to avoid general banning of chemicals by the class to which they belong, but to focus on risk assessment programs. This includes complete reconsideration of polymer flame retardants in the direction of ecologically friendly systems, as well as those that are extremely effective and have important roles in saving lives, preventing, or reducing the spreading of fire. Equilibrium should be so achieved by balancing flame-retardant benefits and environmental and human risks. The results of an investigation by REACH for the main halogenated flame retardants are summarized next.

- Pentabromodiphenyl ether (PentaBDE) (Structure 10.4(a): The risk has been confirmed.
- Octabromodiphenyl ether (OctaBDE) (Structure 10.4(b): Some risks have been shown. Production was stopped.
- Decabromodiphenyl ether (Deca-BDE) (Structure 10.4(c): No risk was identified.
- Hexabromocyclododecane (HBCD) (Structure 10.4(d): Substance is of very high concern for authorization.
- Short chain (C_{10}–C_{13}) chloroparaffins (SCCPs): These are used in metal working and leather processing and pose a risk to the aquatic environment. No significant risks to human health were identified. In all FR applications, no risk of secondary poisoning through accumulation in the environment or the food chain was found.
- Medium-long-chain chlorinated paraffin: In 2005, the EU Risk Assessment identified a risk of accumulation in the food chain, and suggested risk reduction measures for all applications.
- Tetrabromobisphenol A (TBBPA): TBBPA [Chapter 5, Figure 5.1(b)] presents no risk to human health. Therefore, TBBPA is not subject to any classification for health. TBBPA is classified in the EU as an R50/53 substance for the environment: "toxic to aquatic organisms" and "may cause long-term adverse effects in the aquatic environment".

(a)

(b)

(c)

(d)

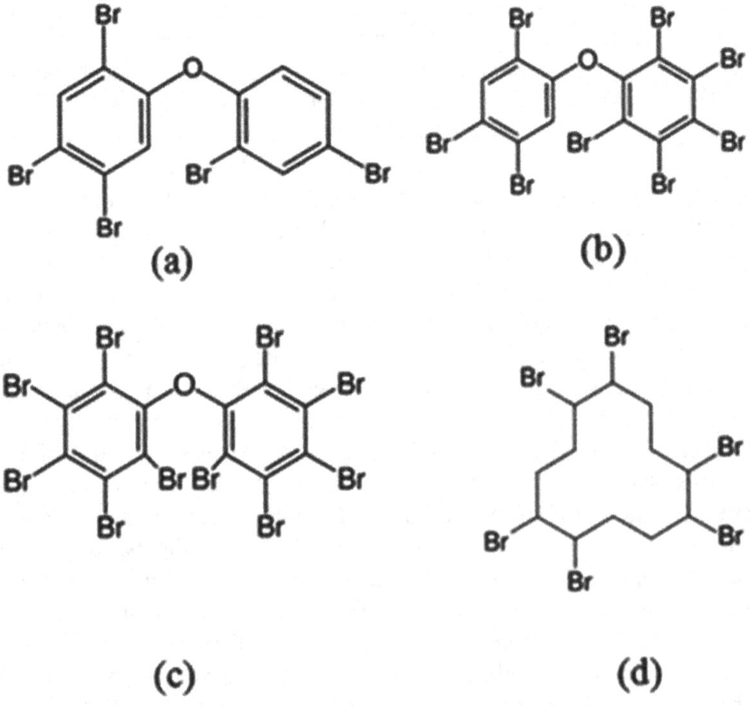

STRUCTURE 10.4   PBDEs

The production of two major class bromine-based retardants, polybrominated diphenyl ethers (PBDEs), and polybrominated biphenyl (PBBs), has been banned in Europe and the US. State legislative activity has focused on three types of PBDEs: pentaBDE, octaBDE, and decaBDE. Evidence has shown that PBDEs would persist in the environment and accumulate in the living organisms. Toxicological testing results indicated that these chemicals could cause liver, thyroid, and neurodevelopmental toxicity. Moreover, the Department of Health and Human Services and the International Agency for Research on Cancer suggested that PBBs could be carcinogenic to human health (Gu et al., 2015).

Considerable concern has been shown regarding the possible formation of polybrominated dioxins associated with incineration of organobromine compounds, especially those based on polybrominated diphenyl oxides (PBDPO) (Horrocks, 2003).

Neither PBDE nor PBB has-been been found mutagenic. For decaBDE, a few studies have shown increased tumor frequency among experimental animals (liver tumors) while other studies have shown no effect (WHO, 1994). The doses were very high in the experiments, which indicated increased cancer frequency: 1–2 g/kg/d (NTP, 1985). The increases were not statistically significant. The result has been judged as ambiguous (EU, 1999). In the United States, decaBDE is classified as possibly carcinogenic for humans (class C) based on this study (Federal Register, 1996). IARC (1990) has on the other hand, classified the substance to group 3, i.e., not carcinogenic for humans, but has declared that there is limited evidence for being carcinogenic to experimental animals. The United States has classified octaBDE, pentaBDE, and tetraBDE as noncarcinogenic for humans (class D) (EPA, 1984).

Carcinogenic effects have been observed for PBB, especially hexaBB. Other congeners may also be carcinogenic, e.g., pentaBB, heptaBB, and nonaBB, since they were present in the tested commercial product, which mainly consisted of hexaBB. The lowest dosage promoting cancer (liver tumors in rats) was 0.5 mg/kg/d (WHO, 1994;0.15 mg/kg/d caused no effect (test period 2 years). IARC has classified commercial hexaBB (also including pentaBB and heptaBB) as possibly carcinogenic for humans (class 2B) (Hellström, 2000).

The most extensively used BFRs have been tetrabromobisphenol A (TBBPA), hexabromocyclododecane (HBCD), and three commercial mixtures of polybrominated diphenyl ethers (PBDEs): decabromodiphenylether (decaBDE), octabromodipheylether (octaBDE), and pentabromodiphenylether (pentaBDE) (Shaw et al., 2010). PBDEs and polybrominated biphenyls (PBBs) have been restricted in the EU since 2004 (European Parliament and Council, 2000) and the production of pentaBDE and octaBDE has been phased out in the United Statessince 2004 (US-EPA, 2008). PentaBDE and octa-BDE have been identified as persistent organic pollutants [POPs (SFT,

2009)], and pentaBDE is already listed as a hazardous substance under the Water Framework Directive (WFD), while octaBDE and decaBDE are listed among the substances to be monitored (European Parliament and Council, 2000). DecaBDE and HBCD were included by the European Chemicals Agency (ECHA) in the list of candidates for authorization under REACH Article 57d as substances of very high concern (SVHC) based on PBT (persistent, bioaccumulating, toxic) properties. TBBPA is registered under REACH and is currently not subject to any REACH restriction processes. Tris(2-chloroethyl) phosphate (TCEP) is representative of the chlorinated FRs, and has been included in Annex XIV of REACH ("Authorisation List") because of concerns over its reproductive toxicity (Cat 1B). List of Banned FRs (Nicola, 2015)

The European Union's (EU) legislative framework on chemicals has been streamlined by REACH and the following FRs are completely banned:

- Tris(2,3 dibromopropyl) phosphate (TRIS)
- Tris(1-aziridinyl)phosphine oxide (TEPA)
- Polybromobiphenyls (PBB)
- PentaBDE
- OctaBDE
- HBCD (unless authorized)

TCEP and HBCD have been listed. Assessment of decaBDE is ongoing and is not yet banned in the EU. It is banned in electronics but permitted in other products.

The following halogenated flame retardants for fireproofing garments are in the restricted list (European Commission, 2008).

10.2.2   TRIS(2,3-DIBROMO-1-PROPYL)-PHOSPHATE

It is known as Tris or Fyrol HB 32. Tris is mutagenic and is listed as an IARC Group 2A carcinogen. It is one of the chemicals covered by the Rotterdam Convention in 2004. In the United States, the Consumer Product Safety Commission banned the sale of children's garments containing TRIS in 1977.

10.2.3   TRIS-(AZIRIDINYL)-PHOSPHINOXIDE

Metepa or 1-[Bis(2-methyl-1-aziridinyl)phosphoryl]-2-methylaziridineis a chemosterilant, with the capability to restrict ovarian development. Metepa can also result in carcinogenesis, particularly the formation of teratomas.

10.2.4   POLYBROMINATED BIPHENYLS.

Polybrominated biphenyls (PBBs), also called brominated biphenyls or polybromobiphenyls, are a group of manufactured chemicals that consist of polyhalogenated derivatives of a biphenyl core. Their chlorine analogs are the

PCBs. While once widely used commercially, PBBs are now controlled substances under the Restriction of Hazardous Substances Directive, which limits their use in electrical and electronic products sold in the EU.

10.2.5 Antimony Trioxide

The use of antimony trioxide as synergist, together with halogenated-based flame retardants (e.g., in textile back coatings, flame retarded paints, rubber, textiles) is also under review within the European Union Commission. The International Antimony Association opposes the ban of antimony oxides under RoHS regulation because there is no conclusive evidence regarding the toxicity or environmental impact of antimony oxides (International Antimony Association, 2009).

Local legislations in some countries may not be able to induce a broad sustainable protection of mankind and the environment as many chemicals remain unchanged in nature for a long period of time, and may also spread over large regions of the planet. The Stockholm Convention with more than 160 member states decided on the possible ban of POPs, which have shown toxicity and environmental risks (Secretariat of the Stockholm Convention, 2009). Thus, stronger cooperation between countries and international organizations is required to address the problems associated with the use of harmful chemicals (Gaan et al., 2011).

PBB was banned in textile materials that come into direct contact with the skin in the United States and part of Europe in 1979, and in Switzerland and Austria in 1983. In 1984, according to Directive 76/769/EEC, the use of PBB was prohibited in textiles, such as clothing, underwear, and bedding. Similarly, pentaBDE and octaBDE were banned in Europe in 2004, and decaBDE was banned in 2008. At present, PBB and PBDEs are on the prohibition list of REACH (Horrocks, 2013).

10.3 ECO-FRIENDLY FRs

Replacement of existing flame retardants (FRs) with sustainable and environmentally friendly alternatives for textiles in domestic, safety, transport (automotive, rail, aerospace and marine), civil emergency and military, construction and other industries requires a multidisciplinary approach from textile technology to the physics and chemistry of fire namely:

* Development of new innovative FR with low fire toxicity, environmental impacts and free of halogen.
* Analysis of their effectiveness, durability, (smoke) toxicity and particularly environmental impact by life cycle analysis (LCA).
* Improved surface treatment (plasma, enzymes, ultrasound, UV, etc.) and application processes (coating, spinning, sol-gel, micro-encapsulation, (photo) chemical, etc.) for FRs.
* The synergistic effect of combining nanomaterials with conventional FRs.

Exposure to flame retardants once widely used in consumer products has been falling, according to a new study. The researchers are the first to show that levels of polybrominated diphenyl ethers (PBDEs) measured in children significantly decreased over a 15-year period between 1998 and 2013, although the chemicals were present in all children tested (Sciencedaily, 2018).

Phosphorous-halogen–containing FRs combine the properties of both halogen and phosphorus components, which act synergistically in the polymer system by combining the gas-phase action of halogen species with the condensed-phase action of phosphorus-based compounds such as tris (2,3-dibromopropyl) phosphate (tris-BP) and tris-(1,3-dlchloro-2-propyl) phosphate (tris-CP) (Horrocks, 2008). The halogenated phosphorous FRs were introduced as alternatives for the restricted halogenated FRs, but some of these compounds have been reported to be toxic (Anonymous, 2016).

In general, the toxicity potential of triaryl-phosphates is related to the content of o-phenolic residues in commercial products. The most pressing health and environmental problem associated with the use of triaryl-phosphates arise from their ability to induce delayed polyneuropathy. For TPhP, the contact allergenic effects on humans and possible toxicity to water organisms have been documented (Anonymous, 2016).

There is no data in the literature on the toxicological, immunological, neurological, or carcinogenic and mutagenic effects of oral exposure to ammonium polyphosphate (APP), ammonium ions, or polyphosphates. The only documented negative environmental effect of APPs is the algal toxicity. As a result of low environmental risks, APPs and MPP (melamine polyphosphate) could represent a good alternative to brominated FRs (Horrocks, 2001).

The phosphonate ester is the reactive, functional phosphonamide, which is chemically di-alkylphosphonocarboxylic acid amide (e.g., PYROVATEX® CP NEW). Its main limitation and health and environmental risk represent the formaldehyde generation due to the presence of the reactive methylol group in the structure.

DOPO and its derivatives are reported to be nonneurotoxic with no inflammatory activation potential. Therefore, they are very promising candidates for the replacement of currently applied flame retardants.

The NFRs are melamine and various melamine derivatives separately, they have little effect on flame retardancy; therefore, they are often used in combination with phosphorus FRs, such as MPP. The main advantages of these compounds are their low toxicity, solid state, and in the case of fire, the absence of dioxin and halogen acids and their low evolution of smoke (Isbasara and Hacaloglu, 2012).

There is limited information about the environmental issues associated with any mineral flame retardant compounds. Some scientists have concerns about carbon nanotubes due to unknown harmful impacts to the human body by inhalation, and some suggest that carbon nanotubes have similar toxicity to asbestos fibers. Nanoparticles have higher surface areas than bulk materials. Hence, the mobility of nanoparticles causes more damage to the human body and the environment, compared to the bulk particles (Elbasuney, 2015).

Decabromodiphenyl ether (Structure 10.5) in commercial mixtures is known as c-decaBDE. The amount of c-decaBDE used in plastics and textiles varies globally, but up to 90% of c-decaBDE ends up in plastic and electronics, while the remaining ends up in coated textiles, upholstered furniture, and mattresses. The aviation industry uses c-decaBDE in electrical wiring and cables, interior components, and in electrical and electronic equipment (EEE) in older airplanes and spacecraft. C-decaBDE is expected to be present in plastics and textiles in several waste streams such as *End of Life Vehicles* (ELV), e-waste, textile waste, and mixed waste. For plastics in EEE, substitution strategies range from exchange of the resin system and FR, to complete redesigns of the product itself.

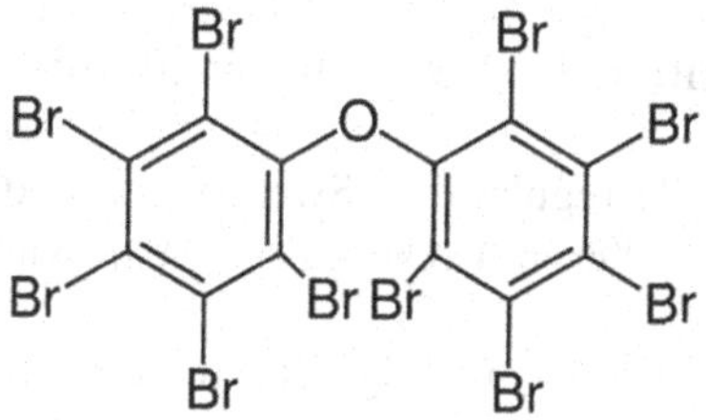

STRUCTURE 10.5   Decabromodiphenyl ether(DecaBDE)

Deca-BDE has long been characterized as an environmentally stable and inert product that is incapable of degradation in the environment, nontoxic, and therefore of no concern. However, some scientists had not particularly believed that decaBDE was so benign, particularly as evidence to this effect came largely from the industry itself. One problem in studying the chemical was that "the detection of decaBDE in environmental samples is difficult and problematic"; only in the late 1990s did analytical advances allow detection at much lower concentrations. DecaBDE is released by different processes into the environment, such as emissions from manufacture of decaBDE-containing products and from the products themselves. Elevated concentrations can be found in air, water, soil, food, sediment, sludge, and dust. A 2006 study concluded "in general, environmental concentrations of BDE-209 [i.e., decaBDE] appear to be increasing" (Alcock and Busby, 2006).

According to the EU restriction proposal, which assessed different alternatives to c-decaBDE, eight possible alternative chemicals that appear to be possible substitutes for c-decaBDE in plastic polymers and textiles are (Stockholm Convention, 2008):

1 Decabromodiphenyl ethane (DBDPE),
2 Bisphenol A bis (diphenyl phosphate) (BDP/BAPP),
3 Resorcinol bis (diphenylphosphate) (RDP),
4 Ethylene bis (tetrabromophthalimide) (EBTBP),

5 Magnesium hydroxide (MDH),
6 Triphenyl phosphate (TPP),
7 Aluminum trihydroxide (ATH), and
8 Red phosphorous.

Norwegian legislation generally follows EU legislation and implements REACH regulations. However, Norway also banned the use, production, import, and export of decaBDE in 2008, and now also plans to eliminate all brominated flame retardants by 2020, having already banned pentaBDE, octaBDE, decaBDE, and PBB. A priority list has been prepared to specify chemicals to be phased out, both in terms of emissions and use, by 2020. The list includes pentaBDE, octaBDE, decaBDE, HBCD, TBBPA, bisphenol A, DEHP, SCCPs, PFO A, PFO S, perchloroethylene, trichloroethylene, and triclosan (http://www.pic.int/Portals/5/download.aspx?d=UNEP-FAO-RC-FRA-NOTIF-decaBDE-1163195-Norway-20140502.En.pdf, accessed on 13.5.2020).

In addition to REACH regulations, Sweden imposed a ban on decaBDE in textiles and furniture in 2006, but reversed this ban when it was challenged by the EU.

No US federal regulations currently exist to restrict flame retardants; however, individual states impose varying restrictions. The United States signed the Stockholm Treaty in 2001, but it has not been ratified yet. Many states have banned pentaBDE and octaBDE. Surprisingly, decaBDE is largely untouched, despite the fact that it actually degrades to form lower brominated flame retardants such as octaBDE, and has been listed as a chemical of high concern by GreenScreen for Safer Chemicals. In the absence of national standards, some US states, including Washington, Maine, Hawaii, Maryland, Vermont, and Oregon, have independently banned the use of decaBDE.

According to the American Chemistry Council (ACC), The US Environmental Protection Agency (EPA) issued a significant new use regulation (SNUR) for the use in textiles of six polybrominated diphenyl ethers (PBDEs) in 2005 and proposed a SNUR for decaBDE and HBCD in 2012, but has not yet finalized the proposal. Several Asian nations have similar restrictions to those of US states.

10.4   INTUMESCENT FRs FROM RENEWABLE SOURCES

During their decomposition (combustion), char-forming substances often swell and expand. One approach to optimization thus lies in promoting a flame-retardant action through the formation of intumescent char. The char layer is formed at the expense of combustible gases. This layer prevents further flame-spread by surrounding the unburned material like a thermal barrier. The tendency of a polymer to char can be increased with chemical additives, as well as by altering its molecular structure. These additives generally produce highly conjugated systems and aromatic structures which char when exposed to heat and/or are converted into cross-linking agents at high temperatures.

Intumescent coatings when heated form a thick, porous, carbon-containing layer, providing ideal insulation of the surface against excessive rises in temperature and oxygen ingress. Thermal degradation is thus prevented, which plays a decisive part in delaying the degradation process. Intumescent compounds increase their volume by a factor of 200 or more when heated (Vandersall, 1971).

The oxidative modification of polysaccharides is a widespread practice for preparing a charring agent. The oxidation leads to salts of polyoxyacids. Copper salts, in conjunction with alkalis, are the most effective catalysts for the oxidation of polysaccharides. In the absence of a catalyst, virtually no oxidation of starch and cellulose takes place at temperatures below 100°C. Lignin, however, contains many highly reactive phenol groups, which can be oxidized directly with oxygen at measurable rates, even without a catalyst.

As in the oxidation of low molecular weight polyols, the anionic form of a polysaccharide reacts with Cu^{2+} ions to form an adduct $[Cu^{2+} \ldots A\text{-}]$ (A--deprotonated polysaccharide). The function of the catalyst, i.e., the Cu^{2+} ions, is based on activating the deprotonated substrate for oxygen attack. It may be assumed that the divalent copper ions oxidize the substrate anion, and the anion radicals, or radicals forming as intermediates, then react with O_2. Not only does starch on its own act as a starting material for the production of the polyoxyacid salts, but any other raw materials containing starch can also be used. These include not only maize kernels, oats, and rice, but also substandard materials such as cereals affected by fungal diseases, waste from rice cultivation, and waste from mills. This enables initial costs to be kept low. Furthermore, for the oxidation of lignin in alkaline media, with or without a catalyst, a similar mode of reaction was observed to that of the polysaccharides. Lignin is one of the most widely occurring biopolymers. The current world total production of lignin is estimated to be 3×10^{11} tons. New production by biosynthesis is approximately 2×10^{10} tons. Lignin makes up about 30% of the dry weight of softwoods and about 20% of that of hardwoods. Itis a random copolymer of phenylpropane units with characteristic side chains. Lignin is readily crosslinked and, in the solid state, has no regular structure. The hydroxyl group in lignin plays an important part in the interaction with water. Lignin is generally regarded as an amorphous polyphenolic material that is formed in enzyme-initiated, dehydrogenative, free-radical polymerization from phenylpropanoid monomers p-coumaryl, coniferyl, and sinapyl alcohol. The basic structure of lignin consists of just two components, an aromatic section and a C_3 chain. The only reactive sites in lignin are the OH groups in the aromatic framework (phenolic) and those in the C chain (alcoholic). If lignin is regarded as an amorphous, three-dimensional network, then it has three main elements as shown in Structure 10.6 namely:

STRUCTURE 10.6 Functional groups of lignin:

1. 4-hydroxyphenyl,
2. Guaiacyl, and
3. Syringyl groups.

Lignin comes from renewable sources such as trees, grasses, and agricultural crops. Lignin makes up over 30% of the components of wood. They are non-toxic and extremely versatile in their capabilities. Worldwide production of lignin as a by-product of pulp and paper manufacturing is over 30 million tons per annum (Lomakin et al., 2012).

The initial formation of the char foam took place in the temperature range of 150°–280°C. It was the result of synchronous processes: on the one hand, the decrease in viscosity of the polymer with its transition from a glassy to a viscous state, and on the other hand, the chemical reactions such as decarboxylation and dehydration. The foam is produced by gaseous decomposition products–water and carbon dioxide.

The intermolecular dehydration reactions promote the formation of the spatially linked network structure and stabilizing and strengthening of the char foam that forms.

The high effectiveness of intumescent additives with regard to flame retardancy was demonstrated in the context of the combustibility tests according to ASTM E136-09 on pinewood, which had been treated with an aqueous solution of oxidized starch for protection. Fire leads to significantly lower mass losses in the wood treated with oxidized lignin or oxidized starch than in the untreated wood. The maximum loss as a function of time is 41% lower for lignin, and 33% lower for starch than that of pure pinewood. The improved flame resistance of the sample coated with oxidized lignin, compared with the uncoated sample, is the result of the development of a char layer that provides a temporary protective barrier of polyphenolic, charring lignin (Lomakin et al., 2012).

Recently, a biobased phosphorus-containing compound phytic acid (PA) (Structure 10.7) has provoked interest in textile treatment, particularly in FR treatment. Phytic acid is a sixfold dihydrogenphosphate ester of inositol (specifically, of the myo isomer), also called inositol hexakisphosphate (IP6) or inositol polyphosphate. At physiological pH, the phosphates are partially ionized, resulting in the phytate anion.

STRUCTURE 10.7 Phytic acid (PA)

PA is a unique natural substance found in plant seeds. It has received considerable attention due to its effects on mineral absorption. PA impairs the absorption of iron, zinc, and calcium, and may promote mineral deficiencies. Therefore, it is often referred to as an anti-nutrient.

PA is known as inositol hexakisphosphate acid or phytate in the salt form, and is regarded as a *green* molecule because it is found in abundance in plant tissues, such as beans, cereal grains, and oil seeds. As a biocompatible, environmentally friendly, nontoxic and easily obtained organic acid (Ye et al., 2012), PA has already been widely applied in antioxidant, anticancer, biosensor, cation exchange, nanomaterial, and other fields because of its special inositol hexaphosphate structure (Laufer et al., 2012). PA contains 28 wt % phosphorus based upon molecular weight, and is promising as a possible and effective FR material. PA has been used to reduce the flammability of cellulose-based materials. PA has been employed as a doping acid to greatly improve the FR performance of polyaniline-deposited paper composites. PA/chitosan and PA/nitrogen-modified silane hybrids have been used via layer-by-layer assembly to fabricate FR thin films on cotton fabric (Laufer et al., 2012). In addition, the potential FR effect of different metallic phytates has been evaluated as biosourced phosphorus additives for poly(lactic acid) composites (Costes et al., 2015).

PA consists of six negatively charged phosphate groups and has a strong tendency to combine or interact with positively charged metal ions or proteins (Yang et al., 2007). This means it is possible that PA can combine with wool fibers (natural protein fiber) by means of the electrostatic interaction between the positively charged amino groups in wool and the negatively phosphate groups in PA. In a study by Cheng et al., (2016), PA was applied to wool fabric through an exhaustion process with the aim of improving the flame resistance of woolen textiles. The important factors affecting the adsorption of PA, such as pH, temperature and PA concentration were considered. The combustion and thermal properties of the treated fabrics were evaluated via limiting oxygen index (LOI), vertical burning test, pyrolysis combustion flow calorimetry (PCFC), and thermogravimetry (TG). Scanning electron microscopy (SEM) coupled with energy dispersive X-ray spectroscopy (EDS) was used to investigate the morphology of the treated fibers. Furthermore, the phosphorus content of the unburned and burned wool fabrics was detected by inductively coupled plasma optical emission spectrometry (ICP-OES) tests. PA has been proven to be a potential FR agent because of its high char-forming ability. The P content of the treated wool fibers was much lower than that of the corresponding char residues, indicating that P participates in the formation of the char layer, and the condensed-phase mechanism is suitable for FR wool fabrics. The use of PA provides an opportunity for producing FR wool fabrics using a green FR agent. In future work, some measures should be taken to improve the wash-resistance of FR wool fabrics.

10.5   BIOMACROMOLECULAR FRs

Biomacromolecules such as proteins (whey proteins, caseins, hydrophobins) and deoxyribonucleic acid (DNA) showed unexpected flame retardant/flame-suppressant features when deposited on cellulosic or synthetic substrates, such as cotton, polyester, or cotton–polyester blends (Alongi et al., 2013). The use of some of these biomacromolecules (e.g., caseins and whey proteins) as flame retardants is a significant advantage, since they can be considered as waste or by-products from the cheese and milk industry; on the other hand, despite the current high cost of DNA, its availability has become competitive with those of other chemicals, due to the recently developed large scale production method, based on the extraction and purification of DNA from salmon milt and roe sacs (Wang et al., 2001).

Caseins are phosphorylated proteins that represent the main fraction of milk proteins (around 80%) and, possibly, the most widely investigated food proteins, obtained as coproducts during the production of skimmed milk. Notwithstanding their standard cheese production uses, these proteins have been mainly employed as a food ingredient for improving such physical properties as foaming and whipping, water binding and thickening, texture, and emulsification. Furthermore, they have been exploited as coatings, specifically referring to papermaking, leather finishing, printing, and manufacturing of synthetic fibers (Liu et al., 2013).

In the context of fire retardancy, cotton, polyester and cotton polyester blend fabrics (polyester content: 65 wt. %) were treated with a casein aqueous suspension (5 wt. %) in a climatic chamber ($30°C$ and 30% R.H.): the suspension was spread onto the samples with a spatula and the excess was removed by gently pressing with a rotary drum; then, the coated samples were dried to constant weight. The final dry add-on was 20 wt % (Carosio et al., 2014). The presence of the casein coatings promoted a strong anticipation of both cellulose and polyester decomposition; this behavior, observed for all the types of treated fabrics, was attributed to the phosphate groups located on the shell of casein micelles, which, upon heating, release phosphoric acid, favoring the degradation of cellulose or polyester toward the formation of a stable char. The latter exerts a protective effect on the underlying textile, limiting the oxygen diffusion and absorbing the heat evolved during combustion (Alongi et al., 2015).

It is very well known that deoxyribonucleic acid (DNA) consists of two long polymer chains of nitrogen-containing bases–namely, adenine (A), guanine (G), cytosine (C) and thymine (T)–with backbones made of five-carbon sugars (i.e., deoxyribose units), as well as of phosphate groups tied by ester bonds. The chains are rolled-up around the same axis and bonded together. This way, a double helix is formed, which exploits the presence of H-bonds between the bases located side-by-side and paired in a specific mode; cytosine bases are combined with guanine, while adenine bases are paired with thymine.

The use of DNA as a sustainable flame retardant is quite recent, and can usually be attributed to the structure of this biomacromolecule, which represents an all-in-one intumescent system (Carosio et al., 2014).

More specifically, the double helix comprises all three constituents of an intumescent material, namely:

- The phosphate groups that form phosphoric acid,
- The deoxyribose units, which act as a carbon source and as blowing agents, and
- The four nitrogen-containing bases that may give rise to the formation of NH_3.

Like an intumescent material, when DNA is exposed to a heat source, it develops a multicellular foamed char that limits the heat and mass transfer between flame and polymer, hence providing flame extinction.

The first pioneering work dealing with the flame retardant properties of this biomacromolecule was published in 2013 and focused on the use of DNA derived from herring sperm for self-extinguishing cotton fabrics. To this aim, a standard impregnation/exhaustion method for reaching the desired final dry add-on (19 wt. %) was employed (Alongi et al., 2013). Among the most important achievements, improved thermal and thermo-oxidative stability of the treated fabrics in nitrogen and air in terms of char residue formed at high temperatures, was demonstrated. Furthermore, combustion was blocked, and the flame was extinguished within 2 s after the application of a methane flame in horizontal configuration to the fabrics treated with this biomacromolecule for 3 s. The remarkable flame retardant features of DNA were further confirmed by LOI (28% vs. 18% for DNA-treated fabrics and untreated cotton, respectively) and cone calorimetry tests; the latter showed that none of the DNA-treated specimens ignited upon exposure to 35 kW/m^2 heat flux.

The high char-forming character of deoxyribonucleic acid was attributed to its chemical structure, in which the phosphate groups are able to form phosphoric acid that catalyzes the dehydration of cotton, favoring its auto-cross-linking toward the formation of a stable aromatic char (that is also formed by the deoxyribose units) and inhibiting the production of volatile flammable species. This char behaves as a physical protective barrier on the underlying substrate, limiting the heat, fuel, and oxygen transfer between the fabric and the flame. At the same time, the decomposition of pyrimidine and purine bases could promote the formation of azo-compounds, which could further induce the char development and the formation of non-combustible gases, such as nitrogen, carbon monoxide, and carbon dioxide.

The aforementioned biomacromolecules can be applied to fabrics using a conventional impregnation/exhaustion process or a layer-by-layer method, starting from aqueous solution/suspensions, hence exploiting a significantly green technology. The mechanism, through which these biomacromolecules are able to confer flame retardancy to fabrics, is still under investigation. However, the flame retardant effectiveness of these green macromolecules seems to be attributable to their chemical composition, as well as to their interaction with the underlying fabrics that, upon heating or exposure to a flame, favors the formation of a stable and protective char (i.e., a carbonaceous residue), which limits the exchange of oxygen and combustible volatile products, hence enhancing textile flame-resistance.

More specifically, caseins and hydrophobins, which contain phosphate groups and disulphide units, respectively, have been assessed as effective flame-retardant systems for cellulosic substrates, since these components are capable of influencing the cellulose pyrolysis toward the formation of char. Furthermore, whey proteins have shown their suitability to form protective coatings on cotton, which exhibit great water vapor adsorption, possibly justifying the achieved flame resistance of the treated fabrics. Compared to proteins, DNA shows unique and peculiar behavior, since it contains all three principal ingredients of an intumescent formulation in one molecule. The charred and foamed layer at the surface of the burning polymer protects the underlying material from the action of the heat or flame. The phosphate groups produce phosphoric acid, the deoxyribose rings act as a carbon source and blowing agents. Upon heating, they may dehydrate, forming char and releasing water. As a result of nitrogen-containing bases (guanine, adenine, thymine, and cytosine) that may release ammonia, DNA-treated cotton fabrics have even reached outstanding self-extinguishment features.

Despite the high flame retardant efficiency observed for most of the chemical and low- impact FRs, there are still some unresolved issues that deserve further attention. Particularly in the case of biomacromolecules, the possibility of adjusting the green technological approach to a larger scale than the lab scale (at least to preindustrialization) is still under assessment and the final decision will be largely based on the cost-effectiveness of the described biomacromolecules. Definitely, the cost of some of the discussed biomacromolecules, such as DNA, is presently very high: therefore, their use as low-impact FRs can be foreseen only on the basis of significant cost reduction.

Several advantages can be conferred from the exploitation of proteins and DNA in providing FR features to textiles (Malucelli et al., 2014). In particular, their ease of manipulation, the possibility of exploiting application techniques that are already designed and optimized for fabric finishing, such as impregnation/exhaustion, spray, or even layer-by-layer deposition, (Malucelli, 2016) and the setup of low impact/sustainable finishing recipes (thanks to the use of water-based solutions/dispersions).

In addition, caseins and whey proteins are, to some extent, by-products or even wastes recovered from the agro-food industry; therefore, their use as possible FRs may represent a new way toward the valorization of agro-food crops, reducing and/or preventing their landfill confinement. Finally, the high price of DNA with respect to standard chemical FRs is being overcome, because large scale extraction and purification processes have been developed (Wang et al., 2001).

Quite recently, DNA was also exploited as a component of layer-by-layer (LbL) coatings; indeed, its assemblies were found to improve the flame-retardant features of the treated fabrics, keeping, at the same time, a green character. In particular, Carosio et al. (2013) combined negatively charged DNA with positively charged chitosan layers on cotton, thus building up an assembly containing up to 20 bilayers.

At present, some of the proposed P-based FRs (such as the proposed phosphorus-containing biomacromolecules) cannot accomplish this goal. The Bio-FRs developed recently have little or no light fastness or, in some cases, a limited durability up to few washing cycles. This surely represents a current drawback that limits the use of the treated textiles to those restricted applications for which durability to washing treatments is not required. Thus, further research will also have to consider the design of new strategies that can overcome this challenging issue (Salmeia et al., 2016).

10.6 FUTURE TRENDS

Flame retardant products have difficulties in obtaining eco-labels, and some of them cannot obtain them. The most important objective comprises a complete replacement of halogenated FRs with phosphorus and phosphorus-nitrogen *greener* FR formulations for achieving efficient flame retardancy with minimal health and environmental hazards. According to this objective, the introduction of novel intumescent FR systems represents a major scientific and technological challenge; however, this will enable a breakthrough in the production of flame-retarded polymer materials, which would follow the principles of eco-design. Adding two or more flame retardants in products that work in combination and have a synergistic effect is another novel and efficient way of achieving efficient flame retardant polymers (Šehić et al., 2016b).

The phosphorus-based additives allow exploiting an almost infinite number of highly efficient FR systems that are applicable on different fibers and fabrics.

Moreover, the simultaneous presence of phosphorus and other elements (such as nitrogen, silicon, and sulfur) may be very important for developing additive or synergistic effects during the exposure of the treated textiles to a flame or to a heat source.

Although the phosphorus-based flame retardants are being developed as an alternative to toxic halogenated counterparts, they should be thoroughly checked for toxicity before application. New phosphorus-based flame retardants have been evaluated for toxicity, and these have shown a low toxicity profile (Hirsch et al., 2016).

The persistence, bioaccumulation, and toxicity (PBT) of a selection of HFFRs must be studied. Some preliminary studies have already been made (Waaijers et al., 2013).

The future requirements of polymeric materials will most likely be (Morgan, 2017):

- They must pass new flammability tests based on fire risk scenarios and fire safety principles.
- They must be environmentally friendly, recyclable, and sustainable.
- Recyclability could mean energy recovery until separate the additives can be separated and recycled as well.
- They must be inexpensive.

The replacement of halogen-containing FRs by halogen-free FRs is a long-term process. The replacement of halogen-based FRs will be faced with many challenges. A feasible and practical strategy is to follow two routes:

1. Firstly, the existing halogen-based FRs which cannot be replaced should be used selectively and in reduced quantities.
2. In parallel, novel HFFRs are to be developed which may be based on phosphorous and/or nitrogen organic compounds and inorganic compounds, especially nanomaterials.

A wide variety of phosphorus compounds have been explored, both as additives and reactive components, for the flame retardation of a wide range of polymers, both thermoplastics and thermosets. However, the commercial exploitation of phosphorus-based flame retardant systems is still, to some extent, in its infancy, with halogen-containing flame retardants still dominating the market. This is mostly a consequence of the generally higher cost of phosphorus-based materials, especially organophosphorus compounds, and the lower flame retardant efficiency of many phosphorous-based FRs. As environmental pressures to reduce the use of halogenated organic flame retardants increase, it is likely that the commercial exploitation of phosphorus-containing alternatives will increase, bringing with it higher FR efficiencies and reduced costs (Joseph and Ebdon, 2010).

However, perhaps the most exciting developments, are likely to be in the increasing use of various nanoscopic additives, such as nanoclays and carbon nanotubes, in combination with both halogenated- and phosphorus-containing flame retardants. Surface treatments in which a flame retardant barrier layer is laid down by plasma polymerization could be particularly effective for polymeric materials having high surface to-volume ratios, such as fibers and films (Joseph and Ebdon, 2010).

Future research directions and activities in the field of the use of FRs for polymer materials are guided by sustainable development goals, following REACH (EU) and other eco-legislation. It is difficult, if not impossible to achieve eco-levels for products finished with commercial flame retardants. The most important objective is to replace halogenated FRs completely with phosphorus and phosphorus-nitrogen *greener* FR formulations for achieving perfect flame retardancy with minimal health and environmental hazards.

REFERENCES

Alcock R.E. and Busby J. (April 2006). Risk migration and scientific advance: The case of flame-retardant compounds. *Risk Analysis*, 26(2), 369–381. doi:10.1111/j.1539-6924.2006.00739.x. PMID 16573627).

Alongi J., Carletto R.A., Di Blasio A., Cuttica F., Carosio F., Bosco F., and Malucelli G. (2013). Update on Flame Retardant textiles: State of the art, *Environmental Issues and Innovative Solutions, Shawbury, Smithers Rapra*, 2013, 1–348.

Alongi J., Bosco F., Carosio F., Di Blasio A., and Malucelli G. (2014). A new era for flame retardant materials?, *Materials Today*, 17(4), May, 152–153.

Alongi J., Carletto R.A., Di Blasio A., Cuttica F., Carosio F., Bosco F., and Malucelli G. (2013b). Intrinsic intumescent-like flame retardant properties of DNA-treated cotton fabrics, *Carbohydrate Polymers*, 96, 296–304.

Alongi J. and Malucelli G. (2015). Thermal degradation of cellulose and cellulosic substrates, in: A. Tiwari and B. Raj, Eds., *Reactions and Mechanisms in Thermal Analysis of Advanced Materials*. JohnWiley & Sons, Hoboken, NJ, USA, Vol 14, pp. 301–332.

Anonymous (2016). Flame retardants: the case for policy change, environment & human health, *Tekstilec*, 59(3), 196–205. www.ehhi.org/reports/flame/EHHI_FlameRetardants_1113.pdf, accessed on 1.7.2016.

Braun U., Schartel B., Fichera M.A., and Jager C. (2007). *Polymer Degradation and Stability*, 92, 1528–1545.

Buser H.R. (1987). Brominated and brominated/chlorinated dibenzodioxins and dibenzofurans: Potential environmental contaminants, *Chemosphere*, 16(8–9), 1873–1876.

Carosio F., Di Blasio A., Alongi J., and Malucelli G. (2013). Green DNA-based flame retardant coatings assembled through Layer by Layer, *Polymer*, 54, 5148–5153.

Carosio F., Di Blasio A., Cuttica F., Alongi J., and Malucelli G. (2014). Flame retardancy of polyester and polyester-cotton blends treated with caseins, *Industrial & Engineering Chemistry Research*, 53, 3917–3923.

Cheng X.-W., Guan J.P., Tang R.-C., and Kai-Qiang Liu K.-Q. (2016). Improvement of flame retardancy of poly(lactic acid) nonwoven fabric with a phosphorus-containing flame retardant, *Journal of Industrial Textiles*, 46(3), 914–928. DOI: 10.1177/1528083715606105.

Costes L., Laoutid F., Dumazert L., Lopez-cuesta J.M., Brohez S., Delvosalle C., and Dubois P. (2015). Metallic phytates as efficient bio-based phosphorous flame retardant additives for poly (lactic acid), *Polymer Degradation and Stability*, 119, 217–227.

Drinker C.K., Warren M.F., and Bennet G.A. (1937). The problem of possible systemic effects from certain chlorinated hydrocarbons, *Journal of Industrial Hygiene and Toxicology*, 19(7), 283–311.

D'Silva K., Fernandes A., and Rose M. (2004). Brominated organic micropollutants – Igniting the flame retardant issue, *Critical Reviews in Environmental Science and Technology*, 34, 141–207.

EHP (Environmental Health Perspectives). (1978). Final report of the subcommittee on health effects of PCBs and PBBs, 24, 133–198.

Elbasuney S. (2015). Surface engineering of layered double hydroxide (LDH) nanoparticles for polymer flame retardancy, *Powder Technology*, 277, 63–73. DOI: 10.1016/j.powtec.2015.02.044.

EPA (US Environmental Protection Agency) (1984). *Health and Environmental Effects Profile for Brominated Diphenyl Ethers*. Environ Criteria and Assessment Office, Cincinnati, USA.

EU. (1999). Draft Risk assessment of diphenyl ether, pentabromo derivative (pentabromodiphenyl ether, of diphenyl ether, octabromo derivative, of bis(pentabromophenyl) ether (decabromodiphenyl ether), https://echa.europa.eu/documents/10162/da9bc4c4-8e5b-4562-964c-5b4cf59d2432, accessed on 12.5.2020.

European Commission (2008).Textiles Background Product Report, Brussels, European Commission, DG Environment-G2, B-1049.

European Parliament and Council (2000). Directive 2000/60/EC of the European Parliamentand of the Council of 23 October 2000 establishing a framework for community action in the field of water policy. *OJ. L* 327, 92.

Federal Register (1996). Proposed Guidelines for Carcinogen Risk Assessment, *Federal Register*, 61(79), 17960–18011, 23 April, https://www.govinfo.gov/content/pkg/FR-1996-04-23/html/96-9711.htm, accessed on 12.5.2020.

Gaan S., Salimova V., Rupper P., Ritter A., and Schmid H. (2011). *Chapter 5 Flame Retardant Functional Textiles in Functional Textiles for Improved Performance*. Protection and Health, 1st Edition, N Pan G. Sun, Woodhead, UK.

Gu L., Ge Z., Huang M., and Luo Y. (2015). Halogen-free flame-retardant waterborne polyurethane with a novel cyclic structure of phosphorus-nitrogen synergistic flame retardant, Journal of Applied Polymer Science. DOI: 10.1002/APP.41288.

Hardy M.L., Banasik M., and Stedeford T. (2009). Toxicology and human health assessment of decabromodiphenyl ether, *Critical Reviews in Toxicology*, 39, 1–44.

Hellström T. (2000) *Brominated Flame Retardants (PBDE and PBB) in Sludge – a Problem?*, The Swedish Water and Wastewater Association Report No M 113 (eng), April.

Hirsch C., Striegl B., Mathes S., Adlhart C., Edelmann M., Bono E., Gaan S., Salmeia K. A., Hoelting L., Krebs A., et al. (2016). Multiparameter toxicity assessment of novel DOPO derived organo-PFRs. *Archives of Toxicology*, 90, 1–19. DOI: 10.1007/s00204-016-1680-4.

Horrocks A.R. (1997). Environmental consequences of using Flame-retardant textiles – Asimple life cycle analytical model, *Fire and Materials*, 21(5), 229–234. doi:10.1002/(SICI)1099-1018(199709/10)21:5229::AID-FAM6143.0.CO;2-U.

Horrocks A.R. (2001). *Textiles. Fire Retardant Materials*. A.R. Horrocks and D. Price. Woodhead Publishing, Cambridge, pp. 128–181.

Horrocks A.R. (2003). Flame retardant finishes and finishing, in: D. Heywood, Ed., *Textile Finishing, Volume 2*. Society of Dyers and Colourists, Bradford, UK, pp. 214–250.

Horrocks A.R. (2008) *Textiles. Advances in Fire Retardant Materials.* A.R. Horrocks and D. Price, Eds., Woodhead Publishing, Cambridge, pp. 188–233.

Horrocks A.R. (2011). Flame retardant challenges for textiles and fibres: New chemistry versusinnovatory solutions, *Polymer Degradation and Stability*, 96, 377–392.

Horrocks R.A. (2013). Flame retardant and environmental issues, in: J. Alongi, A. R. Horrocks, F. Carosio, and G. Malucelli, Eds., *Update on % Flame Retardant Textiles: State of the Art, Environmental Issues and Innovative Solutions.* Smithers Rapra Technology Ltd, Shawbury, pp. 207–239.

IARC (International Agency for Research on Cancer) (1990). *Some Flame Retardants and Textile Chemicals, and Exposures in the Textile Manufacturing Industry.* IARC Monographs on the Evaluation of Carcinogenic Risks to Humans, Volume 48, ISBN-13 (Print Book), 978-92-832-1248-5, ISBN-13 (PDF,978-92-832-1248-5.

International Antimony Association (2009). Downstream user exposure scenarios being prepared for REACH, Brussels, International Antimony Association VZW. Available at: www.antimony.be/newsletter/docs/i2a/i2a-newsletter-december-2009.pdf, accessed on 18.2.2020.

Isbasara C. and Hacaloglu J. (2012). Investigation of thermal degradation characteristics of polyamide-6 containing melamine or melamine cyanurate via direct pyrolysis mass spectrometry, *Journal of Analytical and Applied Pyrolysis*, 98, 221–230. DOI: 10.1016/j.jaap.2012.09.002.

Jensen S. (1966). Report of a new chemical hazard, *New Scientist*, 32, 612.

Joseph P. and Ebdon J.R. (2010). *Phosphorus-Based Flame Retardants in Fire Retardancy of Polymeric Materials*, 2nd edition, C.A. Wilkie and A.B. Morgan, CRC Press, USA.

Jurgen T.H. (1998). Overview of flame retardants, fire & environment protection service, *Chemica Oggi*, 16, 1–19.

Kemmlein S., Herzke D. et al. (2009). Brominated flame retardants in the European chemicals policy of REACH – Regulation and determination in materials, *Journal of Chromatography A*, 1216, 320–333.

Laufer G., Kirkland C., Morgan A.B., and Grunlan J.C. (2012). Intumescent multilayer nanocoating, made with renewable polyelectrolytes, for flame-retardant cotton, *Biomacromolecules*, 13, 2843–2848.

Liu Y., Liu L., Yuan M., and Guo R. (2013). Preparation and characterization of casein-stabilized gold nanoparticles for catalytic applications, *Colloids and Surfaces A*, 417, 18–25.

Lomakin S.M., Sakharov A.M., Sakharov P.A., and Zaikov G.E. (2012). Environmentally friendly flame retardants based on renewable raw materials, *International Polymer Science and Technology*, 39(7).

Malucelli G. (2016). Layer-by-Layer nanostructured assemblies for the fire protection of fabrics, *Materials Letters*, 166, 339–342.

Malucelli G., Bosco F., Alongi J., Carosio F., Di Blasio A., Mollea C., Cuttica F., and Casale A. (2014). Biomacromolecules as novel green flame retardant systems for textiles: An overview, *RSC Advances*, 4, 46024–46039.

Morgan A.B. (2017). *Polymer Flame Retardant Chemistry, FR Chemistry*, 09/30/09, http://www.nist.gov/el/fire_research/upload/2-Morgan.pdf, accessed on 12.11.17.

Nicola (2015). Use and regulation of flame retardants in textiles, *AATCC Review*, 15(6).

NTP (1985). NTP technical report on the toxicology and carcinogenesis studies of dekabromodiphenyl oxide in F344/N rats and B6C3F1 mice (feed studies). National Toxicology Program, NTP-TR-309, NIH-85-2565.

Reemtsma O., Quintana J.B.-N., Rodil R., Garci'A-López M.-C., and Rodriguez I., (2008). Organophosphorus flame retardants and plasticizers in water and air I. Occuraence and fate. *Trends in Analytical Chemistry*, 27(9), 727–737.

Salmeia K.A., Hoelting L., Krebs A., et al. (2016). Multiparameter toxicity assessment of novel DOPO derived organo-PFRs. *Archives of Toxicology*, 1–19. DOI: 10.1007/s00204-016-1680-4.

Schecter A., Shah N. et al. (2009). PBDEs in US and German clothes dryer lint: A potential source of indoor contamination and exposure, *Chemosphere*, 75, 623–628.

Schindler W.D. and Hauser P.J. (2004). *Chemical Finishing of Textiles*. Woodhead, Cambridge, England.

Sciencedaily (2018). www.sciencedaily.com/releases/2018/04/180404093947.h, accessed on 10.5.18. (Source: Columbia University's Mailman School of Public Health, Date: April 4, 2018).

Šehić A., Tavčer P.F., and Simončič B. (2016). Flame Retardants and Environmental Issues, *Tekstilec*, 59(3), 196–205. doi:10.14502/Tekstilec2016.59.196-205.

SFT (2009). *Guidance on Alternative Flame Retardants to the Use of Commercial Pentabromodiphenylether (c-PentaBDE)*. Oslo, http://chm.pops.int/Portals/0/docs/POPRC4/intersession/Substitution/pentaBDE_revised_Stefan_Posner_final%20version.pdf.

Shaw S.D., Blum A., Weber R., Kurunthachalam K., Rich D., Lucas D., Koshland C.P., Dobraca D., Hanson S., and Birnbaum L.S. (2010). Halogenated flame retardants: do the fire safety benefits justify the risks? *Reviews on Environmental Health*, 25(5), 261–305.

Stockholm Convention (2008). Decabromodiphenyl ether (commercial mixture, c-decaBDE), www.chm.pops.int/

US-EPA (2008). Tracking progress on U.S. EPA's Polybrominated Diphenyl Ethers (PBDEs), www.epa.gov/sites/production/files/2015-09/documents/pbdestatus1208.pdf.

USGS (2018). *Phosphorus and Water*. US Dept. of Interior, http://water.usgs.gov/edu/phosphorus.html, accessed on 27.3.18.

Vandersall H.L. (1971). Intumescent coating systems, their development and chemistry, *Journal of Fire and Flammability*, 2, 97.

Waaijers S.L., Kong D., and Hendriks H.S. (2013). Persistence, bioaccumulation, and toxicity of HFFRs, *Reviews of Environmental Contamination and Toxicology*, 222, 171.

Wang L., Yoshida J., and Ogata N. (2001). Self-assembled supramolecular films derived from marine deoxyribonucleic acid (DNA)-surfactant complexes: Large scale preparation and optical and thermal properties, *Chemistry of Materials*, 13, 1273–1281.

WHO (World Health Organization). (1994). Brominated diphenyl ethers. International programme on substance safety, *Environmental Health Criteria*, 162. http://www.inchem.org/documents/ehc/ehc/ehc162.htm, accesed on 12.5.2020.

Yang L.Z., Liu H.Y., and Hu N.F. (2007). Assembly of electroactive layer-by-layer films of myoglobin and small-molecular phytic acid, *Electrochemistry Communications*, 9, 1057–1061.

Ye C.H., Zheng Y.F., Wang S.Q., Xi T.F., and Li Y.D. (2012). In vitro corrosion and biocompatibility study of phytic acid modified WE43 magnesium alloy, *Applied Surface Science*, 258, 3420–3427.

Index

A

Abramov reaction, 248
Acrylic fibers, 6, 16, 44–47, 154, 242, 261, 353, 368
  modacrylic, 28, 42, 130, 372
    FR, 154, 159–160
    oxidized, 145–152
Acrylonitrile-butadiene-styrene copolymers (ABS), 171, 200–236, 252
Additive flame retardants, 31, 154–166, 207, 215
After-flame time, 41, 59, 64, 68, 88
Afterglow, 21, 68
Aircraft, fire safety of, 112, 146, 151, 269, 299–300, 383
Aluminum hydroxide, alumina trihydrate (ATH), 31, 159, 165–170, 184, 201, 216–218, 333, 345, 366
Alumina, 169, 170
  fibers e.g. Saffil®, 151
  alumina-silica fibers e.g. Nextel®, 151
Aluminosilicate
  layered or clays, 175, 327–329
Amino resins, 16
Ammonium salts, 30, 167, 172, 229, 252–253, 293, 336
  AEDTMPA, AHDTMPA, 253
  bromide, 149, 167, 200, 369, 378
  polyphosphate, 167–174, 217–240, 293–308, 353–359, 376, 392
  polyphosphate-pentaerythryol, 294
  sulfamate, 159, 167–172, 217, 228, 364–373
Antimony trioxide
  bromine systems, 32, 33, 165–170, 200, 204, 367, 372, 391
  zinc borate, 345
Application-based flammability tests, 51, 73, 83
Aramid fiber, 130, 190, 301, 344
  meta-aramid, Nomex, 17, 47, 132–153
  para-aramid, Kevler, 17, 47, 132–153, 378
Aramid-arimid fiber, Kermel, 148
Arimid or polyimide fibers, 146, 148
Aryl polyphosphonates (ArPPN), 241
ASTM Standards
  1930, 95, 97
  D 635, D 4804, 65
  D1230-17, 52, 54
  D 2859-16, 62
  D2863, 109, 120
  D 3801, 66
  D 4108-87, 93
  D 4804, 65, 69
  D 5048, 66, 69
  D6413-11, 61
  D6413/15, 52
  D6545-18, 52
  D 7309, 112, 117, 122
  F1506, 99, 100
  F1891-06, 99
  F2700, 94
  E 1321-97a, 112
  E1354, 75, 119
  ATH, see Alumina trihydrate, 119, 169, 170, 217. 237, 345, 394
Atherton–Todd reaction, 250, 251, 275
Auto-crosslinking, 400

B

Back-coating, 291
Basofil® (BASF), (see melamine-formaldehyde fiber), 134, 138–141
Benzofurans, 385, 386
  di-, 213, 387
Biocomposite, 343, 359
Biodegradation, 247
Black phosphorus or phosphorene, 320
Boron compounds, 31
  boric acid and borax, 167, 186, 265, 293, 322
  boric oxide, 185
  Fluoboric acid, 363
Boron-based FR, 31
  flame retardancy mechanism, 185
Boronic acid, 160
British standard
  BS 476, Parts 4, 6, 7 and 11, reaction to fire, 62
  EN 367:1992, Protective clothing, 68
  BS 5852, BS EN 1021-1 or BS EN, 1021–2, 68
Bromine, 168, 199, 217
  aliphatic derivatives, 166, 171, 193, 199, 369
  antimony synergism, 202–204
  aromatic derivatives, 166, 170, 195, 199, 203, 358, 369
  classification of, 204
  FR activity, 201–204
  phosphorus synergism, 33

polymeric, 32, 158
   with chlorine derivatives, 204
Burn injury, 7, 51, 88–98
Burning process (*see* combustion), 9, 40
   textiles, 48, 121, 160
      burning rate, 40–43, 48, 66, 115, 159,
         179, 239, 373

C

Candle flame, 23
Carbon fiber, *also* graphite, grapheme, 119, 140,
      160, 265, 320, 377–378
   Reinforced epoxy composites, 341
   Semicarbon, 145
Carbon monoxide, 2, 9, 10, 21, 22, 114, 179,
      183, 213, 225, 307, 384, 385, 400
Carbon-based nanocomposites, 326, 330
Carbon-based nanoparticles (NPs), 318
   Fullerenes, 318
   nanotubes (CNTs), 318
   multi-wall carbon nanotubes (MWNTs),
      318, 319, 331–333, 338–341
      FR properties, 319
      mechanical and electrical properties, 318
Carcinogenicity, 211, 245, 246, 383
Carpets, 4–6, 62, 63, 129–131, 194, 208, 266,
      363–369
Cellulose
   cross-linking and esterification with
      phosphorous, 187
   dehydration and cross-linking, 188
   flame retardant of, 167, 227
   flame retardant viscose, 154
   flame-spreading mechanisms, 46
   Smoldering, 102
   thermal degradation, 181
   volatile pyrolytic products, 182
Char
   formation and characterization, 20, 42,
      115, 173–176, 188–191, 208, 223,
      227–231, 252–296, 302, 339, 353,
      365, 373
   intumescent, 173, 292, 300, 301, 373, 394
   length, 41, 48–52, 59–63, 101, 253,
      256, 374
   soft and hard, 174, 293
Chlorendic acid, 32, 170, 209
Chlorine, 168, 170, 171, 199, 201
   Bleach, 132
   Containing FRs, 205, 368, 369
   organo-as paraffin, 32
CNT buckypaper, 20, 341
Coatings, 29, 30, 148–150, 160, 167, 235, 267,
      291, 292, 322, 371, 391, 398

Combustion (*see* burning), 2, 8–17, 19–26, 9,
      42–52, 65, 66, 102–120, 151–185,
      202, 273
   products, 8–16, 22, 24, 75, 115, 366
Composites (*see also* nanocomposites), 9, 48,
      81, 83, 119, 137–143, 151, 315
   Bio, 359
   epoxy, 307
   fiber-reinforced, 114
   flammability of, 119
   flame retardancy
   nano-, 148, 168, 175, 316–320
      FR mechanism, 325
   Clay-based nano, 327
   Exfoliated nano-, 330
   polymer-nano, 323–325
   Polymer-layered silicate nano, 334, 344
   test, 102, 129
   textile-reinforced, 315
Condensed phase activity (*see also* flame
      Retardant), 20, 75, 102, 116, 155,
      166–176, 187–194, 201–208, 224–238,
      249, 260, 269–280, 3010, 303, 325,
      340–344, 383, 392
Cone calorimeter test (ISO 5660), 73–83, 112,
      119, 269, 276, 277, 301, 306, 307,
      332, 336
Congener, 207
Cotton
   flame retardant, 186, 262
   polyester blends, 371, 374, 398
Cross-linking, 12, 20, 139, 160, 182–188, 228,
      258, 322, 353, 354, 374, 394

D

Decabromodiphenyl ether (DECABDE), 32,
      170, 207, 367, 370, 389–394
Decabromodiphenyl ethane, 32, 170
Decabromodiphenyl oxide (or ether),
      (DBDPO), 194, 206, 358, 367, 369,
      372, 384
Dechlorane Plus, 199, 203, 369
Dichlorotribromophenyl phosphate (DCTBPP),
      356
Degradation, 114, 115, 129, 144, 153, 170, 175,
      179, 181–192, 210–235, 244, 272
DGEBA, 273, 277
Differential thermal analysis (DTA), 204, 228
Dilution of fuel, 184, 185, 191
Diammonium phosphate, 167, 172, 186, 225,
      229, 230, 371
Dibenzodioxins, 213, 387
Dioxins (*see* environmental issues and toxicity),
      171, 194, 209, 384, 386, 387, 389

Dioxin-like effects, 211
Diphenylchloro phosphate (DPCP), 155
Doping, 153, 397
9,10-dihydro-9-oxa-10-phosphaphenanthrene-
        10-oxide (DOPO), 248, 271–278, 361
    MPL-DOPO, 275, 276
    DDM-DOPO, 275, 276
DOPO based phosphinate flame retardants, 24
DOPO-T, 273
DOPO-VTS, 272, 273
DPBAEP, 277
DP-DDM, 274
DPO, 278
Durability to laundering
    durable finishes, 167, 168, 177, 232, 244,
        253–258
    non-durable finishes, 167, 168, 172, 229, 354
    semi-durable finishes, 167, 169, 172, 195,
        229, 254, 354

E

Elastomers, 306, 333
    flame retarded, 205, 233, 300
    Thermoplastic poly(ester ether) elastomers
        (TPEEs)
Endocrine disruption, 210
Endothermic Degradation, 184
Environmental issues (*see also* toxicity), 175,
        200, 215
Epoxy resin, 13
    Flame retardants, 205, 217, 225, 231, 233,
        252, 267–279, 301, 374–375
Ethylene-vinyl acetate (EVA), 252, 342
Expandable graphite (EG), 278–300, 338, 359

F

Fibers (*see also* (high performance fibers)/
        fabrics/polymers,
    Thermoplastic, 9, 43–62, 119, 130, 190, 204,
        205, 217, 230, 298, 306, 358
    glass transition temperature, 9, 44, 136, 241,
        269, 277
    melting temperature, 9, 133
    pyrolysis temperature, 9, 10, 44, 169, 184,
        188, 227, 253.
Fire
    textile-related, 4
Fire growth, 73
Fire growth rate (FIGRA), 74, 82, 11
Fire hazard statistics
    Textiles, 3
        Precautions, 4
    fatalities, 1

non-war related infernos in history, 1
Fire-resistant clothing (FRC), 24
Fire-resistant polymer, fire safe polymer (*see*
        inherent FR polymer), 10, 129–162
Fire toxicity, 391
Flame
    extinction, 116, 118, 399
    heat release, 10, 13, 17, 21, 43, 46, 51, 54, 55,
        74–83
    heat release rate, 74–82, 107–119, 153, 179,
        225, 269, 276, 306, 309, 319, 364, 365
        peak, 249, 254, 322
    mechanism, 17
    spread, 5, 2, 46–58, 65–107, 112, 119, 132,
        196, 266, 273
        ignition, 5–29, 39–121, 129–196, 244,
            265, 322, 356
        index, 71
        smoke and toxic gas, 292, 307, 308, 375
        surface flame spread, 54
    spreading mechanism, 46
    temperatures, 21, 24, 43, 45, 46, 115,
        202, 302
Flame-retardant (FR), flame retardancy
    Additives, 130, 156–159, 174, 176, 278, 279,
        354, 368, 383
    Application methods, 194
    back-coating, 160, 184, 195, 240, 291, 301,
        354, 355, 378
    blanketing, 193
    char formation, 20, 42, 115, 173–176,
        188–191, 208, 223, 227–231, 252–296,
        302, 339, 353, 365, 373
    classifications, 31, 166
    coating, 29, 30, 130, 149, 150, 167, 169, 180,
        183, 194, 235, 259, 262, 273, 356, 375
    condensed phase, 20, 75, 102, 116, 155,
        166–176, 187–194, 201–208, 224–238,
        249, 260, 269–280, 300, 303, 325,
        340–344, 383, 392
    durable, 168
    expandable graphite, 278, 279, 294, 300,
        359, 376
    gas (or vapor) phase, 33, 159, 161, 166, 171,
        191–199, 225, 291, 366–371
    glassy layer, 30, 186, 190
    grafting, 143, 149, 177, 207, 259, 261, 262,
        369, 373
    halogen, 199–215, 268
    halogen-free, 159, 169, 171, 180, 196,
        215–218, 223–235, 250–259, 270–280,
        303–307, 359–363, 402.
    mechanisms, 201
historical development, 31

inorganic/mineral, 31, 32, 166, 168
intumescent coating., 291, 296, 363, 395
mechanisms for synthetic polymers, 183
mechanisms in general, 184
melamine and derivatives, 172, 293, 294,
    298, 302, 305, 308, 362, 366,
    376, 392
nanocoating, 339, 341, 344
phosphorous (PFRs) additives, 171, 172,
    223–250
    mechanisms, 227
phosphorous-nitrogen additives
    mechanisms, 191
viscose e.g. Lenzing FR, 29, 47, 130, 147,
    148, 152, 154
hazards from halogenated
hazards from phosphorus based
hybrid organic–inorganic
intumescent, 173
mechanisms
    reduction of flammable volatiles, 186
    thermal shielding, 185
minerals, 29, 32, 166, 168, 175, 184, 296,
    299–307, 306, 392, 397
nano-composite, 174
nitrogen-based, 168, 216, 218, 373
nono-durable, 167
phosphorus, 171
phosphorus and nitrogen-, 186
radical trap, 154, 193, 202–206, 385
reactive, 176
    TDHTPP, 161
silicone, 179
synergism, 33, 194, 200, 202, 228, 332,
    370, 372
toxicity, 17, 179, 194, 203–218, 224–248,
    292, 343–366, 372
Flammability
and LOI, 47
brief specifications, 52
fabric properties, 40
factors for apparel, 27
of composite materials, 11
Flammability tests
Application-based tests, protective
    clothing, 83
Horizontal, floor covering, 62
ignitability
    cone calorimeter, 74
    glow wire test, 21
ignition test, 51, 102
inclined plane, 54
large-scale, small scale, 49
LOI test, 110

micro-scale Combustion Calorimetry
    (MCC), 112
NASA STD 6001 upward, 111
Radiant Panel Flame Spread Test, 71
reaction-to-fire test, 51, 73–79
room-corner test, single burning item (SBI)
    test, 74, 79–82
textile fibers, 42
Thermal Protective Performance (TPP) Test,
    91, 93, 94, 101
upholstered furniture, smoldering cigarette
    test, 52, 103
UL94 test, 53, 64
vertical, 52, 59
Flocking, 6
Floor covering, 52, 62
Friedel–Crafts reactions, 249
Fuel, 5–25, 41, 46, 50, 74–78, 95, 102, 112–116,
    173, 183–202, 224–238, 267, 316, 326,
    355, 365, 400
Furans (see environmental issues and Toxicity),
    171, 194, 209, 213, 384
Furnishing fabrics, 194, 195, 291
Furniture, 4, 21, 63, 102–106, 120, 122, 236,
    375, 394

G

Gas phase radical quenching (see also flame
    retardant mechanisms), 187, 192, 193
Gas singeing, 6
Glass fiber, 137, 151
Graphene and its oxide, 295, 296, 320, 321, 326,
    327, 338, 365, 376
    Intercalation, 295, 300
Grafting, (copolymerisation), 143, 149, 177,
    207, 259, 261, 262, 369, 373
Guanidine sulfamate, 294, 365

H

Halogen flame retardants 199–215, 268
    and environment, 154, 168, 171, 200,
        209–213, 267, 385, 388, 389
    classification, 204
    gas phase activity, 192
    mechanisms, 201
    physical effects
    halogen synergisms, 194, 200, 202
    hazards
        carcinogenicity, 211
        endocrine disruption, 210
Halogenated phosphorus FRs
    brominated: tris(2,3-dibromopropyl)
        phosphate (TRIS), 172, 214, 390

chlorinated: tris(1,3-dichloro-2-propyl) phosphate (TDCPP), 32, 171, 172
Halogen- free FRs, 159, 169, 171, 180, 196, 215–218, 223–235, 250–259, 270–280, 303–307, 359–363, 402.
Hazards
  Fire, 1
    in textile industry, 3
    fire-prone area, 5
    precautionary steps, 5
Heat of combustion, 14, 44, 74, 112–115, 119, 131, 189
Heat release rate (HRR), 74–82, 107–119, 153, 179, 225, 269, 276, 306, 309, 319, 364, 365
  peak, 249, 254, 322
    Hexabromocyclododecane, HBCD, 205
Hexafluorotitanates and hexafluorozirconates, 208, 266
Hexaglycidyl tris(3-(bis(oxiran-2-ylmethyl) amino)phenyl)phosphine oxide (HGE), 270
High performance fibers, 17, 119, 301
Horizontal flammability test, 62
Hybrid organic-inorganic flame retardants, 177

I

Incineration, 389
Inclined plane flammability test, 5
Ignition, 5–29, 39–121, 129–196, 244, 265, 322, 356
  auto, piloted, 17
  ease of, 5, 41–55, 119, 322
  edge and surface, e, 42
  process, 16
Inherently FR fibres, polymers, 10, 129–162
Intrinsic flammability, 44
Intumescence and intumescent systems, 173, 291
  Advantages, 293
  application, 291
  char
    soft, 174, 293
    hard, 174, 293
  components, 174
    acid sources, 293
    blowing agents, 293
    carbonization agent, 293
  mechanisms, 294

K

Kevlar (*see* para aramid), 10, 17, 28, 45, 132, 134, 135, 138, 144, 152
Kabachnik–Fields reaction, 248

L

Lateral Ignition and Flame Spread Tests (LIFT), 112
Levoglucosan, 181–183, 186–190, 200, 227, 262, 353
Lignocellulose, 260
Limiting oxygen index, 11, 19, 48–51, 109–118, 129–139, 147–159, 177, 183, 188, 204, 228, 248–260, 265–279, 298–323, 328–341, 345, 357–374, 399
  and Flammability, 7

M

Magnesium hydroxide (MDH), 31, 168, 169, 394
Maleic anhydride, 332
Manikin flammability test, 8, 51, 88, 95–102
Melamine, 130, 138, 139, 172–174, 195, 216, 218, 224, 240, 254, 292–294, 305, 308, 362, 392
Melamine-formaldehyde fibers, 133, 134, 138, 139, 140
Melamine phosphate, 172, 217, 234–235, 279, 293–294
Melamine pyrophosphate, 234, 235, 376
Melamine resin
  Polyphosphate, 216–217, 235–240, 392
Melt blending, 143, 332, 333, 337
Melt dripping, 159, 171, 216, 226, 238, 303, 366
Melt viscosity, 156, 182, 250, 319
Meta aramid (*see* Nomex), 16, 132–136, 147, 152
Metaboric acid, 185
Metal hydroxides, 169, 216, 218, 342, 366
Michaelis–Arbuzov reaction, 248
Michaelis–Becker reaction, 248
Micro-scale Combustion Calorimetry (MCC), 112–128, 252, 263
Mineral or inorganic flame retardants, 29, 32, 166, 168, 175, 184, 296, 299–307, 306, 392, 397
  aluminum hydroxide (ATH), 31, 159, 165–170, 184, 201, 216–218, 333, 345, 366
  magnesium hydroxide (MDH), 31, 168, 169, 394
  Chalk (calcium carbonate), 168, 317
Modacrylic, 28, 45, 47, 130, 154, 160, 354, 368, 372
Montmorillonite, 159, 175, 296, 307, 328–336, 366, 375
  Intercalation, 329

N

Nanoclay, 159, 175, 178, 304, 318, 327–342,
    365, 366, 402
Nanocoating, 339, 341, 344
    FR mechanisms, 340
Nanocomposite, 148, 168, 175, 316–320
    carbon-based, 318
    clay-based, 327
    exfoliation and delamination, 330
    FR mechanisms, 176, 325, 330
    layered silicates, 334, 344
    polymer, 323–325
    preparation, 335
    thermal stability, 324
    toxicity, 343-
    types, 317
Natural fibers, 6, 9, 30, 44, 45, 119, 122, 129,
    139, 301, 343, 345
Natural fiber reinforced (NFR) composites,
    343, 344, 345, 359
Nightwear, 4, 46
Nitrogen, nitrogen compounds, 9, 10, 16,
    19, 23, 109, 129, 154, 188–191,
    268–280, 298, 353–375,
    393–403
Nitrogen Flame Retardants, 168, 216,
    218, 373
    Mechanism, 173
Nomex (*see* Aramid), 10, 132–137,
    144. 147
Novoloid fibers, 134–140

O

Octabromo bisphenyl, 358
OEKO-TEX® standard, 100, 244
Organomodified clays, 329
Organophosphorus compounds, 29, 32, 154,
    156, 165, 166, 223–238, 239, 242, 247,
    248, 268, 293, 301, 353, 369, 373,
    377, 383, 402
    Triphenyl phosphate (TPP), 32
    Resorcinol bis(diphenylphosphate)
        (RDP), 32
    Bisphenol A diphenyl phosphate (BADP),
        32, 172
    Tricresyl phosphate (TCP), 32
Oxidized PAN fibers, 134, 145
Oxygen index (*see* limiting oxygen index), 11,
    19, 48–51, 109–118, 129–139,
    147–159, 177, 183, 188, 204, 228,
    248–260, 265–279, 298–323, 328–341,
    345, 357–374, 399
Organo–sulfur compounds, 129

P

Para aramid (*see* Kevlar), 10, 17, 28, 45, 132,
    134, 135, 138, 144, 152
PBDEs, 205
Pentaerythritol, 253
    phosphate, 233
Phenol-formaldehyde (*see* Novoloid), 134–140
Phenolic resin, 16
phosphate esters, 191, 224–234, 237
    effects on environment, 238
Phosphonates, 32, 239, 275
    cyclic oligomeric, 240
Phosphonium salt (THPC), 242–247, 263
    Toxicity, 245
Phosphorus
    condensed phase activity, 187, 190
    flame retardants, 171, 172, 223–250
        for cotton, 261
        for epoxy, 267
        for wool, 266
    black, phosphorene, 320–322
    red, 190, 230, 394
Phosphorus-nitrogen FRs (PNFRs), 186
Phosphorylamide, 242
PINFA, 173, 216, 217
Poly(1,2-dicarboxyl ethylene spirocyclic
    pentaerythritol bisphosphonate)
    (PDESPB), 255, 256
Polyacrylics, Polyacrylonitrile, PAN (*see also*
    acrylic Fibres), 12, 130, 143–146, 159,
    160, 186, 200, 242, 353, 354, 369, 372
Polyamide (nylon), 42–47, 53, 130, 133, 135,
    153, 173, 175, 200, 205, 207, 230,
    240, 264, 273, 304, 333, 363–372
    Flame retardant, 158, 159
Poly(amide–imide) (PAI), 148, 149
Polybenzimidazole, PBI, fibre, 10, 45, 47, 134,
    143, 144–152
Polybenzoxazole fibre, PBO, 133, 134, 143–153
PBDPO, 389
Polybrominated biphenyls (PBB), 206
Polybrominated Diphenylethers (PBDE),
    205–207, 389
Poly(butylene terephthalate), PBT, 133, 214,
    252, 298, 299, 304, 334, 390, 402
Polycarbonate, PC, 200, 249, 293
Polychlorinated biphenyls (PCB)
Polydimethyl siloxane, PDMS, 179, 180
Polyhaloalkenes, 149
Polyhedral oligomeric silsesquioxane (POSS),
    257, 335
Polyimide fibers P84® (Evonik), 134
Poly(methylmethacrylate) (PMMA), 134
Polyphenylene sulfide (PPS), 134, 141, 142

Polysiloxanes, 179
Polyester, poly(ethylene terephthalate), PET, 42, 45, 46, 154, 155, 355, 360
 cotton blends, 42, 43, 45, 363
 flame retardant for, 355, 359
 IFR, (e.g. Trevira CS,), 130, 154, 155, 156, 355
 resins, 200, 235, 240, 242, 279, 362, 374
Polyethylene, PE and UHMPE, 135, 357, 370
Polytetrafluoroethylene (PTFE), 42, 45, 48, 134, 150, 199, 334
Polymer
 Thermoplastic (*see* fibers), 9, 43–62, 119, 130, 190, 204, 205, 217, 230, 298, 306, 358
 Thermoset, 19, 119, 138, 139, 140, 148, 169, 205, 240–242, 260, 271–277
 Combustion products, 115
 Thermal decomposition, 11
 Inherently flame retarded, 10, 129–162
Polymeric FRs, 32, 158
Polymer-layered silicate nano-composites, 334, 344
Polymethyl-methacrylate (PMMA), 48, 159, 175, 369
Polyolefins, 169, 184, 188, 194, 205, 225, 233, 240, 284, 304, 370
Polypropylene, PP, 12, 42, 44, 47, 153, 154, 171–175, 190, 216–218, 235, 255, 369
 Flame retardant, 161, 304–306, 369
Polystyrene, PS, 22, 32, 158, 160, 279
 high impact, HIPS, 205, 207, 236
 brominated, 170, 367
 phosphorous, 278, 279
Poly(tetrafluoroethylene), PTFE, 42, 45, 48, 134, 150, 199, 334
Polyurethane foam, 48
 Waterborne, 309
Polyvinyl alcohol, PVA, 336, 372, 374
Polyvinyl chloride, PVC, 16, 48, 133, 149, 150, 200, 205, 294, 35, 369, 370
 flame retardant for, 294
Polyvinylidene chloride, PVDC, 149
Protective clothing, 83, 86
 flammability test, 88, 90
Pudovik reaction, 248
Pyrolysis, 11, 12, 113, 115, 184, 194
Pyrolysis combustion flow calorimetry (PCFC), 112–115
Pyrolysis of cellulose, 11
Pyrolysis of synthetic fibers, 353
PYROVATEX CP, 254, 255, 262

R

Radiant Panel Flame Spread Test, 71, 72
Rate
 of burning, 44, 64, 123, 165, 323, 357
 heat release rate, 74–82, 107–119, 153, 179, 225, 269, 276, 306, 309, 319, 364, 365
 peak, 249, 254, 322
REACH, 214, 242, 367, 387, 390, 391, 394, 403
Reaction-to-fire test, 51, 73, 78, 79
Reactive flame retardants, 176
TDHTPP, 161
Resins
 epoxy, 13, 205, 218, 219, 225, 252, 267, 271–279
 maleimide, 273
 phenolic, 16, 133, 139, 377
 polyester, 218, 235, 240, 242, 279, 364, 372, 374
 thermoplastic, 345
Resorcinol bis(diphenyl phosphate) (RDP), 32, 172, 236, 393
Rice husk (RH), 260, 272, 322
Risk assessment, 71, 96, 97, 387, 388
Room corner test, 80
Rubber, 5, 16, 138, 165, 179, 237, 300, 360, 391

S

Sandmeyer-type reaction, 248
Single burning item (SBI) test, 51, 73–83
Scaffolding effect, 371
Self-extinguishing, 12, 21, 22, 28, 41, 61, 107, 130, 149, 160, 165, 230, 237, 268, 323, 399
Silica-based fibers e.g. Quartzel®, 151
Silica nanoparticles, 322
Silicon carbide fibers, 151
Silicon flame retardants, 256
silicone thermoplastic elastomer (Si-TPE), 235
Smoke, 22
Smoke and toxicity, 73
 cone calorimeter
 smoke chamber
Smoke Growth Rate (SMOGRA)
Smoke suppression
Smouldering
 inhibition
Sodium polyphosphates, 363
Sol–gel process
Spirocyclic pentaerythritol bisphosphorate disphosphoryl chloride (SPBDC)
Stenter
Sulfur compounds
Synergism, 33

antimony for halogen FRs, 194, 200, 202
halogen-phosphorus, 370
nanoclay and IFR, 332
phosphorus-nitrogen, P-N, 194, 228, 372
Synthetic fibers (*also* man-made)
    flame retardant treatments, 353
    inherently flame retardant, 134–150

T

Tris(2-chloroethyl) phosphate (TCEP), 214, 242,
    375, 390
Tris(aziridinyl)phosphinoxide, 214
Trixylylphosphate (TXP), 214
Tetrabromobisphenol–A (TBBPA), 32, 170,
    205, 210, 211, 214, 217, 388–394
Tetrabromophthalic anhydride (TBPA), 207,
    208, 209
Tetrakis(hydroxymethyl)phosphonium chloride
    (THPC), 242–247, 263, 2
Tetrakis(hydroxymethyl)phosphonium
        Hydroxide (THPOH), 243, 247, 263
Textile
    fire-prone substances, 5
    thermal properties, 44, 71, 93, 136, 144
    testing, 47
Thermal decomposition of polymers, 9, 11,
    187, 188
    Mechanisms, 12
Thermal response parameter (TRP), 108
Thermal shielding, 185
Thermal transitions of fibers, 9, 10
    glass transition temperature, 9
    melting temperature, 9
    pyrolysis temperature, 9
Thermoplasticity, 130, 153, 353
Thermogravimetric Analysis (TGA), 106–108,
    114–115, 204, 233, 253, 258, 306–310,
    374
Thermoplastics and thermosets, 169, 235, 325
    flame retarded, 207, 329, 402
Thermoplastic poly(ester ether) elastomers
    (TPEEs), 333
Toxicity
    of flame retardants, 194, 210, 211, 218, 224,
    383–391

dioxins and furans, 194, 213, 384–387
of NPs, 343
of smoke, 22
of THPC, 245, 246
Toxic potency, 150
Triethyl phosphate, 235, 376
Triphenyl phosphate, 32, 172, 190, 268,
    385, 394
Tris(chloroethyl)phosphate, 214
Trixylylphosphate (TXP), 214
Triphenylphosphine oxide (TPPO),
    155, 156

U

UL94 flammability test, 64
Upholstered furniture, fire regulations,
    102–105
Urea and thiourea, 173, 174, 191, 195, 217, 228,
    243–245, 263–265

V

Vapor (or gas) phase, 33, 159, 161, 166, 171,
    191–199, 225, 291, 366–371
Vertical flammability tests, 59
Viscose, 47
    flame retardant, 29, 113, 147, 153, 154, 155,
    264, 265
Viscosity (*see* melt viscosity), 156, 182, 242, 250,
    318, 396

W

Waste, 9, 11, 117, 154, 226, 291, 322
Wool
    blends
    flame retardancy of, 208
    Zirpro process, 208, 266

Z

Zinc compounds, 165, 169, 203, 204, 216, 217,
    300, 305, 345, 397
Zirconates (and titanates), 208, 209, 266
Zirpro process, 208, 266

Printed in the United States
by Baker & Taylor Publisher Services